7th International Workshop on Multi-Rate Processes and Hysteresis and the 2nd International Workshop on Hysteresis and Slow-Fast Systems (MURPHYS-HSFS 2014)

Journal of Physics: Conference Series Volume 727

Berlin, Germany
7 – 11 April 2014

Editors:

O. Klein
M. Dimian
P. Gurevich
D. Knees
D. Rachinskii
S. Tikhomirov

ISBN: 978-1-5108-2613-7
ISSN: 1742-6588

Printed from e-media with permission by:

Curran Associates, Inc.
57 Morehouse Lane
Red Hook, NY 12571

Some format issues inherent in the e-media version may also appear in this print version.

Printed by Curran Associates, Inc. (2016)

For permission requests, please contact the Institute of Physics
at the address below.

Institute of Physics
Dirac House, Temple Back
Bristol BS1 6BE UK

Phone: 44 1 17 929 7481
Fax: 44 1 17 920 0979

techtracking@iop.org

Additional copies of this publication are available from:

Curran Associates, Inc.
57 Morehouse Lane
Red Hook, NY 12571 USA
Phone: 845-758-0400
Fax: 845-758-2633
Email: curran@proceedings.com
Web: www.proceedings.com

Table of contents

Volume 727

MURPHYS-HSFS-2014: 7th International Workshop on MUlti-Rate Processes & HYSteresis (MURPHYS) & the 2nd International Workshop on Hysteresis and Slow-Fast Systems (HSFS)

7–11 April 2014, Berlin, Germany

Accepted papers received: 7 June 2016
Published online: 1 July 2016

Preface

Papers

Journal of Physics: Conference Series **727** (2016) 011001 doi:10.1088/1742-6596/727/1/011001

Foreword

MURPHYS-HSFS-2014 was the *7th International Workshop on **MU**lti-**R**ate **P**rocesses & **HYS**teresis (MURPHYS)* in conjunction with the *2nd International Workshop on **H**ysteresis and **S**low-**F**ast **S**ystems (HSFS)*. It took place at the Weierstrass Institute for Applied Analysis and Stochastics (WIAS), Berlin, Germany, from April 7 to April 11 in 2014.

The international workshop on "Multi-Rate Processes & Hysteresis" continued a series of biennial conferences (Cork, Ireland, 2002–2008; Pecs, Hungary, 2010; Suceava, Romania, 2012) and the international workshop on "Hysteresis and Slow-Fast Systems" was the follow-up of the HSFS-workshop that had taken place in Lutherstadt Wittenberg, Germany, in 2011.

More then 60 scientists from nine European countries and from the USA participated in MURPHYS-HSFS-2014. The program of the workshop featured 49 talks, including 15 main lectures and 15 invited talks. Recent mathematical results for systems with hysteresis operators, multiple scale systems, rate-independent systems, systems with energetic solutions, singularly perturbed systems, and systems with stochastic effects were presented. The considered applications included magnetization dynamics, biological systems, smart materials, networks, ferroelectric and ferroelastic hysteresis, fatigue in materials, market models with hysteresis, biomedical applications, chemical reactions, noise-induced phenomena, partially saturated soils, colloidal films and evaporation of automotive fuel droplets.

Statement of Peer Review: All papers published in this volume of Journal of Physics: Conference Series have been peer reviewed through processes administered by the Editors. Reviews were conducted by expert referees to the professional and scientific standards expected of a proceedings journal published by IOP Publishing.

International steering committee: E. Benoit (France), M. Brokate (Germany), R. Cross (UK), K. Dahmen (USA), M. Dimian (Romania), M. Eleuteri (Italy), G. Friedman (USA), P. Gurevich (Germany), A. Ivanyi (Hungary), L. Kalachev (USA), O. Klein (Germany), D. Knees (Germany), P. Krejčí (Czech Republic), A. Neishtadt (UK), R. O'Malley (USA), N. Popovic (UK), D. Rachinskii (Ireland/USA), S. Sazhin (England), A. Shchepakina (Russia), V. Sobolev (Russia), S. Tikhomirov (Germany), C. Visone (Italy)

International scientific program committee: S. Alonso (Germany), A. Amann (Ireland), P. Andrei (USA), Z. Balanov (USA), M. Brøns (Denmark), D. Flockerzi (Germany), V. Goldshtein (Israel), R. Iyer (USA), A. Korobeinikov (Spain), I. Mayergoyz (USA), A. Mielke (Germany), P. O'Kane (Ireland), J. Sprekels (Germany), A. Stancu (Romania), M. Thomas (Germany), A. Visintin (Italy)

Organization committee: O. Klein (WIAS Berlin, Germany), M. Dimian (Stefan cel Mare University, Suceava, Romania), P. Gurevich (Free University Berlin, Germany), D. Knees (WIAS Berlin/ University of Kassel, Germany), D. Rachinskii (University College Cork, Ireland/ University of Texas, Dallas, USA), S. Tikhomirov (Max Planck Institute for Mathematics in the Sciences, Leipzig, Germany).

Financial support: The workshop was supported by the DFG Research Center MATHEON "Mathematics for key technologies", by the DFG Collaborative Research Center 910: "Control of self-organizing nonlinear systems: Theoretical methods and concepts of application" and by the Weierstrass Institute for Applied Analysis and Stochastics, Berlin.

Collaborative Research Center 910: Control of self-organizing nonlinear systems: Theoretical methods and concepts of application

[1]

O. Klein, M. Dimian, P. Gurevich, D. Knees, D. Rachinskii, S. Tikhomirov
Guest Editors for Journal of Physics: Conference Series

[1] Some logos are omitted because of licensing issues.

Journal of Physics: Conference Series **727** (2016) 011002 doi:10.1088/1742-6596/727/1/011002

Peer review statement

All papers published in this volume of *Journal of Physics: Conference Series* have been peer reviewed through processes administered by the proceedings Editors. Reviews were conducted by expert referees to the professional and scientific standards expected of a proceedings journal published by IOP Publishing.

On the influence of the hysteretic behavior of the capillary pressure on the wave propagation in partially saturated soils

Bettina Albers

Einstein Junior Fellow, TU Berlin, Soil Mechanics and Geotechnical Engineering, Sekr. TIB1-B7, Gustav-Meyer-Allee 25, 13355 Berlin, Germany

E-mail: albers@grundbau.tu-berlin.de

Abstract. It is well known that the capillary pressure curve of partially saturated soils exhibits a hysteresis. For the same degree of saturation it has different values depending on the initial state of the soil, thus for drying of a wet soil or wetting of a dry soil. The influence of these different values of the capillary pressure on the propagation of sound waves is studied by use of a linear hyperbolic model. Even if the model does not contain a hysteresis operator, the effect of hysteresis in the capillary pressure curve is accounted for. In order to obtain the limits of phase speeds and attenuations for the two processes the correspondent values for main drying and main wetting are inserted into the model separately. This is done for two examples of soils, namely for Del Monte sand and for a silt loam both filled by an air-water mixture. The wave analysis reveals four waves: one transversal wave and three longitudinal waves. The waves which are driven by the immiscible pore fluids are influenced by the hysteresis in the capillary pressure curve while the waves which are mainly driven by the solid are not.

1. Introduction

The capillary pressure or suction curve of partially saturated soils (i.e. of soils whose pore space is filled by a mixture of immiscible fluids) exhibits different values for the same degree of saturation depending on whether water is drained or supplied. This phenomenon, known as soil-moisture hysteresis, is discussed in details in [1]. Measuring methods, the prediction of capillary pressure curves from measured data and first results on the influence of drying and wetting on the propagation of sound waves are addressed in [2].

In the present paper the examination of the influence of the hysteresis on the wave propagation is deepened by comparing the behavior of two soil types whose capillary pressure curves along with experimentally observed values have been presented in [3]. The experimental and predicted values of Del Monte sand and "EFEDA" silt loam are illustrated in [3].

For the sand the main drying curve (MDC) has been measured. The main wetting curve (MWC) has been predicted using the measured data. Measurements of the capillary pressure curves are time consuming and often one measurement technique is not sufficient to cover the large range of capillary pressures which mostly spans certain orders of magnitude. For the silt loam even both main curves have been predicted from measurements of the primary wetting curve (PWC) and scanning drying curves (SDC). The latter are inner hysteresis curves which appear upon redrying and rewetting. The prediction method, proposed by Haverkamp et al. [3]

implies that all drying and wetting curves, regardless of the scanning order, have the shape of the van Genuchten equation [4] in normalized form

$$\theta^* = \frac{\theta - \theta_r}{\theta_S - \theta_r} = \left[1 + \left(\frac{h}{h_g}\right)^n\right]^{-m}, \tag{1}$$

where θ is the volumetric water content, θ_r the residual water content, θ_s the water content at natural saturation, h the soil water pressure head or matric potential, h_g a van Genuchten pressure head scale parameter and n and m are van Genuchten shape parameters.

This leads to the following equations for the MWC and the MDC

$$\theta^*_{mw} \equiv \frac{\theta_{mw}}{\theta_{Smw}} = \left[1 + \left(\frac{h}{h_{gmw}}\right)^{n_{mw}}\right]^{-m_{mw}}, \quad \theta^*_{md} \equiv \frac{\theta_{md}}{\theta_{Smd}} = \left[1 + \left(\frac{h}{h_{gmd}}\right)^{n_{md}}\right]^{-m_{md}}, \tag{2}$$

where the subscripts mw and md refer to main wetting and main drying, respectively. The relations between the specific wetting parameters m_{mw}, n_{mw} and h_{gmw} and the drying parameters m_{md}, n_{md} and h_{gmd} are specified in [3] in the following way

$$\begin{matrix} m_{mw} = m_{md}, \\ n_{mw} = n_{md}, \end{matrix} \quad \text{and} \quad h_{gmd} = 2h_{gmw}, \quad \text{for} \quad \begin{cases} m_{mw}n_{mw} \geq 1, \\ m_{md}n_{md} \geq 1. \end{cases} \tag{3}$$

The prediction of several curves of different scanning order from measured data has been demonstrated in [3] for 22 different soil samples. Two of them (sand and silt loam) are used here to calculate the wave speeds and attenuations of the waves occurring if the pore space of the soil contains the immiscible pore fluids water and air. The possible range of the acoustic properties is determined by calculating them for the limit cases. This means, that separately the values of the MDC and of the MWC are used in the relation between capillary pressure and saturation.

The material properties of the two soil types are summarized in Table 1. Some of them are given in [3]. Others have been calculated using e.g. the value of the earth acceleration $g = 9.81$ m s^{-2} and the conversion of the pressure head in cm H_2O (used in [3]) into the capillary pressure in Pa (used in the continuum model)

$$\text{pressure head } h \text{ [m H}_2\text{O]} = \frac{\text{capillary pressure } p_c \text{ [Pa]}}{\rho^{FR} \text{ [kg m}^{-3}] \; g \text{ [m s}^{-2}]} \quad \Rightarrow \quad 1[\text{cm H}_2\text{O}] = 100[\text{Pa}].$$

The dimensional van Genuchten scale parameters α for drying and wetting, used in the continuum model, have been calculated from the head scale parameters given in [3] by

$$\alpha_{d/w} \text{ [Pa]} = \frac{1}{10 \; g \text{ [m s}^{-2}] \; h_{gd/gw} \text{ [cm]}}.$$

The material parameters connected with the properties of the solid, as e.g. porosity, mass density and Poisson's ratio, have been ascertained in the literature.

The example of Del Monte sand is widely studied and the material parameters for this soil type, given in [5], have been used even if they deviate from experienced data for sandy soils classified in the German standard DIN 4220 (compare [6]). For the porosity of Del Monte sand the authors of [5] refer to measurements of Liakopoulos who measured also the MDC data [7]. The reported value of the initial porosity $n_0 = 0.2975$ is used also here even if in the literature more plausible bigger values between 34.5% and 37% are mentioned. Also the Poisson ratio of 0.4 seems rather high and the solid grain density of 2000 kg m^{-3} rather low.

Table 1. Material properties of Del Monte sand and loam silt filled by an air-water mixture.

	Del Monte sand	silt loam
real compressibility grains K_s [GPa]	35	35
real compressibility fluid K_f [GPa]	2.25	2.25
real compressibility gas K_g [MPa]	0.101	0.101
Poisson's ratio ν [-]	0.4	0.35
shear modulus μ^S [GPa]	0.4724	0.4487
solid grain density ρ^{SR} [kg m^{-3}]	2000	2650
fluid density ρ^{FR} [kg m^{-3}]	1000	1000
initial porosity n_0 [-]	0.2975	0.5
intrinsic permeability k [m^2]	$4.5 \cdot 10^{-13}$	$1.92 \cdot 10^{-13}$
water conductivity $\mathrm{K} = k\frac{\rho^{FR}g}{\mu_w}$ [m/s]	$4.44 \cdot 10^{-6}$	$1.888 \cdot 10^{-6}$
water resistance $\pi^F = \frac{n_0\rho^{FR}g}{\mathrm{K}}$ [kg m^{-3}s^{-1}]	$6.573 \cdot 10^8$	$2.598 \cdot 10^9$
air resistance π^G [kg m^{-3}s^{-1}]	$1.82 \cdot 10^5$	$1.82 \cdot 10^5$
water viscosity μ_w [Pa s]	$1 \cdot 10^{-3}$	$1 \cdot 10^{-3}$
air viscosity μ_a [Pa s]	$1.82 \cdot 10^{-5}$	$1.82 \cdot 10^{-5}$
van Genuchten parameter n [-]	4.150	4.556
van Genuchten parameter m [-]	0.518	0.561
h_g for drying [cm]	116.56	720.520
h_g for wetting [cm]	58.28	360.26
α for drying [Pa]	0.000087454	0.000014147
α for wetting [Pa]	0.000174909	0.000028295

The material parameters of the silt loam (see Table 1) are in better accordance to the values given for similar soil types classified in DIN 4220.

Two different approaches for the estimation of the shear modulus have been compared in [8]. Here, the expression stemming from classical elasticity is used

$$\mu^S = \frac{3}{2}\frac{1-2\nu}{1+\nu}\frac{K_s}{1+50n_0}. \tag{4}$$

The last term is the widely used Geertsma formula for the drained compressibility modulus $K_d = K_s/(1+50n_0)$ (see e.g. [9]).

Also at high values of capillary pressure a certain quantity of water (the amount depends on the soil type) remains in the sample. For both soil types this so-called irreducible saturation is very small. Therefore, nearly the whole range of initial saturations $0.01 \leq S_0 \leq 1$ is studied in this paper.

2. Linear model for partially saturated soils

The book [6] deals with the modeling of partially saturated soils and the analysis of sound wave propagation in one-, two- and three-component media. A hyperbolic linear model which is able to describe the wave propagation is introduced and analyzed. However, this model only accounts for one branch of the capillary pressure curve. The saturation-capillary pressure-relation enters the model in form of the van Genuchten equation [4] as a constitutive law. Since this relation and also the measurable material parameters are microscopic but the model is macroscopic, all quantities have to be transferred from the micro- to the macro-scale. The macroscopic material parameters $\left\{\lambda^S + \frac{2}{3}\mu^S, \kappa^F, \kappa^G, Q^F, Q^G, Q^{FG}\right\}$ follow from the initial porosity n_0, the real compressibilities of solid, fluid and gas, K_s, K_f and K_g, respectively, together with the above quoted Geertsma formula and the van Genuchten law (1) in the notation of the continuum model

$$p_c = \frac{1}{\alpha}\left[S^{(-1/m)} - 1\right]^{1/n}, \tag{5}$$

where S denotes the degree of saturation, p_c the capillary pressure and α, n and m the van Genuchten parameters. λ^S and μ^S are Lamé parameters, κ^F and κ^G are the compressibilities of fluid and gas. Q^F, Q^G and Q^{FG} are coupling parameters between solid-fluid, solid-gas and fluid-gas, respectively. They appear in the macroscopic set of field equations

$$\begin{aligned}
\rho_0^S \frac{\partial \mathbf{v}^S}{\partial t} &= \operatorname{div}\left\{\lambda^S e\mathbf{1} + 2\mu^S \mathbf{e}^S + Q^F \varepsilon^F \mathbf{1} + Q^G \varepsilon^G \mathbf{1}\right\} + \\
&\quad + \pi^{FS}\left(\mathbf{v}^F - \mathbf{v}^S\right) + \pi^{GS}\left(\mathbf{v}^G - \mathbf{v}^S\right), \\
\rho_0^F \frac{\partial \mathbf{v}^F}{\partial t} &= \operatorname{grad}\left\{\rho_0^F \kappa^F \varepsilon^F + Q^F e + Q^{FG}\varepsilon^G\right\} - \pi^{FS}\left(\mathbf{v}^F - \mathbf{v}^S\right), \\
\rho_0^G \frac{\partial \mathbf{v}^G}{\partial t} &= \operatorname{grad}\left\{\rho_0^G \kappa^G \varepsilon^G + Q^G e + Q^{FG}\varepsilon^F\right\} - \pi^{GS}\left(\mathbf{v}^G - \mathbf{v}^S\right), \\
\frac{\partial \mathbf{e}^S}{\partial t} = \operatorname{sym}\operatorname{grad}\mathbf{v}^S, &\quad \frac{\partial \varepsilon^F}{\partial t} = \operatorname{div}\mathbf{v}^F, \quad \frac{\partial \varepsilon^G}{\partial t} = \operatorname{div}\mathbf{v}^G, \quad e \equiv \operatorname{tr}\mathbf{e}^S.
\end{aligned} \tag{6}$$

The fields $\left\{\mathbf{v}^S, \mathbf{v}^F, \mathbf{v}^G, \mathbf{e}^S, \varepsilon^F, \varepsilon^G\right\}$ are the velocities of the three components, the macroscopic deformation tensor $\mathbf{e}^S$ and the volume changes of fluid and gas, respectively.

The volume changes of the solid, fluid and gas, $e, \varepsilon^F, \varepsilon^G$, are defined by

$$e = \frac{\rho_0^S - \rho^S}{\rho_0^S}, \quad \varepsilon^F = \frac{\rho_0^F - \rho^F}{\rho_0^F}, \quad \varepsilon^G = \frac{\rho_0^G - \rho^G}{\rho_0^G}, \tag{7}$$

where ρ^S, ρ^F and ρ^G are the partial mass densities of the components. Quantities with subindex zero are initial values of the corresponding current quantities.

Moreover, in (6) the relative resistances π^{FS} and π^{GS} of the flow of the pore fluids through the channels of the skeleton occur. Taking into account formulae for the relative permeabilities k_f and k_g proposed by van Genuchten [4]

$$k_f = S^{\frac{1}{2}}\left[1 - \left(1 - S^{\frac{1}{m}}\right)^m\right]^2, \qquad k_g = (1-S)^{\frac{1}{3}}\left(1 - S^{\frac{1}{m}}\right)^{2m}. \tag{8}$$

and the parameters π^F and π^G incorporating the permeability of the solid and the viscosity of the pore fluid (see Table 1) the resistances are calculated by

$$\pi^{FS} = \frac{\pi^F}{k_f}, \qquad \pi^{GS} = \frac{\pi^G}{k_g}. \tag{9}$$

3. Propagation of sound waves

By means of the set (6) the propagation of sound waves in partially saturated sand and silt loam is studied. In order to describe the propagation of plane monochromatic waves in an infinite medium whose fronts are perpendicular to the unit vector in the direction of propagation $\mathbf{n}$, we suppose

$$\begin{aligned}
\varepsilon^F = E^F\mathcal{E}, \quad \varepsilon^G = E^G\mathcal{E}, \quad \mathbf{e}^S = \mathbf{E}^S\mathcal{E}, \\
\mathbf{v}^F = \mathbf{V}^F\mathcal{E}, \quad \mathbf{v}^G = \mathbf{V}^G\mathcal{E}, \quad \mathbf{v}^S = \mathbf{V}^S\mathcal{E}, \\
\mathcal{E} := \exp i\left(\mathbf{k}\cdot\mathbf{x} - \omega t\right),
\end{aligned} \tag{10}$$

where $\mathbf{E}^S, E^F, E^G, \mathbf{V}^S, \mathbf{V}^F, \mathbf{V}^G$ are constant amplitudes, ω is a given frequency and $\mathbf{k} = k\mathbf{n}$ is the wave vector with the complex wave number k.

The compatibility relations

$$E^F = -\frac{1}{\omega}k\mathbf{n}\cdot\mathbf{V}^F, \quad E^G = -\frac{1}{\omega}k\mathbf{n}\cdot\mathbf{V}^G,$$
$$\mathbf{E}^S = -\frac{1}{2\omega}k\left(\mathbf{n}\otimes\mathbf{V}^S + \mathbf{V}^S\otimes\mathbf{n}\right), \quad \text{i.e.} \quad e = -\frac{1}{\omega}k\mathbf{n}\cdot\mathbf{V}^S\mathcal{E}, \tag{11}$$

follow from substitution of (10) into the last line of (6). Using (11) in the remaining field equations leads to

$$\begin{aligned}\omega^2\mathbf{V}^S = &\frac{\lambda^S}{\rho_0^S}k^2\left(\mathbf{V}^S\cdot\mathbf{n}\right)\mathbf{n}+\frac{\mu^S}{\rho_0^S}k^2\left(\left(\mathbf{V}^S\cdot\mathbf{n}\right)\mathbf{n}+\mathbf{V}^S\right)+\\&+\frac{Q^F}{\rho_0^S}k^2\left(\mathbf{V}^F\cdot\mathbf{n}\right)\mathbf{n}+\frac{Q^G}{\rho_0^S}k^2\left(\mathbf{V}^G\cdot\mathbf{n}\right)\mathbf{n}+\\&+i\frac{\pi^{FS}\omega}{\rho_0^S}\left(\mathbf{V}^F-\mathbf{V}^S\right)+i\frac{\pi^{GS}\omega}{\rho_0^S}\left(\mathbf{V}^G-\mathbf{V}^S\right)=0,\end{aligned} \tag{12}$$

$$\begin{aligned}\omega^2\mathbf{V}^F = &\kappa^F k^2\left(\mathbf{V}^F\cdot\mathbf{n}\right)\mathbf{n}+\frac{Q^F}{\rho_0^F}k^2\left(\mathbf{V}^S\cdot\mathbf{n}\right)\mathbf{n}+\\&+\frac{Q^{FG}}{\rho_0^F}k^2\left(\mathbf{V}^G\cdot\mathbf{n}\right)\mathbf{n}-i\frac{\pi^{FS}\omega}{\rho_0^F}\left(\mathbf{V}^F-\mathbf{V}^S\right)=0,\end{aligned} \tag{13}$$

$$\begin{aligned}\omega^2\mathbf{V}^G = &\kappa^G k^2\left(\mathbf{V}^G\cdot\mathbf{n}\right)\mathbf{n}+\frac{Q^G}{\rho_0^G}k^2\left(\mathbf{V}^S\cdot\mathbf{n}\right)\mathbf{n}+\\&+\frac{Q^{FG}}{\rho_0^G}k^2\left(\mathbf{V}^F\cdot\mathbf{n}\right)\mathbf{n}-i\frac{\pi^{GS}\omega}{\rho_0^G}\left(\mathbf{V}^G-\mathbf{V}^S\right)=0.\end{aligned} \tag{14}$$

The transversal wave follows, if the scalar product of (12)-(14) with an arbitrary unit vector $\mathbf{n}_\perp$ perpendicular to $\mathbf{n}$, i.e. $\mathbf{n}\cdot\mathbf{n}_\perp = 0$, is taken. This yields

$$\begin{aligned}&\omega^2 V_\perp^S = \frac{\mu^S}{\rho_0^S}k^2V_\perp^S + i\frac{\pi^{FS}\omega}{\rho_0^S}\left(V_\perp^F - V_\perp^S\right)+i\frac{\pi^{GS}\omega}{\rho_0^S}\left(V_\perp^G-V_\perp^S\right),\\&\omega^2V_\perp^F = -i\frac{\pi^{FS}\omega}{\rho_0^F}\left(V_\perp^F-V_\perp^S\right), \qquad \omega^2V_\perp^G = -i\frac{\pi^{GS}\omega}{\rho_0^G}\left(V_\perp^G-V_\perp^S\right),\\&V_\perp^S := \mathbf{V}^S\cdot\mathbf{n}_\perp, \quad V_\perp^F := \mathbf{V}^F\cdot\mathbf{n}_\perp, \quad V_\perp^G := \mathbf{V}^G\cdot\mathbf{n}_\perp,\end{aligned} \tag{15}$$

and finally the dispersion relation for the shear wave.

In order to obtain relations for longitudinal waves the scalar product of equations (12)-(14) with the vector $\mathbf{n}$ is taken:

$$\begin{aligned}&\left\{\omega^2-\frac{\lambda^S+2\mu^S}{\rho_0^S}k^2+i\frac{\left(\pi^{FS}+\pi^{GS}\right)\omega}{\rho_0^S}\right\}V_\parallel^S-\\&-\left\{\frac{Q^F}{\rho_0^S}k^2+i\frac{\pi^{FS}\omega}{\rho_0^S}\right\}V_\parallel^F-\left\{\frac{Q^G}{\rho_0^S}k^2+i\frac{\pi^{GS}\omega}{\rho_0^S}\right\}V_\parallel^G=0,\\&-\left\{\frac{Q^F}{\rho_0^F}k^2+i\frac{\pi^{FS}\omega}{\rho_0^F}\right\}V_\parallel^S+\left\{\omega^2-\kappa^Fk^2+i\frac{\pi^{FS}\omega}{\rho_0^F}\right\}V_\parallel^F-\frac{Q^{FG}}{\rho_0^F}k^2V_\parallel^G=0,\\&-\left\{\frac{Q^G}{\rho_0^G}k^2+i\frac{\pi^{GS}\omega}{\rho_0^G}\right\}V_\parallel^S-\frac{Q^{FG}}{\rho_0^G}k^2V_\parallel^F+\left\{\omega^2-\kappa^Gk^2+i\frac{\pi^{GS}\omega}{\rho_0^G}\right\}V_\parallel^G=0,\end{aligned} \tag{16}$$

Journal of Physics: Conference Series **727** (2016) 012001 doi:10.1088/1742-6596/727/1/012001

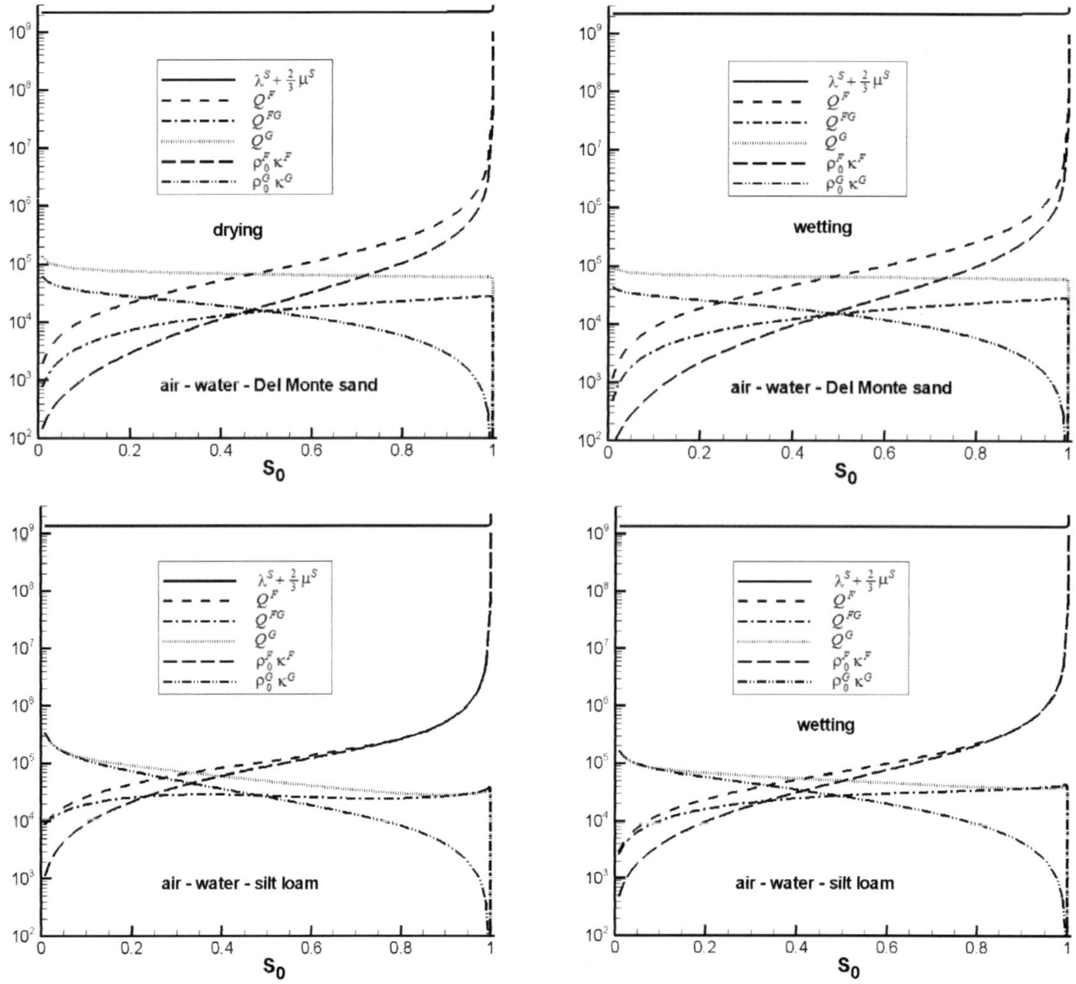

Figure 1. Macroscopic material parameters appearing in (6) in dependence on the initial saturation. Top row: Del Monte sand, bottom row: silt loam; for the parameters on the left hand side the MDC data, for those on the right hand side the MWC data have been used, respectively.

where

$$V_{\|}^{S} = \mathbf{V}^{S} \cdot \mathbf{n}, \quad V_{\|}^{F} = \mathbf{V}^{F} \cdot \mathbf{n}, \quad V_{\|}^{G} = \mathbf{V}^{G} \cdot \mathbf{n}. \tag{17}$$

The dispersion relations which, due to their lengthy form are not shown here, are solved for the complex wave number k. The results specify both the phase velocities $c_{ph} = \omega/\mathrm{Re}(k)$ and the attenuations $\mathrm{Im}(k)$.

4. Numerical results

The behavior of the waves propagating in Del Monte sand filled by an air-water mixture has been studied already in [10]. These results are compared here to those of a silt loam. As mentioned above, both soil types belong to those for which the capillary pressure curves are presented in [3].

4.1. Material parameters

The macroscopic material parameters of the model introduced in Section 2 have been calculated for both soil types from the microscopic material parameters summarized in Table 1. Once MDC data and once MWC data are used. The results are illustrated in Figure 1.

While for the sand a difference between the drying and the wetting data is hardly discernible, for the silt loam slight differences are visible. This does not only concern the numerical values but also the shape and, thus, the points of intersection of the curves.

We are interested in the influence of the consideration of different drying and wetting capillary pressure data on the propagation of sound waves. The behavior of the phase speeds and of the attenuations of the four appearing waves is examined in dependence on the frequency and on the initial saturation.

4.2. Frequency dependence

Figures 2 and 3 show for a single value of the initial saturation ($S_0 = 0.8$) the dependence of the phase speeds $\omega/\mathrm{Re}(k)$ and the attenuations $\mathrm{Im}(k)$ on the frequency. It is obvious that for

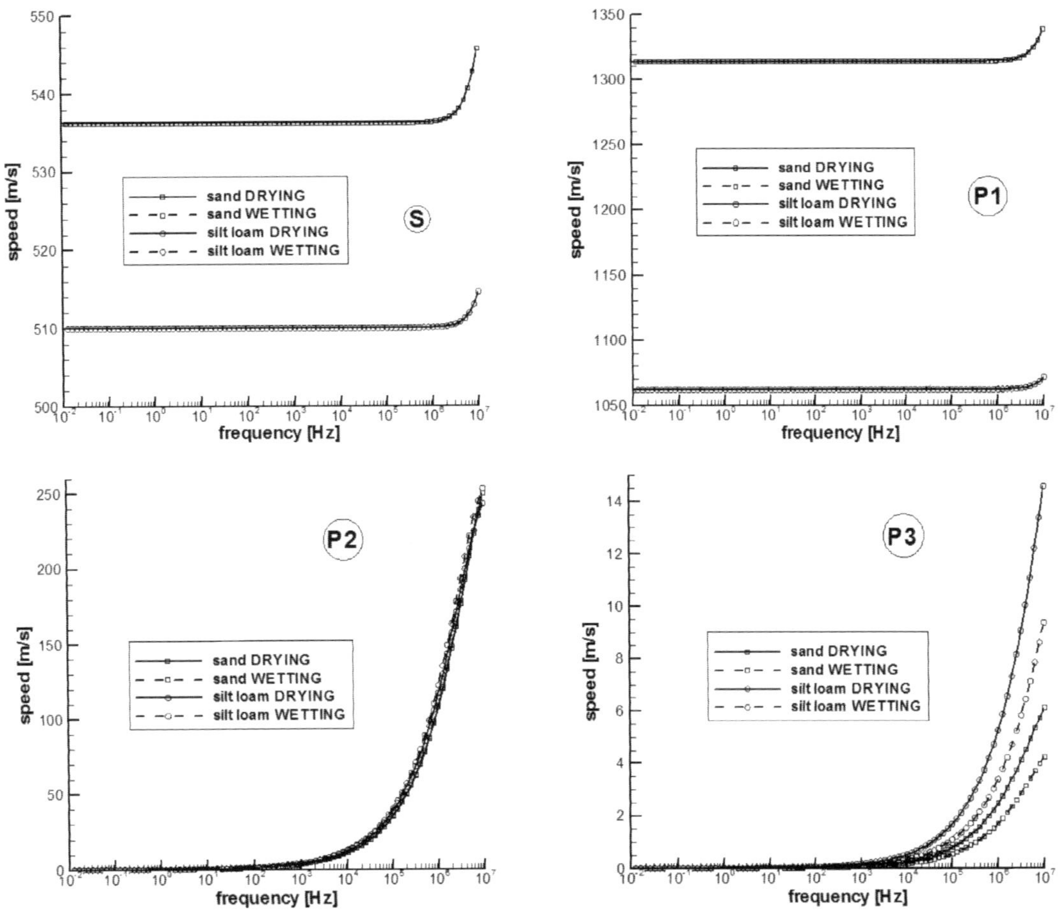

Figure 2. Phase speeds of the transversal wave S and the three longitudinal waves $P1$, $P2$ and $P3$ in dependence on the frequency ω using the parameters of the MDC (solid lines) and MWC (dashed lines) for Del Monte sand (squares) and silt loam (circles); $S_0 = 0.8$.

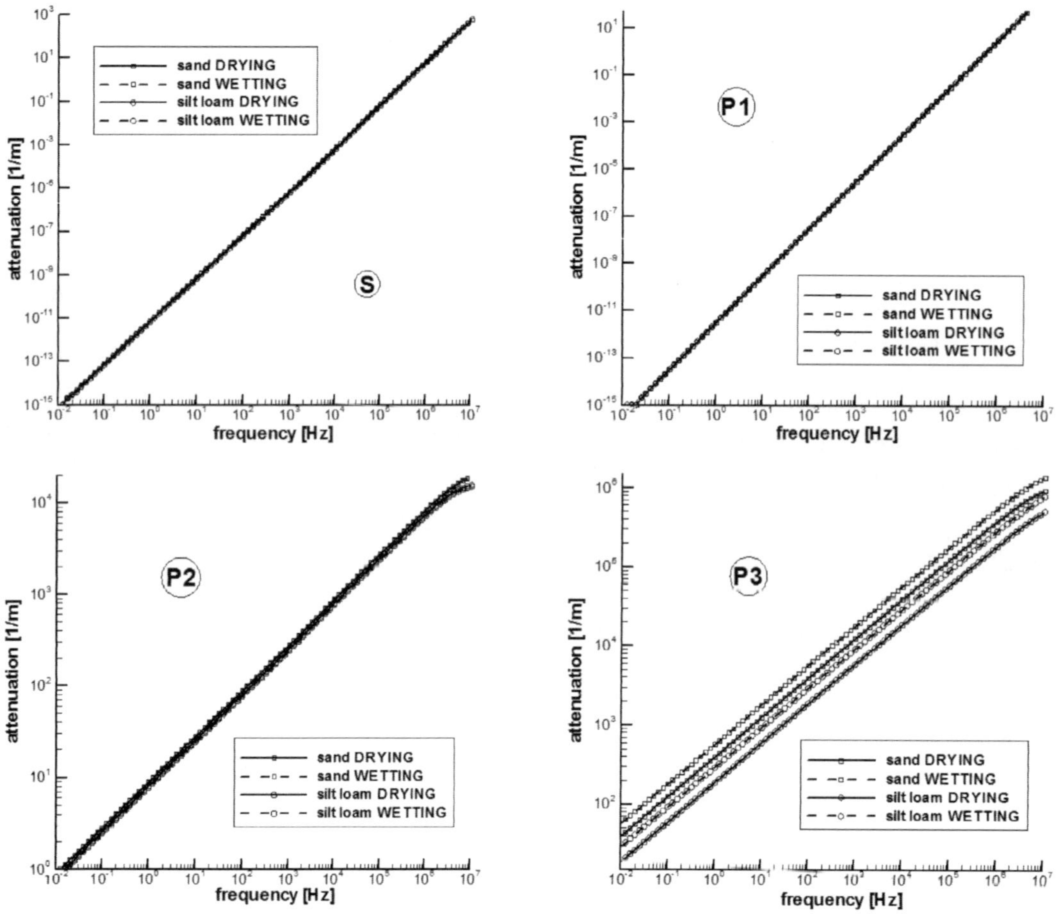

Figure 3. Attenuations of the transversal wave S and the three longitudinal waves $P1$, $P2$ and $P3$ in dependence on the frequency ω using the parameters of the MDC (solid lines) and MWC (dashed lines) for Del Monte sand (squares) and silt loam (circles); $S_0 = 0.8$.

the waves which are mainly influenced by the solid (the shear wave and the fast longitudinal wave $P1$) there is no difference at all between wetting and drying curves. Also for the waves which are driven by the pore fluids a difference occurs only for frequencies which do not appear in geophysical applications. In the next subsection we will have a closer look on what happens in dependence on the saturation for a smaller value of the frequency. The $P2$-wave appears also in saturated porous media and its speed depends on the compressibility of the pore fluid. It is slower than the first longitudinal wave. The third longitudinal wave appears due to capillary effects, i.e. it only appears if at least two immiscible pore fluids are present. Its speed is, again, much smaller than this of the $P2$-wave. Comparing the two soil types it can be observed that the speeds of the S- and $P1$-waves are higher for sand (due to the higher shear modulus and the smaller true solid mass density) while the speeds of the remaining two waves are higher in silt loam. However, the differences are small.

The same behavior becomes obvious inspecting the frequency dependence of the attenuations illustrated in Figure 3. Attenuations for drying and wetting data are the same for S and $P1$, barely distinguishable for $P2$ and different for $P3$. Presumably, the reason for the differences of the latter wave becoming evident is that the attenuation of this wave is extremely high. While

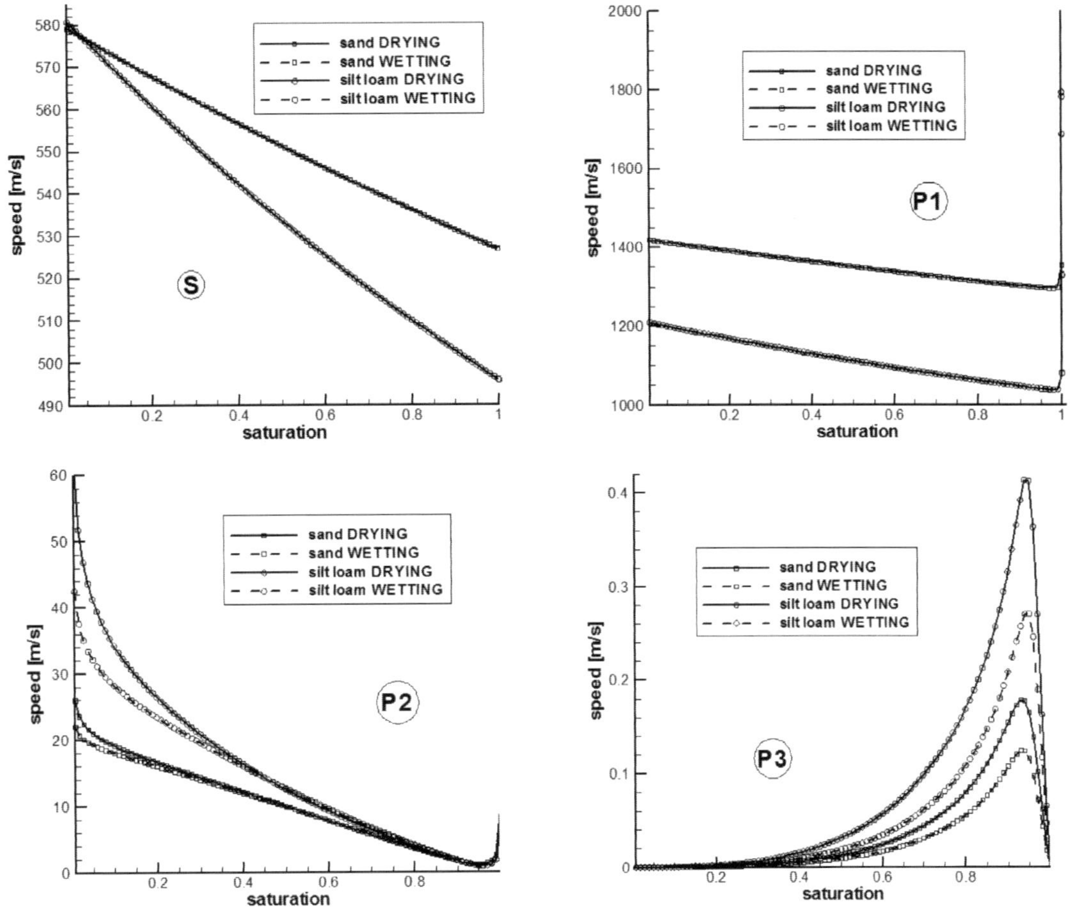

Figure 4. Phase speeds of the transversal wave S and the three longitudinal waves $P1$, $P2$ and $P3$ in dependence on the initial saturation S_0 using the parameters of the MDC (solid lines) and MWC (dashed lines) for Del Monte sand (squares) and silt loam (circles); $\omega = 1000$ Hz.

in the region of frequencies appearing in geophysical applications the attenuations of S- and $P1$-wave are of the order of 10^{-15} to 10^{-5} 1/m, for the $P3$-waves values of 10^1 to 10^4 1/m are observable. Such high values are barely measurable in practice. Already for the $P2$-wave whose attenuations for such frequencies are in the range of 1 to 100 1/m measurements were sophisticated but finally succeeded [11]. Attempts to measure speeds (very low) or attenuations (very high) of the $P3$-wave are not known. The theoretical values of the attenuation are bigger for drying than for wetting and higher for silt loam than for sand.

4.3. Dependence on the initial saturation

More interesting than the frequency dependence is the dependence of wave speeds and attenuations on the initial degree of saturation. To analyze this behavior we choose a frequency of 1000 Hz which is typical for geophysical applications.

The shear wave speed shows an almost linear behavior. It decreases for increasing degree of saturation. For a small initial saturation the speed for drying and wetting nearly coincides. It has a value of around 580 m/s. The decrease for bigger values of saturation is steeper for silt loam than it is for sand. While for silt loam it is approximately 14%, for sand around 9% are observable.

The speed of the $P1$-wave shows a behavior which may be useful for the development of nondestructive testing methods for partially saturated soils. In dependence on the saturation it slightly linearly decreases and approximately doubles its value for high degrees of saturation. The speeds do not differ for drying and wetting. For small saturation the $P1$-speed for sand is approximately 1400 m/s, for silt loam around 1200 m/s. The two curves for increasing values of the saturation are nearly parallel.

An influence of the hysteresis in the capillary pressure curve is visible for the speeds of $P2$- and $P3$-wave. These speeds behave oppositely. While the $P2$-wave has a minimum for a certain degree of saturation (for both soil types in the region of 95%), the $P3$-wave shows a maximum for the same degree of saturation. The existence of a minimum in the speed of sound in air-water-mixtures is well known for a long time (e.g. [12]) and occurs in a comparable way also in partially saturated porous media.

Generally, the speeds of the pore-fluid driven waves are much less than those of the solid driven waves. The $P2$-wave speed lies in the range between 0 and 60 m/s, this of the $P3$-wave between 0 and 0.5 m/s. The maximum values are higher for silt loam than for sand and also

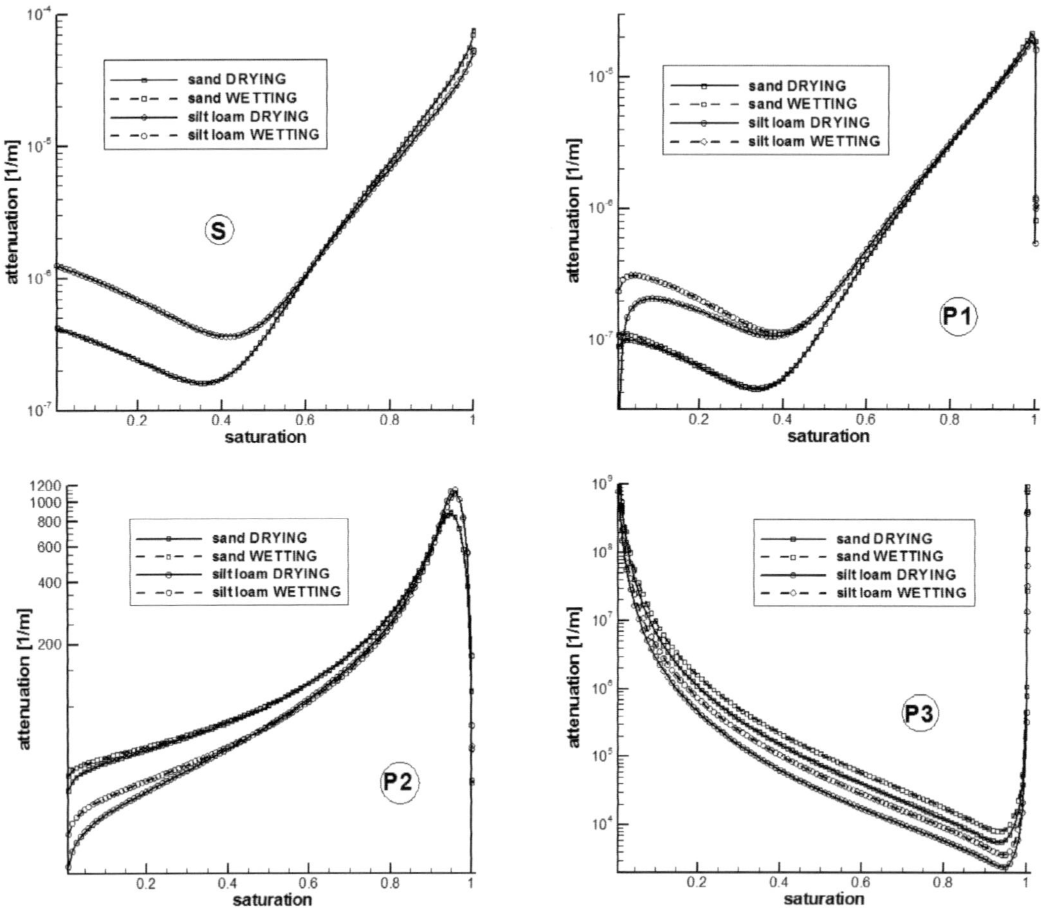

Figure 5. Attenuations of the transversal wave S and the three longitudinal waves $P1$, $P2$ and $P3$ in dependence on the initial saturation S_0 using the parameters of the MDC (solid lines) and MWC (dashed lines) for Del Monte sand (squares) and silt loam (circles); $\omega = 1000$ Hz.

the differences between drying and wetting are more pronounced for silt loam than for sand. Namely, the maximum value for the $P2$-speed for drying in silt loam is approximately 60 m/s, for wetting 42 m/s. Thus, the difference is around 30% whereas the values of 26 and 22 m/s in sand result in a difference of approximately 15%. Also for the speed of the $P3$-wave a difference nearly twice as big as for sand is observed for silt loam. While the differences occur for the $P2$-wave in the region of low saturations, for the $P3$-wave it happens in the region of high degrees of saturation.

The differences in the attenuations are not that obvious as for the speeds. Again, S- and $P1$-wave are little attenuated. Values between 10^{-7} and 10^{-4} 1/m appear. All curves exhibit a minimum at a degree of saturation of approximately 40%. For the shear wave drying and wetting curves do not show a difference, for low values of the saturation the attenuation of silt loam is higher than this for sand; for saturations higher than approximately 50% the attenuations of the two soil types are nearly the same. This behavior can be also observed for the attenuations of the $P1$-wave. Interestingly, for small degrees of saturation, for this wave in silt loam a difference between drying and wetting is visible (for the speed this was not the case). For the sand this behavior is barely observable.

As mentioned above, the $P2$-wave and, particularly, the $P3$-wave are strongly attenuated. The attenuations of these waves exhibit maxima for degrees of saturation at which the speed shows minima and vice versa. The attenuations of silt loam are slightly smaller than those of sand. While for the $P3$-wave the attenuation for drying and wetting differs for both soil types in nearly the whole range of saturations (except for the limit values), for the $P2$-wave this is only the case for small degrees of saturation.

Conclusions and outlook

In [10] the influence of the hysteresis in the capillary pressure curve already had been studied on the example of Del Monte sand filled by an air-water mixture. Four waves, one transversal wave and three longitudinal waves, had been analyzed. For the waves driven mainly by the skeleton (the shear wave and the fastest longitudinal wave) it could be shown that – as expected – the influence of the use of either drying or wetting data is negligible. However, for the pore fluid driven waves a smaller influence than expected had been emerged. Several possible reasons for this behavior had been mentioned: that the porosity had been chosen rather small, that only one soil type had been investigated and that the prediction of the main wetting curve had not been proven by experimental data.

By investigation of a second soil type and with a choice of material parameters which seems much more plausible, in this paper, a significant influence of drying and wetting on the propagation of sound waves in silt loam could be observed. It could be imagined that the influence in silt loam, indeed, is considerably larger than in sand. However, more likely, the reason of the small influence in sand is the choice of the material parameters.

In order to clarify the above mentioned questions, measurements of the main drying and wetting curves of the twelve soil types studied in [6] have been requested in a laboratory and will be interpreted theoretically once available. Then, several soil types for which significant material parameters are available can be compared, the quality of the prediction of the main curves can be checked and a sufficient number of soil types can be analyzed.

Acknowledgments

Financial support as an Einstein Junior Fellow by the Einstein Foundation Berlin is highly appreciated.

Journal of Physics: Conference Series **727** (2016) 012001 doi:10.1088/1742-6596/727/1/012001

References

[1] Albers B 2014 Modeling the hysteretic behavior of the capillary pressure in partially saturated porous media – a review, *Acta Mechanica* **225**(8), 2163–89

[2] Albers B 2015 Main drying and wetting curves of soils - on measurements, prediction and influence on wave propagation, accepted for publication in *Engineering Transactions*

[3] Haverkamp R, Reggiani P, Ross PJ and Parlange, J-Y 2002 Soil water hysteresis prediction model based on theory and geometric scaling. In *Environmental Mechanics, Water, Mass and Engergy Transfer in the Biosphere*, PAC Raats, D Smiles, AW Warrick, Ed. 213–46

[4] Van Genuchten MT 1980 A closed-form equation for predicting the hydraulic conductivity of unsaturated soils. *Soil Sci. Soc. Am. J.* **44**:892–8

[5] Lewis RW and Schrefler BA 1998 *The finite element method in the static and dynamic deformation and consolidation of porous media.* Wiley, Chichester

[6] Albers B 2010 *Modeling and Numerical Analysis of Wave Propagation in Saturated and Partially Saturated Porous Media.* Habilitation thesis, Shaker Verlag, Aachen

[7] Liakopoulos AC 1966 Theoretical approach to the solution of the infiltration problem. *International Association of Scientific Hydrology* **11**(1) 69–110

[8] Albers B 2010 Micro-Macro Transition and Linear Wave Propagation in Three-Component Compacted Granular Materials, AIP Conference Proceedings of the Joint IUTAM-ISIMM Symposium on Mathematical Modeling and Physical Instances of Granular Flows, vol. 1227, JD Goddard, JT Jenkins, P Giovine (Eds.), Melville, New York, 391-404

[9] White JE 1983 *Underground sound. Application of seismic waves, Methods in Geochemistry and Geophysics.* Vol. 18. Elsevier, Amsterdam, New York

[10] Albers B 2014 On modeling three-component porous media incorporating hysteresis, Proceedings of the 11th. World Congress on Computational Mechanics (WCCM XI), paper 2568

[11] Kelder O and Smeulders D 1996 Measurement of ultrasonic bulk properties of water-saturated porous media, in: EAGE, 58th Conference and Technical Exhibition, Amsterdam

[12] Wood, AB 1957 *A Textbook of Sound.* G. Bell and Sons, London

Reentry produced by small-scale heterogeneities in a discrete model of cardiac tissue

Sergio Alonso
Department of Applied Physics, Universitat Politcnica de Catalunya - BarcelonaTech, Barcelona, Spain
E-mail: s.alonso@upc.edu

Markus Bär
Physikalisch-Technische Bundesanstalt, Abbestrasse 2-12, 10587 Berlin, Germany
E-mail: Markus.Baer@ptb.de

Abstract. Reentries are reexcitations of cardiac tissue after the passing of an excitation wave which can cause dangerous arrhythmias like tachycardia or life-threatening heart failures like fibrillation. The heart is formed by a network of cells connected by gap junctions. Under ischemic conditions some of the cells lose their connections, because gap junctions are blocked and the excitability is decreased. We model a circular region of the tissue where a fraction of connections among individual cells are removed and substituted by non-conducting material in a two-dimensional (2D) discrete model of a heterogeneous excitable medium with local kinetics based on electrophysiology. Thus, two neighbouring cells are connected (disconnected) with a probability ϕ $(1-\phi)$. Such a region is assumed to be surrounded by homogeneous tissue. The circular heterogeneous area is shown to act as a source of new waves which reenter into the tissue and reexcitate the whole domain. We employ the Fenton-Karma equations to model the action potential for the local kinetics of the discrete nodes to study the statistics of the reentries in two dimensional networks with different topologies. We conclude that the probability of reentry is determined by the proximity of the fraction of disrupted connections between neighboring nodes ("cells") in the heterogeneous region to the percolation threshold.

1. Introduction

Cardiac tissue is an example of an excitable media where electrical travelling waves propagate. The propagation of such electrical signal through the tissue synchronizes the contraction of all the fibers of the heart. The electrical signal corresponds to a pulse-like wave of membrane action potential, which is around 5 cm wide, that propagates at high speed, 0.5 m/s [1]. The general propagation properties of such waves are similar to chemical waves in the Belousov-Zhabotinsky reaction or the oxidation of CO on catalytic surfaces [2]. In excitable media waves may break up giving rise to spiral waves of excitation [3]. In cardiac tissue such spiral waves have been experimentally observed [4] and are associated with tachycardia [1]. The presence of multiple spirals and spatiotemporally chaotic dynamics is related to ventricular fibrillation [5].

The electrical wave is generated by the depolarization of the cellular membrane of the myocytes. The depolarization permits the influx of different types of ions and in particular of calcium, which induces the contraction of the cardiac fibers. Myocytes are electrically active cells 100-120 μm long and 10-20 μm wide and are coupled by gap junctions. Such junctions permit the propagation of the membrane depolarization to the neighbouring cells and maintain wave propagation. Research on modeling of the electrical activity of the myocytes has produced models with a large number of equations whose details depend on the type of animal or the level of detail in the description. For a survey of physiological models see for example [6, 7].

The large spatial scale of the electrical propagating waves in comparison with the size of the myocytes, has usually permitted the use of homogenization theories to generate continuous models of action potential propagation in cardiac tissue [3]. However, during the last years more detailed models of cardiac tissue including the individual cells have appeared in two [8] and three [9] dimensional tissue. It permits also the inclusion of other types of cells like fibroblast [10] or non-conducting tissue originated by fibrosis [11].

The inclusion of heterogeneities at the cellular level affects the dynamics of the macroscopic waves. The velocity, for example, decreases with the number of included heterogeneities [12, 13, 14] because the connectivity between the cells is reduced and the communication is slower [15]. The decrease of the velocity implies a reduction of the excitability of the medium and it may preclude spiral breakup in two-dimensional tissue [12] or induce negative filament tension [16, 17] in three dimensional slabs. Cardiac propagation can be perturbed by a so-called reentry of the excitation, e.g. by spontaneous formation of spirals due to ectopic beats. The accumulation of the heterogeneities in a particular region of the system may generate reentries [18, 19] due to the formation of a source-sink mismatch [20, 21]. On the other hand, heterogeneities can be also employed as virtual electrodes to generate a large number of new waves to suppress fibrillatory electrical activity [22, 23].

Here, we study the generation of reentries by a heterogeneous circular region and we systematically classify the different types of reentries found in the numerical simulations for square and hexagonal topologies of the discrete network of connected cells. For all the cases we find similar relation between proximity of the heterogeneity fraction to the critical value for percolation [24] and the probability to obtain reentrant activity [19]. The formation of reentries due to small heterogeneities have been shown previously for generic models of excitable media [12, 14], in models of heterogeneous cardiac tissue [13, 18, 19] and in models of cardiac cell cultures [25, 26]. While previous models consider only square networks of cells, here, we perform a detailed analysis for different topologies, and in particular simulate excitations in heterogeneous hexagonal networks, which are closer to the topology of real cardiac tissue where the elongation of the myocytes creates typically a hexagonal distribution of cells as has been already included in previous models of the microstructure of cardiac tissue [27, 28].

The paper is organized as follows: in Section 2 we introduce the three modeling approaches considered and the types of network topologies employed in the numerical simulations. In Section 3 we show the main results obtained from more than 20000 simulations. The paper ends with the main conclusions from the statistical analysis of the numerical results.

2. Model

2.1. Continuous tissue model

The propagation of the action potential is described by the cable equation [3]. Such equation relates the variation of the membrane potential (V) with the total current due to the ion channels across the membrane (I), and the conduction of potential along the cell membrane. The cable equation reads:

$$\frac{\partial V}{\partial t} = -I + \nabla \cdot (D \nabla V), \quad (1)$$

where $D = 0.001\ cm^2/ms$ is the effective diffusion coefficient of the action potential, which is proportional to the conductivity tensor.

The total current I is a sum of different types of ion currents $I = \sum_i I_i$. The number and complexity of these currents determines the particular ionic model employed [6]. Here we use a simplified model, which involves in only three ion currents [5, 29]:

$$\frac{\partial V}{\partial t} = -(I_{fi} + I_{so} + I_{si}) + \nabla \cdot (D\nabla V), \quad (2)$$

where the currents do not exactly correspond to an explicit ion channel, but are controlled by parameters which can be fitted to particular experimental measurements or to more complex models. The ionic currents read:

$$\begin{aligned} I_{fi} &= \frac{-vp(V - V_c)(1 - V)}{\tau_d}, \\ I_{so} &= \frac{V(1-p)}{\tau_o} + \frac{p}{\tau_r}, \\ I_{si} &= \frac{-w(1 + tanh(k(V - V_c^{si})))}{2\tau_{si}}, \end{aligned} \quad (3)$$

corresponding, respectively, to the fast inward, slow outward and slow inward currents. These currents are controlled by the gating variables v and w. The evolution of such gate variables depends on the action potential:

$$\begin{aligned} \frac{\partial v}{\partial t} &= \frac{(1-p)(1-v)}{\tau_v^-(V)} - \frac{pv}{\tau_v^+}, \\ \frac{\partial w}{\partial t} &= \frac{(1-p)(1-w)}{\tau_w^-} - \frac{pw}{\tau_w^+}; \end{aligned} \quad (4)$$

where $\tau_v^-(V) = (1-q)\tau_{v1}^- + q\tau_{v2}^-$ with $p = \theta(V - V_c)$ and $q = \theta(V - V_v)$, where $\theta(x)$ is the Heaviside step function.

During our study we vary the parameter τ_d keeping constant the values of the rest of parameters: $\tau_v^+ = 3.33ms$, $\tau_{v1}^- = 19.6ms$, $\tau_{v2}^- = 1000ms$, $\tau_w^+ = 667ms$, $\tau_w^- = 11ms$, $\tau_o = 8.3ms$, $\tau_r = 50ms$, $\tau_{si} = 45ms$, $k = 10$, $V_c^{si} = 0.85$, $V_c = 0.13$ and $V_v = 0.055$, for more details see [5]. For $\tau_d = 0.25$ the propagating waves obtained in the numerical integration of eq.(2) together with eq.(3, 4) reproduce the main properties of the waves obtained by the numerical integration of the Beeler-Reuter model [30]

We employ finite differences for the spatial discretization ($\Delta x = 250\ \mu m$) and a second order Runge-Kutta method for the temporal integration of the equations ($\Delta t = 0.15\ ms$) for the numerical integration of eqs.(2-4).

The combination of the nonlinealities controlling the total current I with the transport of the action potential determined by the parameter D, produces the stationary propagation of a wave through the cardiac tissue which coordinates the contraction of the heart. Waves with free ends produce the formation of spiral waves. Spirals are related with reentry in cardiac tissue which leads to tachycardia.

The action potential duration (APD) decreases with the parameter τ_d, see Fig. 1. Thus, waves are smaller and slower for larger values of τ_d, meaning that excitability decreases with this parameter. As the value of τ_d is increased, the APD decreases, see Fig.1(a), the velocity of the traveling waves (C) decreases practically linearly, see Fig.1(b), and the wave size (λ), defined by the product of these two quantities ($\lambda = C \cdot APD$), also decreases, see Fig.1(c).

In Fig. 2 we show the spiral wave dynamics for three different values of τ_d. For each case, we show first the action potential for a single point, second four snapshots showing the evolution of a spiral wave and finally the integration of the value of V for the whole area $< V >= A^{-1} \int V dA$.

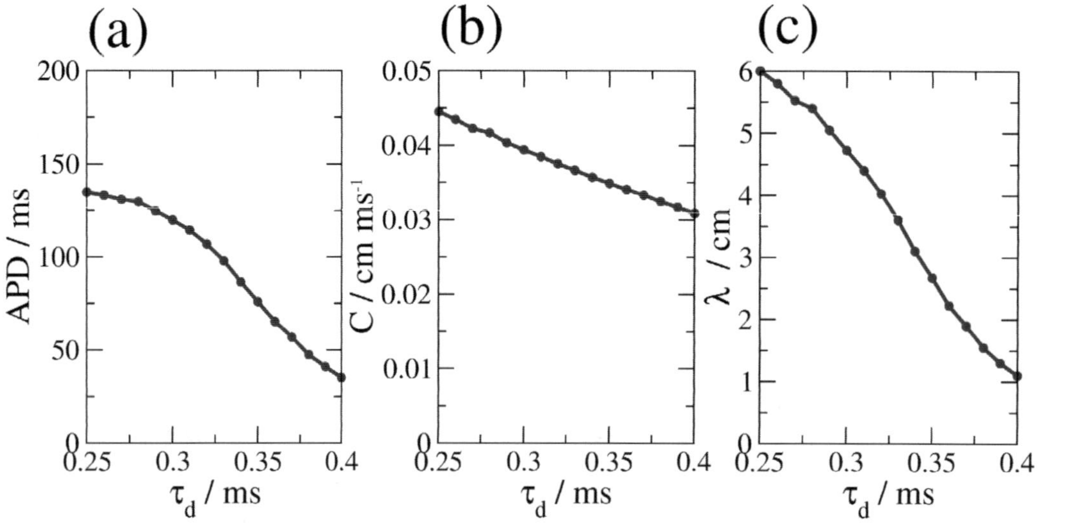

Figure 1. Dependence of the propagation parameters on the medium excitability. Monotonous decrease of the APD (a), linear decrease of the wave velocity (b) and decrease of the wave size (c) for increasing values of the model parameter τ_d.

For $\tau_d = 0.25$ and $\tau_d = 0.30$ large action potentials are obtained and the interaction of the waves of the spiral with the refractory tail of the previous wave produces complex behaviours and even spiral breakup as it is shown in Fig. 2(b). The development of the spiral breakup depends on the initial condition and the size of the system, it is also obtained under some conditions for $\tau_d = 0.25$. Waves for large values of τ_d behave similar to pulses in FitzHugh-Nagumo models [31] and zones with weak excitability are reached. Further increase of τ_d brings the system to the non-excitable region and precludes the propagation of waves [5].

2.2. Discrete network of cells

In contrast with the continuous description of the cardiac tissue, here we consider a two-dimensional network of cells connected by gap junctions. The excitability of the cells are governed by the ion equations described above. The electrical connectivity between the cells is assumed to be at first neighbours and it corresponds to diffusion in the continuous limit. An equivalent equation to eq.(2) can be calculated for each cell i as follow:

$$\frac{\partial V_i}{\partial t} = -(I_{fi} + I_{so} + I_{si}) + \xi \sum_{j}^{N} (V_j - V_i), \qquad (5)$$

where N is the number of first neighbours: four for a square network, see Fig.3(a), and six for a cubic network, see Fig.3(b). The parameter ξ is chosen in order to recover the homogeneous limit of eq.(2) and reads $\xi = D/\Delta x^2$; where $\Delta x = 0.01\ cm$ is the distance between cells centers or, i. e., size of the cells.

2.3. Heterogeneous discrete network of cells

Inactive gap junctions, inert tissue among the cardiac cells or the presence of unexcitable cells, e.g. fibroblast, in the tissue breaks the continuous and homogeneous propagation of action potential through the cardiac tissue, and the resulting waves are rough and noisy. We model

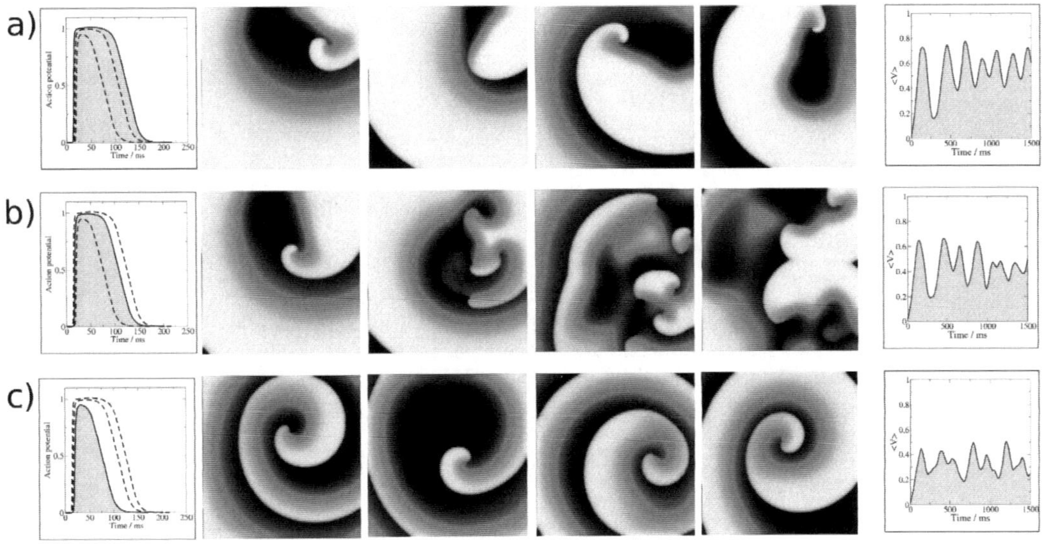

Figure 2. Dynamics of two-dimensional wave propagation in a homogeneous continuous model for three different values of $\tau_d =: 0.25ms$ (a), $0.30ms$ (b) and $0.35ms$ (c). For each value of $\tau_d =$ we show the evolution of the action potential of an individual cell (left panel), four snapshots corresponding to times 480 ms, 660 ms, 960 ms and 1200 ms (middle panel), and the average of the action potential ($< V >$) in the whole domain (right panel). Square grid 10×10 cm^2 with $\Delta t = 0.15$ ms and $\Delta x = 0.025$ cm.

this disruption in the propagation by the introduction of random heterogeneities in the network of cells using eq.(5) together with a connectivity parameter η_{ij}:

$$\frac{\partial V_i}{\partial t} = -(I_{fi} + I_{so} + I_{si}) + \xi \sum_j^N \eta_{ij}(V_j - V_i), \tag{6}$$

if the connectivity parameter $\eta_{ij} = 1$ for all i and j we recover the homogeneous discrete network limit of eq.(5). We consider here than the value of η_{ij} can take only two values: $\eta_{ij} = 1$ for normal connections between two cells and $\eta_{ij} = 0$ for abnormal connections. These abnormal connections are randomly scattered in the tissue, thus, we do not consider any correlation.

We study two types of heterogeneities:

- *Non-conducting bonds.* The connection between two cells (i and j) is removed ($\eta_{ij} = 0$) with certain probability ϕ which is independent of the connectivity with the rest of first neighbours, see Fig. 3(c,d).
- *Non-conducting sites.* For a certain probability ϕ a cell (i) is abnormally connected with all the first neighbours ($\eta_{ij} = 0, \forall\ j$), see Fig. 3(e,f).

These two interpretations gives rise to different types of discrete networks of cells with different topological properties.

The introduction in the medium of a randomly distributed fraction of heterogeneities affects the dynamics of the propagating waves. As example we first consider the case of non-conducting bonds in a square network. The propagation of waves in the medium depends on the fraction of heterogeneities ϕ. As the heterogeneities do not modify the ion currents, the APD in each cell

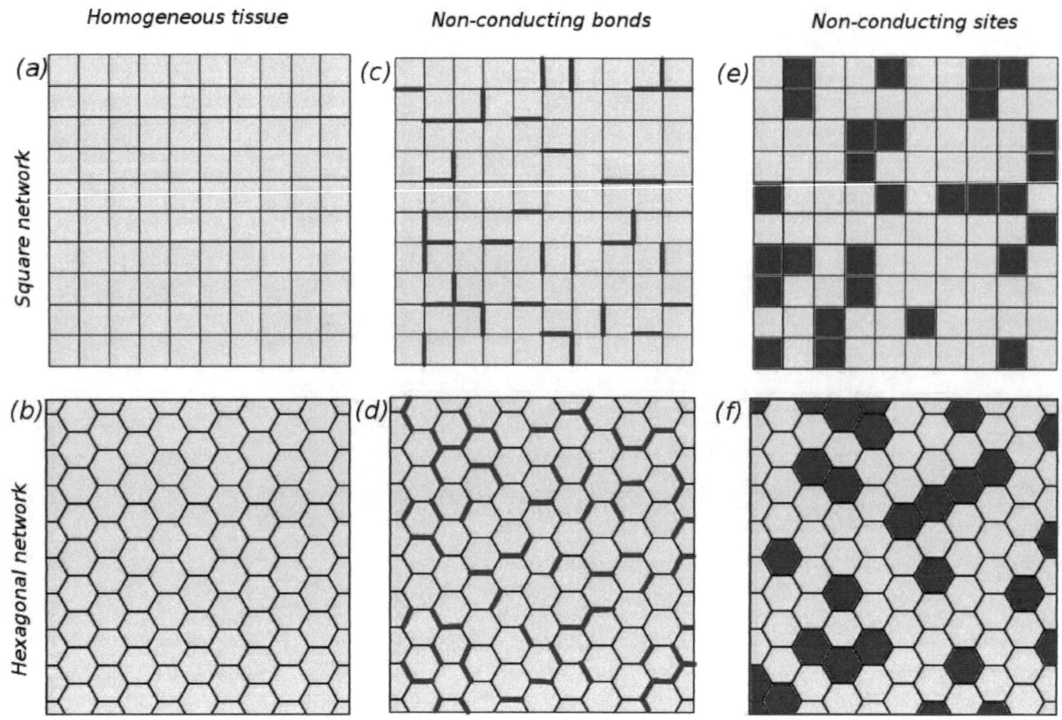

Figure 3. Types of networks: Homogeneous Square (a), homogeneous hexagonal (b), heterogeneous square with non-conducting bonds (c), heterogeneous hexagonal with non-conducting bonds (d), heterogeneous square with non-conducting sites (e), heterogeneous hexagonal with non-conducting sites (f).

remains independent on the fraction ϕ, see Fig.4(a). The velocity of wave propagation, however, clearly depends on ϕ, and it decreases for increasing ϕ, see Fig.4(b). The decrease of the velocity produces an equivalent decrease of the wave size, see Fig.4(c).

The increase of the fraction of the heterogeneities produces the appearance of clusters of isolated regions surrounded by removed links or cells. For a particular fraction ϕ_c the network percolates, i.e. a cluster of the system size separates the whole medium in two disconnected parts [24]. The critical fraction ϕ_c for non-conducting bonds and sites is different and gives rise to two different percolation thresholds. Above the percolation threshold of the cell network wave propagation is not more possible. Close to such value, the velocity rapidly decrease to zero [14] and the wave size approaches the size of the heterogeneities, inducing wave breakup, and giving rise to complex dynamics [19].

2.4. Localized heterogeneous region inside a discrete network of cells

In the numerical simulations we divide the system in two parts, see Fig.5 for a sketch of the system:

- *Damaged region:* a circular heterogeneous region corresponding to a damaged area of the tissue. The heterogeneities are introduced with a fraction ϕ in a circular region of the tissue. This region represents a damaged area of the tissue where the parameter τ_d is also modified by the decrease of the excitability.

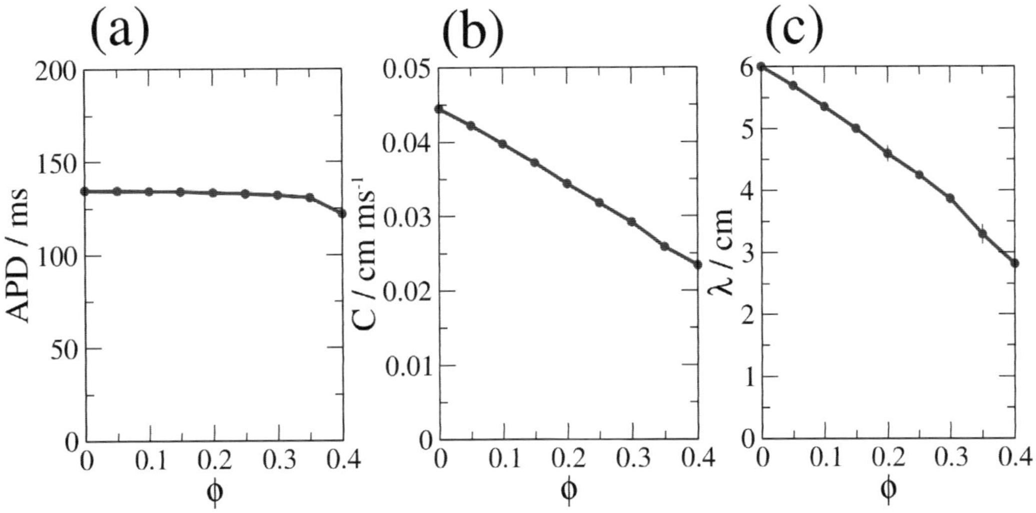

Figure 4. Dependence of the propagation parameters on the fraction of heterogeneities ϕ. Independence of the APD (a), linear decrease of the wave velocity (b) and decrease of the wave size (c) on the parameter ϕ.

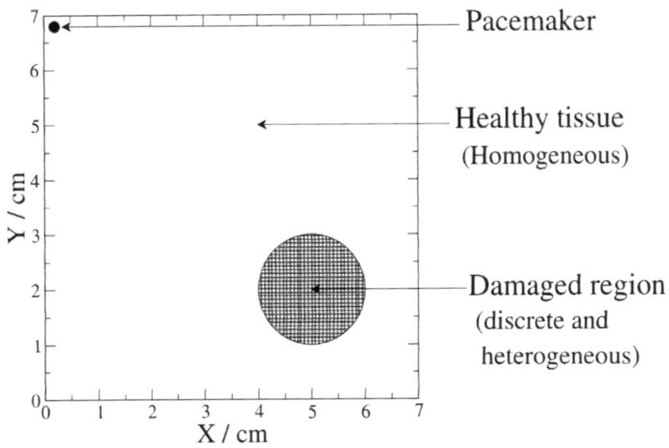

Figure 5. Sketch of the system employed for the numerical simulations. Waves are initially generated at the pacemaker and propagate normally in the homogeneous tissue until the wave interacts with the damaged region, which is highly heterogeneous and where reentries are formed.

- *Healthy tissue:* a normal homogeneous region corresponding to the healthy tissue. The propagation of the action potential recovers the standard properties.

We evaluate the interaction between the wave propagating in an homogeneous region of the tissue with the damaged region with a fraction of heterogeneities. An excitation wave of action potential is initially induced in one corner of the two-dimensional system, mimicking a pacemaker, see Fig.5. The excitation propagates through the medium until it interacts with the circular region. Depending on the relation between ϕ and ϕ_c (percolation threshold) three behaviours are observed:

- *For low fraction ($\phi << \phi_c$):* the wave propagates inside the damaged region. The wave deforms and propagates inside slower than outside, however, wave propagation is stable.

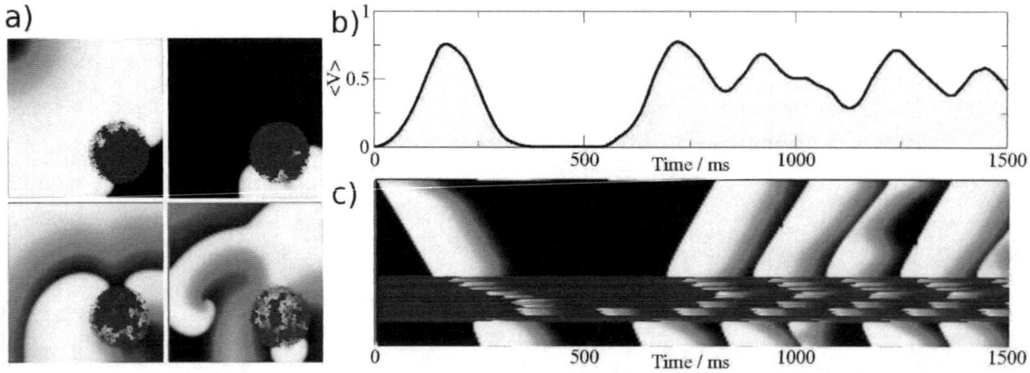

Figure 6. Sustained reentry induced in a damaged region of the tissue. (a) Four snapshots of the evolution of the reentry for times: 184 ms, 598 ms, 1002 ms and 1398 ms. (b) Evolution of the global spatial average of the action potential showing the activity of the tissue. (c) Spatio-temporal dynamics from a diagonal cut of the system which crosses the damaged area and illustrates the activity inside and outside the damaged region. Parameter values: Radius of the circular inhomogeneity: 1.4 cm; inside the damaged area $\phi = 0.49$ and $\tau_d = 0.35ms$; and square network $7 \times 7\ cm^2$.

- *For large fraction ($\phi >> \phi_c$):* the wave does not enter in the damaged region and the it propagates around this region.
- *For intermediate fraction ($\phi \sim \phi_c$):* the wave propagates in the circular region, however the propagation inside is highly irregular. The different possibilities obtained from the numerical simulations under such condition are discussed in the next section.

3. Results

3.1. Types of reentries

As already stated in the previous section, wave propagation is irregular for $\phi \sim \phi_c$. The most frequent behaviour corresponding to the interaction between the wave and the damaged heterogeneous region for $\phi \sim \phi_c$ is a transient propagation of the wave inside the damaged region that rapidly disappears. There are, however, other dynamics when the propagation inside the heterogeneous region persists. We classify the new behaviours into three different groups:

- *Sustained reentry:* The pieces of waves inside the circular region survive after the wave passes and repeatedly propagate to the boundary of the circle giving rise to reentry. An example of such behaviour is shown in Fig.6. The moving pieces produce a new wave which excites the whole medium again; see Fig.6(a). These pieces produce a pacing of the homogeneous region and if the reentry path inside the circular region is fast enough, it can even break the waves by the interaction with the refractory tail of the previous wave as shown in Fig.6(a), and produce the formation of spiral waves. The evolution of the total excitation of the medium ($< V >$) is also plotted and shows the intense excitation of the system from the reentry, see Fig.6(b). Finally, the spatio-temporal plot shown in Fig.6(c) permits the observation of the interaction among the waves induced by the circular region. When the path inside the heterogeneous region is large the pacing is slow and it produces a stable pacemaker of new waves.
- *Non-sustained reentry:* The first wave generated by the previous mechanism interacts with the wave inside the circular region and eliminates the activity, see Fig.7. A single event is produced which basically resets the whole heterogeneous system again. Different cases with

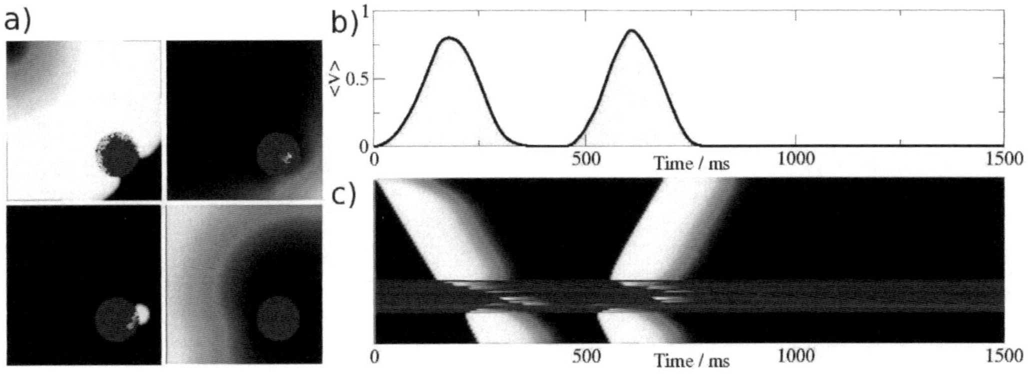

Figure 7. Non-sustained reentry induced in a damaged region of the tissue. (a) Four snapshots of the evolution of the reentry for times: 184 ms, 345 ms, 469 ms and 717 ms. (b) Evolution of the global spatial average of the action potential showing the activity of the tissue. (c) Spatio-temporal dynamics from a diagonal cut of the system which crosses the damaged area and illustrates the activity inside and outside the damaged region. Parameter values: Radius of the circular inhomogeneity: 1 cm; inside the damaged area $\phi = 0.495$ and $\tau_d = 0.35ms$; and square network 7×7 cm^2.

two or three reentries and the final reset to the rest state have been also observed in the numerical simulations.

- *Residual patterns:* If the excitability of the medium inside the circular region is very low in comparison to the outside medium, the wave may not be able to reenter into the homogeneous highly excitable medium. Then the pieces of waves are restricted to propagate inside the heterogeneous region. An example of such dynamics is shown in Fig.8. The average activity ($< V >$) of the medium is not zero but small, see Fig.8(b), and there is a periodic dynamics restricted to the circular region produced by the continuous repetition of the reentry path inside the heterogeneous system, see Fig.8(c).

The three types of dynamics have been observed in the numerical simulations. They can appear in different random realizations under exactly the same conditions and they can appear simultaneously, for instance a non-sustained reentry can finish in a residual pattern. As we show below, the probability of reentry depends not only on ϕ but also on the size, excitability and network topology of the circular heterogeneous region.

3.2. Relation of the reentry probability with the percolation threshold

We study the probability of appearance of the different dynamical patterns by increasing systematically the fraction of non-conducting bonds ϕ in a square network. We increase ϕ in small steps ($\Delta\phi = 0.005$) and for each value of ϕ we perform 100 realizations of the same numerical experiment with different random distribution of the heterogeneities.

The dependence of the probability of the three types of reentries on ϕ is shown in Fig.9. At low and large values of ϕ reentries do not appear. For a window of intermediate values ($\phi \sim \phi_c$) the three types of reentries are observed. The higher probability of reentry is obtained at $\phi = 0.48$ while the percolation threshold for the square network with non-conducting bonds is $\phi_c = 0.5$.

The most common reentry is the sustained reentry (74%) while non-sustained (13%) and the residual patterns (13%) are less frequent. The relative probability of the non-sustained reentries and residual patterns is higher for large values of ϕ. A possible reason for such effect is that

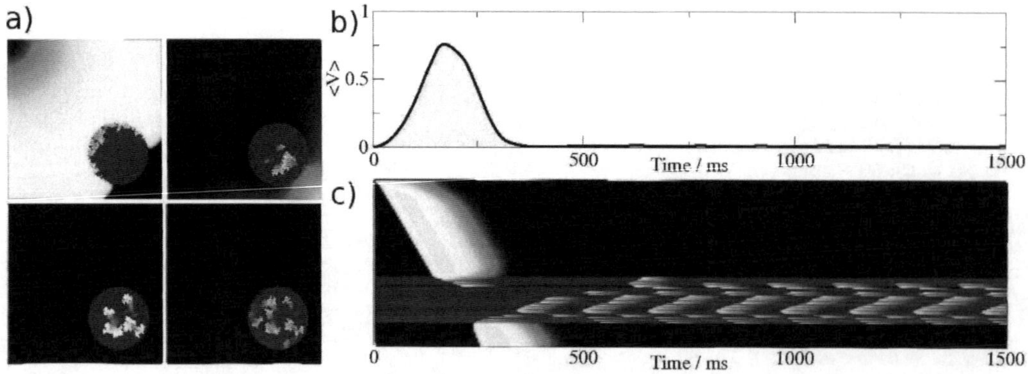

Figure 8. Residual reentry induced in a damaged region of the tissue. (a) Four snapshots of the evolution of the reentry for times: 184 ms, 349 ms, 483 ms and 763 ms. (b) Evolution of the global spatial average of the action potential showing the activity of the tissue. (c) Spatio-temporal dynamics from a diagonal cut of the system which crosses the damaged area and illustrates the activity inside and outside the damaged region. Parameter values: Radius of the circular inhomogeneity: 1.4 cm; inside the damaged area $\phi = 0.49$ and $\tau_d = 0.35ms$; and square network 7×7 cm^2.

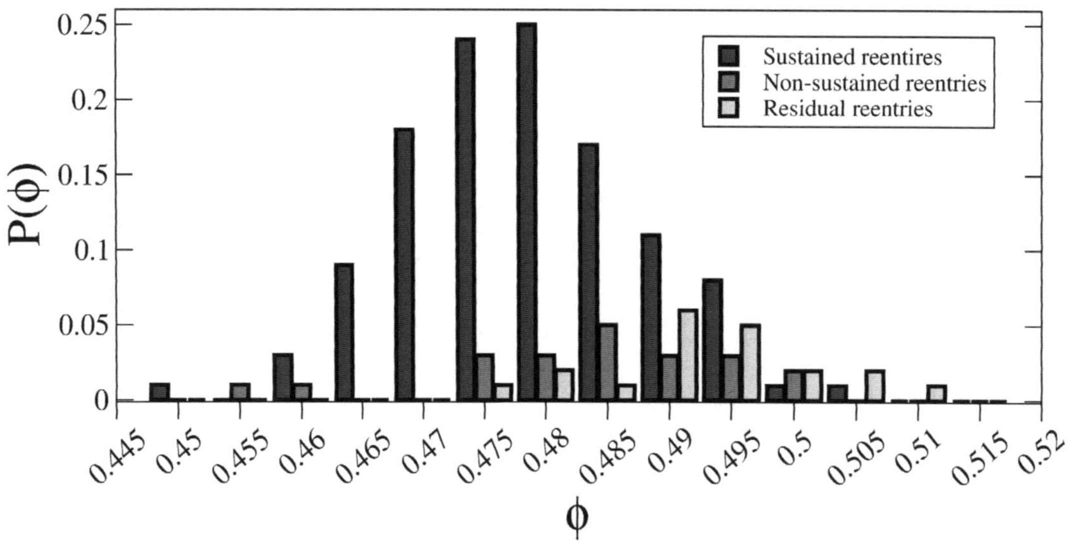

Figure 9. Probability of the different types of reentries produced by a circular heterogeneous region for different values of ϕ. Three types of interactions are observed: sustained reentry, non-sustained reentry and residual patterns. The radius of the inhomogeneity is kept constant $R = 1.4$ cm. The parameter τ_d is spatially inhomogeneous: $\tau_d = 0.25ms$ outside and $\tau_d = 0.35ms$ inside the damaged region. A total of 100 realizations have been done for each value of ϕ.

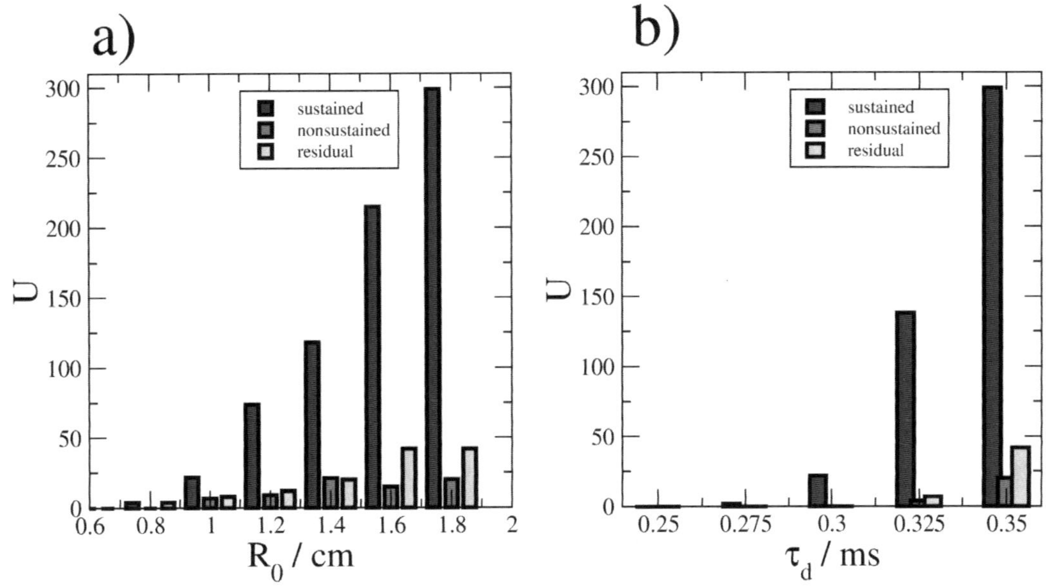

Figure 10. a) Total reentry probability for the three different types of reentry for different radius keeping $\tau_d = 0.35ms$ (a) and for different values of τ_d keeping $R_0 = 1.8cm$ (b). Parameter τ_d outside the damaged region is kept $\tau_d = 0.25ms$.

at high values of ϕ the propagation inside the damaged region is slower. Such reduction of the velocity is equivalent to an effective reduction of the excitability. The difference of excitability between inside and outside the region is, therefore, larger. A gap in excitability hinders the exit of the wave, as it is known for generic excitable media.

3.3. Dependence on the size of the damaged region

We also study the effect of the size of the damaged region. If the region is small the entrance of the wave in the damaged region rapidly excites the active cells. A possible reentrance into the homogeneous healthy region would rapidly interact with the posterior part of the initial wave, which hinders the appearance of the reentry. Given a value of τ_d there is a minimal size of the circular area which can induce reentries.

We define the total number of reentries for all the possible heterogeneity fraction values ($0 < \phi < 1$) as control parameter: $U = \sum_i NP(\phi_i)$ where $NP(\phi_i)$ is the number of reentries observed after N=100 realizations with ϕ_i. As already noted in the previous subsection we increase the fraction of heterogeneities in a discrete way: $\phi_i = i\Delta\phi$. In Fig. 10(a) the dependence of this quantity U on the radius of the circular damaged region is shown. As the radius increases the number of reentries grows.

The relative fractions of the different types of reentries change also with the size of the damaged region. For low radius the distribution is centered around the percolation threshold. However, for larger radius the center of the distribution moves to lower values of ϕ [19]. This displacement indicates that the majority of reentries appear at higher excitability (low values of ϕ) and the case of sustained reentry is the most probable. For smaller regions the distribution of the three types of reentries is more homogeneous, for example for $R_0 = 0.8cm$ we observe 50% sustained reentries and 50% of residual patterns, in comparison with the results shown in Fig. 9.

Table 1. Distribution of the different types of reentries corresponding to Fig. 11

Topology	Threshold	Total	Sustained	Non-sustained	Residual
Square Bond	0.5	361	299 (83%)	20 (5%)	42 (12%)
Square Site	0.41	220	167 (76%)	22 (10%)	31 (15%)
Hexagonal Bond	0.65	227	82 (36%)	12 (5%)	133 (59%)
Hexagonal Site	0.5	58	11 (19%)	4 (7%)	43 (74%)

3.4. Dependence on the excitability of the damaged region

We perform simulations under different excitability conditions. We change the value of the parameter τ_d inside the damaged region. The numerical results show a clear dependence of the probability on the value of this parameter, see Fig. 10(b). For standard conditions of the model ($\tau_d = 0.25$) reentries are not observed in the numerical simulations. The increase of the parameter τ_d decreases the velocity and APD of the waves inside the damaged region and permits the appearance of the reentries, see Fig. 10(b). The probability of reentry increases with τ_d.

The distribution of the three types of reentries depends also on the excitability of the medium, see Fig. 10(b). For larger excitability, i.e small τ_d, the probability of reentry is small and all the cases are associated with sustained reentries. For smaller excitability, i.e large τ_d, the other two types of reentries also are observed. The relative probability of non-sustained and residual reentries increases with τ_d.

3.5. Probability of reentry on different topologies of the networks of cells

The previous results are obtained by changing the fraction of non-conducting bonds in a square grid. Next, we consider other types of topologies and heterogeneities, keeping the size of the damaged region and the excitability inside this region. We consider square or hexagonal grids with non-conducting bonds or sites, see Fig.3.

In Fig. 11 the dependence of the probability of the three types of reentries on the fraction ϕ is plotted for different network topologies keeping the values of the parameters. The interaction of a travelling wave with the heterogeneous region in a square or in a hexagonal grid produces the onset of reentries for fraction values close to the particular percolation threshold of the corresponding network. In this figure the distribution of the type of reentries as function of ϕ is also plotted. For all network topologies, while sustained reentries appear typically for small values of ϕ, residual patterns dominate at large fractions.

Although for the four cases the probability of reentry is related to the percolation threshold, the results shown in Fig. 11 indicate a dependence of the relative probability of the reentries on the network topology:

- *Square vs. hexagonal:* Square grids produce higher probability of reentry than hexagonal grids under the same conditions. Furthermore, the distribution of the reentries is different, while in square grids around 80% of reentries correspond to sustained, in hexagonal grids the major part of reentries (59% and 74%, see Table 1) are residual.
- *Bond vs. Sites:* While the distributions of the different types of reentries does not depend on the type of heterogeneity, site or bond, there is a decrease of the total number of reentries in the case of site heterogeneities in comparison with bond heterogeneities, see Table 1 and Fig. 11.

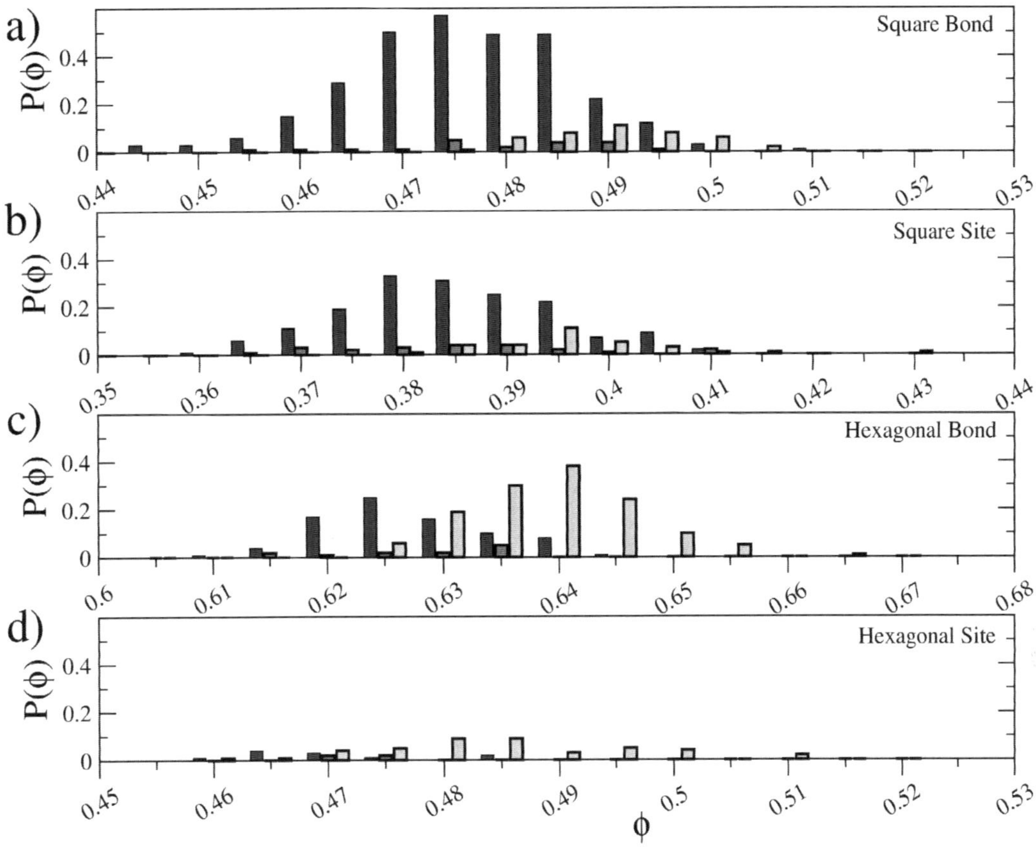

Figure 11. Probability of the different types of reentries produced by a circular heterogeneous region for different values of ϕ for different topologies and heterogeneity types: a) Square network with site heterogeneities, b) Square network with site heterogeneities, c) Hexagonal network with bond heterogeneities, and d) Hexagonal network with site heterogeneities. The radius of the inhomogeneity is kept constant $R = 1.8\ cm$. The parameter τ_d is spatially inhomogeneous: $\tau_d = 0.25$ outside and $\tau_d = 0.35$ inside the damaged region.

4. Conclusions

We employ a heterogeneous discrete model of cardiac tissue to numerically study the reentry produced by a damaged region (modelled as a heterogeneous discrete system) inside the healthy tissue (homogeneous). Following, we list the main conclusions after the realization of more than 20000 numerical simulations:

- Heterogeneous tissue may produce reentry if the fraction of non-conducting heterogeneities is close to the percolation threshold of the network.
- The reentries are classified in three main types depending on their capability to survive and to leave from the damaged region to the healthy area: sustained (survive in the long run and leave), non-sustained (do not survive but leave at least once) and residual (survive but do not leave).
- The probability of reentry depends on the size and excitability of the damaged region.
- The probability of reentry depends on the type of heterogeneities, reentries produced by non-conducting bonds are more frequent than by non-conducting sites.

- The type of the network changes the predominant type of the reentry. While for square networks sustained reentry are more common, for hexagonal networks residual reentries are more frequent.

These conclusions are relevant for the understanding of anatomical reentry in cardiac tissue. Extensions of the employed model have to consider the three-dimensional structure of the tissue, which changes the topological properties, and more concrete models of cellular networks in real cardiac tissue.

An additional biological example of heterogeneous excitable media is myocyte cell cultures. In such case heterogeneities play a crucial role in the pattern formation processes [25, 26, 32], because the lack of orientation of myocytes. In summary, the use of well established concepts in physics like percolation threshold in discrete networks can improve the general understanding of reentries in cardiac tissue which may lead to dangerous arrhythmia.

Acknowledgements
We acknowledge financial support from the German Science Foundation DFG within the framework of SFB 910 "Control of self-organizing nonlinear systems" and from MINECO of Spain under the Ramon y Cajal program with the grant number RYC-2012-11265.

References

[1] D. P. Zipes, and J. Jalife, *Cardiac Electrophysiology: From Cell to Bedside* (Philadelphia, PA: Saunders, 2004).
[2] *Chemical Waves and Patterns*, edited by R. Kapral and K. Showalter (Kluwer, Dordrecht, 1994).
[3] J. P. Keener, and J. Sneyd, *Mathematical Physiology* (New York: Springer, 1998).
[4] Davidenko J M, Pertsov A M, Salomonz R, Baxter W, and Jalife J, 1992 *Nature* **335** 349
[5] Fenton F H, Cherry E M, Hastings H M, and Evans S J, 2002 *Chaos* **12** 852
[6] Clayton R H, Bernus O, Cherry E M, Dierckx H, Fenton F H, Mirabella L, Panfilov A V, Sachse F B, Seemann G, and Zhang H, 2011 *Prog. Bio. Phys. Mol. Biol.* **104** 22-48
[7] Fenton F H, and Cherry E M, 2008 *Scholarpedia* **3** 1868
[8] Gouvea de Barros B, Sachetto Oliveira R, Meira W, Lobosco M, and Weber dos Santos R, 2012 *Comp. Math. Meth. Med.* 824569
[9] Stinstra J G, MacLeod R, and Henriquez C S, 2010 *Ann Biomed Eng* **38** 1390-1414.
[10] Xie Y, Garfinkel A, Camelliti P, Kohl P, Weiss J N, and Qu Z, 2009 *Am J Physiol Heart Circ Physiol* **297** H775-H784
[11] Qu Z, Karagueuzian H S, Garfinkel A, and Weiss J N, 2004 *Am. J. Physiol. Heart Circ. Physiol.* **286** H1310
[12] Panfilov A V, 2002 *Phys. Rev. Lett.* **88** 118101
[13] Ten Tusscher K H W J, and Panfilov A V, 2007 *Europace* **9**, vi38
[14] Alonso S, Kapral R, and Bär M, 2009 *Phys. Rev. Lett.* **102** 238302
[15] Rohr S, 2004 *Cardiovas. Res.* **62** 309-322
[16] Alonso S, Bär M, and Panfilov A V, 2011 *Chaos* **21** 013118
[17] Alonso S, Bär M, and Panfilov A V, 2013 *Bull. Math. Biol.* **75** 1351-1376
[18] Cherry E M, Ehrlich J R, Nattel S, and Fenton F H, 2007 *Heart Rhythm* **4** 1553-1562
[19] Alonso S, and Bär M, 2013 *Phys. Rev. Lett.* **110** 158101
[20] Kleber A G, and Rudy Y, 2004 *Physiol. Rev.* **84** 431-488
[21] Kinoshita S, Iwamoto M, Tateishi K, Suematsu N J, and Ueyama D, 2013 *Phys. Rev. E* **87** 062815
[22] Pumir A, Nikolski V, Hörning M, Isomura A, Agladze K, Yoshikawa K, Gilmour R, Bodenschatz E, and Krinsky V, 2007 *Phys. Rev. Lett.* **99** 208101
[23] Luther S, Fenton F H, Kornreich B G, Squires A, Bittihn P, Hornung D, Zabel M, Flanders J, Gladuli A, Campoy L, Cherry E M, Luther G, Hasenfuss G, Krinsky V, Pumir A, Gilmour R F Jr and Bodenschatz E, 2011 *Nature* **475** 235
[24] Torquato S, *Random Heterogeneous Materials* (Springer, New York, 2002).
[25] Bub G, Shrier A, and Glass L, 2002 *Phys. Rev. Lett.* **88** 058101
[26] Steinberg B E, Glass L, Shrier A, and Bub G, 2006 *Phil. Trans. R. Soc. A* **364** 1299-1311
[27] Hubbard M L, Xu J, Henriquez C S, 2014 *Comp. in Card.* **41** 65
[28] Gouvea de Barros B, Weber dos Santos R, Lobosco M., Alonso S, 2015 *Submitted for publication*
[29] Fenton F H, and Karma A, 1998 *Chaos* **8** 22
[30] Beeler, G W, Reuter H, 1977 *J. Physiol* **268** 177-210

[31] FitzHugh R, 1961 *Biophys. J.* **1** 445
[32] Shajahan T K, Borek B, Shrier A, and Glass L, 2011 *Phys. Rev. E* **84** 046208.

Investigation of energy dissipation due to contact angle hysteresis in capillary effect

Bhagya Athukorallage[1] and Ram Iyer[2]
Department of Mathematics and Statistics, Texas Tech University.
E-mail: bhagya.athukorala@ttu.edu[1] and ram.iyer@ttu.edu[2]

Abstract. Capillary action or Capillarity is the ability of a liquid to flow in narrow spaces without the assistance of, and in opposition to, external forces like gravity. Three effects contribute to capillary action, namely, adhesion of the liquid to the walls of the confining solid; meniscus formation; and low Reynolds number fluid flow. We investigate the dissipation of energy during one cycle of capillary action, when the liquid volume inside a capillary tube first increases and subsequently decreases while assuming quasi-static motion. The quasi-static assumption allows us to focus on the wetting phenomenon of the solid wall by the liquid and the formation of the meniscus. It is well known that the motion of a liquid on an non-ideal surface involves the expenditure of energy due to contact angle hysteresis. In this paper, we derive the equations for the menisci and the flow rules for the change of the contact angles for a liquid column in a capillary tube at a constant temperature and volume by minimizing the Helmholtz free energy using calculus of variations. We describe the numerical solution of these equations and present results from computations for the case of a capillary tube with 1 mm diameter.

1. Introduction

Capillarity and wetting phenomena play prominent roles in soil science, plant biology, surface physics, and hence, both fields gain more attention from researchers in various areas like chemistry, physics, and engineering. Studies on wetting phenomenon mainly look at how liquid spreads on a solid surface. Industrial processes, such as cleaning, painting, coating, and adhesion [1] widely use the key concepts of wetting. As an example, in the automobile industry, surfaces are prepared prior to painting, and tires are treated to promote adhesion on wet roadways.

The surface chemistry of a solid substrate is the key factor in determining its wetting behavior; hence, the specific wetting properties can be obtained by modifying the surface chemistry. At large scales, research on wetting properties of a liquid is used to increase the deposition efficiency of pesticides on plant leaves and the cooling of industrial reactors. On smaller scales, it is used to improve the efficiency of microfluidic and nanoprinting devices [2].

Capillarity is the study of the interface between two immiscible fluids, or between liquid and air. These interfaces change their shape to minimize the surface energy [3]. The two fundamental equations that describe the *macroscopic* theory of capillary surfaces are the Young-Laplace equation and the Kelvin equation. The wetting of a solid surface by a liquid in the presence of vapor is described by Young's equation [4].

- The Young-Laplace equation describes the shape of a capillary interface. When a curved capillary interface is in equilibrium, a pressure difference, δp, forms across the interface. Young-

Laplace equation relates the pressure deficiency to the curvature of the interface, and it reads

$$\delta p = 2\gamma_{LG} H, \tag{1}$$

where H denotes the mean curvature of the interface [3].

- Lord Kelvin's equation describes the evaporation or condensation of liquid from a capillary surface. In particular, this equation describes the change in the pressure of the vapor as a function of the temperature. This change in the vapor pressure leads to a change in the shape of the capillary surface in accordance with the Young-Laplace equation.
- Young's equation describes the boundary condition at the three phase line where a capillary surface meets a solid surface.

  Consider a liquid drop on a solid substrate, where one encounters three different interfaces: solid-liquid (SL), solid-gas (SG), and liquid-gas (LG). We assume that the contact line is a differentiable function so that one may assign a tangent direction at any point. We also assume that the solid and liquid surfaces are differentiable so that we may assign normal directions at any point. The *contact angle* θ between the solid and the liquid is defined to be (please refer to Figure 1) as the angle between the normal vectors to the solid and the liquid at a point on the contact line [5].

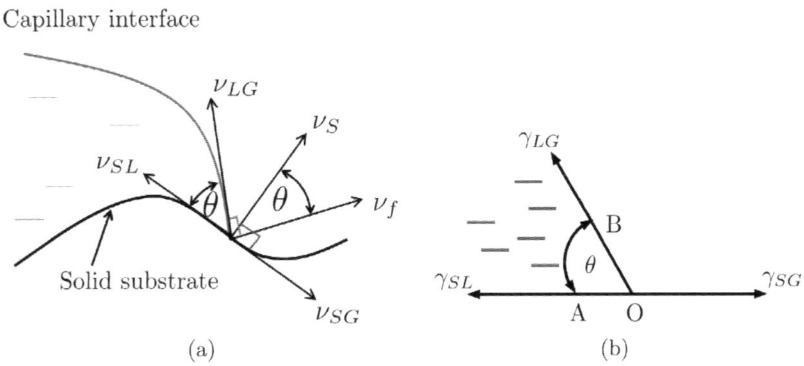

Figure 1: (a) Geometric definition of the contact angle for a 3-D droplet on a solid substrate [5]: ν_S is the normal to the solid substrate at the contact line, and ν_f is the normal to the droplet surface at the contact line. The angle between ν_S and ν_f is the contact angle. On the tangent plane to the solid substrate, let ν_{SG} be the unit vector perpendicular to the contact line pointing away from the wetted region. The vector $\nu_{SL} = -\nu_{SG}$ is the unit vector pointing into the wetted region. On the tangent plane to the capillary interface at the contact point, let ν_{LG} be the unit vector perpendicular to the contact line pointing away from the solid. (b) Forces per unit length of magnitude γ_{SG}, γ_{SL}, and γ_{LG} respectively act along the unit directions ν_{SG}, ν_{SL}, and ν_{LG}. Force balance in the ν_{SG} direction yields the Young equation.

Forces per unit length of magnitude γ_{SL}, γ_{SG}, and γ_{LG} act along certain directions perpendicular to the contact line as described in Figure 1(b). Balancing the forces on the tangent plane to the solid surface yields Young's equation [3]:

$$\gamma_{SG} - \gamma_{SL} = \gamma_{LG} \cos\theta. \tag{2}$$

The contact angle, θ, is not simply the tangential angle at the point of contact. In the region of the triple-point line (the line of contact between the substrate, air, and liquid), the liquid exhibits three distinct regions: the molecular region, the transition region, and the capillary region [6]. It is at the transition region where one should measure the angle of contact. The angle that the transition region makes with the substrate is the contact angle to be used in the Young's law [7]. This subtlety in the notion of the contact angle will be important in Section 2.2. The scalar quantities γ_{SL}, γ_{SG}, and γ_{LG} can also be understood as the energies of the interfaces per unit area.

In this article, we continue the work presented in [8] and compute the energy that must be expended by an external agent while changing the volume of a liquid column in a capillary tube through one cycle. The liquid volume itself could be changed slowly by changing the temperature of the system through Kelvin's effect described earlier. The very fact that a liquid column exists in a vertical capillary tube implies that the angles at the upper and lower menisci have different contact angle values. This is only possible due to the contact angle hysteresis phenomenon described next.

1.1. Contact angle hysteresis

Consider a liquid droplet on a solid surface with a contact angle of θ (Fig. 2). Experiments show that if the liquid is carefully added to the droplet via a syringe, the volume and contact angle of the droplet will increase without changing its initial contact area. Further increase of its volume results in an increase in the contact area with the contact angle fixed at θ_A (refer to Fig. 2(a)). Similarly, if the liquid is removed from a droplet, volume and contact angle of the droplet decrease but retain the same contact area. Continuing this process results in a recession of the contact area at a contact angle of θ_R. These two limiting values, θ_A and θ_R, are referred to as *advancing* and *receding angles*. For a symmetric droplet, one can obtain a hysteresis diagram for the contact angle θ verses droplet diameter D as depicted in Figure 2 (b).

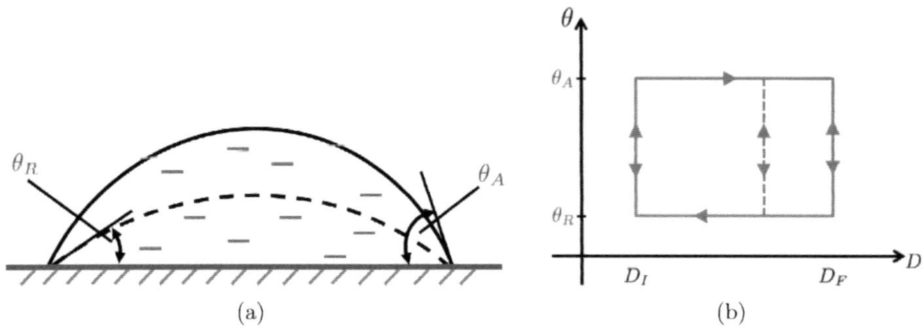

Figure 2: (a) Contact angle hysteresis effect: advancing θ_A, and receding θ_R contact angles of a liquid drop. (b) Plot of the contact angle θ verses contact diameter D for a drop on a solid surface.

1.2. Effect of surface chemistry and roughness

The contact angle θ at the contact line may be found by solving (2). It has been known since the work of H. L. Sulman and H. Picard [9] that there is a difference in the contact angle at the solid-liquid-gas contact line between rising drops and falling drops. They called this phenomenon *contact angle hysteresis*. The amount of hysteresis depends on the chemical compositions of the solid and the liquid, and the physical roughness of the solid surface [10]. Several researchers have shown that even for solids with surface height variation in the nanometers and drop sizes that are 2 or 3 orders of magnitude larger, there is still significant contact angle hysteresis [11, 12, 13, 14]. These results show that the solid-liquid chemistry is the primary reason for contact angle hysteresis. The same experiments have also shown that increase of surface roughness changes the amount of contact angle hysteresis but in a consistent manner [10, 15].

As experiments have pointed out that the primary reason for contact angle hysteresis is the chemistry between the solid and liquid, the practice of assigning only a single value for the interfacial energy γ_{SL} was questioned by Penn and Miller [12]. Recently, Snoeijer and Andreotti [16] have shown through a theoretical analysis at the microscopic level of a liquid drop on a smooth surface with no surface roughness and with no chemical impurities that the macroscopic (or apparent) contact angle cannot have a single value. Extrand[17] started with the assumption that the advancing and receding contact angles are known and concluded that the interfacial energy of the solid-liquid must be set valued if the angles are different.

Due to the above mentioned reasons, we assume γ_{SL} takes values in the set $[\gamma_{SL_{\min}}, \gamma_{SL_{\max}}]$. In the next section, we follow the theory outlined in [8] to derive the equations satisfied by a finite column of liquid in a capillary tube.

1.3. Derivation of equations

We obtain the governing equations for a capillary surface by minimizing the total energy of the three-phase system (solid-liquid-gas) in a capillary tube for a constant liquid volume and temperature, while subject that $\gamma_{SL} \in [\gamma_{SL_{\min}}, \gamma_{SL_{\max}}]$ [8]. Consider a capillary surface formed between a three-phase system: liquid, gas, and a solid boundary. Let the domain occupied by the liquid be Ω and the domain of the capillary interface be S. Further, S_w denotes the area of the wetted part of the solid boundary. The total energy of this system consists of three main terms [18]:

(i) Free surface energy,
(ii) Wetting energy, and
(iii) Gravitational potential energy.

The free surface energy, $\mathcal{E}_f$, of a capillary interface is proportional to the area of the liquid surface that is not in contact with the solid. Then, if the energy per unit area (surface tension) is γ_{LG},

$$\mathcal{E}_f = \gamma_{LG} \int_S dS. \tag{3}$$

In equation (3), dS represents the surface area element on the capillary interface. In equation (4), wetting energy, $\mathcal{E}_w$, is the adhesion energy between the liquid and the solid substrate, and

$$\mathcal{E}_w = \int_{S_w} (\gamma_{SL} - \gamma_{SG})\, dS_w. \tag{4}$$

Note that in this quantity, γ_{SL} is not a constant function. Finally, the gravitational potential energy, $\mathcal{E}_g$, in terms of the height coordinate z and volume element dV is

$$\mathcal{E}_g = \int_\Omega \rho g\, z\, dV, \tag{5}$$

where ρ and g are the liquid density and gravitational acceleration, respectively. The total Helmholtz energy at constant temperature of the system is then given by:

$$\mathcal{E}_{tot} = \mathcal{E}_f + \mathcal{E}_w + \mathcal{E}_g. \tag{6}$$

One obtains equations for the capillary surfaces and the boundary conditions at the contact line from first-order necessary conditions arising from the minimization of the total energy functional subject to a constant liquid volume constraint and $\gamma_{SL} \in [\gamma_{SL_{\min}}, \gamma_{SL_{\max}}]$.

2. Liquid column in a capillary tube: Effect due to the contact angle hysteresis

Experiments suggest that a thin, vertical capillary can hold a finite column of liquid. Consider a liquid column trapped in a capillary tube with radius R. Let θ_1 and θ_2 be the equilibrium contact angles of the upper and lower menisci, respectively. Hence, for the static equilibrium of the liquid column, it satisfies the equation:

$$2\pi R\gamma_{LG}(\cos\theta_1 - \cos\theta_2) = \rho\, g\, V, \tag{7}$$

where γ_{LG} is the interfacial tension between the liquid-gas interface, and ρ, V denote the density and volume of the liquid respectively. Thus, the existence of such a capillary column implies that the equilibrium contact angles θ_1 and θ_2 satisfy $\theta_1 < \theta_2$.

Given V, to find the angles θ_1 and θ_2, we mathematically model the capillary menisci formed in a liquid column trapped in a cylindrical tube. As our goal is to find concrete values for energy losses for comparison purposes, we restrict ourselves to the case where the capillary interface is axisymmmetric. Governing equations for the upper and lower menisci are obtained using calculus of variations [18, 19, 20] after minimizing the total energy functional at a constant temperature subject to the constraint that $\gamma_{SL} \in [\gamma_{SL_{\min}}, \gamma_{SL_{\max}}]$.

Consider an axisymmetric capillary column in a tube with radius R. Let the liquid density be ρ. The profiles of the upper and lower menisci are $h_1(r)$ and $h_2(r)$ with contact angles θ_1 and θ_2, respectively. Consider the Cartesian coordinate system as depicted in Figure 3. The distance between the origin with the minimum of the upper meniscus and maximum of the lower meniscus are H_1 and H_2; $\bar{p}_0$ is the capillary pressure at the origin, and $\bar{p}_0 \equiv p(0) - p_{atm}$. Interfacial tensions between the solid-gas, solid-liquid, and liquid-gas are γ_{SG}, γ_{SL}, and γ_{LG}.

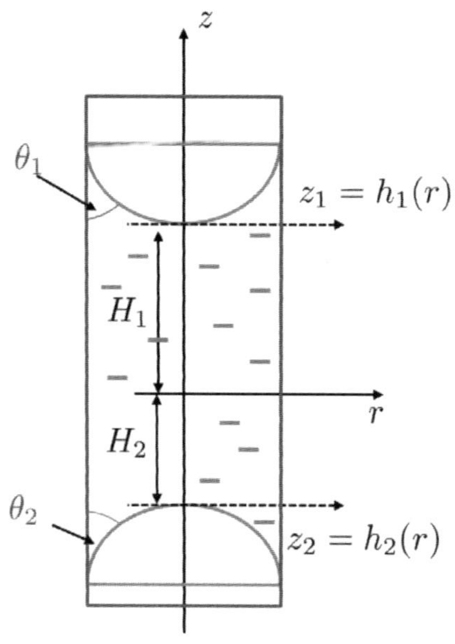

Figure 3: Schematic drawing of a liquid column in a capillary tube with gravity acting along the $-z$ direction. Upper and lower contact angles satisfy $\theta_1 < \theta_2$. The capillary surface heights $z_i = h_i(r)$, $i = 1, 2$ are measured from the base of the surface as shown in figure with z direction as positive – hence, h_2 is a non-positive function whilst h_1 is a non-negative function.

Let the total surface area of the tube be A_t. The total energy functional, $\mathcal{E}_{tot}$, is

$$\begin{aligned}\mathcal{E}_{tot} &= \left(A_t - 2\pi R \int_{H_2+h_2(R)}^{H_1+h_1(R)} dz\right)\gamma_{SG} + 2\pi R \int_{H_2+h_2(R)}^{H_1+h_1(R)} \gamma_{SL}(z)\, dz + \\ &\gamma_{LG} \int_0^R 2\pi r \sqrt{1+h_1'^2(r)}\, dr + \gamma_{LG} \int_0^R 2\pi r \sqrt{1+h_2'^2(r)}\, dr + \\ &\int_0^R \int_{H_2+h_2(r)}^{H_1+h_1(r)} 2\pi r\, \rho g\, z\, dz\, dr. \end{aligned} \tag{8}$$

In (8), the first two terms are the energy contribution due to the non-wetting and wetting surfaces, on the capillary tube, respectively. These two energy terms are described by the integrals over the wetting and non-wetting area of the capillary tube. In the second term, γ_{SL} takes values in the interval $[\gamma_{SL_{\min}}, \gamma_{SL_{\max}}]$ and it is considered to be z-dependent. The third and the fourth terms describe the capillary surface energy of the upper and lower meniscus profiles. This energy term is related to the capillary-free surface area. Gravitational potential energy term is represented by the last term and is defined over the liquid volume; liquid density is ρ, and g is the gravitational acceleration.

Expression (8) may be simplified to

$$\begin{aligned}\mathcal{E}_{tot} &= A_t\gamma_{SG} + 2\pi R \int_{H_2+h_2(R)}^{H_1+h_1(R)} (\gamma_{SL}(z) - \gamma_{SG})\, dz + \gamma_{LG} \int_0^R 2\pi r \sqrt{1+h_1'^2(r)}\, dr + \\ &\gamma_{LG} \int_0^R 2\pi r \sqrt{1+h_2'^2(r)}\, dr + 2\pi\rho g \int_0^R \frac{r}{2}[(h_1(r)+H_1)^2 - (h_2(r)+H_2)^2]\, dr. \end{aligned} \tag{9}$$

2.1. Derivation of the equilibrium conditions: First variation of the total energy functional

By minimizing the total energy of the system at constant volume and temperature subject to the constraint $\gamma_{SL} \in [\gamma_{SL_{\min}}, \gamma_{SL_{\max}}]$, we find the equilibrium meniscus profiles: $h_1(r)$ and $h_2(r)$ and the liquid column heights: H_1 and H_2. This approach results in the associated Euler-Lagrange equation for $h_1(r)$, $h_2(r)$, H_1, and H_2 together with the boundary conditions at the three-phase contact line [21, 18, 22]. These boundary conditions yield the corresponding Young's equation at the upper and lower contact lines. Furthermore, we also obtain flow rules for the change in the contact angles when the contact lines move.

Although volume V is considered to be constant, the wetting area is allowed to vary, and therefore, we take the first variation of (9) with respect to the variables: H_1, H_2, $h_1(r)$, and $h_2(r)$. Denote the Lagrange multiplier corresponding to the constant volume constraint as $\bar{p}$. Then,

$$\begin{aligned}\delta\mathcal{E}_{tot} - \bar{p}\,\delta V \;=\;& 2\pi R[\gamma_{SL}(H_1 + h_1(R)) - \gamma_{SG}](\delta H_1 + \delta h_1(R)) - \\ & 2\pi R[\gamma_{SL}(H_2 + h_2(R)) - \gamma_{SG}](\delta H_2 + \delta h_2(R)) + \\ & \gamma_{LG}\int_0^R 2\pi r\left[\frac{h_1'(r)}{\sqrt{1+h_1'^2(r)}}\delta h_1' + \frac{h_2'(r)}{\sqrt{1+h_2'^2(r)}}\delta h_2'\right] dr + \\ & 2\pi\rho g\int_0^R r[(h_1(r)+H_1)(\delta h_1(r)+\delta H_1) - (h_2(r)+H_2)(\delta h_2(r)+\delta H_2)]\,dr - \\ & 2\pi\bar{p}\int_0^R r[\delta h_1(r) + \delta H_1 - \delta h_2(r) - \delta H_2]\,dr \\ \le\;& 0, \end{aligned} \quad (10)$$

where the inequality is due to the fact that equality may not be attained due to the constraint $\gamma_{SL} \in [\gamma_{SL_{\min}}, \gamma_{SL_{\max}}]$. By using integration by parts, the first integral in (10) may be recast into

$$\int_0^R r\frac{h_i'(r)}{\sqrt{1+h_i'^2(r)}}\delta h_i'\,dr = r\frac{h_i'(r)}{\sqrt{1+h_i'^2(r)}}\delta h_i(r)\Big|_0^R - \int_0^R \left(\frac{r h_i'(r)}{\sqrt{1+h_i'^2(r)}}\right)'\delta h_i(r)\,dr \quad \text{for}\, i \in \{1,2\}, \quad (11)$$

and hence, (10) can be rewritten in the form:

$$\begin{aligned}\delta\mathcal{E}_{tot} - \bar{p}\,\delta V \;=\;& R[\gamma_{SL}(H_1 + h_1(R)) - \gamma_{SG}](\delta H_1 + \delta h_1(R)) - \\ & R[\gamma_{SL}(H_2 + h_2(R)) - \gamma_{SG}](\delta H_2 + \delta h_2(R)) + \\ & \gamma_{LG}\frac{r h_1'(r)}{\sqrt{1+h_1'^2(r)}}\delta h_1(r)\Big|_0^R + \gamma_{LG}\frac{r h_2'(r)}{\sqrt{1+h_2'^2(r)}}\delta h_2(r)\Big|_0^R - \\ & \gamma_{LG}\int_0^R\left(\frac{r h_1'(r)}{\sqrt{1+h_1'^2(r)}}\right)'\delta h_1(r)\,dr - \gamma_{LG}\int_0^R\left(\frac{r h_2'(r)}{\sqrt{1+h_2'^2(r)}}\right)'\delta h_2(r)\,dr + \\ & \rho g\int_0^R r[(h_1(r)+H_1)(\delta h_1(r)+\delta H_1) - (h_2(r)+H_2)(\delta h_2(r)+\delta H_2)]\,dr - \\ & \bar{p}\int_0^R r[\delta h_1(r) + \delta H_1 - \delta h_2(r) - \delta H_2]\,dr. \end{aligned} \quad (12)$$

The necessary condition for $\mathcal{E}_{tot}$ to have an extremum is $\delta\mathcal{E}_{tot} - \bar{p}\,\delta V \le 0$. Now, $\delta\, h_1(r)$ and $\delta\, h_2(r)$ have no restriction on sign and so we get:

$$-\gamma_{LG}\left(\frac{rh_1'(r)}{\sqrt{1+h_1'^2(r)}}\right)' - \bar{p}r + \rho g r(h_1(r)+H_1) = 0, \tag{13}$$

$$-\gamma_{LG}\left(\frac{rh_2'(r)}{\sqrt{1+h_2'^2(r)}}\right)' + \bar{p}r - \rho g r(h_2(r)+H_2) = 0. \tag{14}$$

Similarly, δH_1 and δH_2 have no restriction on sign and this leads to:

$$\Big(\gamma_{SL}(H_1+h_1(R)) - \gamma_{SG}\Big)R - \bar{p}\frac{R^2}{2} + \rho g\int_0^R rh_1(r)\,dr + \rho g H_1\frac{R^2}{2} = 0, \tag{15}$$

$$-\Big(\gamma_{SL}(H_2+h_2(R)) - \gamma_{SG}\Big)R + \bar{p}\frac{R^2}{2} - \rho g\int_0^R rh_2(r)\,dr - \rho g H_2\frac{R^2}{2} = 0. \tag{16}$$

These equations are satisfied for $r \in [0, R]$. With these deductions, the inequality $\delta\mathcal{E}_{tot} - \bar{p}\,\delta V \leq 0$ reduces to:

$$\begin{aligned}\delta\mathcal{E}_{tot} - \bar{p}\,\delta V &= R[\gamma_{SL}(H_1+h_1(R)) - \gamma_{SG}]\,\delta h_1(R) - R[\gamma_{SL}(H_2+h_2(R)) - \gamma_{SG}]\,\delta h_2(R) + \\ &\quad \gamma_{LG}\frac{rh_1'(r)}{\sqrt{1+h_1'^2(r)}}\delta h_1(r)\Big|_0^R + \gamma_{LG}\frac{rh_2'(r)}{\sqrt{1+h_2'^2(r)}}\delta h_2(r)\Big|_0^R \\ &\leq 0.\end{aligned} \tag{17}$$

Note that the contact line variation $\delta h_1(R)$ and $\delta h_2(R)$ are related to θ_1 and θ_2 respectively. For a particular θ_i value, $\delta h_i(R)$ is determined by the inequality (17). To simplify notation, denote:

$$\frac{\gamma_{SG} - \gamma_{SL}(H_1+h_1(R))}{\gamma_{LG}} = \cos\theta_{Y_1}, \tag{18}$$

$$\frac{\gamma_{SG} - \gamma_{SL}(H_2+h_2(R))}{\gamma_{LG}} = \cos\theta_{Y_2}, \tag{19}$$

Note that for $i = 1, 2$, $\cos\theta_{Y_i} \in [\cos\theta_A, \cos\theta_R]$. Then, using the fact that $h_1'(R) = \cot\theta_1$ and $h_2'(R) = -\cot\theta_2$, inequality (17) becomes:

$$(\cos\theta_1 - \cos\theta_{Y_1})\,\delta h_1(R) + (\cos\theta_2 - \cos\theta_{Y_2})\,\delta h_2(R) \leq 0 \tag{20}$$

Define: $\cos\theta_A := \frac{\gamma_{SG}-\gamma_{SL_{\max}}}{\gamma_{LG}}$ and $\cos\theta_R := \frac{\gamma_{SG}-\gamma_{SL_{\min}}}{\gamma_{LG}}$. We have the following cases. For $i = 1, 2$,

(i) If $\cos\theta_i \in (\cos\theta_A, \cos\theta_R)$, then $\forall\,\theta_{Y_i}$, $(\cos\theta_i - \cos\theta_{Y_i}) \lesseqgtr 0$. Hence by (20) $\delta h_i(R) = 0$.

(ii) If $\cos\theta_i = \cos\theta_A$, then $\forall\,\theta_{Y_i}$, $(\cos\theta_i - cos\theta_{Y_i}) \leq 0$. Thus, by (20) $\delta h_i(R) \geq 0$.

(iii) If $\cos\theta_i = \cos\theta_R$, then $\forall\,\theta_{Y_i}$, $(\cos\theta_i - \cos\theta_{Y_i}) \geq 0$. Hence, by (20) $\delta h_i(R) \leq 0$.

The above inequalities yield the *flow rules* for the contact line motion and correspond to observed results from experiment described in Section 1.1.

2.2. Concept of capillary pressure and solution of the system of equations

First, we provide a physical interpretation to the Lagrange multiplier term $\bar{p}$. The left hand side of Inequality (10) may be thought of as the first variation of the *augmented* energy functional $\mathcal{E}_{aug} = \mathcal{E}_{tot} - \bar{p}\,V$. From Equation (8), we see that the last (gravitational energy) term of $\mathcal{E}_{aug}$ is

$$\int_0^R \int_{H_2+h_2(r)}^{H_1+h_1(r)} 2\pi r\,(-\bar{p} + \rho\, g\, z)\, dz\, dr.$$

This expression allows us to interpret $\bar{p}$ as the pressure at the origin $z = 0$. We term this pressure as the *Capillary pressure.*

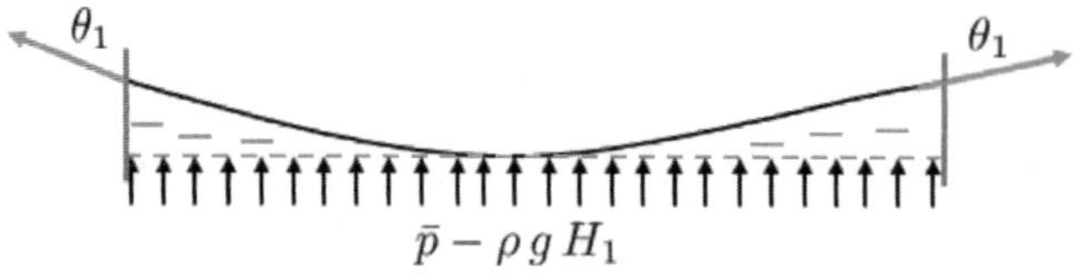

Figure 4: Schematic drawing showing the top meniscus with a liquid column underneath. Force balance on this meniscus is given by Equation (21).

Equation (15) may be rewritten after multiplication by $2\,\pi$ as:

$$\cos\theta_{Y_1}\,\gamma_{LG}\,(2\,\pi\,R) + (\bar{p} - \rho\, g\, H_1)\,(\pi\,R^2) = \rho g \int_0^R 2\,\pi\, r h_1(r)\, dr. \tag{21}$$

Equation (21) represents the balance of forces for the liquid column under the meniscus $z_1 = h_1(r)$ (see Figure 4). It is interesting to note the first term on the left hand side. As the real contact angle is θ_1, the difference $(\cos\theta_{Y_1} - \cos\theta_1)\,\gamma_{LG}\,(2\,\pi\,R)$ represents the additional force due to the adhesion between the solid wall and the liquid column – a similar or same interpretation may be found in Joanny and de Gennes [23].

Now, to facilitate computation, we make the *assumption* that the height of the liquid column under the meniscus as shown in Figure 4 is much smaller than the height of the entire liquid in the capillary tube. This assumption is a minor one and facilitates the computation of the quantities easily as we can re-state the assumption as $\theta_{Y_1} = \theta_1$. Similarly, we assume that $\theta_{Y_2} = \theta_2$ by considering the bottom meniscus. Note that, with this assumption, (18) yields: $\gamma_{SG} - \gamma_{SL}(H_1 + h_1(R)) = \gamma_{LG}\,\cos\theta_1$ which is Young's law applied to the contact line of the top meniscus.

Denote by $\alpha = \rho g R^2$. Solving (15) and (16) for H_1 and H_2, we get

$$H_1 = \frac{2\gamma_{LG}\cos\theta_1 R}{\alpha} + \frac{\bar{p}R^2}{\alpha} - \frac{2\rho g}{\alpha}\int_0^R r h_1(r)\, dr, \tag{22}$$

$$H_2 = \frac{2\gamma_{LG}\cos\theta_2 R}{\alpha} + \frac{\bar{p}R^2}{\alpha} - \frac{2\rho g}{\alpha}\int_0^R r h_2(r)\, dr. \tag{23}$$

Substituting H_1 from (22) into Equation (13), we obtain a partial differential equation for $h_1(r)$ that depends on the parameter $\bar{p}$. Similarly, we may obtain a partial differential equation for $h_2(r)$. We solve these PDEs using COMSOL® Multiphysics software.

2.3. Energy dissipation due to hysteresis

In [8], it is shown that the energy that must be input by an external source over the time interval $[0, T]$ for the volume to change according to the profile given by $V(t)$ (assumed to be differentiable with respect to t), where $t \in [0, T]$ is given by:

$$\mathcal{W} = \int_{t_0}^{t_f} \bar{p}(t)\,\dot{V}(t)\,dt, \tag{24}$$

where $\bar{p}(t)$ is the capillary pressure associated with the volume $V(t)$. The above formula states that if the capillary pressure $\bar{p}$ is plotted against the volume then the area of the region under the curve yields the energy input from the external source that is causing the volume to change. As the relation between $\bar{p}$ and V shows hysteresis (see Section 3), Equation (24) equivalently represents the energy loss due to contact angle hysteresis when the volume of the liquid is changed in a cycle. If the initial and final volumes are not the same, then Equation (24) also includes the internal (potential) energy change in addition to the energy loss. Although capillary effect is a complex phenomenon, it is interesting to note that the energetics of the system may be described by two scalar variables - the volume of the liquid and the capillary pressure!

In the next section, we consider a specific profile for the volume change and compute the energy that must be input by an external source to complete a cycle.

3. Numerical results and Discussion

The formation of an axisymmetric liquid column in a capillary tube is depicted in Figure 3. In the present study, we investigate the dissipation of energy during one cycle of capillary action, when the liquid volume inside the capillary tube first increases and subsequently decreases. Figure 5 shows a hysteresis diagram obtained by considering the variation of wetting area with the liquid volume. Let θ_1 and θ_2 be the upper and lower contact angle values and assume both angles are invariant along the corresponding contact lines. A_{Low} is the initial wetting area. Recall the necessary condition: $\theta_1 < \theta_2$ for the static equilibrium of the liquid column. We first consider the increase in the liquid volume at $\theta_1 = \theta_R$ and $\theta_R < \theta_2 < \theta_A$ (refer Figure 5). During this process, assume both θ_1 and θ_2 increasing until θ_2 reaches its advancing angle θ_A, and $\theta_R < \theta_1 < \theta_A$ (segment 1–2). Once θ_2 equals θ_A, further continuation of liquid addition results in a downward motion of the lower contact line with a constant contact angle θ_A and a decrease in the upper contact angle θ_1(segment 2–3). At point 3, let the wetting area be A_{High}, and θ_1 is still in the interval $[\theta_R, \theta_A]$. Paths 3–4 and 4–1 refer to the decrease in the liquid volume. As volume is quasi-statically removed, assume both θ_1 and θ_2 decrease until θ_1 approaches its receding value, and let θ_2 be in the interval $[\theta_R, \theta_A]$ at the point 4. Hence, segment 3–4 depicts a constant wetting area. Since $\theta_1 = \theta_R$, now the lower meniscus begins to recede with constant contact angle θ_R. Liquid is removed until the column arrives at the initial configuration: that is, wetting area A_{Low} and liquid volume V_1. This formulation assumes that both upper and lower contact lines be pinned during the segments 1–2 and 3–4.

On vertical segments 1–2 and 3–4 the wetting area is fixed at A_{Low} and A_{High} respectively. On segments 2–3 and 4–1, $\theta_2 = \theta_A$ and $\theta_1 = \theta_R$, respectively. Thus, we solve the governing equations (13) – (16), for the upper and lower menisci with appropriate boundary conditions (described below) using the COMSOL® Multiphysics software.

Upper and lower capillary surfaces are numerically computed by using the boundary conditions: $h_1'(0) = h_2'(0) = 0$, $h_1'(R) = \cot\theta_1$, and $h_2'(R) = -\cot\theta_2$ with prescribed θ_1 and θ_2. Additional wetting area constraint is imposed to calculate the capillary surfaces that correspond to the line segments 1–2 and 3–4. For the numerical computations, we consider capillary tubes with a diameter of 5mm and 1mm. All the numerical computations are made with $\gamma_{LG} = 73\,\mathrm{mJ/m^2}$, $\rho = 1000\,\mathrm{kg/m^3}$, and $g = 9.82\,\mathrm{m/s^2}$ [3].

Wetting area, liquid volume, and area of the upper and lower interfaces are numerically computed for each solution.

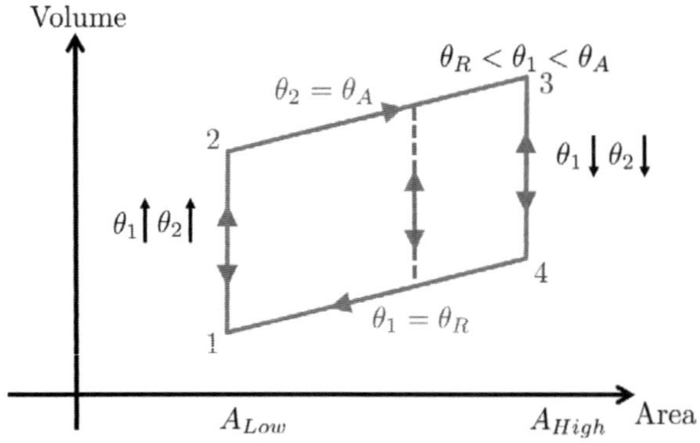

Figure 5: A hysteresis diagram for a capillary column: the liquid volume variation with respect to the wetting area of the liquid column.

Results of numerical computations of upper and lower menisci in a capillary tube of diameter $2R = 5\,\text{mm}$ are displayed in Figure 6. Upper and lower contact angles are at 35° and 65°, respectively. The liquid column volume is $4.63 \times 10^{-8}\,\text{m}^3$. For any pair of $\{\theta_1, \theta_2\}$, the volume V of the liquid may be calculated using the static equilibrium condition:

$$2\pi\gamma_{LG}R(\cos\theta_1 - \cos\theta_2) = \rho g V, \tag{25}$$

hence, we use the criterion (25) to validate our meniscus profiles.

We calculate the energy losses when the liquid volume inside the capillary tube first increases and subsequently decreases. The corresponding energy diagram is in Figures 7. Area under each closed loop denotes the energy dissipation due to the capillary pressure. Table 1 presents the variations of the energy terms for different upper and lower contact angle (θ_1 and θ_2) values.

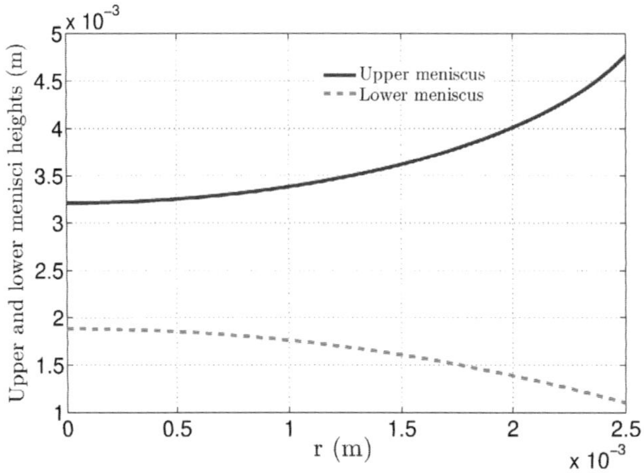

Figure 6: Upper and lower menisci profiles. The capillary tube diameter is 5mm, and the capillary pressure at the origin $\bar{p}(0) = -10\,\text{N/m}^2$. In this numerical simulation, the upper and lower contact angles are $\theta_1 = 35°$ and $\theta_2 = 65°$, respectively. The liquid bridge has the wetting area $5.76 \times 10^{-5}\,\text{m}^2$ with volume $4.63 \times 10^{-8}\,\text{m}^3$.

Journal of Physics: Conference Series **727** (2016) 012003 doi:10.1088/1742-6596/727/1/012003

Table 1: Energy variations with capillary tube with diameter of 1 mm. Initial and final wetting areas are $4.00 \times 10^{-5}\,\text{m}^2$ and $4.18 \times 10^{-5}\,\text{m}^2$. All the numerical computations are performed under the assumptions of advancing angle $\theta_A = 66°$ and receding angle $\theta_R = 30°$.

θ_1	θ_2	Wetting area $\times 10^{-5}(\text{m}^2)$	Volume $\times 10^{-8}(\text{m}^3)$	Capillary Pressure (N/m^2)
30.0	62.0	3.99	1.85	-7.05
30.8	62.6	4.00	1.86	-11.32
33.6	64.4	4.00	1.87	-15.28
36.0	66.0	4.00	1.88	-18.94
35.8	66.0	4.02	1.89	-18.51
35.6	66.0	4.04	1.90	-18.20
35.4	66.0	4.06	1.91	-17.88
35.2	66.0	4.08	1.92	-17.57
35.0	66.0	4.10	1.93	-17.25
34.8	66.0	4.12	1.94	-16.93
34.6	66.0	4.14	1.95	-16.62
34.2	66.0	4.18	1.97	-16.07
33.0	65.2	4.18	1.96	-14.22
31.2	64.0	4.18	1.95	-11.52
30.0	63.2	4.18	1.94	-9.6
30.0	63.0	4.13	1.93	-9.23
30.0	62.6	4.07	1.90	-8.37
30.0	62.4	4.04	1.88	-7.90

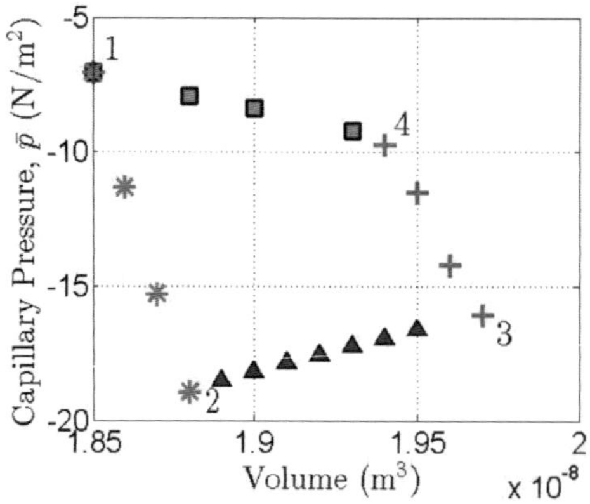

Figure 7: Energy diagram for capillary pressure versus liquid volume. Capillary tube diameter is 1 mm, and in this numerical simulation $\theta_R = 30°$ and $\theta_A = 66°$. The low and high wetting areas are 4.00×10^{-5} m^2 and 4.18×10^{-5} m^2, respectively. Energy required to complete the cycle 1–2–3–4–1 is approximately 8.21 nJ.

Table 1 shows the capillary energy variations with volume of the liquid column. Rows 1–4, 5–11, 12–15, and 16–18 of the table correspond to the volume variations that also correspond to line segments 1–2, 2–3, 3–4, and 4–1, respectively (see the hysteresis diagram: Figure 5). Figure 7 depicts the capillary pressure variation with volume of the liquid column with diameter of 1 mm. Energy required to complete the cycle 1–2–3–4–1 is approximately 8.21 nJ.

3.1. Energy losses due to viscosity of the liquid

A study of the energy mechanisms involved in wetting and drying of water in soil is very important in the field of climate change, and also for determining optimal irrigation procedures for agriculture [24]. Soil pore spaces are micro capillary tubes which fill-up and dry-out due to rainfall and/or irrigation. Completion of one cycle of wetting and drying involves motion of the contact line. There are two different mechanisms for energy lost in this process – (1) hysteresis that originates due to contact angle hysteresis, which is rate-independent; (2) viscous friction opposing fluid motion.

Numerical computation presented in Section 3 shows that the energy demand due to rate-independent hysteresis for a capillary tube with diameter of 0.1 cm, approximate wetting area of 0.4 cm^2, and capillary surface motion of 0.006 cm (corresponding to wetting area change of 0.018 cm^2) is 0.0821 ergs. Next, we consider viscous friction during contact line motion so that one may compare numbers and get an idea of the more dominant mechanism of entropy increase (or increase in internal energy of the liquid and surroundings) in capillary effect. To enable analysis, and for clarity, a simpler configuration of a liquid bridge is chosen as shown in Figure 8 for analysis [25, 26].

In the previous research work [25, 26], we investigated the viscous energy dissipation of a fluid flow that results as a consequence of the deformation of a capillary interface. In particular, we considered a capillary surface between two parallel, non-ideal solid surfaces and assumed the invariance of the interface in $z-$direction (see Figure 8). The fluid flow was analyzed using the Navier-Stokes and Continuity equations, and the viscous energy dissipation was numerically computed during the deformation of the interface. The distance between two plates is 0.1 cm and the wetting area change is approximately 0.01 cm^2. During the motion of the capillary interface, we found the energy dissipation due to viscosity to be 3.64×10^{-4} ergs. For this reason, hysteresis energy loss seems to be significantly greater (by a factor of more than 200) than energy loss due to viscosity for comparable dimensions.

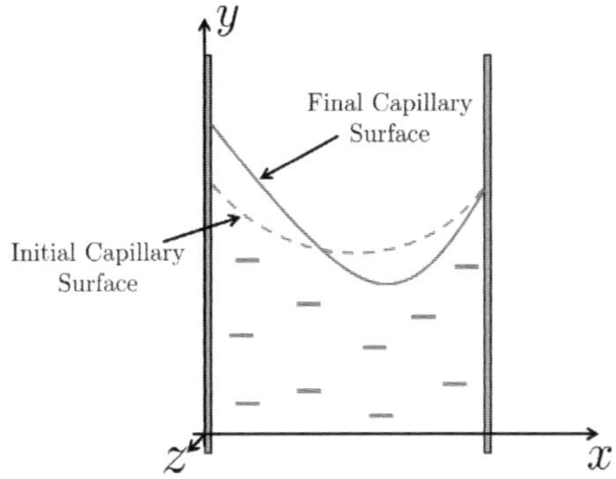

Figure 8: Two vertical plates, which are immersed in a liquid. Capillary interface is invariant in the $z-$direction.

4. Conclusions

In this study, we investigated the energy dissipation due to the effect of contact angle hysteresis using a liquid column formed in a capillary tube in which volume of the column first increases and subsequently decreases. Equations for the upper and lower menisci were obtained by minimizing the Helmholtz energy at a constant volume and temperature subject to the constraint that the energy of the interface between the solid and the liquid takes values in an interval. We present numerical solutions of the derived equations for the case of a capillary tube with 1 mm diameter. The energy required to overcome contact angle hysteresis seems to be 2 orders of magnitude greater than that required to overcome fluid viscosity for a comparable liquid bridge. This suggests that contact angle hysteresis is the dominant mechanism of energy loss in nano-fluidics.

References

[1] Extrand C W 2002 *Langmuir* **1** 7991–7999

[2] Bonn D, Eggers J, Indekeu J, Meunier J and Rolley E 2009 *Reviews of Modern Physics* **81** 739–805 ISSN 0034-6861

[3] de Gennes P G, Brochard-Wyart F and Quere D 2003 *Capillarity and Wetting Phenomena: Drops, Bubbles, Pearls, Waves* (Springer)

[4] Butt H and Kappl M 2009 *Surface and interfacial forces* (John Wiley & Sons)

[5] Alberti G and DeSimone A 2011 *Archive for rational mechanics and analysis* **202** 295–348

[6] Lin S Y, Chang H C, Lin L W and Huang P Y 1996 *Review of Scientific Instruments* **67** 2852–2858

[7] Merchant G J and Keller J B 1992 *Physics of Fluids A: Fluid Dynamics* **4** 477–485

[8] Athukorallage B, Aulisa E, Iyer R and Zhang L 2015 *Langmuir* **31** 2390–2397

[9] Sulman H L and Picard H K 1920 *The theory of concentration processes involving surface-tension* (Original typewritten notes, Columbia University, NY)

[10] de Gennes P G 1985 *Reviews Modern Physics* **57** 828–861

[11] Dettre R and Johnson Jr R 1964 Contact angle hysteresis ii: Contact angle measurements on rough surfaces *Contact Angle, Wettability, and Adhesion* ed Fowkes F (American Chemical Society) pp 136–144

[12] Penn L and Miller B 1980 *Journal of Colloid and Interface Science* **1** 238–241

[13] Ramos S, Charlaix E and Benyagoub A 2003 *Surface Science* **540** 355–362

[14] Chibowski E and Malgorzata J 2013 *Colloid Polymer Science* **291** 391–399

[15] Marmur A 2010 A guide to the equilibrium contact angles maze *Contact Angle, Wettability, and Adhesion* ed Mittal K L (Brill, Leiden) pp 3–18

[16] Snoeijer J and Andreotti B 2008 *Physics of Fluids* **20** 057101

[17] Extrand C W 1998 *Journal of Colloid and Interface Science* **207** 11–19

[18] Finn R 1986 *Equilibrium capillary surfaces* (Springer-Verlag)

[19] Frankel T 1997 *The Geometry of Physics: An Introduction* (Cambridge University Press) ISBN 9780521383349

[20] Evans L 2010 *Partial Differential Equations* Graduate studies in mathematics (American Mathematical Society) ISBN 9780821849743

[21] Bullard J W and Garboczi E J 2009 *Phys. Rev. E* **79**(1) 011604

[22] Gelfand I and Fomin S 2000 *Calculus of Variations* (Dover Publications)

[23] Joanny J F and de Gennes P G 1984 *The Journal of Chemical Physics* **81** 552–562 ISSN 00219606

[24] Appelbe B, Flynn D, McNamara H, O'Kane P, Pimenov A, Pokrovskii A, Rachinskii D and Zhezherun A 2009 *IEEE Control Systems Magazine* **29** 44–69

[25] Athukorallage B and Iyer R 2014 *Physica B: Condensed Matter* **435** 28 – 30 ISSN 0921-4526 9th International Symposium on Hysteresis Modeling and Micromagnetics (HMM 2013)

[26] Athukorallage B 2014 *Capillarity and Elastic Membrane Theory from an Energy Point of View* Ph.D. thesis Texas Tech University

Self-Sustained Relaxation Oscillations in Time-Delay Neural Systems

S D Glyzin[1], A Yu Kolesov[1] and N Kh Rozov[3]

[1] Faculty of Mathematics, Yarostavl State University, Sovetskaya str., 14, Yaroslavl, 150000, Russia

[2] Faculty of Mechanics and Mathematics, Moscow State University, Leninskiye Gory, Main Building, Moscow, 119991, Russia

E-mail: glyzin@uniyar.ac.ru; kolesov@uniyar.ac.ru; fpo.mgu@mail.ru

Abstract. A new method to model the phenomena 'bursting' and 'buffering' in neural systems is represented. Namely, a singularly perturbed nonlinear scalar differential difference equation with two delays is introduced, which is a mathematical model of a single neuron. It is shown that for suitably chosen parameters this equation has a stable periodic solution with an arbitrary prescribed number of asymptotically high impulses (spikes) on a period interval. It is also shown that the buffering phenomenon occurs in a one-dimensional chain of diffusively coupled neurons of this type: as the number of components in the chain grows in a way compatible with a decrease of the diffusion coefficient, the number of co-existing stable periodic motions increases indefinitely.

1. Introduction

Self-sustained oscillations in neural systems have two characteristic features, the so-called bursting phenomenon and the buffering phenomenon. The first consists of the alternation of packets of pulses (strings of several spikes) and pieces of relatively moderate variation of the membrane potentials. The second term is commonly used in situations when an arbitrary prescribed number of co-existing attractors can be realized by an appropriate choice of parameters of the corresponding dynamical system.

The bursting phenomenon has been considered by many authors (see, for instance, [1, 2, 3, 4, 5] and the literature in these papers). To construct a mathematical model of this phenomenon, authors usually use singularly perturbed systems of ordinary differential equations with one slow and two fast variables, in which stable bursting-cycles (that is, periodic motions exhibiting the bursting phenomenon) can exist under certain assumptions. However, there is also another possible approach to this problem which is based on taking into account time delays.

The situation with the buffering phenomenon is somewhat different. Although it is quite common for nonlinear models in various fields of natural sciences [6, 7, 8, 9, 10], it has never got the appropriate attention in the neurodynamics literature. On the other hand this feature is important for neural sciences: it reflects a competitive interaction between different concepts and ideas in the neocortex part of human brain and can be used to explain the mechanisms behind associative memory. Thus we arrive at the problem of realizing both the phenomena described above in the framework of a single mathematical model. We put forward such a model in this paper.

2. Description of the Mathematical Model

First models of the dynamics of the electric potential of neurons are due to A.L. Hodgkin and A.F. Huxley [11, 12], who presented a phenomenological model based on balance-type relations such that (if the parameters have been suitably chosen) its dynamics is consistent with experimental data about the oscillations of the membrane potentials of neurons.

Their model is quite complicated and many authors attempted to simplify it while preserving the main characteristic features of neural dynamics. The surveys [3, 4] present a list of requirements for a pulse neuron model and enumerated many model systems. The most important of these requirements is that a model has a stable periodic pulse-type regime. Another important condition is that the model must display the bursting phenomenon (for some values of parameters).

To describe the new mathematical model of an spiking neuron we use the fundamental idea in [11, 12] that a biological neuron can be replaced by an equivalent generator of electric oscillations. We shall only take into account the potassium and sodium currents, and take the maximum polarization level of the membrane for a reference point by assuming that $u(t) \geq 0$ is the deviation of the potential from this level. Then, discounting the leakage current, we can write the current balance equations:

$$c\dot{u} = -I_{\mathrm{Na}} - I_{\mathrm{K}}, \tag{1}$$

where $c > 0$ is usually called the membrane capacity.

To construct a reasonable model we make several additional assumptions.

Assumption 1. *We assume that the currents I_{Na} and I_K can be represented as follows:*

$$I_{Na} = -\chi_{Na}(u) \cdot u, \qquad I_K = -\chi_K(u) \cdot u, \tag{2}$$

where $\chi_{Na}(u)$ and $\chi_K(u)$ are functions characterizing the sodium and potassium conductance.

Assumption 2. *We assume that $\chi_{Na}(0) = -\alpha_0$ and $\chi_{Na}(u) \to -\beta_0$ as $u \to +\infty$, where α_0 and β_0 are positive constants. We also assume that $\beta_0 > \alpha_0$.*

Assumption 3. *We assume that $\chi_K(0) = \alpha_1$ and $\chi_K(u) \to -\beta_1$ as $u \to \infty$, where α_1 and β_1 are positive constants, as above.*

Assumption 4. *We assume that the values of the potassium and sodium conductance lag behind the current value of the membrane potential. We take the value of the first, potassium lag for a unit of time and assume that the second lag is $h \in (0, 1)$. Thus we have $\chi_K = \chi_K(u(t-1))$ and $\chi_{Na} = \chi_{Na}(u(t-h))$.*

Assumption 5. *We assume that*

$$\chi_{\mathrm{Na}}(0) + \chi_{\mathrm{K}}(0) > 0. \tag{3}$$

We can present many arguments, biophysical and mathematical alike, in favour of these assumptions. We start with biophysical considerations. Relations (2) ensure that $u(t)$ is positive (as it must be) and reflect the traditional approach to use Volterra-type equations for simulating biophysical and ecological processes. We make the assumption that $\chi_{\mathrm{Na}}(0)$ is negative because for strong polarization ($u \ll 1$) there is a surplus of sodium ions on the inner surface of the membrane. Pumping them out of the cell is what ion transporters are for (see [12]). Since sodium ions are positively charged, this process reduces the membrane potential, so $\chi_{\mathrm{Na}}(u) < 0$ for $u \ll 1$. Now we go over to the behaviour of $\chi_{\mathrm{K}}(u)$. When polarization is strong, the flow of potassium ions is directed inside the cell thus contributing to the growth of the membrane potential, so $\chi_{\mathrm{K}}(u) > 0$ for $u \ll 1$. However, after passing the peak of the potential, the flow of potassium ions changes direction. Hence there exists a level of the potential such that $\chi_{\mathrm{K}}(u) < 0$

for values of u higher than this level. Thus we arrive at Assumption 3. As regards Assumption 4, we can only say that a conductance time lag is an important feature of ion channels, which must therefore be taken into account. And as regards inequality (3), it reflects the fact that when polarization is strong, the membrane potential is increasing. On the mathematics level, Assumptions 1–5 ensure that all the requirements for neural models stated above are fulfilled. Namely, we show below that equation (1) has periodic pulse regimes with an arbitrary prescribed number of pulses on a period interval.

Under the Assumptions 1–5, from (1) we obtain

$$c\dot{u} = [\chi_{\mathrm{Na}}(u(t-h)) + \chi_{\mathrm{K}}(u(t-1))]u. \tag{4}$$

It is easy to see that for $h = 1$ the model (4) reduces to the generalized Hutchinson equation, which was thoroughly investigated in [13]. Note that a model with one delay was constructed in [14] in a similar way.

For further analysis we reduce (4) to a more convenient form. To do this we set

$$\begin{gathered}\chi_{\mathrm{Na}}(u) = (\chi_{\mathrm{K}}(0) + \chi_{\mathrm{Na}}(0))f(u) - \chi_{\mathrm{K}}(0),\ \chi_{\mathrm{K}}(u) = \chi_{\mathrm{K}}(0) - (\chi_{\mathrm{K}}(0) + \chi_{\mathrm{Na}}(0))g(u),\\ \lambda = (\chi_{\mathrm{K}}(0) + \chi_{\mathrm{Na}}(0))/c.\end{gathered} \tag{5}$$

Then we obtain the equation

$$\dot{u} = \lambda[f(u(t-h)) - g(u(t-1))]u. \tag{6}$$

As previously, here $u(t) > 0$ is the membrane potential of a neuron. The positive parameter λ characterizes the speed of electric processes in the system, so it must be large; and h is a fixed parameter in the interval $(0, 1)$. As concerns the functions $f(u)$, $g(u) \in C^1(\mathbb{R}_+)$, where $\mathbb{R}_+ = \{u \in \mathbb{R} : u \geq 0\}$, in accordance with Assumptions 1–5 and equalities (5) we assume that they have the following properties:

$$\begin{gathered}f(0) = 1,\ \ g(0) = 0;\ \ f(u) = -a_0 + O(1/u),\ \ uf'(u) = O(1/u),\ \ u^2 f''(u) = O(1/u),\\ g(u) = b_0 + O(1/u),\ \ ug'(u) = O(1/u),\ \ u^2 g''(u) = O(1/u)\ \text{ as }\ u \to +\infty,\end{gathered} \tag{7}$$

where $a_0 = -\dfrac{\alpha_1 - \beta_0}{\alpha_1 - \alpha_0}$ and $b_0 = \dfrac{\alpha_1 + \beta_1}{\alpha_1 - \alpha_0}$ are positive constants. Examples of such functions are given by

$$f(u) = (1-u)/(1+c_1 u) \text{ and } g(u) = c_2 u/(1+u),\ \ c_1, c_2 = \mathrm{const} > 0. \tag{8}$$

The main results of this paper concern relaxation properties of equations (6) and systems of coupled equations of this type. It is important to note that the model we obtain is quite meaningful: for parameters chosen appropriately it has regimes with a single spike on a period interval (for example, for $h = 1$), as well as regimes with any prescribed number of such spikes. In particular, we shall see that for any fixed positive integer n we can pick parameters h, a_0, and b_0 in (6) and (7) such that for any sufficiently large λ equation (6) has an exponentially orbitally stable cycle $u = u_*(t, \lambda)$ with period $T_*(\lambda)$, where $T_*(\lambda)$ approaches a finite limit $T_* > 0$ as $\lambda \to \infty$. Furthermore, on an interval of length $T_*(\lambda)$ the function $u_*(t, \lambda)$ shows precisely n asymptotically high impulses (of order $\exp(\lambda h)$) with duration $\Delta t = (1 + 1/a_0)h$ following one another, and u_* is asymptotically small for the rest of time. In other words, $u_*(t, \lambda)$ is a bursting-cycle for this choice of parameters.

We get a graphic impression of the relaxation properties of the bursting-cycle $u_*(t, \lambda)$ from its graph in the plane (t, u) (see Figure 1), which is plotted to scale $1 : 25$ in the case of $h = 1/26$ and $\lambda = 130$ for the functions in (8) with $c_1 = 0.5$ and $c_2 = 4$.

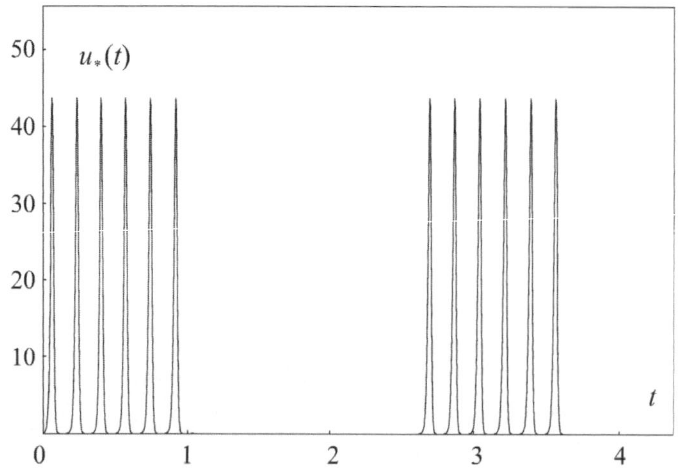

Figure 1. Stable periodic solution $u_*(t, \lambda)$ of equation (6) for $h = 1/26$ and $\lambda = 130$ for the functions in (8) with $c_1 = 0.5$ and $c_2 = 4$

As we have already said, (6) is a mathematical model of an isolated neuron. Now let us look at a one-dimensional chain of m such neurons, $m \geq 2$, each of which interacts with its two immediate neighbours. Then instead of (6) we obtain the system

$$\dot{u}_j = d\,(u_{j+1} - 2u_j + u_{j-1}) + \lambda[f(u_j(t-h)) - g(u_j(t-1))]u_j, \;\; j = 1, \ldots, m, \tag{9}$$

where $u_0 = u_1$, $u_{m+1} = u_m$, and the parameter $d > 0$, which has the order of 1, characterizes the strength of coupling between neurons. Clearly, system (9) possesses the so-called homogeneous (or synchronous) cycle

$$u_1 \equiv \ldots \equiv u_m = u_*(t, \lambda), \tag{10}$$

where $u_*(t, \lambda)$ is the stable periodic solution of (6). Our central result claims that for suitably decreased d and all $\lambda \gg 1$ this system has at least m exponentially orbitally stable inhomogeneous periodic motions (apart from the stable cycle (10); recall that m is the order of the system). We call each of these motions a discrete autowave process, or simply an autowave.

System (9) is the required mathematical model, in which buffering and the bursting phenomenon occur at the same time. In fact, we show below that all the m stable autowaves in this system are bursting-cycles. This means that each component u_j of an autowave, $j = 1, \ldots, m$, displays the same asymptotic behaviour as $u_*(t, \lambda)$ on a period interval. For each positive integer n we can pick parameters a_0, b_0, and h such that each of these components shows precisely n asymptotically high impulses on a period interval for these values of the parameters.

3. Relaxation Properties of a Single Neuron

In this section we investigate the question whether equation (6) has a stable relaxation bursting-cycle. In (6) we make the substitution $u = \exp(\lambda x)$, which takes this equation to the form

$$\dot{x} = F(x(t-h), \varepsilon) - G(x(t-1), \varepsilon), \tag{11}$$

where $F(x,\varepsilon) = f(\exp(x/\varepsilon))$ and $G(x,\varepsilon) = g(\exp(x/\varepsilon))$, $\varepsilon = 1/\lambda \ll 1$. It follows from the properties (7) of the functions $f(u)$ and $g(u)$ that

$$\lim_{\varepsilon\to 0} F(x,\varepsilon) = R(x),\ \lim_{\varepsilon\to 0} G(x,\varepsilon) = H(x),\ R(x) = \begin{cases} 1 & \text{for } x < 0, \\ -a_0 & \text{for } x > 0, \end{cases}$$
$$H(x) = \begin{cases} 0 & \text{for } x < 0, \\ b_0 & \text{for } x > 0. \end{cases} \tag{12}$$

Relations (12) allow us to proceed from (11) to the limiting relay equation with delays

$$\dot{x} = R(x(t-h)) - H(x(t-1)). \tag{13}$$

As in [13, 15, 16], we define the notion of a solution to (13) in a constructive way. To do this we fix an arbitrary positive integer n and assume that the parameters h, a_0, and b_0 in (6) and (7) satisfy

$$\frac{1}{(n+1)(2+a_0+1/a_0)} < h < \frac{1}{n(2+a_0+1/a_0)+2+1/a_0}, \tag{14}$$

$$b_0 > 1 + a_0. \tag{15}$$

Next we fix a sufficiently small $\sigma_0 > 0$ (we indicate an upper bound for σ_0 in what follows), consider the set of functions

$$\begin{aligned} &\varphi \in C[-1-\sigma_0, -\sigma_0],\ \varphi(t) < 0 \\ &\text{for } t \in [-1-\sigma_0, -\sigma_0],\ \varphi(-\sigma_0) = -\sigma_0 \end{aligned} \tag{16}$$

and let $x_\varphi(t)$, where $t \geq -\sigma_0$, denote the solution of (13) with an arbitrary initial function (16).

Concerning the integration of (13) we note that the right-hand side of this equation is a piecewise constant function, which changes value only when $x(t-h)$ or $x(t-1)$ changes sign. In particular, for $-\sigma_0 \leq t \leq -\sigma_0 + h$ we have simultaneously $\varphi(t-h) < 0$ and $\varphi(t-1) < 0$. Hence by (13) and (16) the function $x_\varphi(t)$ solves the Cauchy problem $\dot{x} = 1$, $x(-\sigma_0) = -\sigma_0$ on this interval of time, and therefore

$$x_\varphi(t) = t. \tag{17}$$

It is also clear that (17) holds as long as $x_\varphi(t-h) < 0$ and $x_\varphi(t-1) < 0$. Thus, it also holds for $-\sigma_0 \leq t \leq 0$.

In view of our constructions, for $0 \leq t < 1$ we have $x_\varphi(t-1) < 0$, so that $H(x_\varphi(t-1)) = 0$. Hence the solution $x_\varphi(t)$ satisfies the auxiliary equation

$$\dot{x} = R(x(t-h)) \tag{18}$$

on this interval. As regards equation (18), its properties were thoroughly analyzed in [13], where it was shown, in particular, that if a solution $x(t)$ of this equation satisfies $x(0) = 0$ and $x(t) < 0$ for $-h \leq t < 0$ then for $t \geq 0$ it must coincide with the periodic function

$$x_0(t) = \begin{cases} t & \text{for } 0 \leq t \leq h, \\ h - a_0(t-h) & \text{for } h \leq t \leq t_0 + h, \\ -a_0 h + t - t_0 - h & \text{for } t_0 + h \leq t \leq T_0, \end{cases} \qquad x_0(t+T_0) \equiv x_0(t), \tag{19}$$

where $t_0 = h(1+1/a_0)$ and $T_0 = h(2+a_0+1/a_0)$. Returning to (13) and taking into account all the above we arrive at the equality

$$x_\varphi(t) = x_0(t),\ 0 \leq t \leq 1. \tag{20}$$

For further analysis we require the special function $y_0(t)$ solving the Cauchy problem

$$\dot{x} = 1 - H(x_0(t)), \quad x|_{t=0} = 0. \tag{21}$$

In view of (19), it is easy to see that for $t \geq 0$ this function is given by the relations

$$y_0(t) = \begin{cases} -(b_0 - 1)t & \text{for } 0 \leq t \leq t_0, \\ t - b_0 t_0 & \text{for } t_0 \leq t \leq T_0, \end{cases} \quad y_0(t) = (k-1)(T_0 - b_0 t_0) + \tag{22}$$
$$+ y_0(t - (k-1)T_0) \ \text{ for } (k-1)T_0 \leq t \leq kT_0, \ k \in \mathbb{N}, \ k \geq 2.$$

We note that condition (15) ensures the following properties of y_0:

$$y_0(t) < 0 \ \forall t \geq 0, \quad \lim_{t \to +\infty} y_0(t) = -\infty. \tag{23}$$

Now we consider the next time interval $1 \leq t \leq 1 + h$. Note that because of (14), $t = 1$ lies in the interval $(nT_0 + t_0 + h, (n+1)T_0)$. Hence, from (19) and (20) we obtain $x_\varphi(t-h) < 0$ for $t \in [1, 1+h]$, so that the function $x_\varphi(t)$ solves a Cauchy problem similar to (21) on this interval:

$$\dot{x} = 1 - H(x_0(t-1)), \quad x|_{t=1} = x_0(1).$$

Hence, for $1 \leq t \leq 1 + h$ we conclude that

$$x_\varphi(t) = x_0(1) + y_0(t-1). \tag{24}$$

In the next step we observe that if we know a priori that

$$x_\varphi(t-h) < 0 \tag{25}$$

then (24) also holds for $1 \leq t \leq 2$. However, (25) holds indeed because, by (14) and (19)

$$x_0(1) = x_0(1 - nT_0) = 1 - (n+1)T_0 < 0 \tag{26}$$

and from (23) we obtain $y_0(t-1) < 0$ for $t \geq 1$. Thus (24) holds for $1 \leq t \leq 2$.

For $t \geq 2$ we shall assume by (25) that we have the a priori estimate

$$x_\varphi(t-1) < 0. \tag{27}$$

Then the solution $x_\varphi(t)$ in question solves the Cauchy problem

$$\dot{x} = 1, \quad x|_{t=2} = x_0(1) + y_0(1),$$

so that it can be defined by the formula

$$x_\varphi(t) = t - T_*, \quad T_* = (n+1)(T_0 + b_0 t_0), \tag{28}$$

which follows from (22) and (26). It remains to add that by (24) and (28) the a priori assumptions (25) and (27) are certainly valid on the interval $2 \leq t \leq T_*$, and by the inequality $T_* - 2 > 0$ (following from (14) and (15)) this interval has positive length.

Now we will assume that the parameter σ_0 (see (16)) satisfies

$$\sigma_0 < (n+1)T_0 - 1. \tag{29}$$

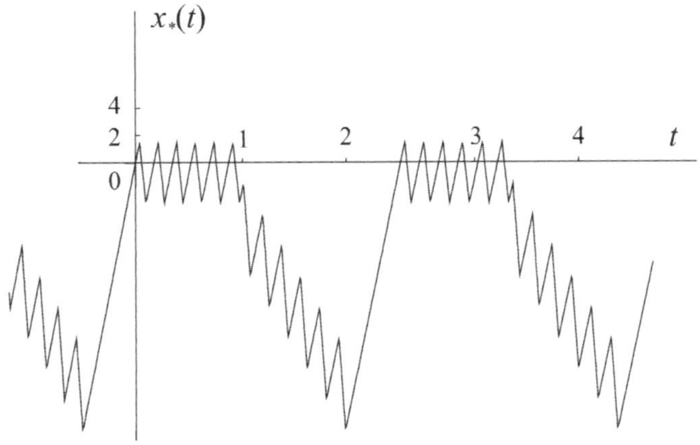

Figure 2. Function $x_*(t)$ for $a_0 = 2$, $b_0 = 4$, and $h = 1/26$ (inequalities (14) with $n = 5$ hold for these values of the parameters)

Then it follows from the above constructions that $x_\varphi(t + T_*)$, $-1 - \sigma_0 \le t \le -\sigma_0$, is a function of the set (16) and the equation $x_\varphi(t - \sigma_0) = -\sigma_0$ has precisely $2n + 2$ zeros on the interval $(0, T_*]$. Thus for $t \ge -\sigma_0$ each solution $x_\varphi(t)$ with initial condition (16) coincides with the same T_*-periodic function

$$x_*(t) = \begin{cases} x_0(t) & \text{for } 0 \le t \le 1, \\ x_0(1) + y_0(t-1) & \text{for } 1 \le t \le 2, \\ t - T_* & \text{for } 2 \le t \le T_*, \end{cases} \qquad x_*(t + T_*) \equiv x_*(t). \tag{30}$$

We show the graph of this function for $a_0 = 2$, $b_0 = 4$, and $h = 1/26$ in Figure 2 (inequalities (14) with $n = 5$ hold for these values of the parameters).

Now we consider the connections between periodic solutions of (11) and (13). We have the following result.

Theorem 3.1. *For all sufficiently small $\varepsilon > 0$ equation (11) has a unique orbitally exponentially stable cycle $x_*(t, \varepsilon)$, $x_*(-\sigma_0, \varepsilon) \equiv -\sigma_0$, whose period $T_*(\varepsilon)$ satisfies the limit relations*

$$\lim_{\varepsilon \to 0} T_*(\varepsilon) = T_*, \qquad \lim_{\varepsilon \to 0} \max_{0 \le t \le T_*(\varepsilon)} |x_*(t, \varepsilon) - x_*(t)| = 0.$$

We omit the full proof of this theorem and refer the reader to [13, 16] for technical details. We only present the scheme of the proof.

Apart from the constant σ_0 satisfying (29), we also fix $q_1 > \sigma_0$ and $q_2 \in (0, \sigma_0)$ and let $S(\sigma_0, q_1, q_2) \subset C[-1 - \sigma_0, -\sigma_0]$ denote the closed bounded set of functions φ satisfying the conditions $-q_1 \le \varphi(t) \le -q_2$ and $\varphi(-\sigma_0) = -\sigma_0$. Then for an arbitrary function $\varphi \in S(\sigma_0, q_1, q_2)$ we look at the solution $x = x_\varphi(t, \varepsilon)$, $t \ge -\sigma_0$, of equation (11) with the initial condition $x = \varphi(t)$ for $-1 - \sigma_0 \le t \le -\sigma_0$. We denote by $t = T_\varphi$ the $(2n + 2)$nd positive root of the equation $x_\varphi(t - \sigma_0, \varepsilon) = -\sigma_0$ (we assume that this equation has at least $2n + 2$ roots on the half-axis $t > 0$, which are numbered in ascending order). Finally, we define the Poincaré first return operator $\Pi_\varepsilon : S(\sigma_0, q_1, q_2) \to C[-1 - \sigma_0, -\sigma_0]$ by

$$\Pi_\varepsilon(\varphi) = x_\varphi(t + T_\varphi, \varepsilon), \quad -1 - \sigma_0 \le t \le -\sigma_0. \tag{31}$$

Journal of Physics: Conference Series **727** (2016) 012004 doi:10.1088/1742-6596/727/1/012004

The rest of the scheme is standard (see [13] and [16]). First we prove that the solution $x_\varphi(t,\varepsilon)$, is asymptotically close to $x_*(t)$ as $\varepsilon \to 0$ uniformly with respect to $\varphi \in S(\sigma_0, q_1, q_2)$ and $t \in [-\sigma_0, T_* - \sigma_0/2]$. This implies that for appropriately chosen parameters σ_0, q_1, and q_2 the operator (31) is defined on the set $S(\sigma_0, q_1, q_2)$ and transforms it into itself. It remains to analyse the variational equation along the solution $x_\varphi(t,\varepsilon)$ and to verify that Π_ε is a contraction operator.

Additionally to Theorem 3.1, we point out that the relaxation cycle

$$u_*(t,\lambda) = \exp\left(x_*(t,\varepsilon)/\varepsilon\right)\big|_{\varepsilon=1/\lambda} \tag{32}$$

of equation (6) has the required asymptotic features of a bursting-cycle. In fact, its period $T_*(\lambda)$ satisfies $\lim\limits_{\lambda\to\infty} T_*(\lambda) = T_*$. Moreover, the cycle (32) shows $n+1$ successive asymptotically high impulses (of the order of $\exp(\lambda h)$) on the interval $0 \le t \le T_*(\lambda)$. These spikes correspond to the intervals $kT_0 < t < t_0 + kT_0$, $k = 0, 1, \dots, n$, on which the function in (30) is positive. On the other hand, if t is a fixed moment of time in the set $[0, T_*) \setminus \bigcup_{k=0}^{n}[kT_0, t_0 + kT_0]$, then $u_*(t,\lambda)$ has order $\exp(-\lambda q)$, where $q = \text{const} > 0$.

4. Relaxation Properties of Neural Chains

Now we turn to system (9), where we assume, as before, that the parameters a_0, b_0, and h satisfy (14) and (15). For the analysis of this system we introduce new variables x, $y_1, \dots, y_{m-1}$:

$$u_1 = \exp(x/\varepsilon), \;\; u_j = \exp\left(x/\varepsilon + \sum_{k=1}^{j-1} y_k\right), \;\; j = 2, \dots, m, \;\; \varepsilon = 1/\lambda \ll 1. \tag{33}$$

Substituting (33) into (9) we obtain the relaxation system

$$\begin{gathered}\dot{x} = \varepsilon d\,(\exp y_1 - 1) + F(x(t-h),\varepsilon) - G(x(t-1),\varepsilon),\\ \dot{y}_j = d\,[\exp y_{j+1} + \exp(-y_j) - \exp y_j - \exp(-y_{j-1})] + \\ +F_j(x(t-h), y_1(t-h), \dots, y_j(t-h), \varepsilon) - G_j(x(t-1), y_1(t-1), \dots, y_j(t-1), \varepsilon),\\ j = 1, \dots, m-1,\end{gathered} \tag{34}$$

where $y_0 = y_m = 0$, the functions F and G are as in (11), and the functions F_j and G_j have the following form:

$$F_j(x, y_1, \dots, y_j) = \frac{1}{\varepsilon}\left[f\left(\exp\left(x/\varepsilon + \sum_{k=1}^{j} y_k\right)\right) - f\left(\exp\left(x/\varepsilon + \sum_{k=1}^{j-1} y_k\right)\right)\right],$$

$$G_j(x, y_1, \dots, y_j) = \frac{1}{\varepsilon}\left[g\left(\exp\left(x/\varepsilon + \sum_{k=1}^{j} y_k\right)\right) - g\left(\exp\left(x/\varepsilon + \sum_{k=1}^{j-1} y_k\right)\right)\right].$$

Let us fix a positive constant σ_0 satisfying (29). Next we introduce at the interval $-\sigma_0 \le t \le T_* - \sigma_0$, where T_* is the quantity in (28). For $z = (z_1, \dots, z_{m-1}) \in \mathbb{R}^{m-1}$ let $y_1^0(t,z)$, …, $y_{m-1}^0(t,z)$ denote the components of the solution of the impulse system

$$\begin{gathered}\dot{y}_j = d\,[\exp y_{j+1} + \exp(-y_j) - \exp y_j - \exp(-y_{j-1})],\\ j = 1, \dots, m-1, \;\; y_0 = y_m = 0;\end{gathered} \tag{35}$$

$$\begin{gathered} y_j(h+kT_0+0)=y_j(h+kT_0-0)-(1+a_0)y_j(kT_0), \\ y_j(t_0+h+kT_0+0)=y_j(t_0+h+kT_0-0)-(1+1/a_0)y_j(t_0+kT_0), \\ y_j(1+kT_0+0)=y_j(1+kT_0-0)-b_0y_j(kT_0), \\ y_j(1+t_0+kT_0+0)=y_j(1+t_0+kT_0-0)-(b_0/a_0)y_j(t_0+kT_0), \\ k=0,1,\ldots,n, \quad j=1,\ldots,m-1, \end{gathered} \tag{36}$$

which satisfies the initial condition

$$(y_1,\ldots,y_{m-1})\big|_{t=-\sigma_0}=(z_1,\ldots,z_{m-1}) \tag{37}$$

and is defined on this interval. Finally, we consider the map

$$z\to\Phi(z)\overset{\text{def}}{=}(y_1^0(t,z),\ldots,y_{m-1}^0(t,z))\big|_{t=T_*-\sigma_0} \tag{38}$$

from $\mathbb{R}^{m-1}$ to $\mathbb{R}^{m-1}$.

Theorem 4.1. *To each fixed point $z=z_*$ of the map (38) which is exponentially stable or dichotomous there corresponds for sufficiently small $\varepsilon>0$ a relaxation cycle $(x(t,\varepsilon),y_1(t,\varepsilon),\ldots,y_{m-1}(t,\varepsilon))$ of system (34) with the same stability properties and which satisfies $x(-\sigma_0,\varepsilon)\equiv-\sigma_0$ and has period $T(\varepsilon)$. Furthermore,*

$$\begin{gathered} \lim_{\varepsilon\to0}T(\varepsilon)=T_*, \qquad \max_{-\sigma_0\le t\le T(\varepsilon)-\sigma_0}|x(t,\varepsilon)-x_*(t)|=O(\varepsilon), \\ \lim_{\varepsilon\to0}\max_{t\in\Sigma(\varepsilon)}|y_j(t,\varepsilon)-y_j^0(t,z_*)|=0, \qquad \max_{-\sigma_0\le t\le T(\varepsilon)-\sigma_0}|y_j(t,\varepsilon)|\le M,\ j=1,\ldots,m-1, \end{gathered} \tag{39}$$

where $x_*(t)$ is the function in (30), $M=\text{const}>0$, and $\Sigma(\varepsilon)$ is the interval $[-\sigma_0,T(\varepsilon)-\sigma_0]$ from which the subintervals

$$\begin{gathered} (h+kT_0-\varepsilon^\delta,h+kT_0+\varepsilon^\delta),\ (t_0+h+kT_0-\varepsilon^\delta,t_0+h+kT_0+\varepsilon^\delta), \\ (1+kT_0-\varepsilon^\delta,1+kT_0+\varepsilon^\delta),\ (1+t_0+kT_0-\varepsilon^\delta,1+t_0+kT_0+\varepsilon^\delta),\ k=0,1,\ldots,n, \\ \delta=const\in(0,1) \end{gathered}$$

have been removed.

We omit the proof of this theorem since for quite similar situations it was presented in [17] with details.

Theorem 4.1 is of fundamental nature as it reduces the problem of finding discrete autowave processes in system (9) to the problem of identifying stable fixed points of the map (38). We can represent this map in an invariant form, independent of the choice of σ_0. Let $P^t(z)$, $z\in\mathbb{R}^{m-1}$, $P^0(z)=z$, be the shift operator along the trajectories of system (35). It is easy to see that after the change of variables $P^{\sigma_0}(z)\to z$ the map (38) takes the invariant form

$$z\to\Phi_0(z)\overset{\text{def}}{=}(y_1^0(t,z),\ldots,y_{m-1}^0(t,z))\big|_{t=T_*}, \tag{40}$$

where $(y_1^0(t,z),\ldots,y_{m-1}^0(t,z))$ is the solution of the impulse system (35), (36) with the initial condition $(z_1,\ldots,z_{m-1})$ for $t=0$ (analogous to (37)).

We start the investigation of attractors of the map (40) by analysing the stability properties of its fixed point $z=0$. We establish the following result by simple verification.

Theorem 4.2. *There exists a sufficiently small $d_0>0$ such that for all $d\in(0,d_0]$ the fixed point $z=0$ of the map (40) is exponentially stable.*

Corresponding to the fixed point $z = 0$ there is a cycle of system (34) with components $x = x_*(t, \varepsilon)$, $y_j \equiv 0$, $j = 1, \dots, m-1$, where $x_*(t, \varepsilon)$ is the periodic solution of (11) established by Theorem 2.1. In system (9) the same point corresponds to the homogeneous cycle (10), where $u_*(t, \lambda)$ is the function (32). Theorems 4.1 and 4.2 imply that this cycle is exponentially orbitally stable for each fixed $d \in (0, d_0]$ and all sufficiently large λ.

Now we look for stable fixed points of the map (40) which are distinct from $z = 0$ under the assumption $d \ll 1$. In this case we can asymptotically integrate (35), (36) for $0 \le t \le T_*$ and

$$(a_0, b_0) \in \mathcal{U}_1 \cup \mathcal{U}_2 \cup \mathcal{U}_3, \tag{41}$$

where

$$\mathcal{U}_1 = \{(a_0, b_0): \ a_0 > 1, \ a_0 + 1 < b_0 < 2a_0\}, \ \mathcal{U}_2 = \{(a_0, b_0): \ a_0 > 1, \ b_0 > 2a_0\},$$
$$\mathcal{U}_3 = \{(a_0, b_0): \ 0 < a_0 < 1, \ b_0 > a_0 + 1\}.$$

Then we get the following result.

Theorem 4.3. *Let h, a_0, and b_0 be fixed parameters satisfying* (14) *and* (41). *Then for all sufficiently small $d > 0$ the map* (40) *has m exponentially stable fixed points*

$$O_{r_0}(d) = (z_{1,r_0}(d), z_{2,r_0}(d), \dots, z_{m-1,r_0}(d)), \ \ r_0 = 0, 1, \dots, m-1, \tag{42}$$

such that for $(a_0, b_0) \in \mathcal{U}_1$ their location is determined by the asymptotic behaviour

$$\begin{aligned} z_{j,r_0} &= -\frac{b_0 - a_0}{a_0} \ln \frac{1}{d} + O(1), \ \ j = 1, \dots, r_0; \\ z_{j,r_0} &= \frac{b_0 - a_0}{a_0} \ln \frac{1}{d} + O(1), \ \ j = r_0 + 1, \dots, m-1 \end{aligned} \tag{43}$$

as $d \to 0$, while for $(a_0, b_0) \in \mathcal{U}_2 \cup \mathcal{U}_3$ their asymptotic behaviour is as follows:

$$z_{j,r_0} = -\ln \frac{1}{d} + O(1), \ \ j = 1, \dots, r_0; \ \ z_{j,r_0} = \ln \frac{1}{d} + O(1), \ \ j = r_0 + 1, \dots, m-1, \tag{44}$$

as $d \to 0$.

We do not justify this result here and refer the reader to [18], which contains the proof in a quite similar case.

Theorem 4.1 leads in combination with the asymptotic properties of the map (40) (see Theorems 4.2 and 4.3) to the main result of this paper.

Theorem 4.4. *Let h, a_0, and b_0 be fixed parameters satisfying* (14) *and* (41). *Then for any sufficiently small d_1 and d_2, $d_2 > d_1 > 0$, there exists a sufficiently large $\lambda_0 = \lambda_0(d_1, d_2) > 0$ such that for $d_1 \le d \le d_2$ and $\lambda \ge \lambda_0$ the system of m equations* (9) *has at least m exponentially orbitally stable spatially inhomogeneous cycles co-existing with the stable homogeneous cycle* (10).

It follows from this theorem that when we decrease d and when we increase λ and the number m of neurons in an appropriate way then the buffer phenomenon occurs in (9): the number of co-existing stable cycles increases unbounded. Moreover, all these cycles display the bursting phenomenon. In fact, by (39) each of these cycle has components u_j, $j = 1, \dots, m$, with the asymptotic representation $u_j = \exp(\lambda x_*(t) + O(1))$ as $\lambda \to \infty$, $j = 1, \dots, m$, which holds uniformly with respect to $t \in [0, T(\varepsilon)]$, where $x_*(t)$ is the function in (30). Hence, the graphs of the u_j are similar in shape to the graph in Figure 1, that is, they contain a string of $n + 1$ spikes. To any given natural number n the parameters h, a_0, and b_0, we can be chosen select appropriately.

For example, on Figure 3 for $m = 5$ we show a graphs of components of one of the five coexistent stable periodic solutions of system (9) (parameter values are the same as above, $h = 1/26$, $\lambda = 130$, $c_1 = 0.5$ and $c_2 = 4$).

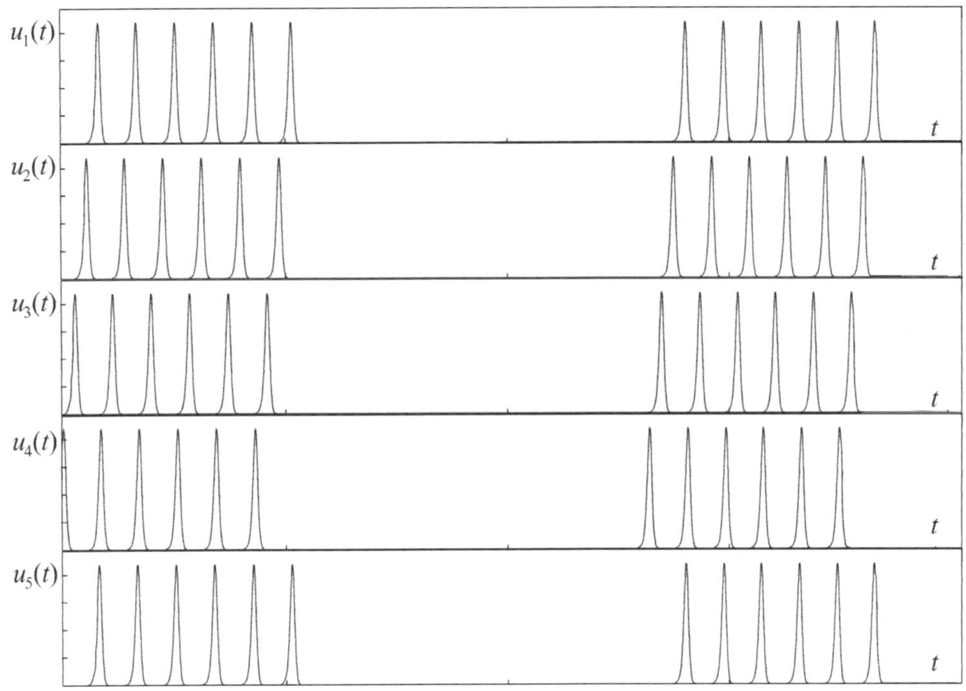

Figure 3. Components $u_1(t), \ldots, u_5(t)$ of the stable periodic solution of system (9) for $m = 5$, $h = 1/26$ and $\lambda = 130$ for the functions in (8) with $c_1 = 0.5$ and $c_2 = 4$

5. Conclusion

Mathematical results presented in the paper show that our new model (9) of single neuron operation combines simplicity of its form with rich dynamical properties. We show that this model can describe such fundamental neurodynamical phenomena as bursting-effect and buffer phenomenon.

Historically main attention of researchers was paid to "bursting behavior". This is not the case for buffer phenomenon. Although this phenomenon is one of the fundamental laws of operation of nonlinear systems from different fields [9, 10], it is still not properly described in neurodynamical literature. At the same time it is clear that buffer phenomenon is important for neuron subjects as it can be useful in describing the associative memory. We shall discuss this in detail.

According to [19, 20] in the simplest case associative memory operates like is shown on fig. 4. That is, there exist some storage medium implemented as a neuron system and storing blocks of data R_n, $n = 1, 2, \ldots, n_0$. Next, in order to obtain from system a certain result R_n it is necessary to feed to its input the corresponding symbolic key K_n and a certain additional data C_n referred to as context.

Now assume that storage medium mentioned above is described by equation (6) or by system of coupled equations of the form (9). Next, let parameters of this system be taken so that it permits n_0 coexistent stable cycles enumerated with $n = 1, 2, \ldots, n_0$. Finally, assume that with every such cycle connected is a corresponding data block R_n.

The operation of mathematical model described above is clear: in order to receive the necessary data R_n, we have to keep the system in a stable periodic regime with number n. This can be fulfilled by introducing the key $K_n = n$ and additional context C_n. We take as C_n the initial conditions of mentioned periodic regime.

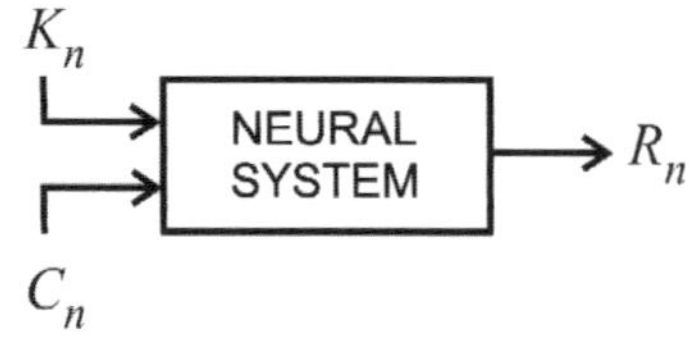

Figure 4. Model of memory with associative access

We should particularly mention the following feature. All the results presented in the paper hold for more general forms of $f(u)$, $g(u)$ from (9). That is, we can assume that as $u \to +\infty$, instead of (7) the following asymptotic equalities are valid

$$f(u) = -a_0 + O(1/u^{\gamma_1}), \quad uf'(u) = O(1/u^{\gamma_1}), \quad u^2 f''(u) = O(1/u^{\gamma_1}),$$
$$g(u) = b_0 + O(1/u^{\gamma_2}), \quad ug'(u) = O(1/u^{\gamma_2}), \quad u^2 g''(u) = O(1/u^{\gamma_2})$$

for arbitrarily fixed $\gamma_1, \gamma_2 > 0$.

Now we dwell on several unsolved problems. It might be interesting to look at the neural chain (9) for other boundary conditions. For example, we can set $u_0 = u_{m+1} = 0$ (Dirichlet-type conditions) or $u_0 = u_m$, $u_{m+1} = u_1$ (periodicity conditions). The question of autowave regimes in 2-dimensional lattices of diffusively coupled neurons of the form (6) is also of interest. Finally, it is still an open question whether the buffering phenomenon survives the transition from discrete chains (9) to the corresponding distributed model, when we set $d = m^2 D$ with $D = \text{const} > 0$ and let $m \to \infty$.

Acknowledgments

This work was supported by Ministry of Education and Science of Russian Federation (project 2014/258-1875) and the Russian Foundation for Basic Research (projects 15-01-04066a)

References

[1] Chay T R. and Rinzel J 1985 *Biophys. J.* **47** 357
[2] Ermentrout G B and Kopell N 1986 *SIAM J. Appl. Math.* **46** 233
[3] Izhikevich E 2000 *International Journal of Bifurcation and Chaos* **10(6)** 1171
[4] Rabinovich M I, Varona P, Selverston A I and Abarbanel H D I 2006. **Rev. Mod. Phys. 78** 1213–65.
[5] Coombes S and Bressloff P C 2005 *Bursting: the genesis of rhythm in the nervous system* (Hackensack, NJ: World Sci. Publ.)
[6] Kolesov A Yu, Mishchenko E F and Rozov N Kh 1998 *Proc. Steklov Inst. Math.* **222.** 1
[7] Kolesov A Yu and Rozov N Kh 2001 *Proc. Steklov Inst. Math.* **233** 143
[8] Kolesov A Yu and Rozov N Kh 1998 *Izv. Math.* **62** 985
[9] Kolesov A Yu and Rozov N Kh 2004 *Invariant tori of nonlinear wave equations* (Moscow: Fizmatlit (Russian))
[10] Mishchenko E F, Sadovnichii V A, Kolesov A Yu and Rozov N Kh *Auto-wave processes in nonlinear media with diffusion* (Moscow: Fizmatlit (Russian))
[11] Hodgkin A L and Huxley A F 1939 *Nature* **144** 710
[12] Hodgkin A L and Huxley A F 1952 *J. Physiol.* **117** 500
[13] Kolesov A Yu, Mishchenko E F and Rozov N Kh 2010 *Comp. Math. and Math. Phys.* **50** 1990
[14] Kashchenko S A and Mayorov V V 2009 *Models of the wave memory* (Moscow: Editorial URSS (Russian))
[15] Kolesov A Yu, Mishchenko E F and Rozov N Kh 1997 *Proc. Steklov Inst. Math.* **216** 119
[16] Glyzin S D, Kolesov A Yu and Rozov N Kh 2011 *Diff. Equat.* **47** 927
[17] Glyzin S D, Kolesov A Yu and Rozov N Kh 2011 *Diff. Equat.* **47** 1697
[18] Glyzin S D, Kolesov A Yu and Rozov N Kh 2012 *Diff. Equat.* **48** 159
[19] Kohonen T. 1977 *Associative memory: a system-theoretical approach* (Berlin: Springer-Verlag)
[20] Kohonen T. 1989 *Self-organization and associative memory* (Berlin Heidelberg: Springer)

Modelling Simple Experimental Platform for In Vitro Study of Drug Elution from Drug Eluting Stents (DES)

L V Kalachev[1]
Department of Mathematical Sciences, University of Montana
Missoula, MT 59812, USA

E-mail: kalachev@mso.umt.edu

Abstract. We present a simple model of experimental setup for in vitro study of drug release from drug eluting stents and drug propagation in artificial tissue samples representing blood vessels. The model is further reduced using the assumption on vastly different characteristic diffusion times in the stent coating and in the artificial tissue. The model is used to derive a relationship between the times at which the measurements have to be taken for two experimental platforms, with corresponding artificial tissue samples made of different materials with different drug diffusion coefficients, to properly compare the drug release characteristics of drug eluting stents.
Key words: model reduction, asymptotic expansion, drug eluting stents, experimental measurements

1. Introduction

Drug Eluting Stents (DES) are widely used for treating coronary artery disease; compared to bare metal stents DES represent more advanced medical devices that are intended to reduce the tissue damage and to speed up the healing of artery walls after stent implantation / expansion within the blocked region of an artery. Modelling of drug elution from DES is a widely expanding area of research. A large number of publications address modelling of drug elution from stents of different geometry, the effect of artery wall conditions, such as, e.g., thrombosis, on drug propagation within vessel wall, etc. (see, e.g., [1] – [13]). In addition to in vivo drug release models for DES the comparison of different possible experimental platforms for study of in vitro release setups, which may mimic the in vivo release processes for their future use in product quality control, regulatory documents filing, etc., is of current interest. One of the ideas is related to using the "artificial tissue" samples, made, e.g., of collagen, or some porous polymer, and shaped as a blood vessel, for modelling and analysis of drug release into real vessels; some materials on "artificial tissue" fabrication for studying drug release and other medical application may be found, e.g., in [14], [15].

In this paper we briefly discuss the basic ideas behind comparison of drug release measurement results obtained for two in vitro release experiments where two "artificial tissue" materials are characterized by different diffusion coefficients for a particular drug. Here we formulate and analyse a simple model addressing some important general features of the process omitting certain details which are not easily quantifiable in real in vivo experiments (such as the difference between propagation of

[1] To whom any correspondence should be addressed.

lipophilic and hydrophilic drugs, the effect of drug partition coefficient of characteristics of drug elution and propagation, the actual tissue constitution which is highly non-uniform, etc.).

The main question that is addressed in this paper is formulated as follows. Assume that we have an experimental system #1 that measures the percentage of drug remnant on stent at an instant of time t^*during drug release from DES into an "artificial tissue" sample, that is a substitution for a real blood vessel, characterized by a drug diffusion coefficient D_1. We want to design an experimental system #2 to check the properties of the same type of DES release, but we now use a different type of an "artificial tissue" sample, playing the role of a real blood vessel, that is characterized by the diffusion coefficient D_2. Once again, we measure the percentage of drug remnant on stent, but now at an instant of time t^{**}. We want to look at a measurement obtained using experimental system #2 and make a conclusion on expected outcome of the measurement (actual or potential) obtained using the experimental system #1. In fact, ideally, if the stents are absolutely the same we want the measured percentages of drug remnant on stent be the same in the case #1 and in the case #2. Here is the question that we address: if for the two experimental systems the measurements are performed at the same instants of time, i.e., $t^{**} = t^*$, will it be possible to get the same results for the percentages of drug remnant on stents? The answer to this question is NO. It is, however, possible to choose the time of measurement t^{**} for the experimental system #2 in such a way that the measurement result will be absolutely the same as that obtained from experimental system #1 at time t^*. The formula that relates the choice of the measurement time t^{**} to the original measurement time t^* for particular experimental platform geometry is presented below (see section 4, formula (40)). Thus, one of the conclusions may be formulated as follows: for the same type of DES to obtain the same measurement results on two different experimental systems, the measurements collected using these systems must be taken at different time points (e.g., if for system #1 we originally take the measurements at $t = 1$ day, 3 days, 7 days, etc., for system #2 corresponding measurements must be taken at $t = t_1$ days, t_2 days, t_3 days, etc., where, in general, $t_1 \neq 1$, $t_2 \neq 2$, $t_3 \neq 3$ may be computed for particular choice of experimental system geometry, choice of materials, etc.).

We note that below we talk about one instant of time at which the measurement is taken (for each experimental system). Since this instant of time is chosen arbitrarily, the same analysis will work for the case where measurements are taken at several time points. Then the analysis just has to be repeated for each individual time point.

The motivation for the current analysis comes from the fact that in numerous practical pharmacokinetics studies performed at pharmaceutical companies the "purely statistical" approach to comparing kinetic drug release curves is used to evaluate different drug eluting devices, where one exponential curve, or a linear combination of exponential curves, are fit to the experimental data. This analysis represents one of the examples where modelling approach allows one to introduce meaningful explanation of where these "exponentials" come from and how they are affected by the parameters in the models describing real experimental setups.

The paper is organized as follows. In section 2 we introduce the description of a simple experimental platform for studying in vitro drug release from DES that is supposed to mimic the in vivo drug release into blood vessel tissue. In section 3 we introduce a model formulation for a simplified version of the experimental setup and perform model reduction. Simplification is needed to clarify the basic ideas and underlying questions that must be addressed when the real experimental systems of this type are built. In section 4 we present the comparison of two experimental setups based on the introduced simplified model formulation. We derive the explicit expressions that relate measurement times for two experimental systems that produce the same measurement results for similar DES. In section 5 we briefly comment on more realistic models and possible extensions of the results derived in sections 3 and 4. Finally, conclusions and discussion of the main results are presented in section 6.

2. Description of a simple experimental platform

We start with the description of a simple (in vitro drug release) experimental platform that is intended to imitate in vivo drug release from DES into a blood vessel. In consists of a hollow cylindrical tube made of collagen or some porous polymer ("artificial tissue") immersed in a buffer solution. The hollow part of the cylinder representing "artificial vessel's" lumen is connected by the hoses to a pump that forces a liquid, representing blood, to constantly move through the lumen. The cross-section of a portion of such "artificial vessel" (without hoses and without a container that holds the buffer solution) is shown in figure 1. The DES is expanded inside the luminal part of the artificial vessel. The drug will be entering the "artificial tissue" sample from DES from the coating pressed against the luminal surface of the sample. At a certain instant of time the elution experiment is stopped, the amount of drug remnant on stent is measured, and the percentage of drug remaining on stent is determined (using the original known drug load).

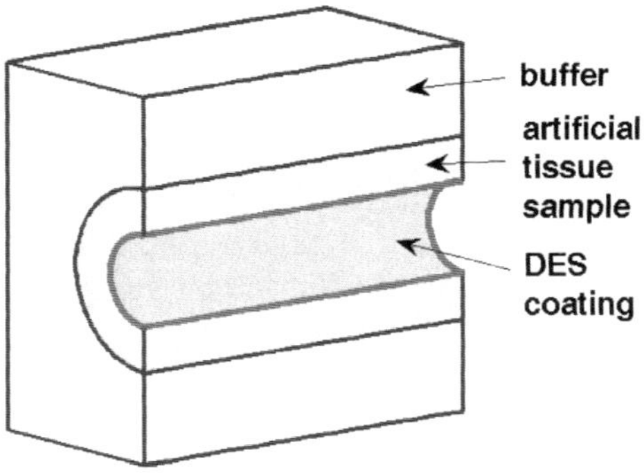

Figure 1. Cross-section of a portion of an in vitro drug release experimental platform.

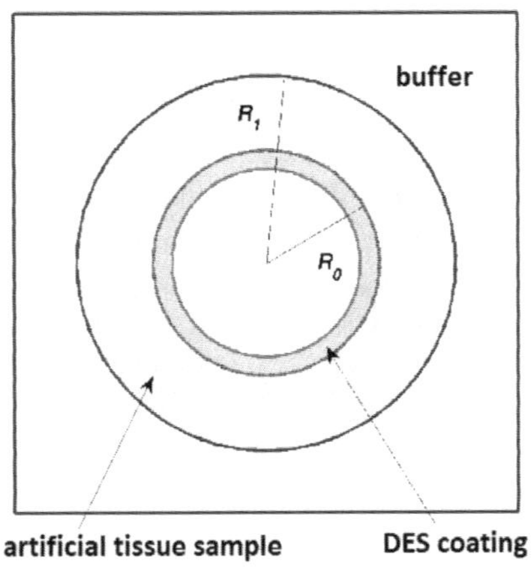

Figure 2. Transversal cross-section corresponding to the case shown in figure 1.

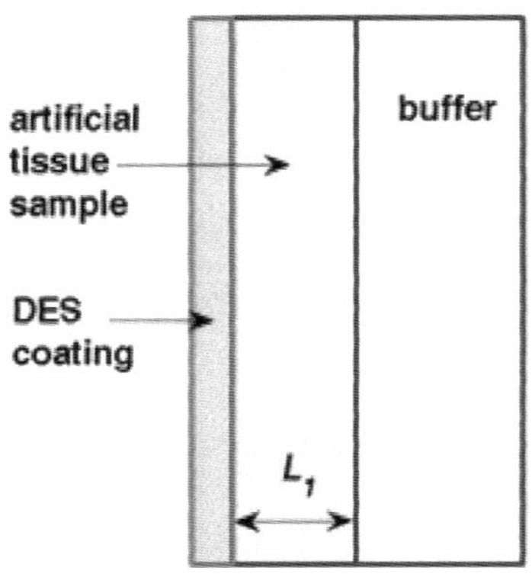

Figure 3. Cross-section representing a simplified model.

3. Model formulation for a simplified experimental setup and model reduction

We start with the description of a simple (in vitro drug release) experimental platform that is intended to imitate in vivo drug release from DES into a blood Instead of the situation characterized by the cross-section shown in figure 2 and corresponding to the experimental setup shown in figure 1, we will consider a simplified experimental setup illustrated in figure 2. The main difference between the geometry of the experimental system represented by cross-section shown in figure 2 and that shown in figure 3 is that in the latter case the coating and the "artificial tissue" sample are planar (not cylindrical) which will allow us to use the symmetry of the simplified case to write corresponding model as spatially one-dimensional. We want to reiterate that this simplification is needed to clarify the conclusions following from the analysis, to simplify notations and to make the formulas in the presentation shorter; similar results will hold for the original geometry as well. Also, for the purpose of model simplification, we consider the homogenized drug eluting coating distribution over the luminal surface of the vessel tissue (i.e., we suppress the dependence of drug distribution on the particular local geometry of stent struts). Such approach was used previously (see, e.g., [5]); it allows one to concentrate on the general characteristics of drug elution / propagation omitting details of local drug distribution in the tissue.

Let us list / re-iterate the assumptions that we make to formulate the model: (a) The coating is assumed to be uniformly distributed over the luminal surface of the "artificial tissue" sample. (b) We assume that the coating is fully hydrated, so that the entire drug in DES is immediately available for release. Mathematically this condition may be formulated in terms of drug diffusion coefficients D_0 (in the coating) and D_1 (in the tissue): $D_0 \gg D_1 > 0$. (c) The drug may leave coating only in the direction of the "artificial tissue" sample, e.g., no flux conditions for drug hold at all the boundaries of the coating except for the boundary facing the sample. This condition is quite realistic since the coating is attached to the stent struts made of metal providing impermeable boundary for drug on the side opposite to that corresponding to vessel's luminal boundary. (d) No membrane slowing down the drug transfer from the coating into the sample is present at the sample's boundary. (e) The problem is spatially one-dimensional, i.e., only one spatial variable x enters the model formulation instead of three spatial variables x, y and z, i.e., we have diffusion present only in one spatial direction. This condition naturally holds in the following situation. We consider originally the "artificial tissue" sample shaped as a parallelepiped with height Z, i.e., $0 \le z \le Z$, width Y, i.e., $0 \le y \le Y$, and thickness L_1, i.e., $0 \le x \le L_1$; the coating also has the shape of a parallelepiped with the same height Z, the same width Y, and some other thickness $-L_0 \le x \le 0$. If the "no flux" boundary conditions (i.e., zero Neumann type boundary conditions) are specified at $z = Z$, $y = 0$ and $y = Y$ for the drug in the sample, the final formulation of the model will not contain y and z. (f) Sufficient amount of buffer is present so that the drug concentration in the buffer is far from saturation; in fact, we assume that concentration of drug in the buffer is zero, i.e., as soon as drug molecules enter the buffer solution through the adventitial boundary of the "artificial tissue" sample, they are immediately removed from the vicinity of this boundary. More realistic situations mentioned in section 4 can be considered as well. Here we concentrate on simple explanation of qualitative results using a simple model. For more complex models the results, and related conclusion, will become even more complicated.

The percentage of drug remnant on stent will depend on the geometry of the "artificial tissue" sample, on drug diffusion coefficient of the "artificial tissue" material, on such properties of the DES coating as its porosity, drug content, shape of stent struts as well as on other factors. While the model for the case of geometry shown in figure 1 may be constructed and analysed, in the next section, to make the explanation and qualitative results of analysis as explicit as possible, we will discuss a somewhat simpler model of drug elution from DES (see figures 2 and 3 as well as detailed explanation in section 3 below).

We start with the model describing concentration $w(x,t)$ of drug in the coating and concentration $u(x,t)$ of drug in the tissue sample can be written as follows (here we note once again that $x = 0$ corresponds to the boundary between drug eluting stent and vessel tissue):

$$\frac{\partial w}{\partial t} = D_0 \frac{\partial^2 w}{\partial x^2}, \qquad -L_0 < x < 0, \tag{1}$$

$$\frac{\partial u}{\partial t} = D_1 \frac{\partial^2 u}{\partial x^2}, \qquad 0 < x < L_1, \tag{2}$$

$$0 < t \le t^*,$$

at $x = 0$:

$$w = u, \qquad D_0 \frac{\partial w}{\partial x} = D_1 \frac{\partial u}{\partial x}, \tag{3}$$

at $x = -L_0$:

$$\frac{\partial w}{\partial x} = 0, \tag{4}$$

at $x = L_1$:

$$u = 0. \tag{5}$$

At $t = 0$:

We rescale the model $(1) - (6)$ using the following changes of variables:

$$\tau = t \cdot (D_1/{L_1}^2), \; W = w/w^*, U = u/w^*, \rho = x/L_1.$$

The non-dimensional model formulation may be written as follows:

$$\frac{\partial W}{\partial \tau} = \frac{D_0}{D_1} \frac{\partial^2 W}{\partial \rho^2}, \qquad -\frac{L_0}{L_1} < \rho < 0, \tag{7}$$

$$\frac{\partial U}{\partial \tau} = \frac{\partial^2 U}{\partial \rho^2}, \qquad 0 < \rho < 1, \tag{8}$$

$$0 < \tau \le \tau^* = t^* \cdot (D_1/{L_1}^2).$$

at $\rho = 0$:

$$W = U, \qquad \frac{\partial W}{\partial \rho} = \frac{D_1}{D_0} \frac{\partial U}{\partial \rho}, \tag{9}$$

at $\rho = -L_0/L_1$:

$$\frac{\partial W}{\partial \rho} = 0, \tag{10}$$

at $\rho = 1$:

$$U = 0. \tag{11}$$

At $\tau = 0$:

$$W = 1, \; U = 0. \tag{12}$$

Taking into account condition (b), $D_0 \gg D_1 > 0$, we can introduce a small parameter $0 < \varepsilon = D_1/D_0 \ll 1$, and use asymptotic approximation

$$W = \widetilde{W} + \varepsilon \cdot \widehat{W} + \ldots, \qquad U = \widetilde{U} + \varepsilon \cdot \widehat{U} + \ldots, \tag{13}$$

to derive a simplified (reduced) formulation of the model; see, e.g., [16] for more detailed description of one of the asymptotic algorithms, the Boundary Function Method, that may be used to construct corresponding approximation. The functions used in the expansions (13) do not contain sub-indices; this is done intentionally to simplify the notation. Also, the resulting rescaled problem turns out to be singularly perturbed. However, in (13) we use the, so-called, "regularly perturbation" ansatz since the experimental data in the experimental setups characteristic for this problem is usually collected on the time scales which are much longer compared to the "fast" mixing time over which the drug is re-distributed within the DES coating. For such "long" characteristic times the, so-called, boundary functions describing the exponentially fast adjustment of the solution to the comparatively slowly varying leading order terms of the regular expansion (13) do not need to be taken into consideration.

We substitute the series (13) into (7) – (12), where equation (7) now has the form

$$\varepsilon \cdot \frac{\partial W}{\partial \tau} = \frac{\partial^2 W}{\partial \rho^2}, \tag{14}$$

and the second boundary condition in (9) has the form

$$\rho = 0: \qquad \frac{\partial W}{\partial \rho} = \varepsilon \cdot \frac{\partial U}{\partial \rho}, \tag{15}$$

and determine consecutively the terms of the asymptotic series in the standard way. In the leading order approximation, we obtain:

$$0 = \frac{\partial^2 \widetilde{W}}{\partial \rho^2}, \tag{16}$$

$$\frac{\partial \widetilde{U}}{\partial \tau} = \frac{\partial^2 \widetilde{U}}{\partial \rho^2}, \tag{17}$$

at $\rho = 0$:

$$\widetilde{W} = \widetilde{U}, \qquad \frac{\partial \widetilde{W}}{\partial \rho} = 0, \tag{18}$$

at $\rho = -L_0/L_1$:

$$\frac{\partial \widetilde{W}}{\partial \rho} = 0, \tag{19}$$

at $\rho = 1$:

$$\widetilde{U} = 0. \tag{20}$$

At $\tau = 0$:

$$\widetilde{W} = 1, \; \widetilde{U} = 0. \tag{21}$$

It follows from (16), the second condition in (18) and (19) that $\widetilde{W} = \widetilde{W}(\tau)$ is, yet unknown, function of only one variable, non-dimensional time τ. In the first order approximation we obtain:

$$\frac{\partial \widetilde{W}}{\partial \tau} = \frac{\partial^2 \widehat{W}}{\partial \rho^2}, \tag{22}$$

$$\frac{\partial \widetilde{U}}{\partial \tau} = \frac{\partial^2 \widehat{U}}{\partial \rho^2}, \tag{23}$$

at $\rho = 0$:

$$\widehat{W} = \widehat{U}, \qquad \frac{\partial \widehat{W}}{\partial \rho} = \frac{\partial \widetilde{U}}{\partial \rho}, \tag{24}$$

at $\rho = -L_0/L_1$:

$$\frac{\partial \widehat{W}}{\partial \rho} = 0, \tag{25}$$

at $\rho = 1$:

$$\widehat{U} = 0. \tag{26}$$

At $\tau = 0$:

$$\widehat{W} = 0, \; \widehat{U} = 0. \tag{27}$$

From solvability condition of equation (22) with the second condition in (24) and condition (25) we obtain the equation for $\widetilde{W} = \widetilde{W}(\tau)$:

$$\frac{\partial \widehat{W}}{\partial \rho}(\rho = 0) - \frac{\partial \widehat{W}}{\partial \rho}(\rho = -L_0/L_1)$$
$$= \frac{\partial \widetilde{U}}{\partial \rho}(\rho = 0) = \int_{-L_0/L_1}^{0} \frac{\partial \widetilde{W}}{\partial \tau}(\tau) d\rho = \frac{L_0}{L_1}\frac{\partial \widetilde{W}}{\partial \tau},$$

and thus,

$$\frac{\partial \widetilde{W}}{\partial \tau} = \frac{L_1}{L_0}\frac{\partial \widetilde{U}}{\partial \rho}(0,\tau). \tag{28}$$

The reduced model in the final form can now be written as (here, without loss of generality, we omit tildes to make notations shorter):

$$\frac{\partial U}{\partial \tau} = \frac{\partial^2 U}{\partial \rho^2}, \tag{29}$$

$$\frac{\partial W(\tau)}{\partial \tau} = \frac{L_1}{L_0}\frac{\partial U}{\partial \rho}(0,\tau), \tag{30}$$

at $\rho = 0$:

$$W(\tau) = U(0,\tau), \tag{31}$$

at $\rho = 1$:

$$U(1,\tau) = 0. \tag{32}$$

At $\tau = 0$:

$$W(0) = 1, \;\; U(\rho, 0) = 0. \tag{33}$$

The original model solution has to be considered in a finite time interval $0 < t \leq t^*$, which corresponds to a non-dimensional time interval $0 < \tau \leq \tau^* = t^* \cdot (D_1/{L_1}^2)$. The non-dimensional spatial domain for (29) is $0 < \rho < 1$. In the next section we compare the measurements taken for two different experimental setups using the reduced model formulation (29) – (33).

Remark: The theorem on the passage to the limit or the theorem on estimation of the remainder term, which relates the reduced model formulation (29) – (33) to the original model formulation (14) – (21) can be proved using standard approaches (see, e.g., [16]). These results will be published later elsewhere.

4. Comparison of results obtained for two different "artificial tissue" materials

In the previous section we discussed the model for experimental setup for which "artificial tissue" sample is characterized by the diffusion coefficient D_1 (experimental system #1). Now consider an "artificial tissue" sample made of a different material and characterized by a different diffusion coefficient D_2 associated with drug propagation through this material (experimental system #2). We want to address the following question: If DES used in both experiments are similar, and if the shape of the "artificial tissue" sample in the second experimental setup is the same as that in the first experimental setup, at which instant of time t^{**} the measurement of the total drug remnant on stent must be taken for the experimental system #2 to produce the same result as that obtained in on experimental system #1 at the instant of time t^*?

To answer this question let us first write down the model system and additional conditions that describe the second experimental setup. Since the stents and the coating are the same, the diffusion coefficient in the coating, D_0, is the same for both experimental systems. We have:

$$\frac{\partial w}{\partial t} = D_0 \frac{\partial^2 w}{\partial x^2}, \qquad -L_0 < x < 0, \tag{34}$$

$$\frac{\partial u}{\partial t} = D_2 \frac{\partial^2 u}{\partial x^2}, \qquad 0 < x < L_1, \tag{35}$$

$$0 < t \leq t^{**},$$

at $x = 0$:

$$w = u, \qquad D_0 \frac{\partial w}{\partial x} = D_2 \frac{\partial u}{\partial x}, \tag{36}$$

at $x = -L_0$:

$$\frac{\partial w}{\partial x} = 0, \tag{37}$$

at $x = L_1$:

$$u = 0. \tag{38}$$

At $t = 0$:

$$w = w^*, \;\; u = 0. \tag{39}$$

Now we rescale the model (34) – (39) using the following changes of variables:

$$\tau = t \cdot (D_2/{L_1}^2),\ W = w/w^*, U = u/w^*, \rho = x/L_1.$$

In the new non-dimensional variables the model formulation for experimental system #2 will have the form (7) – (12) with one exception: the non-dimensional time interval will now be

$$0 < \tau \le \tau^{**} = t^{**} \cdot (D_2/{L_1}^2).$$

We assume that for experimental system #2 the condition (b), $D_0 \gg D_2 > 0$, holds. This means that we can introduce the small parameter $0 < \varepsilon = D_1/D_0 \ll 1$, use asymptotic approximation (13) and apply the analysis similar to that presented in section 3 to obtain the reduced model formulation (29) – (33) with a different expression for the non-dimensional time τ.

If $\tau^* = \tau^{**}$, then the non-dimensional model formulations for experimental systems #1 and #2 will be absolutely the same. So, if the measurements on the two systems are performed for the same DES, then the drug remnant on stent values must be the same at the same non-dimensional instants of time $\tau^* = \tau^{**}$. This means that for identical DES the experimental data values must be the same if the measurements are obtained for the original dimensional instants of time t^* and t^{**} satisfying the following relationship:

$$\tau^* = t^* \cdot \left(\frac{D_1}{{L_1}^2}\right) = t^{**} \cdot \left(\frac{D_2}{{L_1}^2}\right) = \tau^{**},$$

or

$$t^{**} = t^* \cdot \left(\frac{D_1}{D_2}\right). \qquad (40)$$

We illustrate this result qualitatively in figure 4.

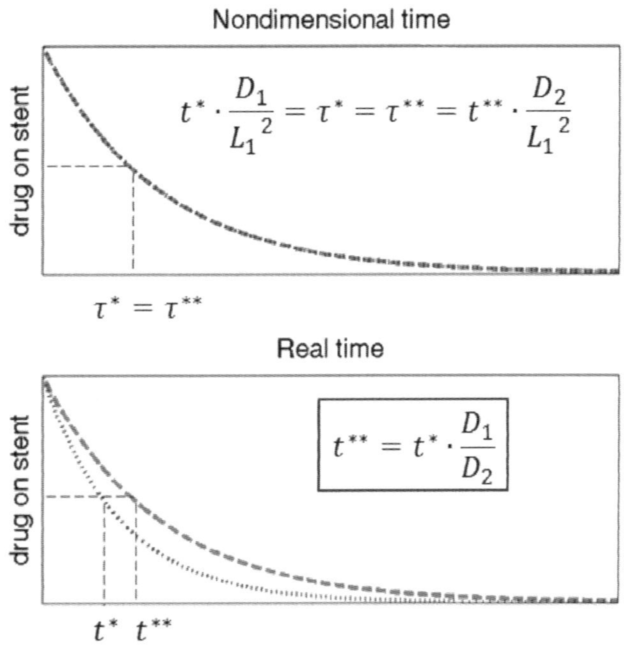

Figure 4. Qualitative illustration of formula (40): the case for which $D_1 > D_2$.

5. Comments on more realistic models

Let us discuss possible extensions of the simple model presented above to more realistic cases and make several important comments.

- In the above model the assumptions were made that the entire drug in the coating is immediately available for release (after instantaneous hydration of coating polymer), the diffusion (drug mixing) within the coating is infinitely fast, and the release is modulated by the finite diffusion within the "artificial tissue" sample. The natural extension of this model is related to introduction of fast release and slow release compartments in the coating. The model will then still be formulated as a partial differential (diffusion) equation for the drug concentration in the sample and two ordinary differential equations describing drug concentrations in "fast" and "slow" release compartments. Within each compartment the drug mixing is fast, however the drug transfer between the compartments is slow (e.g., due to finite rate of coating polymer hydration). It is still assumed that the coating and the "artificial tissue" sample are in direct contact with each other.
- The spatially one-dimensional model (1) – (6) is derived under no flux (zero Neumann type) boundary conditions for drug at the boundaries of the "artificial tissue" sample corresponding to $z = 0$, $z = Z$, $y = 0$, and $y = Y$ (in other words, we have a translational symmetry in the two spatial directions corresponding to z and y). If the Neumann type boundary conditions at these boundaries are substituted by Dirichlet type conditions (e.g., describing drug washout) or by Robin type conditions (describing, e.g., drug exchange between the sample and the buffer through these boundaries), the model will become spatially three-dimensional.
- The dependence of drug concentration in the coating not only on time but also on spatial variable(s) may be taken into account in case the costing polymer is not fully hydrated. We then must use the original partial differential (diffusion) equation for description of drug concentration in the coating (in addition to diffusion equation describing drug distribution in the "artificial tissue" sample) instead of using the reduced model.
- In the presence of a membrane between the coating and the "artificial tissue" sample, the boundary conditions between the coating domain and the tissue sample domain will change (from Dirichlet type to Robin type). Also, depending on whether the concentration of the drug in a buffer is close to saturation or not, the conditions at the boundary of the sample exposed to the buffer may change. These new assumptions will make the model more complicated.
- Other extensions of the above mentioned models are related to possible differences in drug partition in the coating and in the "artificial tissue" sample, presence of drug binding in the sample, as well as different mobility properties of hydrophilic and lipophilic drugs that may be loaded in the coating.
- Different shapes of the "artificial tissue" samples may also be considered (e.g., cylinder shown in figure 1) together with changing in time volume of DES coating (volume changes due to coating polymer degradation).
- Numerical implementations of the model extensions mentioned above may be used for design of particular experiments, estimation of parameters from data, etc.

6. Conclusions

Let us discuss possible extensions of the simple model presented above to more realistic cases and make several important comments. We considered some basic ideas behind the comparison of measurement results (representing percentage of drug remnant on stent) obtained for two experimental setups corresponding to in vitro release from DES into "artificial tissue" samples. In vitro release experiments are usually intended to mimic in vivo release into the real blood vessels. It turns out that for a particular experimental platform that we discussed in this paper the difference in diffusion coefficients between the two "artificial tissue" samples of the same shape makes it necessary to take measurements for the two experimental setups at different instants of time to produce comparable percentage of drug remnant of stent results. The explicit formula connecting the dimensional (real) instants of time when the measurements must be taken for proper comparison of DES characteristics using two experimental systems was derived; see (40).

References

[1] O'Connell BM, McGloughlin T M and Walsh M T 2010 Factors that affect mass transport from drug eluting stents into the artery wall *Biomed. Eng. Online* **9(1)** 15-16

[2] Hwang C-W, Levin A D, Jonas M, Li P H and Edelman E R 2005 Thrombosis Modulates Arterial Drug Distribution for Drug-Eluting Stents *Circulation* **111** 1619-26

[3] Tesfamariam B 2008 Drug release kinetics from stent device-based delivery *systems J. Cardiovasc. Pharmacol.* **51(2)** 118-25

[4] Yang C and Burt H M 2006 Drug-eluting stents: factors governing local pharmacokinetics *Adv. Drug Deliv. Rev.* **58(3)** 402-11

[5] Sakharov D V, Kalachev L V and Rijken D C 2002 Numerical simulation of local pharmacokinetics of a drug after intravascular delivery with an eluting stent *J. Drug Target.* **10(6)** 507-13

[6] Hwang C W, Wu D and Edelman E R 2003 Impact of transport and drug properties on the local pharmacology of drug-eluting stents *Int. J. Cardiovasc. Intervent.* **5(1)** 7-12

[7] Borghi A, Foa E, Balossino R, Migliavacca F and Dubini G 2008 Modelling drug elution from stents: effects of reversible binding in the vascular wall and degradable polymeric matrix *Comput. Methods Biomech. Biomed. Engin.* **11(4)** 367-77

[8] Prabhu S and Hossainy S 2007 Modeling of degradation and drug release from a biodegradable stent coating *J. Biomed. Mater. Res. A* **80(3)** 732-41

[9] Lee C H, Tan H C and Lim Y T 2006 Update on drug-eluting stents for prevention of restenosis. *Asian Cardiovasc. Thorac. Ann.* **14(1)** 75-82

[10] Park K 2009 Better design of drug-eluting stents using computer modelling *J. Control Release* **133(1)** 1

[11] O'Brien C C, Kolachalama V B, Barber T J, Simmons A and Edelman E R 2013 Impact of flow pulsatility on arterial drug distribution in stent-based therapy *J. Control Release* **168(2)** 115-24

[12] Levin A D, Jonas M, Hwang C W and Edelman E R 2005 Local and systemic drug competition in drug-eluting stent tissue deposition properties J. Control Release **109(1-3)** 236-43

[13] Tzafriri A R, Groothuis A, Price G S and Edelman E R 2012 Stent elution rate determines drug deposition and receptor-mediated effects *J. Control Release* **161(3)** 918-26

[14] Sarkar S, Dadhania M, Rourke P, Desai T A and Wong J Y 2005 Vascular tissue engineering: microtextured scaffold templates to control organization of vascular smooth muscle cells and extracellular matrix *Acta Biomater.* **1(1)** 93-100

[15] Sarkar S, Isenberg B C, Hodis E, Leach J B, Desai T A and Wong J Y 2008 Fabrication of a layered microstructured polycaprolactone construct for 3-D tissue *engineering J. Biomater. Sci. Polym. Ed.* **19(10)** 1347-62

[16] Vasil'eva A B, Butuzov V F and Kalachev L V 1995 *The Boundary Function Method for Singular Perturbation Problems* (Philadelphia: SIAM)

MURPHYS-HSFS-2014 IOP Publishing
Journal of Physics: Conference Series **727** (2016) 012006 doi:10.1088/1742-6596/727/1/012006

Hausdorff metric BV discontinuity of sweeping processes

Olaf Klein
Weierstrass Institute for Applied Analysis and Stochastics,
Leibniz Institute in Forschungsverbund Berlin e.V.
Mohrenstr. 39, 10117 Berlin, Germany
E-mail: Olaf.Klein@wias-berlin.de

Vincenzo Recupero
Dipartimento di Scienze Matematiche, Politecnico di Torino
C.so Duca degli Abruzzi 24, 10129 Torino, Italy
E-mail: vincenzo.recupero@polito.it

Abstract. Sweeping processes are a class of evolution differential inclusions arising in elastoplasticity and were introduced by J.J. Moreau in the early seventies. The solution operator of the sweeping processes represents a relevant example of *rate independent operator*. As a particular case we get the so called *play operator*, which is a typical example of a hysteresis operator. The continuity properties of these operators were studied in several works. In this note we address the continuity with respect to the strict metric in the space of functions of bounded variation with values in the metric space of closed convex subsets of a Hilbert space. We provide counterexamples showing that for all BV-formulations of the sweeping process the corresponding solution operator is not continuous when its domain is endowed with the strict topology of BV and its codomain is endowed with the L^1-topology. This is at variance with the play operator which has a BV-extension that is continuous in this case.

1. Introduction

A sweeping process is an evolution problem arising in elastoplasticity that can be described in the following way. Let $\mathcal{H}$ be a real Hilbert space and let $\mathcal{C}(t) \subseteq \mathcal{H}$ be a moving closed convex set, i.e. a family of closed convex sets indexed by the time parameter $t \in [0,T]$, T being the final time of the evolution. One has to find a function $y : [0,T] \longrightarrow \mathcal{H}$ such that

$$y(t) \in \mathcal{C}(t) \qquad \forall t \in [0,T], \tag{1.1}$$

$$-y'(t) \in N_{\mathcal{C}(t)}(y(t)) \qquad \text{for } \mathcal{L}^1\text{-a.e. } t \in [0,T], \tag{1.2}$$

$$y(0) = \mathrm{Proj}_{\mathcal{C}(0)}(y_0), \tag{1.3}$$

where $\mathcal{L}^1$ is the one dimensional Lebesgue measure, $y_0 \in \mathcal{H}$ is prescribed, $\mathrm{Proj}_{\mathcal{C}(0)}(y_0)$ is its projection on $\mathcal{C}(0)$, and $N_{\mathcal{C}(t)}(y(t))$ is the normal cone to $\mathcal{C}(t)$ at $y(t)$ (cf. the definition in formula (2.12) below: precise definitions will be given in the next Section 2). The sweeping process formulated as in (1.1)–(1.3) is well posed in the framework of Lipschitz continuous functions, indeed it can be shown that there exists a unique Lipschitz continuous function $y : [0,T] \longrightarrow \mathcal{H}$ satisfying (1.1)–(1.3), once $y_0 \in \mathcal{H}$ and the mapping $t \longmapsto \mathcal{C}(t)$ is Lipschitz continuous when the class of closed convex subsets of $\mathcal{H}$ is endowed with the Hausdorff metric (see (2.13) below). The proof of this fact can be found in [13]. For an overview on sweeping processes we refer the reader to [12].

If one wants to deal with more general movements of $\mathcal{C}(t)$, for instance when $t \longmapsto \mathcal{C}(t)$ is of bounded variation, then the above formulation has to be modified. In [14] the following generalized formulation is proposed. One has to look for a function $y : [0,T] \longrightarrow \mathcal{H}$ of bounded variation and a positive measure μ such that the distributional derivative $\mathrm{D}y$ of y satisfies the equality $\mathrm{D}y = w\mu$ for some $w \in L^1([0,T], \mu; \mathcal{H})$ and the condition

$$-w(t) \in N_{\mathcal{C}(t)}(y(t)) \qquad \text{for } \mu\text{-a.e. } t \tag{1.4}$$

holds together with (1.3). Such a function y is also called *a solution of the sweeping process in the sense of the differential measures.* We will call $\mathsf{S} : \mathcal{C} \longrightarrow y$ the solution operator of the sweeping process, associating with $\mathcal{C}(t)$ the solution $y(t)$.

The operator S is a very relevant example of a *rate independent operator*, i.e. an operator S such that

$$\mathsf{S}(\mathcal{C} \circ \phi)(t) = (\mathsf{S}(\mathcal{C}) \circ \phi)(t) \quad \forall t \in [0,T]. \tag{1.5}$$

whenever $\phi : [0,T] \longrightarrow [0,T]$ is an increasing surjective function. This fact was already observed by J.J. Moreau in (cf. [14, Proposition 2i]), even if he did not use the term "rate independence".

A relevant particular case of sweeping processes is obtained when $\mathcal{C}(t) = u(t) - \mathcal{Z}$, $\mathcal{Z}$ being a fixed closed convex set in $\mathcal{H}$ and $u : [0,T] \longrightarrow \mathcal{H}$ a given function. In this case the solution operator P mapping u to the solution of the corresponding sweeping process can be equivalently defined on the space of $\mathcal{H}$-valued functions of bounded variation. The operator P is usually called *play operator* and plays an important role in elastoplasticity (cf. e.g. [7, 21, 4, 8, 11]).

The study of the continuity properties of the solution operators S and P has been addressed in several works. For instance in [14] it is shown that S is continuous with respect to the topology of the uniform convergence. Instead in [8] it is proved that P is continuous with respect to the BV strict topology when P is restricted to the space of continuous functions of bounded variation provided $\mathcal{Z}$ is a bounded convex closed set whose interior contains 0, for every boundary point x of $\mathcal{Z}$ there exists a unique outward normal $n(x)$, and the mapping $x \to n(x)$ is continuous. All these assumptions are dropped in [17], where $\mathcal{Z}$ is allowed to be an arbitrary closed convex set. Geometric conditions on $\mathcal{Z}$ are given in Section 3 in order to characterize when P is continuous from the space of left continuous functions of bounded variation into itself when the domain is endowed with the strict topology and the codomain is endowed with the L^1-topology.

The aim of the present note is to show that this continuity property does not hold for the general solution operator S. This is achieved by exhibiting a concrete example in the one dimensional case $\mathcal{H} = \mathbb{R}$.

Here is a brief plan of the paper. In the following Section 2 we present all the technical tools in order to deal with the sweeping processes: BV functions with values in a metric space and convex sets in a Hilbert space. In Section 3 we state the main known existence and continuity results about the sweeping process and in Section 4 we present our counterexample showing the BV-discontinuity of its solution operator. In the final section we make some remarks connecting the BV-discontinuity with the existence of multiple geodesics in the space of closed convex subsets of a Hilbert space.

2. Preliminaries

From now on T will be a fixed strictly positive number and $\mathbb{N}$ is the set of integers that are greater or equal than one. The family of Borel sets in $[0,T]$ will be denoted by $\mathscr{B}([0,T])$.

2.1. Functions with bounded variation
We assume that

$$(\mathcal{X}, d) \text{ is a complete metric space.} \tag{2.1}$$

If $x \in \mathcal{X}$ and $S \subseteq \mathcal{X}$, $S \neq \varnothing$, we set $d(x,S) := \inf_{y \in S} d(x,y)$.

We will mainly deal with spaces of $\mathcal{X}$-valued functions defined on $[0,T]$. As usual the space of continuous functions is denoted by $C([0,T];\mathcal{X})$. In the next definition we recall the most simple space containing discontinuous functions.

Definition 2.1. *Given a function $u : [0,T] \longrightarrow \mathcal{X}$ and a subinterval $J \subseteq [0,T]$, the* (pointwise) variation of u on J *is defined by*

$$\mathrm{V}(u,J) := \sup\left\{\sum_{j=1}^{m} d(u(t_{j-1}),u(t_j)) \ : \ m \in \mathbb{N},\ t_j \in J\ \forall j,\ t_0 < \cdots < t_m\right\}. \tag{2.2}$$

If $\mathrm{V}(u,[0,T]) < \infty$ *we say that* u is of bounded variation on $[0,T]$ *and we set* $BV([0,T];\mathcal{X}) := \{u : [0,T] \longrightarrow \mathcal{X} \ : \ \mathrm{V}(u,[0,T]) < \infty\}$.

It is well-known that every $u \in BV([0,T];\mathcal{X})$ admits one sided limits $u(t-), u(t+)$ at every point $t \in [0,T]$, with the convention that $u(0-) := u(0)$ and $u(T+) := u(T)$. In this note we will limit ourselves to left continuous functions, i.e. we will deal with the space

$$BV_L([0,T];\mathcal{X}) := \{u \in BV([0,T];\mathcal{X}) \ : \ u(t-) = u(t) \quad \forall t \in [0,T]\}. \tag{2.3}$$

When we consider left continuous functions we are essentially dealing with Lebesgue equivalence classes of functions with a special view on the initial point 0, allowing us to take into account Dirac masses at 0. In the next definition we introduce some natural metrics in $BV_L([0,T];\mathcal{X})$.

Definition 2.2. *For every $u, v \in BV_L([0,T];\mathcal{X})$ we set*

$$d_\infty(u,v) := \sup_{t\in[0,T]} d(u(t),v(t)), \tag{2.4}$$

$$d_s(u,v) := \int_0^T d(u(t),v(t))\,\mathrm{d}t + d(u(0),v(0)) + |\,\mathrm{V}(u,[0,T]) - \mathrm{V}(v,[0,T])|. \tag{2.5}$$

We call d_s strict metric *and we say that* $u_n \to u$ strictly on $[0,T]$ *if* $d_s(u_n,u) \to 0$ *as* $n \to \infty$. *The topology induced by* d_s *is called* strict topology.

Of course d_∞ is the distance inducing the topology of uniform convergence. Observe that if $u, v \in BV_L([0,T];\mathcal{X})$ then $t \longmapsto d(u(t),v(t))$ is a measurable integrable function (cf. [6, Section 4.5.10, p. 505]), thus formula (2.5) makes sense.

The strict metric d_s is the natural metric in BV in the metric framework. It is also used when one deals with approximation procedures (see, e.g., [2]). In connection with hysteresis, it has been studied in [21, 4, 8, 16, 19]. Notice that the strict topology is different from the strong (or norm) BV-topology (see (2.9) below): the norm topology is usually too strong for applications (indeed it is often called the $W^{1,1}$-topology, that cannot be adapted to the metric framework).

Usually in the definition of strict metric the term $d(u(0),v(0))$ is missing. The reason why we insert it, is that we are considering left continuous functions on the closed interval $[0,T]$ and we want to take into account the value of these functions at the point $t = 0$. This is equivalent to artificially extend any function $u : [0,T] \longrightarrow \mathcal{X}$ from $[0,T]$ to $[-1,T]$ by setting $u(t) = u(0)$ for every $t < 0$. If we write down the classical notion of strict metric for these extended functions, we get exactly our d_s of Definition 2.2.

Let us recall that the space of Lipschitz continuous functions is defined by

$$Lip([0,T]\,;\mathcal{X}) := \left\{u : [0,T] \longrightarrow \mathcal{X} \ : \ \sup_{t\neq s} \frac{d(u(s),u(t))}{|t-s|} < \infty\right\}. \tag{2.6}$$

It is clear that $Lip([0,T]\,;\mathcal{X}) \subseteq C([0,T]\,;\mathcal{X}) \cap BV([0,T]\,;\mathcal{X})$. In the following definition we recall the notion of geodesic in a metric space.

Definition 2.3. *If $x_0, y_0 \in \mathcal{X}$ and there is a curve $g \in Lip([0,1]\,;\mathcal{X})$ such that $g(0) = x_0$, $g(1) = y_0$ and $d(x_0,y_0) = \mathrm{V}(g,[0,1])$, then g is called* a geodesic connecting x_0 and y_0.

2.2. Convex sets in Hilbert spaces
Let us assume that

$$\begin{cases} \mathcal{H} \text{ is a real Hilbert space with inner product } \langle\cdot,\cdot\rangle, \\ \|x\|_{\mathcal{H}} := \langle x,x\rangle^{1/2}. \end{cases} \tag{2.7}$$

If $\mu : \mathscr{B}([0,T]) \longrightarrow [0,+\infty]$ is a measure and $p \in [1,\infty]$, then the space of functions $u : [0,T] \longrightarrow \mathcal{H}$ such that $t \longmapsto \|u(t)\|_{\mathcal{H}}^p$ is integrable with respect to μ will be denoted by $L^p([0,T]\,,\mu;\mathcal{H})$ or by $L^p(\mu;\mathcal{H})$ if no confusion may arise. For the theory of integration of vector valued functions we refer to [3, Appendix]. When $\mu = \mathcal{L}^1$, the one dimensional Lebesgue measure, we will simply write $L^p(0,T;\mathcal{H}) := L^p([0,T],\mu;\mathcal{H})$. We warn the reader that we do not identify two functions which are equal $\mathcal{L}^1$-almost everywhere ($\mathcal{L}^1$-a.e.). For example, two functions u and v that are identical on $(0,T]$ but satisfy $u(0) \neq v(0)$ are not considered as two representation function of the same equivalence class. We recall that the Sobolev space $W^{1,p}([0,T]\,;\mathcal{H})$ is the space of functions $u \in L^p(0,T;\mathcal{H})$ such that there exists $v \in L^p(0,T;\mathcal{H})$ satisfying the equality $u(t) = u(0) + \int_0^t v(s)\,\mathrm{d}s$ for every $t \in [0,T]$ (cf., e.g., the Appendix of [3]). We also have that $Lip([0,T]\,;\mathcal{H}) = W^{1,\infty}([0,T]\,;\mathcal{H})$. Moreover if $u \in W^{1,1}([0,T]\,;\mathcal{H})$, then there exists the derivative $u'(t)$ for $\mathcal{L}^1$-a.e. $t \in [0,T]$, and we have the equality

$$\mathrm{V}(u,[0,T]) = \int_0^T \|u'(t)\|_{\mathcal{H}}\,\mathrm{d}t, \qquad \forall u \in W^{1,1}([0,T]\,;\mathcal{H}). \tag{2.8}$$

Since $\mathcal{H}$ has a linear structure we can also consider the so called strong BV-metric (or $W^{1,1}$-metric) in $BV_L([0,T]\,;\mathcal{H})$:

$$d_{\mathrm{BV}}(u,v) := \int_0^T \|u(t)-v(t)\|_{\mathcal{H}}\,\mathrm{d}t + \|u(0)-v(0)\|_{\mathcal{H}} + \mathrm{V}(u-v,[0,T]), \qquad u,v \in BV_L([0,T]\,;\mathcal{H}), \tag{2.9}$$

In the regular case we obtain the topology of the Sobolev space $W^{1,1}([0,T]\,;\mathcal{H})$ induced by the norm

$$\|u\|_{W^{1,1}([0,T];\mathcal{H})} := \int_0^T \|u(t)\|_{\mathcal{H}}\,\mathrm{d}t + \int_0^T \|u'(t)\|_{\mathcal{H}}\,\mathrm{d}t, \qquad u \in W^{1,1}([0,T]\,;\mathcal{H}). \tag{2.10}$$

Now we set

$$\mathscr{C}_{\mathcal{H}} := \{\mathcal{K} \subseteq \mathcal{H} \ : \ \mathcal{K} \text{ nonempty, bounded, closed and convex}\}. \tag{2.11}$$

If $\mathcal{K} \in \mathscr{C}_{\mathcal{H}}$ and $x \in \mathcal{H}$, then the projection on $\mathcal{K}$ of x is denoted by $\mathrm{Proj}_{\mathcal{K}}(x)$. If $\mathcal{K} \in \mathscr{C}_{\mathcal{H}}$ and $x \in \mathcal{K}$, we recall that the *(exterior) normal cone to $\mathcal{K}$ at x* is defined by

$$N_{\mathcal{K}}(x) := \{y \in \mathcal{H} \ : \ \langle y, v-x\rangle \le 0 \ \forall v \in \mathcal{K}\}. \tag{2.12}$$

We endow the set $\mathscr{C}_{\mathcal{H}}$ with the Hausdorff distance. Here is the definition.

Definition 2.4. *The* Hausdorff distance $d_{\mathscr{H}} : \mathscr{C}_{\mathcal{H}} \times \mathscr{C}_{\mathcal{H}} \longrightarrow [0, \infty[$ *is defined by*

$$d_{\mathscr{H}}(\mathcal{A}, \mathcal{B}) := \max \left\{ \sup_{a \in \mathcal{A}} d(a, \mathcal{B}), \sup_{b \in \mathcal{B}} d(b, \mathcal{A}) \right\}, \qquad \mathcal{A}, \mathcal{B} \subseteq \mathscr{C}_{\mathcal{H}}. \tag{2.13}$$

The distance $d_{\mathscr{H}}$ makes $\mathscr{C}_{\mathcal{H}}$ a complete metric space (see, e.g., [1, Theorem 3.85, Section 3.17, p. 116]).

For the sake of simplicity we assumed that the elements of $\mathscr{C}_{\mathcal{H}}$ are bounded. If this assumption is dropped, then $d_{\mathscr{H}}$ is a metric that may take on the value ∞, thus some supplementary technical details have to be added. However for our purposes this assumption is not restrictive and in order to prove the BV discontinuity of the sweeping process we can limit ourselves to the bounded case.

2.3. Differential measures

We recall that a *($\mathcal{H}$-valued Borel) vector measure* on $[0, T]$ is a map $\mu : \mathscr{B}([0,T]) \longrightarrow \mathcal{H}$ such that $\mu(\bigcup_{n=1}^{\infty} B_n) = \sum_{n=1}^{\infty} \mu(B_n)$ whenever (B_n) is a sequence of mutually disjoint sets in $\mathscr{B}([0,T])$. Let us also recall that if $\mu : \mathscr{B}([0,T]) \longrightarrow \mathcal{H}$ is a vector measure, then $\|\mu\| : \mathscr{B}([0,T]) \longrightarrow [0, \infty]$ is defined by

$$\|\mu\|(B) := \sup \left\{ \sum_{n=1}^{\infty} \|\mu(B_n)\|_{\mathcal{H}} \ : \ B = \bigcup_{n=1}^{\infty} B_n, \ B_n \in \mathscr{B}([0,T]), \ B_h \cap B_k = \varnothing \text{ if } h \neq k \right\}.$$

The map $\|\mu\|$ is a positive measure which is called *total variation of* μ and the vector measure μ is said to be *with bounded variation* if $\|\mu\|([0,T]) < \infty$ (see, e.g., [5, Chapter I, Section 3.]).

The following proposition (cf. [5, Theorem 1, section III.17.2, p. 358]) provides a connection between functions with bounded variation and vector measures.

Theorem 2.1. *If $f \in BV_L([0,T]; \mathcal{H})$ then there exists a unique vector measure of bounded variation $\mu_f : \mathscr{B}([0,T]) \longrightarrow \mathcal{H}$ such that for every $c, d \in [0,T]$ with $c < d$ we have*

$$\mu_f(]c,d[) = f(d) - f(c+), \qquad \mu_f([c,d]) = f(d+) - f(c),$$
$$\mu_f([c,d[) = f(d) - f(c), \qquad \mu_f(]c,d]) = f(d+) - f(c+).$$

Vice versa if $\mu : \mathscr{B}([0,T]) \longrightarrow \mathcal{H}$ is a vector measure with bounded variation, then the map $f_\mu : [0,T] \longrightarrow \mathcal{H}$ defined by $f_\mu(t) := \mu([a,t[)$ is such that $f_\mu \in BV_L([0,T]; \mathcal{H})$ and $\mu_{f_\mu} = \mu$.

The measure μ_f is called *Lebesgue-Stieltjes measure* or *differential measure* of f. Like in the scalar case if $f \in BV([0,T]; \mathcal{H})$ then $\mu_f = \mathrm{D}f$, with $\mathrm{D}f$ being the distributional derivative of f, i.e.

$$-\int_0^T \varphi'(t) f(t) \,\mathrm{d}t = \int_0^T \varphi \,\mathrm{d}\mathrm{D}f \qquad \forall \varphi \in C_c^1(]0,T[)$$

(cf. [17, Section 2]).

3. The sweeping processes and the play operator

Now we can present the general existence result for the sweeping processes in BV_L. This is the main result in [14]: the existence theorem for right continuous data is [14, Propositions 2a and 3a] and the left continuous case can be deduced from [14, Section 2d].

Theorem 3.1. *Let $\mathcal{C} \in BV_L([0,T]; \mathscr{C}_{\mathcal{H}})$ and $y_0 \in \mathcal{H}$ be given. There exists a unique $y =: \mathsf{S}_{y_0}(\mathcal{C}) \in BV_L([0,T]; \mathcal{H})$ such that there exist a measure $\mu : \mathscr{B}([0,T]) \longrightarrow [0, \infty[$ and* *a function* $w \in L^1(\mu; \mathcal{H})$ *satisfying*

$$\mathrm{D}y = w\mu, \tag{3.1}$$
$$y(t) \in \mathcal{C}(t) \qquad \forall t \in [0,T], \tag{3.2}$$
$$-w(t) \in N_{\mathcal{C}(t)}(y(t)) \qquad \textit{for } \mu\textit{-a.e. } t \in [0,T], \tag{3.3}$$
$$y(0) = \mathrm{Proj}_{\mathcal{C}(0)}(y_0), \tag{3.4}$$

where $w\mu$ *is the vector measure defined by* $w\mu(B) := \int_B w \,\mathrm{d}\mu$, $B \in \mathscr{B}([0,T])$. *If* $\mathcal{C} \in BV([0,T]\,;\mathscr{C}_{\mathcal{H}}) \cap C([0,T]\,;\mathscr{C}_{\mathcal{H}})$, *then* $\mathsf{S}_{y_0}(\mathcal{C}) \in BV([0,T]\,;\mathcal{H}) \cap C([0,T]\,;\mathcal{H})$. *Finally if* $\mathcal{C} \in Lip([0,T]\,;\mathcal{H})$ *then* $\mathsf{S}_{y_0}(\mathcal{C}) \in Lip([0,T]\,;\mathcal{H})$ *and problem* (3.1)–(3.4) *reads as* (1.1)–(1.3).

The two last statements of the previous theorem can be easily deduced by the representation formula of [18, Theorem 3.1] for the continuous case. Thanks to the previous theorem the following solution operator is defined

$$\mathsf{S}_{y_0} : BV_L([0,T]\,;\mathscr{C}_{\mathcal{H}}) \longrightarrow BV_L([0,T]\,;\mathcal{H}).$$

If $\mathcal{Z} \in \mathscr{C}_{\mathcal{H}}$, $0 \in \mathcal{Z}$ and $z_0 \in \mathcal{Z}$, we consider the *play operator*

$$\mathsf{P}_{z_0} : BV_L([0,T]\,;\mathcal{H}) \longrightarrow BV_L([0,T]\,;\mathcal{H})$$

which is defined by

$$\mathsf{P}_{z_0}(u) := \mathsf{S}_{u(0)-z_0}(u - \mathcal{Z}), \qquad u \in BV_L([0,T]\,;\mathcal{H}).$$

The restrictions of S_{y_0} and P_{z_0} to the various spaces subspaces of BV will be denoted by the same symbols S_{y_0} and P_{z_0}. Using this notation, it holds that $\mathsf{P}_{z_0} : Lip([0,T]\,;\mathcal{H}) \longrightarrow Lip([0,T]\,;\mathcal{H})$ is the *classical play operator*, which is the operator associating with $u \in Lip([0,T]\,;\mathcal{H})$ the only function $y \in Lip([0,T]\,;\mathcal{H})$ satisfying (1.1)–(1.3) with $\mathcal{C}(t) = u(t) - \mathcal{Z}$ and $y_0 = u(0) - z_0$.

Let us observe that the operator P_{z_0} here defined coincides with the one introduced in [9]. Indeed, as observed in [9], the two operators coincide on the set of left continuous step functions (cf. formula (3.8)) and they are both continuous with respect to the d_∞-metric (cf. Theorem 3.2 (iii) below and [9, Theorem 2.3]).

Another possible extension of the play operator on $Lip([0,T]\,;\mathcal{H})$ to the space of functions of bounded variation is provided in [17], where an operator $\overline{\mathsf{P}}_{z_0} : BV_L([0,T]\,;\mathcal{H}) \longrightarrow BV_L([0,T]\,;\mathcal{H})$ is obtained as the continuous extension of the classical play operator $\mathsf{P}_{z_0} : Lip([0,T]\,;\mathcal{H}) \longrightarrow Lip([0,T]\,;\mathcal{H})$ with respect to the d_s-topology in the domain and the L^1-topology in the codomain. As shown in [17], the two operators coincide on $BV([0,T]\,;\mathcal{H}) \cap C([0,T]\,;\mathcal{H})$, but in general they are different on $BV_L([0,T]\,;\mathcal{H})$.

A main feature of S_{y_0} and P_{z_0} is *rate independence*, i.e. if $\phi : [0,T] \longrightarrow [0,T]$ is an increasing surjective function, then

$$\mathsf{S}_{y_0}(\mathcal{C} \circ \phi) = \mathsf{S}_{y_0}(\mathcal{C}) \circ \phi, \tag{3.5}$$
$$\mathsf{P}_{z_0}(u \circ \phi) = \mathsf{P}_{z_0}(u) \circ \phi, \tag{3.6}$$

whenever $\mathcal{C} \in BV_L([0,T]\,;\mathscr{C}_{\mathcal{H}})$ and $u \in BV_L([0,T]\,;\mathcal{H})$ (cf. [14, Proposition 2i]). Now let $m \in \mathbb{N}$, $t_0 = 0 < t_1 < \cdots < t_m = T$, and $\mathcal{C}_0, \mathcal{C}_1, \ldots, \mathcal{C}_m \in \mathscr{C}_{\mathcal{H}}$. If $\mathcal{C} \in BV_L([0,T]\,;\mathscr{C}_{\mathcal{H}})$ is the step function defined by

$$\mathcal{C}(t) := \begin{cases} \mathcal{C}_0 & \text{if } t = 0, \\ \mathcal{C}_k & \text{if } t \in \left]t_{k-1}, t_k\right], k \in 1, \ldots, m, \end{cases} \tag{3.7}$$

then for every $y_0 \in \mathcal{H}$ it turns out that (cf. [14, Formulas (1.13)-(1.14)])

$$\mathsf{S}_{y_0}(\mathcal{C})(t) = \begin{cases} \mathrm{Proj}_{\mathcal{C}_0}(y_0) & \text{if } t = 0, \\ \mathrm{Proj}_{\mathcal{C}_k}(y_{k-1}) & \text{if } t \in \left]t_{k-1}, t_k\right], k \in 1, \ldots, m. \end{cases} \tag{3.8}$$

Of course the same kind of result holds for the play operator P_{z_0}. Here we list some of the main continuity properties of S_{y_0}, of P_{z_0} and of the restrictions of these operators to various spaces subspaces of BV. We recall that these restrictions are denoted by the same symbols S_{y_0} and P_{z_0}.

Theorem 3.2. *Assume that $y_0 \in \mathcal{H}$, $\mathcal{Z} \in \mathscr{C}_{\mathcal{H}}$, and $0, z_0 \in \mathcal{Z}$. The following statements holds true.*

(i) $\mathsf{S}_{y_0} : BV_L([0,T];\mathscr{C}_{\mathcal{H}}) \longrightarrow BV_L([0,T];\mathcal{H})$ *is continuous with respect to the* d_∞*-topology.*

(ii) $\mathsf{S}_{y_0} : BV([0,T];\mathscr{C}_{\mathcal{H}}) \cap C([0,T];\mathscr{C}_{\mathcal{H}}) \longrightarrow BV([0,T];\mathcal{H}) \cap C([0,T];\mathcal{H})$ *is continuous if its domain is endowed with the* d_s*-topology and its codomain is endowed with the* d_∞*-topology.*

(iii) $\mathsf{P}_{z_0} : BV_L([0,T];\mathcal{H}) \longrightarrow BV_L([0,T];\mathcal{H})$ *is continuous with respect to the* d_∞*-topology.*

(iv) $\mathsf{P}_{z_0} : BV([0,T];\mathcal{H}) \cap C([0,T];\mathcal{H}) \longrightarrow BV([0,T];\mathcal{H}) \cap C([0,T];\mathcal{H})$ *is continuous with respect to the* d_s*-topology.*

(v) $\overline{\mathsf{P}}_{z_0} : BV_L([0,T];\mathcal{H}) \longrightarrow BV_L([0,T];\mathcal{H})$ *is continuous if the domain is endowed with the strict topology and the codomain with the* L^1*-topology.*

(vi) *If* $\mathcal{H}$ *is of finite dimension and* $\mathcal{Z}$ *is a* non-obtuse polyhedron, *i.e.* $\mathcal{Z} = \{x \in \mathcal{H} : \langle n_j, x\rangle \le c_j,\ j = 1, \ldots, p\}$ *for some* $p \in \mathbb{N}$, $c_j \ge 0$ *and* $n_j \in \mathcal{H}$ *with* $\|n_j\|_{\mathcal{H}} = 1$ *and* $\langle n_j, n_k\rangle \le 0$ *whenever* $1 \le i < j \le p$, *then it follows that* $\mathsf{P}_{z_0} : BV_L([0,T];\mathcal{H}) \longrightarrow BV_L([0,T];\mathcal{H})$ *is continuous if its domain is endowed with the* d_s*-topology and its codomain is endowed with the* L^1*-topology.*

Proof. Part (i) of the previous theorem is proved in [14] while part (ii) is proved in [18]. Concerning the play operator P_{z_0}, part (iii) follows directly from part (i), since $\mathrm{d}_\infty(u_n - Z, u - Z) \to 0$ whenever $u_n \to u$ uniformly on $[0,T]$. Part (iv) is instead proved in [17], where our extra-term $d(u(0), v(0))$ of the strict metric in (2.5) can be handled by means of the reduction method presented in [17, Section 4.4] taking into account the Lipschitz continuity of the projection. In order to prove (v) we recall that in [15] it is proved that $\mathsf{P}_{z_0} : BV([0,T];\mathcal{H}) \cap C([0,T];\mathcal{H}) \longrightarrow BV([0,T];\mathcal{H}) \cap C([0,T];\mathcal{H})$ admits a continuous extension to the above mentioned operator $\overline{\mathsf{P}}_{z_0} : BV_L([0,T];\mathcal{H}) \longrightarrow BV_L([0,T];\mathcal{H})$ if the domain is endowed with the strict topology and the codomain with the L^1-topology. This extension $\overline{\mathsf{P}}_{z_0}$ is however not necessarily equal to the sweeping process $\mathsf{P}_{z_0} = \mathsf{S}_{u(0)-z_0}(u - \mathcal{Z})$ on discontinuous functions considered in (vi): in [10] it is shown for $\mathcal{H}$ being finite dimensional that $\overline{\mathsf{P}}_{z_0} = \mathsf{P}_{z_0}$ on the whole $BV_L([0,T];\mathcal{H})$ if and only if $\mathcal{Z}$ is a non-obtuse polyhedron. □

In the one dimensional case the play operator P_{z_0} is continuous from $BV_L([0,T];\mathbb{R})$ into itself endowed with strict topology, this is proved in [15]. In the vector case this property is in general false: this can be deduced from [17, Theorem 3.7] where it is shown that the restriction $\mathsf{P}_{z_0} : BV([0,T];\mathcal{H}) \cap C([0,T];\mathcal{H}) \longrightarrow BV([0,T];\mathcal{H}) \cap C([0,T];\mathcal{H})$ can be d_s-continuously extended to the whole $BV_L([0,T];\mathcal{H})$ if and only if $\mathcal{Z}$ is a closed vector subspace or $\mathcal{Z} = \{x \in \mathcal{H} : -\alpha \le \langle f, x\rangle \le \beta\}$ for some $\alpha, \beta \in [0, \infty]$ and some $f \in \mathcal{H} \setminus \{0\}$, hence for bounded $\mathcal{Z}$ the operator P_{z_0} is never continuous from the whole BV into itself, both domain and codomain endowed with the strict metric.

4. Metric BV discontinuity

From Theorem 3.2 we can infer that if $\mathcal{Z}$ is a non-obtuse polyhedron and $\mathcal{H}$ has finite dimension, then $\mathsf{P}_{z_0} : BV_L([0,T];\mathcal{H}) \longrightarrow BV_L([0,T];\mathcal{H})$ is continuous if the domain is endowed with the strict topology and the codomain with the L^1-topology, while the operator $\overline{\mathsf{P}}_{z_0} : BV_L([0,T];\mathcal{H}) \longrightarrow BV_L([0,T];\mathcal{H})$ is continuous in this sense even if these assumptions on $\mathcal{Z}$ and $\mathcal{H}$ are not satisfied.

Now we provide the counterexample showing that this last continuity property stated is in general not true for the solution operator S_{y_0} of the sweeping process.

Let us consider $\mathcal{H} = \mathbb{R}$ so that

$$\mathscr{C}_{\mathbb{R}} = \{I \subseteq \mathbb{R} \ : \ I \text{ is a bounded closed interval, } I \neq \varnothing\}. \tag{4.1}$$

If $a, b, c, d \in \mathbb{R}$ and $a \le b$, $c \le d$, then

$$d_{\mathscr{H}}([a,b]\,,[c,d]) = \max\{|a-c|, |b-d|\}. \tag{4.2}$$

Let us set

$$\mathcal{K}_0 := [0,2]\,, \qquad \mathcal{K}_1 := [1,4] \tag{4.3}$$

and fix $t_0 \in \,]0,T[$. So $\mathcal{K}_0, \mathcal{K}_1 \in \mathscr{C}_{\mathbb{R}}$ and we can define $\mathcal{C} : [0,T] \longrightarrow \mathscr{C}_{\mathbb{R}}$ by setting

$$\mathcal{C}(t) := \begin{cases} \mathcal{K}_0 & \text{if } \ 0 \le t \le t_0, \\ \mathcal{K}_1 & \text{if } \ t_0 < t \le T, \end{cases} \qquad t \in [0,T]\,. \tag{4.4}$$

Hence $\mathcal{C} \in BV([0,T];\mathscr{C}_{\mathbb{R}})$ and $\mathrm{V}(\mathcal{C},[0,T]) = 2$. Let us now define $\mathcal{B} : [0,1] \longrightarrow \mathscr{C}_{\mathbb{R}}$ by

$$\mathcal{B}(t) := \begin{cases} [2t, 2+2t] & \text{if } \ 0 \le t \le 2/3, \\ [2-t, 2+2t] & \text{if } \ 2/3 < t \le 1. \end{cases} \tag{4.5}$$

Observe that $\mathcal{B} \in Lip([0,1];\mathscr{C}_{\mathbb{R}})$ and thanks to [18, Proposition 6.1] we have that, up to reparametrization, the restriction of $\mathcal{B}$ to $[0,2/3]$ is a geodesic connecting $B(0) = \mathcal{K}_0$ to $\mathcal{B}(2/3) = [4/3, 10/3]$. On the other hand, up to reparametrization, the restriction of $\mathcal{B}$ to $[2/3,1]$ is a geodesic connecting $\mathcal{B}(2/3) = [4/3,10/3]$ and $B(1) = \mathcal{K}_1$. Hence

$$\mathrm{V}(\mathcal{B},[0,1]) = \mathrm{V}(\mathcal{B},[0,2/3]) + \mathrm{V}(\mathcal{B},[2/3,1]) = 4/3 + 2/3 = 2, \tag{4.6}$$

thus $\mathcal{B}$ is a geodesic connecting $\mathcal{K}_0$ and $\mathcal{K}_1$. Let $\mathcal{B}_n \in Lip([0,T];\mathscr{C}_{\mathbb{R}})$ be the sequence defined (for n large enough) by

$$\mathcal{B}_n(t) := \begin{cases} \mathcal{K}_0 & \text{if } 0 \le t \le t_0, \\ \mathcal{B}(n(t-t_0)) & \text{if } t_0 < t \le t_0 + 1/n, \\ \mathcal{K}_1 & \text{if } t_0 + 1/n < t \le T. \end{cases}$$

We have that

$$\mathrm{V}(\mathcal{B}_n,[0,T]) = \mathrm{V}(\mathcal{C},[0,T]) = d_{\mathscr{H}}(\mathcal{K}_0,\mathcal{K}_1) \quad \forall n \in \mathbb{N},$$

and

$$\lim_{n\to\infty} d_{\mathscr{H}}(\mathcal{B}_n(t),\mathcal{C}(t)) = 0 \qquad \forall t \in [0,T]\,.$$

Hence, as $d_{\mathscr{H}}(\mathcal{B}_n(t),\mathcal{C}(t)) \le d_{\mathscr{H}}(\mathcal{K}_0,\mathcal{K}_1)$, by the dominated convergence theorem we have that

$$\lim_{n\to\infty} \int_0^T d_{\mathscr{H}}(\mathcal{B}_n(t),\mathcal{C}(t))\,\mathrm{d}t = 0.$$

Hence $\mathcal{B}_n \to \mathcal{C}$ as $n \to \infty$ in the strict topology of $BV_L([0,T];\mathscr{C}_\mathbb{R})$. If $y_0 := 0$, then $y_0 \in \mathcal{C}(0) = \mathcal{B}(0) = \mathcal{K}_0$ and, thanks to (3.7)–(3.8) we have

$$\mathsf{S}_0(\mathcal{C})(t) = \begin{cases} 0 & \text{if } 0 \le t \le t_0, \\ \mathrm{Proj}_{\mathcal{K}_1}(0) = 1 & \text{if } t_0 < t \le T. \end{cases} \tag{4.7}$$

It is also easy to check that

$$\mathsf{S}_0(\mathcal{B})(t) = \begin{cases} 2t & \text{if } 0 \le t \le 2/3, \\ 4/3 & \text{if } 2/3 < t \le 1, \end{cases}$$

therefore, using for instance rate independence, it follows that

$$\mathsf{S}_0(\mathcal{B}_n)(t) = \begin{cases} 0 & \text{if } 0 \le t \le t_0, \\ 2n(t-t_0) & \text{if } t_0 < t \le t_0 + 2/3n, \\ 4/3 & \text{if } t_0 + 2/3n < t \le T. \end{cases}$$

Hence, if $z : [0,T] \longrightarrow \mathbb{R}$ is defined by

$$z(t) = \begin{cases} 0 & \text{if } 0 \le t \le t_0, \\ 4/3 & \text{if } t_0 < t \le T, \end{cases} \tag{4.8}$$

we have that $\lim_{n\to\infty} \mathsf{S}_0(\mathcal{B}_n)(t) = z(t)$ for every $t \in [0,T]$, thus, thanks to the dominated convergence theorem, we infer that

$$\mathsf{S}_0(\mathcal{B}_n) \to z \qquad \text{in } L^1(0,T;\mathcal{H}). \tag{4.9}$$

Therefore, as $\mathsf{S}_0(\mathcal{C}) \neq z$, the operator S_0 is not continuous when its domain $BV_L([0,T];\mathcal{H})$ is endowed with the topology induced by the strict convergence and its codomain is endowed with any reasonable topology weaker than or equal to the $L^1(0,T;\mathcal{H})$ topology.

5. Multiple geodesics

The lack of metric BV continuity of the solution operator of the sweeping process in the one dimensional case is somehow connected to the existence of more than one geodesic connecting points (sets) in $\mathscr{C}_\mathbb{R}$. Indeed let us consider the curve $\mathcal{A} : [0,1] \longrightarrow \mathscr{C}_\mathbb{R}$ defined by

$$\mathcal{A}(t) := (1-t)\mathcal{K}_0 + t\mathcal{K}_1, \qquad t \in [0,1]. \tag{5.1}$$

Observe that $\mathcal{A} \in Lip([0,1];\mathscr{C}_\mathbb{R})$ and that $\mathcal{A} \neq \mathcal{B}$. In the case of the one dimensional play operator with characteristic $\mathcal{Z}$, the curve $\mathcal{A}$ corresponds to the input $\mathcal{A}(t) = a(t) - \mathcal{Z}$, $\mathcal{K}_j = \{k_j\}$ for some $k_j \in \mathcal{H}$, $j = 1,2$, with $a(t) = (1-t)k_1 + tk_2$, $t \in [0,1]$. Thanks to [18, Proposition 6.1] or [20, Prop. 1]: we have that $\mathcal{A}$ is a geodesic connecting $\mathcal{K}_0$ and $\mathcal{K}_1$. Let $\mathcal{A}_n \in Lip([0,T];\mathscr{C}_\mathbb{R})$ be the sequence defined (for n large enough) by

$$\mathcal{A}_n(t) := \begin{cases} \mathcal{K}_0 & \text{if } 0 \le t \le t_0, \\ \mathcal{A}(n(t-t_0)) & \text{if } t_0 < t \le t_0 + 1/n, \\ \mathcal{K}_1 & \text{if } t_0 + 1/n < t \le T. \end{cases}$$

We have that

$$\mathrm{V}(\mathcal{A}_n,[0,T]) = \mathrm{V}(\mathcal{C},[0,T]) = d_{\mathscr{H}}(\mathcal{C}_0,\mathcal{C}_1) \quad \forall n \in \mathbb{N},$$

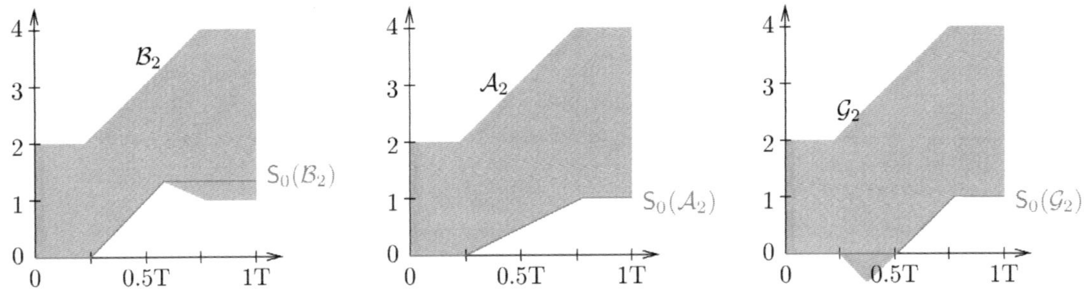

Figure 1. The approximations $\mathcal{B}_2$ (left) and $\mathcal{A}_2$ (center) for $\mathcal{C}$ and the corresponding output of the sweeping process S_0 for $T = 1$ and $t_0 = 0.25$. On the right, the approximation $\mathcal{G}_2$, being analogously derived by using the geodesic $\mathcal{G}$, and the corresponding output of the sweeping process are shown.

and

$$\lim_{n\to\infty} d_{\mathcal{H}}(\mathcal{A}_n(t), \mathcal{C}(t)) = 0 \qquad \forall t \in [0, T].$$

Hence, as $d_{\mathcal{H}}(\mathcal{A}_n(t), \mathcal{C}(t)) \leq d_{\mathcal{H}}(\mathcal{K}_0, \mathcal{K}_1)$, by the dominated convergence theorem we have that

$$\lim_{n\to\infty} \int_0^T d_{\mathcal{H}}(\mathcal{A}_n(t), \mathcal{C}(t))\,\mathrm{d}t = 0.$$

Hence $\mathcal{A}_n \to \mathcal{C}$ as $n \to \infty$ in the strict topology of $BV_L([0,T];\mathscr{C}_{\mathbb{R}})$. Taking again $y_0 := 0$ we find $y_0 \in \mathcal{C}(0) = \mathcal{A}(0) = \mathcal{K}_0$ and

$$\mathsf{S}_0(\mathcal{A})(t) = t,$$

therefore

$$\mathsf{S}_0(\mathcal{A}_n)(t) = \begin{cases} 0 & \text{if } 0 \leq t \leq t_0, \\ n(t - t_0) & \text{if } t_0 < t \leq t_0 + 1/n, \\ 1 & \text{if } t_0 + 1/n < t \leq T. \end{cases}$$

Thus

$$\mathsf{S}_0(\mathcal{A}_n) \to \mathsf{S}_0(\mathcal{C}) \qquad \text{in } L^1(0, T; \mathbb{R}). \tag{5.2}$$

Hence, the sequence $\mathcal{A}_n$ approximating $\mathcal{C}$ by using the geodesic $\mathcal{A}$ does not allow to prove that S_0 is not continuous. [1] But, using this sequences, it follows for **every** extension $\mathsf{E}_{\mathsf{S},0} : BV_L([0,T];\mathscr{C}_{\mathbb{R}}) \to BV_L([0,T];\mathcal{H})$ of the restriction of S_0 to $Lip([0,1];\mathscr{C}_{\mathbb{R}})$ that $\mathsf{E}_{\mathsf{S},0}$ is **not** continuous if the domain is endowed with the strict topology and the codomain with the L^1-topology: If $\mathsf{E}_{\mathsf{S},0}$ were continuous as requested it would follow by the computations for the sequence $\mathcal{A}_n$ that $\mathsf{E}_{\mathsf{S},0}(\mathcal{C}) = \mathsf{S}_0(\mathcal{C})$, while the computations for $\mathcal{B}_n$ would yield that $\mathsf{E}_{\mathsf{S},0}(\mathcal{C}) = z \neq \mathsf{S}_0(\mathcal{C})$.

[1] The same holds for the geodesic $\mathcal{G}$ connecting $\mathcal{K}_0$ and $\mathcal{K}_1$ that is defined as in [20, Theorem 1]:

$$\mathcal{G}(t) := \delta_{t\rho}(\mathcal{K}_1) \cap \delta_{(1-t)\rho}(\mathcal{K}_2)$$

with $\rho := d_{\mathcal{H}}(\mathcal{K}_0, \mathcal{K}_1) = 2$ and $\delta_\lambda(K) := \bigcup_{x\in X} \{y \in X \ : \ d(x,y) \leq \lambda\}$ for $\lambda > 0$ and $K \in \mathscr{C}_{\mathbb{R}}$. We have

$$\mathcal{G}(t) = [-2t, 2+2t] \cap [1 - 2(1-t), 4 + 2(1-t)] = \begin{cases} [-2t, 2t+2] & \text{if } t \leq 1/4, \\ [1 - 2(1-t), 2t+2] & \text{if } t > 1/4. \end{cases}$$

Hence, we see that considering different geodesics connecting two elements in $\mathscr{C}_\mathbb{R}$ yields that there is no extension of the restriction of the sweeping process with the requested continuity properties.

6. Conclusion

We have provided a one dimensional counterexample showing that the solution operator of the sweeping process is not continuous when its domain is endowed with the strict topology of BV and its codomain is endowed with the L^1-topology. For a general Hilbert space $\mathcal{H}$ one obtains a corresponding discontinuity for any BV-formulation of the sweeping process by introducing for any $e \in \mathcal{H}$ with $e \neq 0_H$ the mapping $x \longmapsto \langle x, e\rangle$ and considering the pre-images of the intervals used above intersected with some appropriate bounded convex set. This is at variance with the case of the play operator on $Lip([0,T];\mathcal{H})$ and its continuous BV-extension $\overline{\mathsf{P}}_{z_0}$ (cf. Theorem 3.2 (v)).

Acknowledgment

The authors would like the thank the referees for their valuable comments. The second author is a member of GNAMPA-INDAM, which supported the first author's staying in Torino in April 2013.

References

[1] Aliprantis C D and Border K C, *Infinite Dimensional Analysis - Third Edition*, Springer, Berlin, 2006
[2] Ambrosio L, Fusco N and Pallara D, *Functions of Bounded Variation and Free Discontinuity Problems*, Oxford Mathematical Monographs, Clarendon Press, Oxford, 2000.
[3] Brezis H, *Operateurs maximaux monotones et semi-groupes de contractions dans les espaces de Hilbert*, North-Holland Mathematical Studies, Vol. 5, North-Holland Publishing Company, Amsterdam, 1973.
[4] Brokate M and Sprekels J, *Hysteresis and Phase Transitions*, Applied Mathematical Sciences, Vol. 121, Springer-Verlag, New York, 1996.
[5] Dinculeanu D, *Vector Measures*, International Series of Monographs in Pure and Applied Mathematics, Vol. 95, Pergamon Press, Berlin, 1967.
[6] Federer H, *Geometric Measure Theory*, Springer-Verlag, Berlin-Heidelberg, 1969.
[7] Krasnosel'skiĭ M A and Pokrovskiĭ A V, *Systems with Hysteresis*, Springer-Verlag, Berlin Heidelberg, 1989.
[8] Krejčí P, *Hysteresis, Convexity and Dissipation in Hyperbolic Equations*, Gakuto International Series Mathematical Sciences and Applications, Vol. 8, Gakkōtosho, Tokyo, 1997.
[9] Krejčí P and Laurençot P, *Generalized variational inequalities*, J. Convex Anal. **9** (2002), 159–183.
[10] Krejčí P and Recupero V, *Comparing BV solutions of rate independent processes*, J. Convex Anal., **21** (2014), 121–146.
[11] Mielke A, *Evolution in rate-independent systems*, In "Handbook of Differential Equations, Evolutionary Equations, vol. 2", C. Dafermos, E. Feireisl editors, Elsevier, 2005, 461–559.
[12] Monteiro Marques M D P, *Differential Inclusions in Nonsmooth Mechanical Problems - Shocks and Dry Friction*, Birkhauser Verlag, Basel, 1993.
[13] Moreau J J, *Rafle par un convexe variable, II*, Sém. d'Anal. Convexe, Montpellier, Vol. 2 (1972), Exposé. No. 3, 36 pp.
[14] Moreau J J, *Evolution problem associated with a moving convex set in a Hilbert space*, J. Differential Equations **26** (1977), 347–374.
[15] Recupero V, *BV-extension of rate independent operators*, Math. Nachr., **282** (2009), 86–98.
[16] Recupero V, *Sobolev and strict continuity of general hysteresis operators*, Math. Methods Appl. Sci., **32** (2009), 2003–2018.
[17] Recupero V, *BV solutions of rate independent variational inequalities*, Ann. Sc. Norm. Super. Pisa Cl. Sc. (5), **10** (2011), 269–315.
[18] Recupero V, *A continuity method for sweeping processes*, J. Differential Equations **251** (2011) 2125-2142.
[19] Recupero V, *Hysteresis operators in metric spaces*, Discrete Contin. Dyn. Syst. Ser. S, **8** (2015) 773-792.
[20] Serra J, *Hausdorff distances and interpolations*, Mathematical morphology and its applications to image and signal processing (Amsterdam, 1998), 107–114, Comput. Imaging Vision, 12, Kluwer Acad. Publ., Dordrecht, 1998. 68U10 (54E35)

[21] Visintin A, *Differential Models of Hysteresis*, Applied Mathematical Sciences, Vol. 111, Springer-Verlag, Berlin Heidelberg, 1994.

Multi-scale problem in the model of RNA virus evolution

Andrei Korobeinikov, Aleksei Archibasov, Vladimir Sobolev

Centre de Recerca Matemática, Campus de Bellaterra, Edifici C, 08193 Barcelona, Spain; Departament de Matemàtiques, Universitat Autònoma de Barcelona, Campus de Bellaterra, Edifici C, 08193 Barcelona, Spain
Department of Applied Mathematics Samara State Aerospace University (SSAU), 34, Moskovskoye shosse, Samara 443086, Russia
Department of Technical Cybernetics Samara State Aerospace University (SSAU), 34, Moskovskoye shosse, Samara 443086, Russia

E-mail: akorobeinikov@crm.cat, aarchibasov@gmail.com, hsablem@gmail.com

Abstract. A mathematical or computational model in evolutionary biology should necessary combine several comparatively fast processes, which actually drive natural selection and evolution, with a very slow process of evolution. As a result, several very different time scales are simultaneously present in the model; this makes its analytical study an extremely difficult task. However, the significant difference of the time scales implies the existence of a possibility of the model order reduction through a process of time separation. In this paper we conduct the procedure of model order reduction for a reasonably simple model of RNA virus evolution reducing the original system of three integro-partial derivative equations to a single equation. Computations confirm that there is a good fit between the results for the original and reduced models.

1. Introduction

Due to very high mutation and replication rates combined with high recombination abilities, RNA viruses are able to evolve very fast. This high evolvability makes RNA viruses a convenient experimental model in evolutionary biology and, at the same time, presents a serious problem, as viral evolution is probably the most important single factor accountable for the emergence of new pathogens and the development of drug resistance by existing ones.

Despite its apparent relevance, so far the development of mathematical or computational models of viral evolution which can be used in combination with experimental studies in evolutionary biology, remains a challenge even for such a simple object as RNA virus. In order to be useful in biological research, a mathematical model should be based on explicitly stated basic principles, postulates and hypotheses and allow a transparent interpretation of both results and parameters. In other words, a useful model should be essentially mechanistic. This implies that in evolutionary biology such a model must include a combination of factors responsible for natural selection (which drives evolution) and a mechanism for describing evolution. However, the factors generating the selection pressure act on a time scale comparable with an average generation time, or a life span of a single entity, whereas evolution is a comparatively slow process of accumulation of small changes that progresses through many generations to be evident. As a result, the model unavoidably includes multiple time scales which differ by several orders of magnitude. Computational or analytical handling of such models usually is a highly non-trivial task. In the case of viral evolution this discrepancy is further complicated by the fact that a typical generation times of virus and the hosts differ by orders of magnitude as well.

Besides phenomenological models, such as models suggested by Tsimring, Levin and Kessler [1] or Sasaki [2] and Sasaki and Haraguchi [3, 4], are free of the mentioned drawbacks. However, the use of phenomenological models as a compliment to experimental research is usually questionable, as such models are rather illustrative than explanatory, and interpretation of models' parameters and obtained results in connection with an experimental layout is problematic.

These arguments imply that an ideal model should be mechanistic, include a number of properties of a real–life system and, at the same time, it should be sufficiently simple to allow analysis.

A relatively simple mathematical model of RNA virus evolution was recently suggested by Korobeinikov and Dempsey [5]. This mechanistic model is an extension of Nowak–May HIV model [6], where viral phenotypes (strains) are assumed to be distributed in a continuous phenotype space, and random mutations are described by dispersion. For the sake of simplicity, Korobeinikov and Dempsey formulated the model on a basis of the 2–dimensional Wodarz–Christensen–Thomsen, or Wodarz, model [7]. The latter is a reduction of the original 3–dimensional Nowak–May model under an assumption that the population of free virus particles is proportional to the infected cells population. However, an analytical study of even this comparatively simple model still is a highly challenging task, and it can be expected that for a more complex model, which includes factors, such as immune response or a therapy, the analysis would be even more difficult.

As we mentioned above, to a large extent the model complexity arises as a result of the presence of multiple processes progressing at very different time scales. Thus, the original Nowak–May model combines three processes and three time scales, namely (i) the proliferation of the uninfected target cells, (ii) the infectious process and infected cells' life cycle, and, finally, (iii) free virus life cycle. In the Wodarz model, which was used as a basis for model in [5], one of these processes, namely the free virus life cycle, is omitted and replaced by the assumption that the free virus population is proportional to the infected cells population. The latter makes the model slightly simpler. However, model in [5] includes evolution which occurs on a comparatively slow timescale. Accordingly, the model is a slow–fast system.

The considerable differences of the timescales, which make the model difficult for both computer simulation and analysis, at the same time, indicate that the model can be simplified using the process known as time scales separation (for instance, see [8, 9, 10]). In this paper we carry out such a scale separation. As a result, the model RNA virus evolution in [5] is reduced to a single equation. Simulations demonstrate that apart from a comparatively short transition period there is a good fit of results for both, the full and the reduced models.

We would like to stress that the result is not limited by the specific model introduced in [5]. As the presence of multiple time scales which differ by a several orders of magnitude is typical for mechanistic models in evolutionary biology, the same technique can be applied to any such model.

2. The model

As a basis for modelling, we employ the Nowak–May model of HIV dynamics [6]. This model describes interactions of three populations, namely uninfected target cells (for HIV, the target cells are T helper, or Th cells), infected target cells, and free virus particles, of concentrations $u(t)$, $V(t)$ and $X(t)$, respectively. The model postulates that there is a continuous influx of the target cells (from the thymus, where they mature) at a rate b, that free virus particles infect target cells at a rate $\alpha X u$, that the infected cells produce free virus particles at a per capita rate k, and that average life spans of the uninfected cells, infected cells and the free virus particles are $1/q$, $1/m$ and $1/c$, respectively. Under these assumptions the model equations are

$$\begin{aligned}\frac{du(t)}{dt} &= b - \alpha\, X(t)u(t) - q\, u(t),\\ \frac{dV(t)}{dt} &= \alpha\, X(t)u(t) - m\, V(t),\\ \frac{dX(t)}{dt} &= k\, V(t) - c\, X(t).\end{aligned} \tag{1}$$

The global qualitative behavior of this model is well studied and is known to be completely determined by the basic reproduction number of the virus (or that of infected cells, as for this model these numbers coincide) $R_0 = \alpha bk/cmq$ [11, 12, 13, 14].

Let us assume now that multiple viral strains are possible and are distributed in a continuous phenotype space. For model (1) a viral variant is characterized by parameters α, k, c and m, and hence it could be assumed that the phenotype space is 4–dimensional. However, it is known (and we will demonstrate this further) that for the Nowak-May model (1) these parameters are not independent: as we mentioned, for this model the viral dynamics is completely characterized by the basic reproduction number R_0 which usually serves as a measure of the Darwinian fitness of the virus. In other words, in the framework of the Nowak-May model a viral phenotype is uniquely characterized by the single number R_0, and, therefore, it suffices to consider a 1–dimensional phenotype space $M = \{s \in (0, \infty)\}$. Then fitness R_0 is a function of variable s. (The graph of this function is usually referred to as a "fitness landscape" .) The variables $v(t, s)$ and $x(t, s)$ are strain-specific and hence are the density distributions of the infected cells population and the virus in the phenotype space, whereas the total infected cells population and the total viral load are $V(t) = \int_0^\infty v(t, s)ds$ and $X(t) = \int_0^\infty x(t, s)ds$, respectively. New strains emerge as a result of random mutations which in a continuous phenotype space can be modelled by dispersion.

In the Nowak–May model coefficients α, m, k and c are attributed to a phenotype, and hence these are function of s. We assume that $\alpha = \alpha(s) = \alpha^0\alpha_0(s)$, $m = m(s) = m^0 m_0(s)$, $k = k(s) = k^0 k_0(s)$ and $c = c(s) = c^0 c_0(s)$, where α^0 ,m^0, k^0 and c^0 are coefficients of a wild strain (or a phenotype at the beginning of evolution), $\alpha_0(s)$, $m_0(s)$, $k_0(s)$ and $c_0(s)$ are non–dimensional functions and $m_0(s), k_0(s), c_0(s) = 1$ for this wild phenotype. These assumptions lead to the following system of equations:

$$\begin{aligned}\frac{du(t)}{dt} &= b - u(t)\int_0^\infty \alpha(s)\, x(t, s)\, ds - q\, u(t),\\ \frac{\partial v(t, s)}{\partial t} &= -m(s)\, v(t, s) + \alpha(s)\, u(t)\, x(t, s) + \mu\, \frac{\partial^2 v(t, s)}{\partial s^2},\\ \frac{\partial x(t, s)}{\partial t} &= k(s)\, v(t, s) - c(s)\, x(t, s).\end{aligned} \tag{2}$$

Note that coefficients b and q are characteristics of target cells, and hence these remain constant.

This system of equations should be complemented by initial and boundary conditions. At $s = +\infty$ the natural boundary conditions for $v(t, s)$ and $x(t, s)$ are zeros. At $s = 0$ for convenience we use no-flux conditions $\dfrac{\partial v(t, 0)}{\partial s} = 0$. Initial conditions are $u(0) = u^0$, $v(0, s) = v^0(s)$, $x(0, s) = x^0(s)$, where $u^0, v^0(s), x^0(s) > 0$. Here time t is measured in days and variable s is non–dimensional. Furthermore, $u(t)$ is measured in $cells \cdot mm^{-3}$; $v(t,s)$ and $x(t,s)$ are the densities of cells and free virus particles in the phenotype space, and hence these are measured in $cells \cdot mm^{-3}$ and $virions \cdot mm^{-3}$; cell production rate b is measured in $cells \cdot mm^{-3} \cdot day^{-1}$; per capita mortality rates q, m, k and c are measured in day^{-1}, infectivity α is in $mm^3 \cdot virions^{-1} \cdot day^{-1}$, and dispersion coefficient μ is in day^{-1}.

A model of such type (based on the 2–dimensional Wodarz model, which is, in turn, a reduction of the 3–dimensional Nowak–May model) was introduced in [5], and simulations with this model revealed a number of practically relevant outcomes. In particular, the model demonstrates the behavior which closely resembles the observed development of HIV infection. However, an apparent drawback of this model is that even in its present form its analytical study is a challenging task. A further development of the model, which could involve an incorporation of factors such as immune response and a therapy, would make the model more complicated and its analysis even more difficult. The complexity of this model is a result of an interaction of multiple processes with very different time scales. Thus, the original Nowak–May model (1) combines three processes and three time scales, namely (i) the proliferation of uninfected target cells, (ii) the process of infection and infected cells' life cycle, and, finally, (iii) the free virus particles life cycle. Compared with the Nowak–May model, model (2) includes the process of evolution, which, in comparison, is a very slow process, and hence model (2) is a slow-fast system. The presence of four considerably different timescales indicates that model (2) can be significantly simplified if we separate the timescales.

3. Timescales separation

Let us denote $t = T\bar{t}, s = S\bar{s}$ and

$$u\left(T\bar{t}\right) = U\bar{u}(\bar{t}), \quad v(T\bar{t}, S\bar{s}) = \tilde{V}\bar{v}\left(\bar{t},\bar{s}\right), \quad x(T\bar{t}, S\bar{s}) = \tilde{X}\bar{x}\left(\bar{t},\bar{s}\right),$$

and assume that

$$T = \frac{1}{\mu}, \quad S = 1, \quad U = \frac{b}{q}, \quad \tilde{X} = \frac{k^0}{c^0}\tilde{V}, \quad \tilde{V} = 1\ (cells/s).$$

Substituting these into system (2), we obtain a non–dimensional system

$$\varepsilon\frac{d\bar{u}}{d\bar{t}} = 1 - \int_0^\infty \frac{1}{q}\alpha(\bar{s})\frac{k^0}{c^0}\bar{u}\bar{x}d\bar{s} - \bar{u}, \tag{3}$$

$$\frac{\partial \bar{v}}{\partial \bar{t}} = \alpha(\bar{s})\frac{bk^0}{\mu q c^0}\bar{u}\bar{x} - \frac{m(\bar{s})}{\mu}\bar{v} + \frac{\partial^2 \bar{v}}{\partial \bar{s}^2}, \tag{4}$$

$$\varepsilon\nu\frac{\partial \bar{x}}{\partial \bar{t}} = \bar{k}(\bar{s})\bar{v} - \bar{c}(\bar{s})\bar{x}, \tag{5}$$

where $\varepsilon = \mu/q$ and $\nu = q/c^0$. Initial and boundary conditions for this system are, respectively,

$$\bar{u}(0) = \frac{q}{b}u(0), \quad \bar{v}\left(0,\bar{s}\right) = \frac{1}{\tilde{V}}v(0,s), \quad \bar{x}\left(0,\bar{s}\right) = \frac{1}{\tilde{X}}x(0,s)$$

and

$$\frac{\partial \bar{v}(\bar{t},0)}{\partial \bar{s}} = 0, \quad \lim_{\bar{s}\to\infty} \bar{v}(\bar{t},\bar{s}) = 0.$$

Both numbers, $\varepsilon = \mu/q$ and $\nu = q/c^0$, are small parameters; for instance, for parameters, which were used in [5] and which correspond to a real life HIV case from [15], $\nu \approx 2 \cdot 10^{-3}$ and

ε ranges from $5 \cdot 10^{-5}$ to $5 \cdot 10^{-3}$. That is, system (3)–(5) is a singularly perturbed ("slow-fast") system, and its size can be reduced using the scale separation technique [8]. This procedure corresponds to assumptions that (i) there is a stable slow manifold such that all solutions converge to this manifold, and hence (ii) the values of fast variables $u(t)$ and $x(t,s)$ converge to quasi-equilibrium values which slowly vary as the system progresses along the slow manifold [9].

Let $\nu = 0$, then equality

$$\bar{k}(\bar{s})\bar{v} = \bar{c}(\bar{s})\bar{x}$$

holds. Substituting this equality into (3) and (4), we obtain equations

$$\varepsilon \frac{d\bar{u}}{d\bar{t}} = 1 - \int_0^\infty \frac{1}{q}\beta(\bar{s})\bar{u}\bar{v}d\bar{s} - \bar{u}, \tag{6}$$

$$\frac{\partial \bar{v}}{\partial \bar{t}} = \frac{b}{\mu q}\beta(\bar{s})\bar{u}\bar{v} - \frac{m(\bar{s})}{\mu}\bar{v} + \frac{\partial^2 \bar{v}}{\partial \bar{s}^2}, \tag{7}$$

where $\beta(\bar{s}) = \alpha(\bar{s})k(\bar{s})/c(\bar{s})$. System (6), (7) was suggested in [5].

Now let $\varepsilon = 0$, then

$$\bar{u} = \frac{1}{1 + \int\limits_0^\infty \frac{1}{q}\beta(\bar{s})\bar{v}d\bar{s}} = \frac{1}{1+\bar{\lambda}(t)},$$

where $\bar{\lambda}(t) = \int\limits_0^\infty \frac{1}{q}\beta(\bar{s})\bar{v}d\bar{s}$ is the non-dimensional infection force. (Please recall that the dimensional infection force is usually defined as $\lambda(t) = \int\limits_0^\infty \beta(s)v(t,s)ds$.) Substituting this equality into (7), we finally obtain a single equation

$$\frac{\partial \bar{v}}{\partial \bar{t}} = \bar{m}(\bar{s})\left(\frac{R_0(\bar{s})}{1+\bar{\lambda}} - 1\right)\bar{v} + \frac{\partial^2 \bar{v}}{\partial \bar{s}^2}, \tag{8}$$

where $R_0(\bar{s}) = \dfrac{b\alpha(\bar{s})k(\bar{s})}{qm(\bar{s})c(\bar{s})}$, $\quad \bar{m}(\bar{s}) = \dfrac{m(\bar{s})}{\mu}$.

When $\lambda(t) \ll q$, that is when the reduction of the uninfected population due to infection is small, equation (8) can be further simplified: in this case we can employ the Taylor expansion and use the approximate equality

$$\frac{1}{1+\bar{\lambda}(t)} \approx 1 - \bar{\lambda}(t).$$

Then

$$\frac{\partial \bar{v}}{\partial \bar{t}} = \bar{m}(\bar{s})\left[R_0(s)(1-\bar{\lambda}) - 1\right]\bar{v} + \frac{\partial^2 \bar{v}}{\partial \bar{s}^2}.$$

4. Numerical simulations

Following [5], and for the sake of simplicity, we assume that the diversity of viral phenotypes is described by a single parameter $\alpha(s)$ which can be interpreted as the efficiency of a single virus particle in infecting a target cell; the other parameters are assumed to be constant and the same for all genotypes. To some extent this assumption is justified by the fact that for the Nowak–May model genotypes fitness is defined by a single number $R_0 = \frac{\alpha bk}{cmq}$. Under this hypothesis, and taking into consideration that R_0 is a function of all six parameters, it should make no difference which of these six parameters is varying in the phenotype space. The choice of $\alpha(s)$ is just a matter of convenience, because the direct dependence of R_0 on α simplifies the notation.

Results of numerical analysis are presented at the following figures. (For convenience of comparison, results in these figures are shown for physical dimensional variables.) For these simulations we use the following values for the system parameters: $b = 20\ cells \cdot mm^{-3} \cdot day^{-1}$, $m = 0.8\ day^{-1}$, $q = 0.02\ day^{-1}$, $k = 10^4\ day^{-1}$ and $c = 8\ day^{-1}$. These parameters correspond to patient 2 in [15]. For simplicity we assume a linear landscape, postulating that $\alpha(s) = as$. Here $a = 8 \cdot 10^{-7}\ mm^3 \cdot cells^{-1} \cdot day^{-1}$; this implies that $R_0 = 1.25s$. All simulations are for $\mu = 10^{-6}\ day^{-1}$.

Simulations demonstrate an excellent fit of results obtained for system (6), (7) and equation (8).

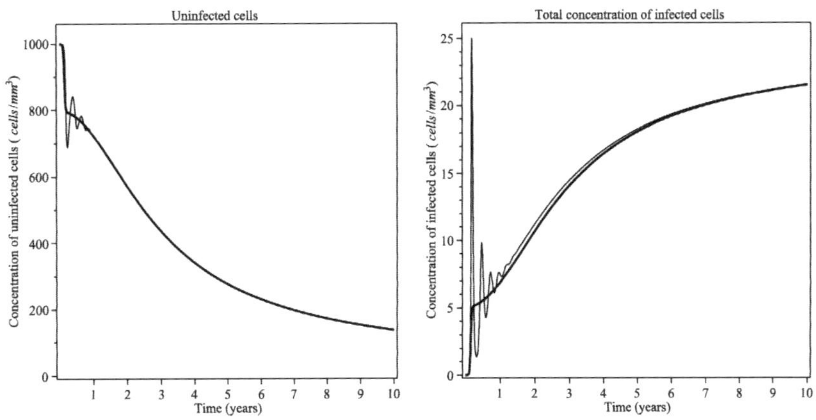

Figure 1. Uninfected cells concentration $u(t)$ and total infected cells concentration $V(t)$ for system (6), (7) (thin line) and equation (8) (bold line).

Figure 1 shows that results for these two systems perfectly coincide everywhere apart from a comparatively short transition period. (We have to remind that equation (8) is unable to describe the transition dynamics at all.) If the initial conditions are taken on or near the slow manifold, then the results for these two systems coincide everywhere (see Fig. 2).

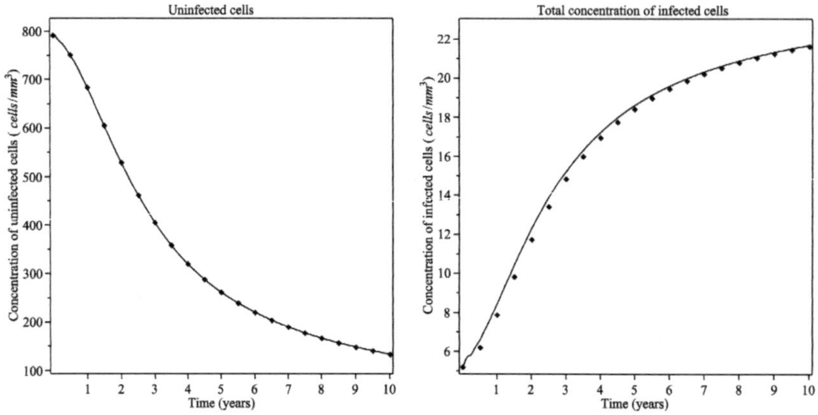

Figure 2. Solutions of system (6), (7) (thin line) and equation (8) (markers), when initial condition is taken on the slow manifold.

However, a match of results for systems (3)–(5) and (6), (7) (and hence for equation (8)) is only qualitative. Figure 3 demonstrates that system (3)–(5) exhibits a slower evolution, producing outcomes which are delayed in time, compared with these for system (6), (7), or equation (8).

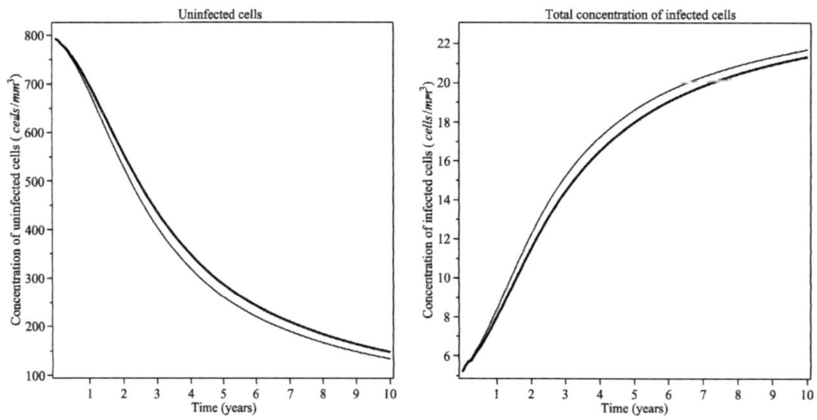

Figure 3. Uninfected cells concentration $u(t)$ and total infected cells concentration $V(t)$ for system (3)–(5) (bold line) and (6), (7) (thin line).

We suppose that this effect is caused by a continuous accumulation of a small delay, caused by the third equation of the Nowak–May model (2), over a long period of evolution.

5. Summary

Due to the fact that a mathematical model in evolutionary biology should necessary combine the factors, which characterize natural selection and drive evolution, and which acts on a comparatively fast time scale, and a very slow process of evolution itself, such a model is complex and its analysis is difficult. However, a combination of several time scales within a model suggests that such a model can be significantly simplified using the scale separation techniques. For ODE such technique is based on the famous Tikhonov's theorem. The mathematical models in evolutionary biology are usually formulated as integro–differential equations and PDE. However, with some reservations the same concepts can be applied to these equations as well (for instance, see [10, 16, 17]).

In this paper we execute the scale separation for a RNA virus evolution model which is based on the 3–dimensional Nowak–May model. As a result of this procedure, the original system of three integro–partial differential equations is reduced to a single integro–partial differential equation, describing the dynamics of infected target cells $v(t, s)$. The other variables, namely the uninfected cells $u(t)$ and free virus particles $x(t, s)$, are expressed as explicit functions of variable $v(t, s)$. Numerical simulations show an acceptable fit of the solution of the original and the reduced models.

A specific feature of this particular model, which is based on the Nowak–May model specifically tailored to describe the HIV dynamics, is that target cells are assumed to be immigrating from the thymus at a constant rate. However, the same concept and the same techniques can be applied to a model of evolution based on any other model of virus dynamics.

Acknowledgments

A.K. is supported by the Ministry of Science and Innovation of Spain via Ramón y Cajal Fellowship RYC-2011-08061 and grant MTM2011-29342, by CONACYT (Mexico) via grant N219614, and by AGAUR (Generalitat de Catalunya) via grant 2014SGR-1307.
A.A. is supported by the Russian Foundation for Basic Research via grant 14-01-97018-p and the Ministry of education and science of the Russian Federation in the framework of the implementation of Program of increasing the competitiveness of SSAU for 2013–2020 years.
V.S. is supported by the Russian Foundation for Basic Research via grant 13–01–97002-p and the Ministry of education and science of the Russian Federation in the framework of the implementation of Program of increasing the competitiveness of SSAU for 2013–2020 years.

References

[1] Tsimring L S, Levine H and Kessler D A 1996 RNA virus evolution via a fitness-space model *Phys. Rev. Lett.* **76** (23) 4440–4443 (doi: 10.1103/PhysRevLett.76.4440)
[2] Sasaki A 1994 Evolution of antigenic drift/switching: continuously evading pathogens *J. Theor. Biol.* **168** 291-308 (doi:10.1006/jtbi.1994.1110)
[3] Haraguchi Y and Sasaki A 1997 Evolutionary pattern of intra–host pathogen antigenic drift: effect of crossreactivity in immune response. *Phil. Trans. R. Soc.* B **352** 11-20 (doi:10.1098/rstb.1997.0002)
[4] Sasaki A and Haraguchi Y 2000 Antigenic drift of viruses within a host: a finite site model with demographic stochasticity *J. Mol. Evol.* **51** 245-255
[5] Korobeinikov A and Dempsey C 2014 A continuous phenotype space model of RNA virus evolution within a host *Math. Biosci. Eng.* **11**(4) 919–927
[6] Nowak M A and May R M 2000 *Virus dynamics: Mathematical principles of Immunology and Virology* (New York: Oxford University Press)
[7] Wodarz D, Christensen J P and Thomsen A R 2002 The importance of lytic and nonlytic immune responses in viral infections *TRENDS in Immunology* **23** 194–200
[8] Khalil H K 1987 Stability analysis of nonlinear multiparameter singularly perturbed systems *IEEE Trans. Aut. Control* AC **32** 260–263
[9] Mortell M P, O'Malley R E, Pokrovskii A and Sobolev V A 2005 *Singular Perturbation and Hysteresis* (Philadelphia: SIAM)
[10] Vasilieva A B, Butuzov V F, Kalachev L V 1995 it The Boundary Function Method for Singular Perturbation Problems (Philadelphia: SIAM)
[11] Korobeinikov A 2004 Global properties of basic virus dynamics models *Bull. Math. Biol.* **66**(4) 879–883
[12] Korobeinikov A 2009 Global asymptotic properties of virus dynamics models with dose dependent parasite reproduction and virulence, and nonlinear incidence rate *Math. Med. Biol.* **26** 225–239
[13] Korobeinikov A 2009 Stability of ecosystem: Global properties of a general prey–predator model *Math. Med. Biol.* **26** 309–321
[14] Vargas-De-León C and Korobeinikov A 2013 Global stability of a population dynamics model with inhibition and negative feedback *Math. Med. Biol.* **30** 65-72

[15] Stafford M A , Corey L, Cao Y , Daar E S, Ho D D, Perlson A S 2000 Modeling plasma virus concentration during primary HIV infection *J. Theor. Biol.* **203** 285-301
[16] Butuzov V F, Nefedov N N, Recke L, Schnieder K R 2012 Global region of attraction of a periodic solution to a singularly perturbed parabolic problem *Applicable Analysis* **91** 1265–1277
[17] Nefedov N N, Nikitin A G 2007 The Cauchy problem for a singularly perturbed integro–differential Fredholm equation *Computational Mathematics and Mathematical Physics* **47** 629–637

MURPHYS-HSFS-2014 IOP Publishing
Journal of Physics: Conference Series **727** (2016) 012008 doi:10.1088/1742-6596/727/1/012008

A remark on geometric desingularization of a non-hyperbolic point using hyperbolic space

Christian Kuehn
Institute for Analysis and Scientific Computing, Vienna University of Technology, 1040 Vienna, Austria
E-mail: ck274@cornell.edu

Abstract. A steady state (or equilibrium point) of a dynamical system is hyperbolic if the Jacobian at the steady state has no eigenvalues with zero real parts. In this case, the linearized system does qualitatively capture the dynamics in a small neighborhood of the hyperbolic steady state. However, one is often forced to consider non-hyperbolic steady states, for example in the context of bifurcation theory. A geometric technique to desingularize non-hyperbolic points is the blow-up method. The classical case of the method is motivated by desingularization techniques arising in algebraic geometry. The idea is to blow up the steady state to a sphere or a cylinder. In the blown-up space, one is then often able to gain additional hyperbolicity at steady states. The method has also turned out to be a key tool to desingularize multiple time scale dynamical systems with singularities. In this paper, we discuss an explicit example of the blow-up method where we replace the sphere in the blow-up by hyperbolic space. It is shown that the calculations work in the hyperbolic space case as for the spherical case. This approach may be even slightly more convenient if one wants to work with directional charts. Hence, it is demonstrated that the sphere should be viewed as an auxiliary object in the blow-up construction. Other smooth manifolds are also natural candidates to be inserted at steady states. Furthermore, we conjecture several problems where replacing the sphere could be particularly useful, i.e., in the context of singularities of geometric flows, for avoiding compactification, and generating 'interior' steady states.

1. Introduction

Consider an ordinary differential equation (ODE) given by

$$\frac{dz}{dt} = z' = f(z), \tag{1}$$

where $z = z(t) \in \mathbb{R}^N$, $N \in \mathbb{N}$, $t \in \mathbb{R}$ and $f : \mathbb{R}^N \to \mathbb{R}^N$ is assumed to be sufficiently smooth. Suppose $z^* \in \mathbb{R}^N$ is a steady state (or equilibrium point) of (1), i.e., $f(z^*) = 0$. Using a translation of coordinates, if necessary, we may assume for the following analysis without loss of generality that $z^* = 0 := (0, 0, \ldots, 0) \in \mathbb{R}^N$. The first standard calculation for steady states is to consider the linearized system in a neighborhood of the steady state

$$Z' = [\mathrm{D}f(0)]Z, \tag{2}$$

where $Z \in \mathbb{R}^N$ and $\mathrm{D}f(0) \in \mathbb{R}^{N\times N}$ denotes the total derivative of f evaluated at $z = 0$. It is also common to refer to $\mathrm{D}f(0)$ as the Jacobian matrix or simply the Jacobian. Let λ_n for

$n \in \{1, 2, \ldots, N\}$ denote the eigenvalues of $\mathrm{D}f(0)$. If the eigenvalues have no zero real parts, $\mathrm{Re}(\lambda_n) \neq 0$ for all n, then the steady state $z^* = 0$ is called hyperbolic. The Hartman-Grobman Theorem (see e.g. [22, p.120-121]) implies that in a neighborhood of a hyperbolic steady state, the flows generated by (1) and (2) are topologically conjugate. For most practical purposes this implies that we may just use the linear ODE (2) to study the dynamics near $z^* = 0$.

However, non-hyperbolic points are unavoidable if we want to analyze bifurcation points [7, 18]. The linearization approach breaks down and one has to carefully consider the influence of nonlinear terms. One possible technique that can be very successful in this context is geometric desingularization; see e.g. [4, p.67-70] for a particular example or [3] for general planar singularities. We are going to introduce geometric desingularization via the blow-up method in more detail in Section 2.

The main geometric idea of the method arose in algebraic geometry in the context of desingularization of algebraic varieties [9, p.29], where one replaces certain singular points by projective space. The resulting variety either has no singular points anymore or one can try to repeat the blow-up. Under certain conditions one may indeed reach a complete desingularization as stated in the celebrated Hironaka Theorem [10, 11].

In the context of ODEs, the classical strategy involves using a spherical blow-up as one works in real space and not in the context of (complex) projective space. The key difference to the algebraic geometry blow-up is that one also has to keep track of the dynamics on the blown-up space. There has been a tremendous amount of work on using the blow-up technique for planar ODEs [3, 4, 2], canard solutions [6, 12, 14, 24], traveling wave problems [20, 5] and a large variety of other problems in the theory of multiple time scale dynamical systems [19, 8, 13, 15, 17]. For fast-slow dynamical systems of the form

$$\begin{aligned} x' &= f_1(x, y, \varepsilon), \\ y' &= \varepsilon g(x, y, \varepsilon), \end{aligned} \tag{3}$$

with $0 < \varepsilon \ll 1$ and $(x, y) \in \mathbb{R}^{m+n}$, it is crucial to observe [6] that the blow-up is frequently very helpful. Indeed, consider the formal limit $\varepsilon \to 0$ in (3) which yields

$$\begin{aligned} x' &= f_1(x, y, 0), \\ y' &= 0. \end{aligned} \tag{4}$$

Linearizing around a steady state, say $(x^*, y^*) = (0, 0) =: 0$ of the fast subsystem (4) means computing the total derivative matrix $D_x f(0) \in \mathbb{R}^{m \times m}$, which is generically not hyperbolic due to the presence of the free parameters (or frozen slow variables) $y \in \mathbb{R}^n$. Therefore, one has to deal with non-hyperbolic points for fast-slow systems [6, 16] and re-write them in the form

$$\begin{aligned} x' &= f_1(x, y, \varepsilon), \\ y' &= \varepsilon g(x, y, \varepsilon), \\ \varepsilon' &= 0. \end{aligned} \tag{5}$$

Setting $z = (x, y, \varepsilon)$, $f_2(x, y, \varepsilon) = \varepsilon g(x, y, \varepsilon)$, $f_3(x, y, \varepsilon) = 0$, and $f := (f_1, f_2, f_3)^\top$ yields precisely a problem of the form (1) with a non-hyperbolic point at the origin.

Using a spherical, or cylindrical, space is currently the standard choice to desingularize non-hyperbolic steady states of ODEs. However, there seems to be no apparent reason why other manifolds could not function equally well, or even better. In this paper, we investigate this idea in more detail and consider a simple example to illustrate the main idea. The spherical case is discussed in Section 2, which is also a fully self-contained introduction to the blow-up method.

In Section 3 we replace the sphere by hyperbolic space, i.e. by using a manifold with constant negative curvature. We emphasize that the word 'hyperbolic' is then used in two distinct ways: (1) for the dynamical type of a steady state and (2) for a smooth manifold which replaces the sphere in the blown-up space. The results in Section 3 confirm the intuition that using a spherical blown-up space is not crucial and hyperbolic space works also for geometric desingularization in the example. This indicates that one should be open-minded about trying to use different manifolds for geometric desingularization. Further directions for this new approach are sketched in Section 4. Three areas are identified, where using manifolds different from spheres could be beneficial.

2. Spherical Blow-Up

In this section a basic test example for the blow-up method is reviewed from [4] and more explicit calculations for this example are provided. The spherical blow-up is constructed in this context, which leads to a geometric desingularization of the problem.

Consider the following planar ODE [4] for $z(t) = (x(t), y(t)) \in \mathbb{R}^2$

$$\begin{array}{ccccccc} \frac{dx}{dt} & = & x' & = & ax^2 - 2xy & =: & f_1(x,y), \\ \frac{dy}{dt} & = & y' & = & y^2 - axy & =: & f_2(x,y), \end{array} \tag{6}$$

where $a > 0$ is a positive parameter, we abbreviate $(x, y) = (x(t), y(t))$ and we denote the vector field by $f := (f_1, f_2)^\top$, where $(\cdot)^\top$ denotes the transpose.

Remark: We note that we are not interested here in trying to solve the problem (6) directly but to use it to illustrate how the blow-up method can resolve the fine-scale geometry of the problem.

We may view the vector field f as a smooth section into the tangent bundle $f : \mathbb{R}^2 \to \mathrm{T}\mathbb{R}^2$. If $p \in \mathbb{R}^2$ is a given point, then we shall usually employ the natural identification of the tangent space $\mathrm{T}_p\mathbb{R}^2 \cong \mathbb{R}^2$.

Observe that $(x, y) = (0, 0) =: 0$ is a steady state, i.e. $f_1(0) = 0 = f_2(0)$, for (6). It is straightforward to compute the linearized system $Z = (X, Y) \in \mathbb{R}^2$ at the origin

$$\begin{aligned} \begin{pmatrix} X' \\ Y' \end{pmatrix} = [\mathrm{D}f(0)] \begin{pmatrix} X \\ Y \end{pmatrix} &= \left.\begin{pmatrix} 2ax - 2y & -2x \\ -ay & 2y - ax \end{pmatrix}\right|_{\{x=0=y\}} \begin{pmatrix} X \\ Y \end{pmatrix} \\ &= \begin{pmatrix} 0 & 0 \\ 0 & 0 \end{pmatrix} \begin{pmatrix} X \\ Y \end{pmatrix}, \end{aligned}$$

where we shall always employ capital variables $Z = (X, Y) \in \mathbb{R}^2$ to emphasize when we work with a linearized problem. We see that the origin is a non-hyperbolic steady state since $\mathrm{D}f(0)$ has two zero eigenvalues; see also Figure 1(a). Hence, further analysis is required and the blow-up method provides one approach to understand the dynamics.

For planar vector fields, the classical approach of the blow-up method is to use a transformation which replaces the point p with a (unit) circle

$$\mathcal{S}^1 = \{(x,y) \in \mathbb{R}^2 : x^2 + y^2 = 1\} = \{(x,y) \in \mathbb{R}^2 : x = \cos\theta, y = \sin\theta, \theta \in [0, 2\pi)\}.$$

In higher-dimensional cases, one usually uses spheres or cylinders. Formally, we fix $r_0 > 0$, consider the interval $\mathcal{I} := [0, r_0]$ and define the manifold

$$\mathcal{B} := \mathcal{S}^1 \times \mathcal{I}. \tag{7}$$

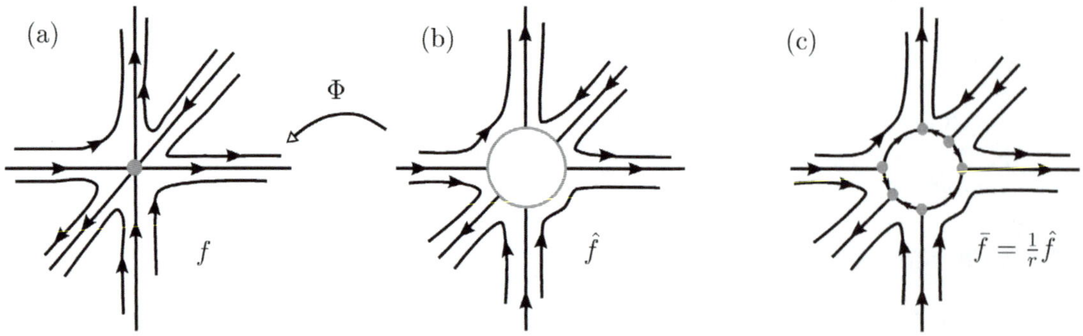

Figure 1. Sketch of the main steps of the (spherical) blow-up method for the example (6). (a) Original vector field f with non-hyperbolic steady state (gray) at the origin. (b) Blown-up vector field $\hat{f}$ on $\mathcal{B}$ with a full circle of steady states (gray) given by $\mathcal{S}^1 \times \{r = 0\}$. (c) Desingularized blown-up vector field $\bar{f}$ with precisely six hyperbolic saddle steady states (gray). The small arrows on $\mathcal{S}^1 \times \{r = 0\}$ indicate the qualitative part of the flow which is different from $\hat{f}$. Observe that the flow directions are compatible with the phase portrait for $\mathcal{S}^1 \times \{r > 0\}$.

Sometimes other choices for $\mathcal{I}$ are convenient such as $\mathcal{I} = \mathbb{R}$, $\mathcal{I} = [-r_0, r_0]$ or $\mathcal{I} = [0, \infty)$ but in our context $\mathcal{I} := [0, r_0]$ will suffice. A spherical blow-up transformation is given by

$$\Phi : \mathcal{B} \to \mathbb{R}^2,$$

where the map Φ will be defined algebraically below. We already note that if Φ is differentiable then the push-forward $\Phi_* : \mathrm{T}\mathcal{B} \to \mathbb{R}^2$ induces a vector field $\hat{f}$ on the blown-up space $\mathcal{B}$ if we require the condition

$$\Phi_*\left(\hat{f}\right) = f.$$

One possibility to define Φ algebraically is to use the weighted polar blow-up. Let $(\theta, r) \in \mathcal{S}^1 \times [0, r_0]$ be coordinates for $\mathcal{B}$ and define

$$\Phi(\theta, r) = (r^\alpha \cos\theta, r^\beta \sin\theta) = (x, y),$$

where $\alpha, \beta \in \mathbb{R}$ are the weights to be chosen below and $\theta \in [0, 2\pi)$. Observe that Φ is a diffeomorphism outside of the circle $\mathcal{S}^1 \times \{r = 0\}$, which corresponds to the steady state $p = (0, 0)$. Hence, the polar blow-up transformation indeed inserts a circle at the non-hyperbolic point and topologically conjugates the dynamics between

$$\mathbb{R}^2 - \{(0,0)\} \qquad \text{and} \qquad \mathcal{B} - \left[\mathcal{S}^1 \times \{r = 0\}\right].$$

To determine good weights α and β one may use quasi-homogeneity of the vector field; recall that f is quasi-homogeneous of type (α, β) and degree $k + 1$ if

$$f(r^\alpha x, r^\beta y) = (r^{\alpha+k} f_1(x, y), r^{\beta+k} f_2(x, y))^\top. \tag{8}$$

Substituting the vector field (6) into (8) yields

$$\begin{aligned} r^{2\alpha} a x^2 - r^{\alpha+\beta} 2xy &= r^{\alpha+k}(ax^2 - 2xy), \\ r^{2\beta} y^2 - r^{\alpha+\beta} axy &= r^{\beta+k}(y^2 - axy). \end{aligned} \tag{9}$$

Therefore, the vector field f is quasi-homogeneous of type $(\alpha, \beta) = (1, 1)$ and degree 2 (with $k = 1$). Then one chooses the blow-up weights as the type of the quasi-homogeneous vector field so that for (6) we just have a polar coordinate change

$$\Phi(\theta, r) = (r \cos\theta, r \sin\theta) = (x, y).$$

Lemma 2.1. *The vector field $\hat{f}$ in polar coordinates is given by*

$$\begin{aligned} \theta' &= r\left(3\cos\theta\sin^2\theta - 2a\sin\theta\cos^2\theta\right), \\ r' &= r^2(a\cos\theta - 2\sin\theta - 2a\cos\theta\sin^2\theta + 3\sin^3\theta). \end{aligned} \tag{10}$$

Proof. One possibility is to note that $\hat{f}(\theta, r) = (\mathrm{D}\Phi)^{-1}f(\Phi(\theta, r))$ and calculate. Alternatively, one may proceed slightly more directly

$$\begin{aligned} ar^2\cos^2\theta - 2r\cos\theta\sin\theta &= x' = r'\cos\theta - r\theta'\sin\theta, \\ r^2\sin^2\theta - ar^2\sin\theta\cos\theta &= y' = r'\sin\theta + r\theta'\cos\theta, \end{aligned} \tag{11}$$

and then solve for θ' and r' in (11). □

The ODE (10) has an entire circle of steady states given by $\mathcal{S}^1 \times \{r = 0\}$; see Figure 1(b). However, it is possible to desingularize the vector field $\hat{f}$ by division by $1/r$, i.e. we define

$$\bar{f} := \frac{1}{r}\hat{f}.$$

The division by $1/r$ does not change the qualitative dynamics on the set $\mathcal{S}^1 \times \{r > 0\}$ up to a time rescaling [1, Sec.1.4.1]. However, the $1/r$ scaling does drastically change the dynamics on the circle $\mathcal{S}^1 \times \{r = 0\}$. The desingularized vector field $\bar{f}$ is given by

$$\begin{aligned} \theta' &= 3\cos\theta\sin^2\theta - 2a\sin\theta\cos^2\theta, \\ r' &= r(a\cos\theta - 2\sin\theta - 2a\cos\theta\sin^2\theta + 3\sin^3\theta). \end{aligned} \tag{12}$$

Having computed (12), the dynamics follows by direct calculation of the steady states and linearization.

Proposition 2.2. *For $a > 0$ fixed, There are six steady states for (12) on $\mathcal{S}^1 \times \{r = 0\}$. Four are given by*

$$\theta = 0, \frac{\pi}{2}, \pi, \frac{3\pi}{2}$$

while the remaining two are defined by the condition $\tan\theta = \frac{2}{3}a$. The six steady states are hyperbolic saddle points as shown in Figure 1(c).

Although the calculations using polar coordinates are easy for our example problem, they become quickly very involved for other problems [6]. In particular, consider the situation when the blow-up has to be used iteratively when new steady states on the sphere associated to $\{r = 0\}$ are also non-hyperbolic.

It is more convenient to use charts for $\mathcal{B}$ in combination with a so-called weighted directional blow-up. Introduce coordinates on $\mathcal{B}$ given by $(\bar{x}, \bar{y}, \bar{r}) \in \mathcal{S}^1 \times [0, r_0]$ with $\bar{x}^2 + \bar{y}^2 = 1$. Then define the weighted directional blow-up map by

$$\Psi : \mathcal{B} \to \mathbb{R}^2, \qquad \Psi(\bar{x}, \bar{y}, \bar{r}) = (\bar{r}\bar{x}, \bar{r}\bar{y}). \tag{13}$$

So how should we define charts $\kappa_i : \mathcal{B} \to \mathbb{R}^2$ to make the calculations as simple as possible? One approach is to require that the induced local coordinate changes

$$\psi_i = \Psi \circ \kappa_i^{-1}$$

are easy to compute and the vector fields $D\psi_i^{-1} \circ f \circ \psi$ have a tractable algebraic form. Let $x_i, y_i \in \mathbb{R}$, $r_i \in [0, r_0]$ and let (r_1, y_1), (r_2, x_2) be coordinates on $\mathbb{R}^2$. One possibility to design the charts is to consider (13) and try to require

$$\psi_1(r_1, y_1) = (r_1, r_1 y_1) \qquad \text{and} \qquad \psi_2(r_2, x_2) = (r_2 x_2, r_2). \tag{14}$$

The following diagram illustrates the main aspects of the weighted directional blow-up:

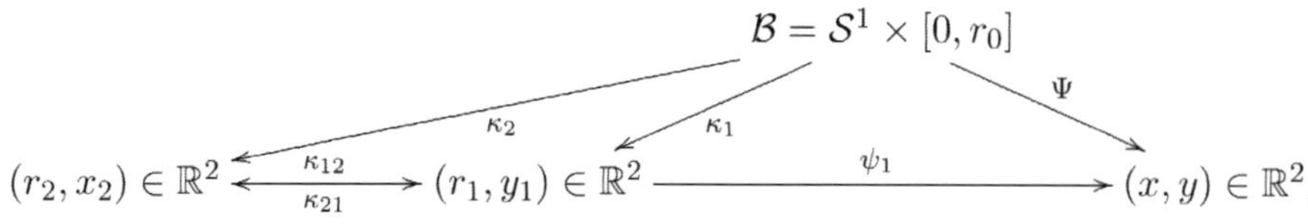

where κ_{12} and κ_{21} denote the transition maps between the two charts κ_1 and κ_2. If (14) holds then this leads to

$$\begin{aligned} \kappa_1(\bar{x}, \bar{y}, \bar{r}) &= \psi_1^{-1} \circ \Psi(\bar{x}, \bar{y}, \bar{r}) = \psi_1^{-1}(\bar{r}\bar{x}, \bar{r}\bar{y}) = (\bar{r}\bar{x}, \bar{r}\bar{y}/(\bar{r}\bar{x})) = (\bar{r}\bar{x}, \bar{y}/\bar{x}), \\ \kappa_2(\bar{x}, \bar{y}, \bar{r}) &= \psi_2^{-1} \circ \Psi(\bar{x}, \bar{y}, \bar{r}) = \psi_2^{-1}(\bar{r}\bar{x}, \bar{r}\bar{y}) = (\bar{r}\bar{x}/(\bar{r}\bar{y}), \bar{r}\bar{y}) = (\bar{x}/\bar{y}, \bar{r}\bar{y}). \end{aligned} \tag{15}$$

Hence we may use (15) as definitions of the charts and obtain that the corresponding coordinate changes on $\mathbb{R}^2$ are given by (14).

Lemma 2.3. *The vector fields using the charts $\kappa_{1,2}$ are given by*

$$\left\{ \begin{array}{rcl} r_1' &=& r_1^2(a - 2y_1), \\ y_1' &=& r_1 y_1 (3y_1 - 2a), \end{array} \right. \qquad \left\{ \begin{array}{rcl} r_2' &=& r_2^2(1 - a x_2), \\ x_2' &=& r_2 x_2 (2a r_2 - 3). \end{array} \right. \tag{16}$$

Proof. As before, we may formally carry out the coordinate change. Or one may use direct calculations, for example, we have

$$r_2' = y' = r_2^2 - a r_2^2 x_2, \qquad x' = r_2' x_2 + r_2 x_2' = a r_2^2 x_1^2 - 2 r_2^2 x_2.$$

From these results, the vector field in (r_2, x_2)-coordinates easily follows. The calculation for the κ_1-chart is similar. □

The ODEs (16) are still polynomial vector fields and algebraically a lot simpler to treat in comparison to long expressions using trigonometric functions. As for the polar case, we may again desingularize the problem using a division by $1/r_i$. For the first chart this yields

$$\begin{array}{rcl} r_1' &=& r_1(a - 2y_1), \\ y_1' &=& y_1(3y_1 - 2a). \end{array} \tag{17}$$

We have that (17) is defined in $(r_1, y_1) \in [0, r_0] \times \mathbb{R}$. We may consider this domain as corresponding to covering the right-half plane of $\mathcal{B} \subset \mathbb{R}^2$ outside of the open half-disc $\{x > 0, x^2 + y^2 < 1\}$; see Figure 2.

There are two steady states for (17) given by

$$(r_1, y_1) = (0, 0), \qquad (r_1, y_1) = \left(0, \frac{2}{3}a\right)$$

which correspond to the steady states with angles $\theta = 0$ and the smallest positive zero of $\tan\theta = \frac{2}{3}a$. In the form (17) it is easier to check the eigenvalues of the linearized system

$$\begin{pmatrix} R_1' \\ Y_1' \end{pmatrix} = \begin{pmatrix} a - 2y_1 & -2r_1 \\ 0 & 6y_1 - 3a \end{pmatrix} \begin{pmatrix} R_1 \\ Y_1 \end{pmatrix}$$

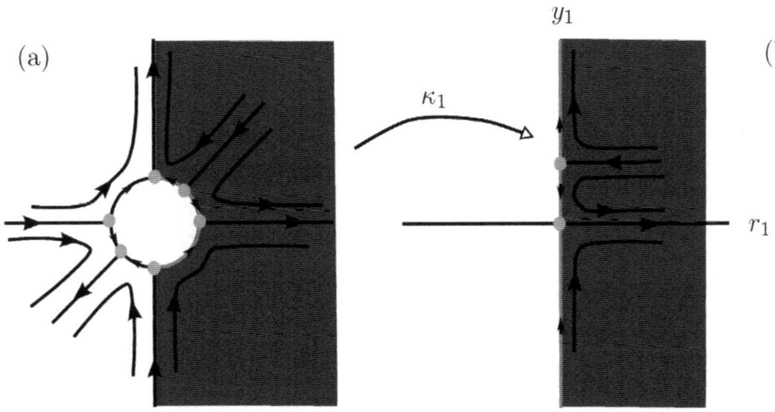

Figure 2. Sketch of the coordinate chart κ_1 associated to the x-directional blow-up. (a) Blown-up space $\mathcal{B}$ with phase portrait (black). (b) Directional coordinates $(r_1, y_1) \in \mathbb{R}^2$; the blue region corresponds to the blue region in (a) using the chart map κ_1, respectively its inverse κ_1^{-1}. Note that the half-circle from (a) is mapped to the vertical y_1-axis.

to conclude that the two steady states are hyperbolic saddle points. The calculations for the second desingularized system

$$\begin{aligned} r_2' &= r_2(1 - ax_2), \\ x_2' &= x_2(2ar_2 - 3), \end{aligned} \tag{18}$$

are similar and we also find two saddle points. The system (17) covers the outside of the open half-disc $\{y > 0, x^2 + y^2 < 1\}$ similar to the case shown in Figure 2 just for the upper half-plane. We can define two more charts, which also cover the left-half plane and the lower half-plane. If we define

$$\begin{aligned} \kappa_3(\bar{x}, \bar{y}, \bar{r}) = (-\bar{r}\bar{x}, \bar{y}/\bar{x}), \\ \kappa_4(\bar{x}, \bar{y}, \bar{r}) = (\bar{x}/\bar{y}, -\bar{r}\bar{y}), \end{aligned} \tag{19}$$

then the local coordinate changes are given by

$$\psi_3(r_3, y_3) = (-r_3, r_3 y_3) \qquad \text{and} \qquad \psi_4(r_4, x_4) = (r_4 x_4, -r_4). \tag{20}$$

With the four charts, one checks that there are six hyperbolic saddle points on $\mathcal{B} \times \{r = 0\}$ and one determines the direction of the flow as shown in Figure 1(c).

As a remaining question we consider the relation between the directional and polar blow-up maps. For example, if we would like to change from polar coordinates (θ, r) to Euclidean coordinates (r_1, y_1), we want the following diagram to commute:

$$\begin{array}{ccccc} & & \mathcal{B} = \mathcal{S}^1 \times [0, r_0] & & \\ & \swarrow^{\alpha_1} & & \searrow^{\Phi} & \\ (r_1, y_1) \in \mathbb{R}^2 & & \xrightarrow{\quad\psi_1\quad} & & (x, y) \in \mathbb{R}^2. \end{array}$$

In particular, this yields the requirement

$$\Phi(\theta, r) = (r\cos\theta, r\sin\theta) = (x, y) = (r_1, r_1 y_1) = \psi_1(r_1, y_1).$$

Therefore, we must have $r_1 = r\cos\theta$ which implies

$$r_1 y_1 = y_1 r\cos\theta = r\sin\theta \quad \Rightarrow \quad y_1 = \tan\theta.$$

The coordinate change

$$\alpha_1(\theta, r) = (r\cos\theta, \tan\theta) = (r_1, y_1) \tag{21}$$

is not well-defined when $\theta = \pi/2, 3\pi/2$ but it is a diffeomorphism otherwise. Note that this implies the polar blow-up is indeed equivalent to the directional blow-up in the x-direction expect on the vertical y_1-axis. This is geometrically clear as we cannot map the circle diffeomorphically, or even homeomorphically, onto the y_1-axis. In some sense, this fact leads one to the viewpoint that using a spherical blow-up, if one eventually wants to calculate in directional coordinates anyway, is not the only choice for the blown-up space. In fact, there may be manifolds that work more naturally with directional coordinate charts.

3. Hyperbolic Space Blow-Up

In this section we address the question whether it is possible to consider a blown-up space other than the sphere to analyze the dynamics. As we shall show below, the answer to this question is positive. The second question is whether other blow-up spaces are more convenient from a practical and/or theoretical perspective. Again, this question has at least a 'non-negative' answer, i.e. we shall show that for our test example, the calculation for hyperbolic space work equally well; in fact, it may be even more convenient to use hyperbolic space if we have distinguished directions and want to work in charts. Furthermore, having the additional freedom to pick a manifold adapted to the problem could be beneficial as outlined in Section 4.

Instead of the sphere, we shall now work with hyperbolic space [23] via the hyperboloid model and define

$$\mathbb{H}_x := \{(x,y) \in \mathbb{R}^2 : x^2 - y^2 = 1\}, \qquad \mathbb{H}_y := \{(x,y) \in \mathbb{R}^2 : y^2 - x^2 = 1\}.$$

Furthermore, we define the associated blow-up spaces

$$\mathcal{B}_x := \mathbb{H}_x \times [0, \rho_0], \qquad \mathcal{B}_y := \mathbb{H}_y \times [0, \rho_0]$$

for some fixed $\rho_0 > 0$; note that ρ_0 plays the same role as r_0 for the spherical case. We start with the blow-up using just the space $\mathcal{B}_x$. Note that we can again use a (weighted) blow-up similar to the polar coordinate map Φ if we recall that $\cosh^2(\varphi) - \sinh^2(\varphi) = 1$. Indeed, we may just define the blow-up map by

$$\Xi : \mathcal{B}_x \to \mathbb{R}^2, \qquad \Xi(\varphi, \rho) = (\rho\cosh\varphi, \rho\sinh\varphi)$$

and apply it to our main example (6). As for the spherical polar blow-up, the map Ξ induces a vector field, which we denote by $\hat{h}$, on $\mathcal{B}_x$ by the requirement

$$\Xi_*\left(\hat{h}\right) = f.$$

Lemma 3.1. *The vector field $\hat{h}$ is given by*

$$\begin{aligned} \varphi' &= \rho(3\sinh^2\varphi\cosh\varphi - 2a\cosh^2\varphi\sinh\varphi), \\ \rho' &= \rho^2(a\cosh\varphi - 2\sinh\varphi - 3\sinh^3\varphi - 2a\cosh\varphi\sinh^2\varphi). \end{aligned} \tag{22}$$

The proof of Lemma 3.1 follows the same approach as Lemma 2.1. As before, we may desingularize the vector field and consider

$$\bar{h} := \frac{1}{\rho}\hat{h}.$$

Then we look for steady states on $\mathbb{H}_x \times \{\rho = 0\}$ and we have to solve

$$\sinh^2 \varphi = \frac{2}{3} a \cosh \varphi \sinh \varphi$$

since $\cosh \varphi \geq 1$.

Proposition 3.2. *For the desingularized vector field $\bar{h}$, there is one steady state at $(\varphi, \rho) = (0, 0)$ and a second one at $(\varphi, \rho) = \left(0, \tanh\left(\frac{2}{3}a\right)\right)$. Both points are hyperbolic saddles.*

The result is expected from the previous computations. Next, we observe that the geometry of the problem for the hyperbolic blow-up space $\mathbb{H}_x$ is similar to the directional blow-up in the x-direction; see Figure 3.

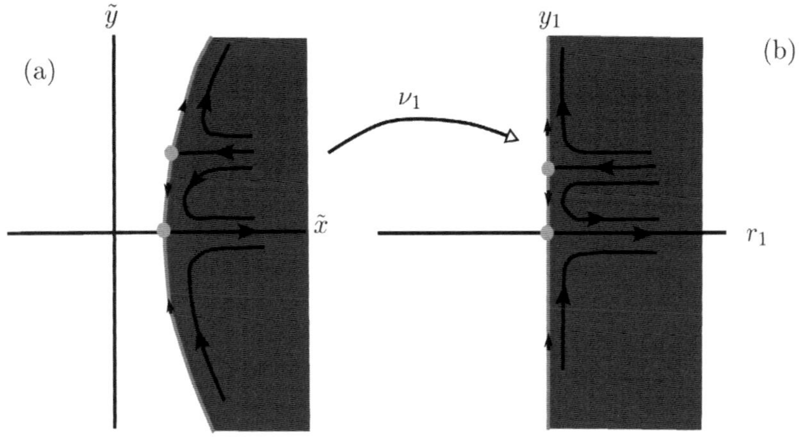

Figure 3. Sketch of the coordinate chart ν_1 associated to the x-directional blow-up. (a) Blown-up space $\mathcal{B}_x = \mathbb{H}_x \times [0, \rho)$ with phase portrait (black). (b) Directional coordinates $(r_1, y_1) \in \mathbb{R}^2$; the blue region corresponds to the blue region in (a) using the chart map ν_1, respectively its inverse ν_1^{-1}. Note that the curve $\{\tilde{x}^2 - \tilde{y}^2 = 1\} \times \{\rho = 0\}$ from (a) is mapped to the vertical y_1-axis.

Next, we check how to define the directional blow-ups based upon $\mathcal{B}_x$. Let $(\tilde{x}, \tilde{y}, \tilde{\rho})$ be coordinates on $\mathcal{B}_x$ with $\tilde{x}^2 - \tilde{y}^2 = 1$ and $\tilde{\rho} \in [0, \rho_0]$. Define the blow-map

$$\Gamma(\tilde{x}, \tilde{y}, \tilde{\rho}) = (\tilde{\rho}\tilde{x}, \tilde{\rho}\tilde{y}).$$

Let $\nu_i : \mathcal{B}_x \to \mathbb{R}^2$ be coordinate charts. As before, we want to construct the charts such that the local coordinate changes are given, as for the spherical case in (14), by

$$\gamma_1(\rho_1, y_1) = (\rho_1, \rho_1 y_1) \qquad \text{and} \qquad \gamma_2(\rho_2, x_2) = (\rho_2 x_2, \rho_2), \tag{23}$$

where $\gamma_i = \Gamma \circ \nu_i^{-1}$. In particular, the following diagram should commute

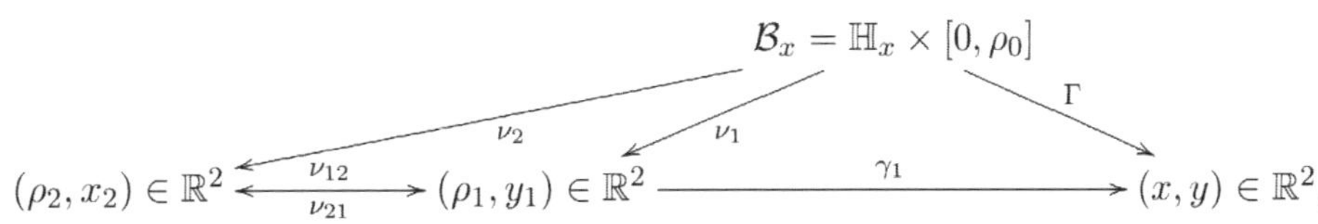

where ν_{12}, ν_{21} denote the transition maps. The conditions (23) yield

$$\begin{aligned} \nu_1(\tilde{x},\tilde{y},\tilde{\rho}) &= \gamma_1^{-1}\circ\Gamma(\tilde{x},\tilde{y},\tilde{\rho}) = \gamma_1^{-1}(\tilde{\rho}\tilde{x},\tilde{\rho}\tilde{y}) = (\tilde{\rho}\tilde{x},\tilde{\rho}\tilde{y}/(\tilde{\rho}\tilde{x})) = (\tilde{\rho}\tilde{x},\tilde{y}/\tilde{x}),\\ \nu_2(\tilde{x},\tilde{y},\tilde{\rho}) &= \gamma_2^{-1}\circ\Gamma(\tilde{x},\tilde{y},\tilde{\rho}) = \gamma_2^{-1}(\tilde{\rho}\tilde{x},\tilde{\rho}\tilde{y}) = (\tilde{r}\tilde{x}/(\tilde{\rho}\tilde{y}),\tilde{\rho}\tilde{y}) = (\tilde{x}/\tilde{y},\tilde{\rho}\tilde{y}), \end{aligned} \tag{24}$$

so the calculations are almost exactly the same as for the spherical case. However, there are some subtle differences when we consider the relation between the directional and hyperbolic polar blow-up maps. If we would like to change from the coordinates (φ,ρ) to Euclidean coordinates (ρ_1,y_1) we get the requirement

$$\Gamma(\varphi,\rho) = (\rho\cosh\varphi,\rho\sinh\varphi) = (x,y) = (\rho_1,\rho_1 y_1) = \gamma_1(\rho_1,y_1).$$

Therefore, it follows that $\rho_1 = \rho\cosh\theta$ which implies

$$\rho_1 y_1 = y_1\rho\cosh\varphi = \rho\sinh\varphi \quad\Rightarrow\quad y_1 = \tanh\varphi.$$

The coordinate change $\beta_1:\mathbb{R}^2\to\mathbb{R}^2$ given by

$$\beta_1(\varphi,\rho) = (\rho\cosh\varphi,\tanh\varphi) = (\rho_1,y_1) \tag{25}$$

is analytic and well-defined everywhere. Geometrically, this is expected since we can easily map the domain

$$\{\tilde{x}:\tilde{x}>0,\tilde{x}^2-\tilde{y}^2=1\}\times[0,\rho_0]$$

diffeomorphically onto a rectangular strip of the form $\{(x,y):x\in[0,\rho_0]\}$; see Figure 3. For the second chart we get

$$\Gamma(\varphi,\rho) = (\rho\cosh\varphi,\rho\sinh\varphi) = (x,y) = (\rho_2 x_2,\rho_2) = \gamma_2(\rho_2,x_2).$$

Therefore, it follows that $\rho_2 = \rho\sinh\theta$ which implies

$$\rho_2 x_2 = x_2\rho\sinh\varphi = \rho\cosh\varphi \quad\Rightarrow\quad x_2 = \frac{1}{\tanh\varphi}.$$

The coordinate change $\beta_2:\mathbb{R}^2\to\mathbb{R}^2$ given by

$$\beta_2(\varphi,\rho) = \left(\rho\sinh\varphi,\frac{1}{\tanh\varphi}\right) = (\rho_2,x_2) \tag{26}$$

is is not defined at $\varphi=0$ as $\tanh(0)=0$. Again, this is expected from the geometry as shown in Figure 3.

4. Conclusion and Outlook

In Section 4.1 a small summary is provided based upon the calculations of the hyperbolic space desingularization example. In Sections 4.2-4.4, three potential applications for blow-up without spheres are suggested, which provide additional motivation for the overall construction proposed here.

4.1. Classical Calculations in Different Coordinates

We may conclude from the example in Section 3 that the space $\mathcal{B}_x$, which is built upon $\mathbb{H}_x$, basically yields immediately a directional blow-up in the x-direction up to the analytic coordinate change β_1. Similarly, one may show that using $\mathcal{B}_y$ corresponds, up to an analytic coordinate change, to a y-direction blow-up. As for the spherical case, we may define charts that also cover the negative half-planes.

In summary, the example demonstrates that the classical choice of a spherical blow-up in $\mathbb{R}^N$ with $\mathcal{S}^{N-1} \times \mathcal{I}$ for some interval $\mathcal{I} \subseteq \mathbb{R}$ is certainly not the only option. In particular, if we already know a certain direction for $z \in \mathbb{R}^N$ where we do not need the directional blow-up, say z_1, then hyperbolic space $\mathbb{H}_{z_1}$ is one good choice as it corresponds via an analytic coordinate change to the respective directional blow-ups. Of course, the intrinsic dynamics of the problem does not change, regardless of the blow-up transformation considered. However, as has been shown in the case of the polar and directional blow-ups, sometimes a certain coordinate system is preferable. Therefore, hyperbolic space, as well as other suitable manifolds, could be viable alternatives to classical methods.

4.2. Geometric Flows and Singularities

Here we briefly outline one viewpoint that motivated the investigation of hyperbolic space blow-up. The basic ODE (1) we started with, as well as the example (6), are vector fields on $\mathbb{R}^N$ given by $f : \mathbb{R}^N \to \mathrm{T}\mathbb{R}^N$. However, what happens if we replace the base manifold $\mathbb{R}^N$ by some more general smooth manifold $\mathcal{M}$? The ODEs under study defined by a vector field $F : \mathcal{M} \to \mathrm{T}\mathcal{M}$ could still have non-hyperbolic points. A natural example is to start considering ODEs arising in differential geometry. Consider the hyperbolic plane $\mathbb{H}$ now viewed in the Poincaré upper half-plane model

$$\mathbb{H} = \{z = x + \mathrm{i}y \in \mathbb{C} : y > 0\}, \qquad \mathrm{d}s = \frac{\sqrt{(\mathrm{d}x)^2 + (\mathrm{d}y)^2}}{y},$$

where $\mathrm{d}s$ is the arclength element [23]. In particular, consider a curve $\gamma(t) = (x(t), y(t)) \in \mathbb{H}$ between two points $p_1, p_2 \in \mathbb{H}$ with $\gamma(0) = p_0$ and $\gamma(1) = p_1$ and define the Lagrangian energy functional

$$L[\gamma] := \frac{1}{2} \int_0^1 \frac{(x')^2 + (y')^2}{y^2} \, \mathrm{d}t. \tag{27}$$

Then it is classical to minimize (27) to determine the differential equations for the geodesics on $\mathbb{H}$. The associated Euler-Lagrange equations are given by

$$q' = f(q), \qquad f(q) := \left(u, v, \frac{2uv}{y}, \frac{v^2 - u^2}{y}\right)^\top$$

with coordinates $q = (x, y, u, v) \in \mathrm{T}\mathbb{H}^2 =: \mathrm{T}\mathcal{M}$. As discussed previously for desingularization of dividing factors it seems natural to consider $F := yf$ so that

$$q' = F(q), \qquad F(q) := \left(uy, vy, 2uv, v^2 - u^2\right)^\top$$

with a suitable time re-parametrization understood. The vector field $F(q)$ has a codimension two submanifold of equilibria $\{u = 0 = v\}$. If we formally linearize $\mathrm{D}_q F(x, y, 0, 0) \in \mathbb{R}^{4\times 4}$ has a quadruple zero eigenvalue and there is no (dynamical) hyperbolicity of the submanifold $\{u = 0 = v\}$. Of course, one could then ask whether the blow-up method can help to desingularize the problem and it seems natural to conjecture that spherical blow-up spaces may not be adequate. Indeed, the entire problem is formulated under the assumption that we work with a hyperbolic metric not with a spherical one. For other geometric flows the situation could be similar since (dynamically) non-hyperbolic points or submanifolds can always appear and we leave this direction as an open problem for future research.

4.3. Point Contraction and Compactness

As outlined in the introduction, a typical situation for the blow-up occurs in the context of fast-slow systems

$$\begin{aligned} x' &= f_1(x,y,\varepsilon), \\ y' &= \varepsilon g(x,y,\varepsilon), \\ \varepsilon' &= 0. \end{aligned} \tag{28}$$

The lowest-dimensional non-trivial case occurs when $z := (x, y, \varepsilon) \in \mathbb{R}^3$. Setting $f_2(x, y, \varepsilon) = \varepsilon g(x, y, \varepsilon)$, $f_3(x, y, \varepsilon) = 0$, and $f := (f_1, f_2, f_3)^\top$ leads to

$$z' = f(z), \qquad z \in \mathbb{R}^3 \tag{29}$$

with a non-hyperbolic steady state at $z = 0$ under the assumption that a non-hyperbolic fast subsystem steady state at the origin exists. Inserting a two-sphere $\mathcal{S}^2$ via a blow-up

$$\Phi : \mathcal{S}^2 \times [0, r_0] \to \mathbb{R}^3 \tag{30}$$

means that all new steady states on $\mathcal{S}^2 \times \{0\}$ can now be analyzed locally. A potential disadvantage is that we now treat all directions of the fast-slow system equally and all orbits on the sphere are compact. To illustrate this problem formally, let us consider the transcritical singularity of the form

$$\begin{aligned} x' &= x(x-y) + \kappa\varepsilon, \\ y' &= \varepsilon, \end{aligned} \tag{31}$$

where $\kappa \in \mathbb{R}$ is a parameter. It is easy to check that $(x, y) = (0, 0)$ is a non-hyperbolic steady state of the fast subsystem. At this transcritical singularity it is well-understood (see e.g. [13, Thm. 2.1]) that depending upon κ it is possible that trajectories starting sufficiently close to the attracting critical manifold $C_0^{a-} := \{(x, y) \in \mathbb{R}^2 : x = 0, y < 0\}$ can jump near the origin towards large negative values of x, the so-called jump case. On the other hand, it is also possible for different values of κ that trajectories stay close to $C_0^r := \{(x, y) \in \mathbb{R}^2 : x = 0, y > 0\}$ (the canard case), or that trajectories stay close to the attracting critical manifold $C_0^{a+} := \{(x, y) \in \mathbb{R}^2 : x = y, y > 0\}$ (the exchange-of-stability case). In those last two cases the motion is slow and bounded on all time scales as $\varepsilon \to 0$. However, for the jump case, the trajectories become unbounded in the singular limit $\varepsilon \to 0$ on the slow time scale. So it seems quite plausible that the x-direction is special in this context as the jumps occur almost parallel to the x-axis. Hence, it could be viewed as more natural to try to represent this fact in the blow-up construction and work with a space such as the hyperboloid

$$\mathbb{H}_x = \{(x, y, \varepsilon) \in \mathbb{R}^3 : y^2 + \varepsilon^2 - x^2 = 1\},$$

or another manifold, which has a distinguished direction. Of course, for the transcritical singularity, this is not necessary as demonstrated in [13]. However, it looks unnecessary to compactify a direction along which one wants to observe unbounded orbits while using a compact blown-up space seems more natural if we want to focus on bounded orbits.

4.4. Tori and Interior Steady States

To illustrate another potential advantage of different manifolds we continue with the fast-slow systems case (28)-(29). Frequently the main problem is to extend a suitable invariant submanifold of phase space near the blown-up sphere [13, 6] from one part of phase space to another part. In this context, a technically challenging problem is that this may involve an ODE without steady states. For example, consider the truncated fold point normal form

$$\begin{aligned} x' &= y - x^2, \\ y' &= \varepsilon. \end{aligned} \tag{32}$$

After a suitable blow-up [13], one obtains the system

$$\begin{aligned} \tilde{x}' &= \tilde{y} - \tilde{x}^2, \\ \tilde{y}' &= 1, \end{aligned} \tag{33}$$

where $(\tilde{x}, \tilde{y}) \in \mathbb{R}^2$ are the coordinates in the so-called classical (or re-scaling) chart. Indeed, one just derives (33) from (32) by applying the scaling

$$x = \varepsilon^{1/3}\tilde{x}, \quad y = \varepsilon^{2/3}\tilde{y}, \quad t = \varepsilon^{-1/3}\tilde{t}$$

where differentiation is with respect to $\tilde{t}$ in (33). To analyze (33) one relies on the classical asymptotic theory of Ricatti equations [21] to find the relevant global solutions of (33). One conjecture is that a fully global analysis could be avoided if there would be additional steady states in $\mathbb{R}^2$ for (33). Instead of using a spherical blow-up of the form (30) a potential choice is

$$\Phi : \mathcal{S}^1 \times \mathcal{S}^1 \times [0, r_0] \to \mathbb{R}^3,$$

where $\mathcal{S}^1 \times \mathcal{S}^1 = \{(x^2 + y^2 + \varepsilon^2 + \mathfrak{R}^2 - \mathfrak{r}^2)^2 = 4\mathfrak{R}^2(x^2 + y^2)\}$ is the usual 2-torus with major and minor radii $\mathfrak{R}$ and $\mathfrak{r}$ that can be used as free parameters. A projection of this torus along the ε-axis yields an annulus in the (x, y)-plane. In particular, if we think about the inner boundary of the annulus in a 'classical chart' then it may potentially contain steady states at which the dynamics could be analyzed locally. Of course, as for the transcritical singularity discussed in the last section, this idea is not necessary for the analysis of the fold point [6, 13] but it may have the potential to avoid difficulties for other singularities.

Acknowledgments

I would like to thank the Austrian Academy of Sciences (ÖAW) for support via an APART fellowship. I also acknowledge the European Commission (EC/REA) for support by a Marie-Curie International Re-integration Grant. I would also like to thank two anonymous referees whose suggestions helped to improve the manuscript.

References

[1] C. Chicone. *Ordinary Differential Equations with Applications.* Texts in Applied Mathematics. Springer, 2nd edition, 2010.
[2] F. Dumortier. Singularities of vector fields on the plane. *J. Differential Equat.*, 23(1):53–106, 1977.
[3] F. Dumortier. *Singularities of Vector Fields.* IMPA, Rio de Janeiro, Brazil, 1978.
[4] F. Dumortier. Techniques in the theory of local bifurcations: Blow-up, normal forms, nilpotent bifurcations, singular perturbations. In D. Schlomiuk, editor, *Bifurcations and Periodic Orbits of Vector Fields*, pages 19–73. Kluwer, Dortrecht, The Netherlands, 1993.
[5] F. Dumortier, N. Popovic, and T.J. Kaper. A geometric approach to bistable front propagation in scalar reaction-diffusion equations with cut-off. *Physica D*, 239(20):1984–1999, 2010.
[6] F. Dumortier and R. Roussarie. *Canard Cycles and Center Manifolds*, volume 121 of *Memoirs Amer. Math. Soc.* AMS, 1996.
[7] J. Guckenheimer and P. Holmes. *Nonlinear Oscillations, Dynamical Systems, and Bifurcations of Vector Fields.* Springer, New York, NY, 1983.
[8] I. Gucwa and P. Szmolyan. Geometric singular perturbation analysis of an autocatalator model. *Discr. Cont. Dyn. Syst. S*, 2(4):783–806, 2009.
[9] Robin Hartshorne. *Algebraic Geometry.* Springer, 1977.
[10] H. Hironaka. Resolution of singularities of an algebraic variety over a field of characteristic zero: I. *Ann. of Math.*, 79(1):109–203, 1964.
[11] H. Hironaka. Resolution of singularities of an algebraic variety over a field of characteristic zero: II. *Ann. of Math.*, 79(2):205–326, 1964.
[12] M. Krupa and P. Szmolyan. Extending geometric singular perturbation theory to nonhyperbolic points - fold and canard points in two dimensions. *SIAM J. Math. Anal.*, 33(2):286–314, 2001.

[13] M. Krupa and P. Szmolyan. Extending slow manifolds near transcritical and pitchfork singularities. *Nonlinearity*, 14:1473–1491, 2001.
[14] M. Krupa and M. Wechselberger. Local analysis near a folded saddle-node singularity. *J. Differential Equat.*, 248(12):2841–2888, 2010.
[15] C. Kuehn. Normal hyperbolicity and unbounded critical manifolds. *Nonlinearity*, 27(6):1351–1366, 2014.
[16] C. Kuehn. *Multiple Time Scale Dynamics*. Springer, 2015.
[17] C. Kuehn and P. Szmolyan. Multiscale geometry of the Olsen model and non-classical relaxation oscillations. *J. Nonlinear Sci.*, 25(3):583–629, 2015.
[18] Yu.A. Kuznetsov. *Elements of Applied Bifurcation Theory*. Springer, New York, NY, 3rd edition, 2004.
[19] P. De Maesschalck and F. Dumortier. Slow-fast Bogdanov-Takens bifurcations. *J. Diff. Eq.*, 250:1000–1025, 2011.
[20] P. De Maesschalck, N. Popovic, and T.J. Kaper. Canards and bifurcation delays of spatially homogeneous and inhomogeneous types in reaction-diffusion equations. *Adv. Differential Equat.*, 14(9):943–962, 2009.
[21] E.F. Mishchenko and N.Kh. Rozov. *Differential Equations with Small Parameters and Relaxation Oscillations (translated from Russian)*. Plenum Press, 1980.
[22] L. Perko. *Differential Equations and Dynamical Systems*. Springer, 2001.
[23] J.G. Ratcliffe. *Foundations of Hyperbolic Manifolds*. Springer, 2006.
[24] M. Wechselberger. Existence and bifurcation of canards in $\mathbb{R}^3$ in the case of a folded node. *SIAM J. Applied Dynamical Systems*, 4(1):101–139, 2005.

Some remarks on a model for rate-independent damage in thermo-visco-elastodynamics

Giuliano Lazzaroni[1], **Riccarda Rossi**[2], **Marita Thomas**[3] **and Rodica Toader**[4]

[1] SISSA, Via Bonomea 265, 34136 Trieste, Italy.
[2] DICATAM - Sezione di Matematica, Università degli studi di Brescia, V. Valotti 9, 25133 Brescia, Italy.
[3] Weierstrass Institute for Applied Analysis and Stochastics, Mohrenstr. 39, 10117 Berlin, Germany.
[4] DIMI, Università degli studi di Udine, Via delle Scienze 206, 33100 Udine, Italy.

E-mail: giuliano.lazzaroni@sissa.it, riccarda.rossi@unibs.it, marita.thomas@wias-berlin.de, rodica.toader@uniud.it

Abstract. This note deals with the analysis of a model for partial damage, where the rate-independent, unidirectional flow rule for the damage variable is coupled with the rate-dependent heat equation, and with the momentum balance featuring inertia and viscosity according to Kelvin-Voigt rheology. The results presented here combine the approach from Roubíček [1, 2] with the methods from Lazzaroni/Rossi/Thomas/Toader [3]. The present analysis encompasses, differently from [2], the monotonicity in time of damage and the dependence of the viscous tensor on damage and temperature, and, unlike [3], a nonconstant heat capacity and a time-dependent Dirichlet loading.

1. Introduction – Energetic solutions for rate-independent processes coupled with rate-dependent effects

In this note we discuss the existence of solutions for an evolutionary model of partial damage, where the rate-independent, unidirectional flow rule for the damage variable is coupled with the rate-dependent heat equation, and with the momentum balance featuring inertia and viscosity according to Kelvin-Voigt rheology. Systems with a mixed rate-independent/rate-dependent character were considered in [1] in the isothermal case, where a suitable notion of weak (*energetic*) solution was introduced, and then in [2] where this notion was extended to thermal processes. In the latter paper, a general existence result for *energetic solutions* was proved, with application to a wide class of thermo-viscoelastic material systems in the frame of *generalized standard solids*, where the flow rule for the internal variable has rate-independent character. The damage model treated here pertains to this class: the internal variable z assesses the soundness of the material, so that one will have $z = 1$ in the fully undamaged, and $z = 0$ in the completely damaged cases, respectively. Additionally, in the model discussed here, z also affects the elastic and the viscous stress tensors.

Here we will further contribute to the analysis initiated in [2] by pointing out that the existence result therein can be extended to the case in which the evolution for the internal variable is *unidirectional* (i.e., monotone nonincreasing), as in the context of the damage model presently considered. Moreover, we will show that time-dependent Dirichlet loadings for the displacement variable can be encompassed in the analysis, whereas in [2, 3] the momentum equation was supplemented with zero Dirichlet boundary data. In this note we will particularly focus on the techniques to treat the difficulties resulting from these additional features of the model. We refer to our previous work [3] for a detailed survey of the literature on rate-independent and rate-dependent damage models in (thermo-)viscoelasticity and for a more detailed discussion of the PDE system under consideration.

We will prove the existence of *energetic solutions* for the damage system using a time discretization, by now standard within the analysis of rate-independent problems. We will also show that, under appropriate conditions on the nonlinear constitutive functions featuring in the PDE system, the time discrete scheme can be *fully decoupled*, which might turn out to be interesting towards the numerical investigation of this model.

The PDE system. We consider the following PDE system which describes the behavior of a thermo-visco-elastic body subject to damage; it consists of the momentum balance (1a), the flow rule (1b), and the heat equation (1c):

$$\rho\ddot{u} - \mathrm{div}\,(\mathbb{D}(z,\theta)e(\dot{u}) + \mathbb{C}(z)e(u) - \theta\,\mathbb{B}) = f_{\mathrm{V}} \qquad \text{in } (0,T)\times\Omega\,, \tag{1a}$$

$$\partial\mathrm{R}_1(\dot{z}) + \mathrm{D}_z G(z,\nabla z) - \mathrm{div}\,(\mathrm{D}_\xi G(z,\nabla z)) + \tfrac{1}{2}\mathbb{C}'(z)e(u):e(u) \ni 0 \qquad \text{in } (0,T)\times\Omega\,, \tag{1b}$$

$$c_{\mathrm{V}}(\theta)\dot{\theta} - \mathrm{div}\,(\mathbb{K}(z,\theta)\nabla\theta) = \mathrm{R}_1(\dot{z}) + (\mathbb{D}(z,\theta)e(\dot{u}) - \theta\,\mathbb{B}):e(\dot{u}) + H \qquad \text{in } (0,T)\times\Omega\,. \tag{1c}$$

Here the unknowns (u,z,θ) stand for the displacement vector field, the damage variable, and the absolute temperature, respectively, $(0,T)$ indicates the time interval, while Ω is a bounded open subset of $\mathbb{R}^3$ with Lipschitz boundary $\partial\Omega$ representing the reference configuration. The strain tensor is $e(u) = \frac{1}{2}(\nabla u + \nabla u^T)$, the constant $\rho > 0$ is the mass density, $\mathbb{D}(z,\theta)$ and $\mathbb{C}(z)$ are the viscous and the elastic tensors, respectively. Thermal stresses are featured by $\theta\,\mathbb{B}$ with $\mathbb{B}$ a fixed symmetric matrix coupling the momentum and the heat equations. In (1c), $c_{\mathrm{V}}(\theta)$ and $\mathbb{K}(z,\theta)$ are, respectively, the heat capacity and the heat conductivity of the material. In (1b) the term $G(z,\nabla z)$ is a regularization for the damage variable as it involves its gradient (e.g. in Sobolev-sense). The term $\mathrm{R}_1(\dot{z})$ is a 1-homogeneous dissipation potential which, in order to encode the rate-independence and the unidirectionality of the damage process, is chosen as

$$\mathrm{R}_1(v) := \begin{cases} |v| & \text{if } v \leq 0\,, \\ +\infty & \text{otherwise,} \end{cases} \tag{2}$$

cf. also e.g. [4] for this ansatz. Note that the unidirectionality is reflected by the fact that $\mathrm{R}_1(\dot{z}) = \infty$ if $\dot{z} > 0$. This ensures that a solution z will be nonincreasing in time in accordance with the definition of z; recall that $z(x) = 1$ in sound and $z(x) = 0$ in completely damaged material points x. In the flow rule (1b) the symbol ∂ indicates the subdifferential in the sense of convex analysis while D_z and D_ξ denote the Gâteaux derivatives. The external forces and the heat source are denoted by f_{V} and H, respectively.

The PDE system (1) is supplemented with Cauchy conditions given on $u(0)$, $\dot{u}(0)$, $z(0)$, and $\theta(0)$, and with the boundary conditions

$$(\mathbb{D}(z,\theta)e(\dot{u}) + \mathbb{C}(z)e(u) - \theta\,\mathbb{B})\,\nu = f_{\mathrm{S}} \quad \text{on } (0,T)\times\partial_{\mathrm{N}}\Omega\,, \tag{3a}$$

$$u = g \quad \text{on } (0,T)\times\partial_{\mathrm{D}}\Omega\,, \tag{3b}$$

$$\mathrm{D}_\xi G(z,\nabla z)\,\nu = 0 \quad \text{on } (0,T)\times\partial\Omega\,, \tag{3c}$$

$$\mathbb{K}(z,\theta)\nabla\theta\cdot\nu = h \quad \text{on } (0,T)\times\partial\Omega\,, \tag{3d}$$

where $\partial_D\Omega$ and $\partial_N\Omega := \partial\Omega\backslash\partial_D\Omega$ are the Dirichlet and the Neumann part of the boundary, ν denotes the outer unit normal vector to $\partial\Omega$, and f_S, g, and h are prescribed external data depending on time and space. In Section 2 we shall detail the assumptions on the constitutive functions and on the given data featuring in system (1).

To some extent, system (1) is a particular case of the general model for rate-independent processes in thermo-viscoelastic materials proposed in [2]. In particular, as in the latter paper the heat equation features a nonconstant heat capacity, depending on θ in a nonlinear way, and accordingly it requires appropriate treatment. Indeed, it will turn out to be useful to switch from the temperature to the *enthalpy* variable, defined as a primitive of the (positive) function c_V. Moreover, as in [2], we shall have to impose specific growth conditions on c_V and on the matrix of heat conduction coefficients $\mathbb{K}$ in order to handle the quadratic terms on the right-hand side of (1c). We refer to [5] for a thermodynamical derivation of the growth assumptions used in what follows.

Nonetheless, let us highlight that, differently from [2] our analysis also encompasses the unidirectionality in the evolution for z, the dependence of the viscous tensor $\mathbb{D}$ on z and θ, and, differently from [3], also a time-dependent Dirichlet loading g.

Indeed, a model for rate-independent, unidirectional, partial damage in thermo-viscoelastic materials with inertia was also proposed in [3] in the case of *constant* heat capacity c_V, which allowed us to avoid the enthalpy transformation. The assumption of a constant heat capacity is valid for large values of the temperature, while a nonconstant heat capacity describes low-temperature regimes. We remark that in [3] it was possible to only *partially* decouple the time-discrete scheme, i.e., only the approximate flow rule for the internal variable could be decoupled from the other equations. Moreover, time-dependent boundary conditions on the displacement u could be accounted for only if the viscous tensor was assumed to be constant.

In the following lines, we briefly sketch the main points of the analysis. First of all, let us remark that, as in [3], we shall have to resort to a weak notion of solution for the initial-boundary value problem for system (1), introduced in [1, 2] and hereafter referred to as *energetic*. While postponing all details to Definition 2.1, we may mention here that this concept consists of the unidirectionality and semistability conditions for the damage variable z combined with a (mechanical) energy balance, and coupled with the weak (distributional-type) formulations of the momentum and enthalpy equations.

After stating our working assumptions on the problem data and introducing energetic solutions, in Section 2 we will state our main existence result, Theorem 2.2. Its proof will be developed throughout Section 3, according to the general strategy suggested in [2]. In fact, in this contribution we will only sketch some parts of the proof, referring for certain details to the latter paper, as well as to [3] for the handling of *unidirectional* processes. Instead, we will dwell on the techniques allowing us to fully decouple the time-discrete scheme and to account for the dynamics of the Dirichlet loading, which causes additional rate-dependent energy terms, see the comments in Section 3.

2. Setup and main result

In this section we collect the conditions on the constitutive functions featuring in system (1), as well as on the loadings and on the prescribed external and initial data. These functions also enter in the free energy ψ of the system which has the structure

$$\psi(e(u), z, \nabla z) = \varphi(e(u), z, \nabla z) + \theta\,\phi(e(u)) - \phi_0(\theta)\,,$$

considered in [2]. The purely thermal contribution ϕ_0 determines the heat capacity function c_V via $c_V(\theta) := \theta\,\phi_0''(\theta)$. As for the mechanical part φ, in the present case we take it in the form

$$\varphi(e(u), z, \nabla z) := \tfrac{1}{2}\mathbb{C}(z)e(u) : e(u) + G(z, \nabla z)\,,$$

while $\phi(e(u)) := -\mathbb{B} : e(u)$.

Next, we derive the version of (1) in terms of the *enthalpy* in place of the temperature, and for the resulting system we recall the notion of *energetic solution* from [2]. The statement of our existence result, Theorem 2.2, closes this section.

Assumptions on the heat capacity and heat conductivity. As mentioned in the introduction, the treatment of the heat equation relies on specific growth conditions on c_{v} and $\mathbb{K}$, adopted from [2] and further tailored to our existence analysis. More precisely, we assume that

$$\begin{aligned}&c_{\mathrm{v}} \in \mathrm{C}^0(\mathbb{R};\mathbb{R}^+) \text{ is such that}\\&\quad \exists\, \alpha_1 \geq \alpha = 1\,,\ c_1 \geq c_0 > 0\ \forall\, \theta \in [0,+\infty): \quad c_0(1+\theta)^{\alpha} \leq c_{\mathrm{v}}(\theta) \leq c_1(1+\theta)^{\alpha_1}\,,\end{aligned} \tag{4a}$$

$$\begin{aligned}&\mathbb{K} \in \mathrm{C}^0(\mathbb{R}\times\mathbb{R};\mathbb{R}^{d\times d}) \text{ is symmetric and}\\&\quad \exists\, c_2,\, c_3 > 0\ \forall\,(z,\theta) \in [0,1]\times[0,+\infty): \quad c_2\, c_{\mathrm{v}}(\theta) \leq |\mathbb{K}(z,\theta)| \leq c_3\, c_{\mathrm{v}}(\theta)\,.\end{aligned} \tag{4b}$$

Let us shortly compare the growth condition in (4a) with the one in [2, formulae (3.12b) and (4.22)]. There, it is required that $\alpha > (d-2)/(d+2)$ with d the space dimension, which would result in $\alpha > 1/5$ in the present three-dimensional context. Hence, $\alpha = 1$ in (4a) is a special case.

This particular choice is made in order to generate a square-root growth of the temperature in dependence of the enthalpy variable, cf. (16), which will be a crucial ingredient to handle the thermal expansion term in combination with time-dependent Dirichlet data. It will also play a key role in the analysis of the time-discrete version of (the enthalpy reformulation of) system (1). In particular, it will be at the core of the proof of Lemma 3.3, by means of which it is possible to have a *fully decoupled* scheme. Let us point out that the linear growth from below imposed on the heat capacity in (4a) is also in accordance with [5, 6] in the context of small-strain thermo-viscoelasticity, where the heat capacity is assumed to be an affine function of temperature. In particular, see [5] for a thermodynamical derivation.

Assumptions on the material tensors. We require that the tensors $\mathbb{B} \in \mathbb{R}^{3\times 3}$, $\mathbb{C}\colon \mathbb{R} \to \mathbb{R}^{3\times3\times3\times3}$, and $\mathbb{D}\colon \mathbb{R}\times\mathbb{R} \to \mathbb{R}^{3\times3\times3\times3}$ fulfill

$$\mathbb{B} \in \mathbb{R}^{3\times3}_{\mathrm{sym}}\,, \text{ and set } C_{\mathbb{B}} := |\mathbb{B}|\,, \tag{5a}$$

$$\mathbb{C} \in \mathrm{C}^{0,1}(\mathbb{R};\mathbb{R}^{3\times3\times3\times3}) \text{ and } \mathbb{D} \in \mathrm{C}^0(\mathbb{R}\times\mathbb{R};\mathbb{R}^{3\times3\times3\times3})\,, \tag{5b}$$

$$\mathbb{C}(z),\, \mathbb{D}(z,\theta) \in \mathbb{R}^{3\times3\times3\times3}_{\mathrm{sym}} \text{ are positive definite for all } z \in \mathbb{R}\,,\ \theta \in \mathbb{R}\,, \tag{5c}$$

$$\exists\, C^1_{\mathbb{C}},\, C^2_{\mathbb{C}} > 0\ \ \forall\, z \in \mathbb{R}\ \ \forall\, A \in \mathbb{R}^{3\times3}_{\mathrm{sym}}: \quad C^1_{\mathbb{C}}\,|A|^2 \leq \mathbb{C}(z)A : A \leq C^2_{\mathbb{C}}\,|A|^2\,, \tag{5d}$$

$$\exists\, C^1_{\mathbb{D}},\, C^2_{\mathbb{D}} > 0\ \ \forall\, z \in \mathbb{R}\ \ \forall\theta \in \mathbb{R}\ \ \forall\, A \in \mathbb{R}^{3\times3}_{\mathrm{sym}}: \quad C^1_{\mathbb{D}}\,|A|^2 \leq \mathbb{D}(z\,,\theta)A : A \leq C^2_{\mathbb{D}}\,|A|^2\,. \tag{5e}$$

In addition, we impose that $\mathbb{C}(\cdot)$ is monotonically nondecreasing, i.e.,

$$\forall\, A \in \mathbb{R}^{3\times3}_{\mathrm{sym}}\ \forall\, 0 \leq z_1 \leq z_2 \leq 1: \quad \mathbb{C}(z_1)A : A \leq \mathbb{C}(z_2)A : A\,. \tag{5f}$$

In the expressions above, $\mathbb{R}^{3\times3}_{\mathrm{sym}}$ denotes the subset of symmetric matrices in $\mathbb{R}^{3\times3}$ and $\mathbb{R}^{3\times3\times3\times3}_{\mathrm{sym}}$ is the subset of symmetric tensors in $\mathbb{R}^{3\times3\times3\times3}$. In particular,

$$\mathbb{C}(z)_{ijkl}{=}\mathbb{C}(z)_{jikl}{=}\mathbb{C}(z)_{ijlk}{=}\mathbb{C}(z)_{klij} \text{ and } \mathbb{D}(z,\theta)_{ijkl}{=}\mathbb{D}(z,\theta)_{jikl}{=}\mathbb{D}(z,\theta)_{ijlk}{=}\mathbb{D}(z,\theta)_{klij}\,.$$

Assumptions on the damage regularization. We require that $G\colon \mathbb{R}\times\mathbb{R}^3 \to \mathbb{R}\cup\{+\infty\}$ fulfills

Indicator: For every $(z,\xi)\in\mathbb{R}\times\mathbb{R}^3$: $\quad G(z,\xi)<+\infty \ \Rightarrow\ z\in[0,1]\,;$ (6a)

Continuity: $G\colon \mathbb{R}\times\mathbb{R}^3\to\mathbb{R}\cup\{+\infty\}$ is continuous on $\mathrm{dom}(G)$, and $G(0,0)=0\,;$ (6b)

Convexity: For every $z\in\mathbb{R}$, $G(z,\cdot)$ is convex; (6c)

Growth: There exist constants $q>1$ and $C_G^1, C_G^2>0$ such that for every $(z,\xi)\in\mathrm{dom}(G)$

$$C_G^1(|\xi|^q-1)\le G(z,\xi)\le C_G^2(|\xi|^q+1)\,. \tag{6d}$$

Accordingly, the state space $\mathcal{Z}$ is defined by

$$\mathcal{Z}:=\{z\in W^{1,q}(\Omega)\colon z\in[0,1]\text{ a.e. in }\Omega\}\,. \tag{7}$$

Assumptions on the given data. With a slight abuse of notation, we will denote by g also the extension into the domain of the non-zero Dirichlet boundary datum on the displacement. We require that

$$g\in H^1(0,T;H^1(\Omega;\mathbb{R}^3))\cap W^{2,2}(0,T;L^2(\Omega;\mathbb{R}^3))\,. \tag{8}$$

On the initial data $(u_0,\dot u_0,z_0)$ we require

$$u_0\in H^1(\Omega;\mathbb{R}^3)\,,\quad u_0=g(0)\quad\text{on }\partial_{\mathrm{D}}\Omega\,,\quad \dot u_0\in L^2(\Omega;\mathbb{R}^3)\,,\quad z_0\in\mathcal{Z}\,, \tag{9}$$

and, in accordance with (4a), we impose on θ_0

$$\theta_0\in L^{\alpha_1+1}(\Omega)\,,\qquad \theta_0\ge 0\quad\text{a.e. in }\Omega\,, \tag{10}$$

where α_1 is the same as in (4a). On the loading and source terms f_{V}, f_{S}, H, and h we require

$$f_{\mathrm{V}}\in H^1(0,T;L^2(\Omega;\mathbb{R}^3))\,,\quad f_{\mathrm{S}}\in H^1(0,T;L^2(\partial_N\Omega;\mathbb{R}^3))\,, \tag{11a}$$

$$H\in L^2(0,T;L^2(\Omega))\,,\qquad H\ge 0\text{ a.e. in }(0,T)\times\Omega\,, \tag{11b}$$

$$h\in L^2(0,T;L^2(\partial\Omega))\,,\qquad h\ge 0\text{ a.e. in }(0,T)\times\partial\Omega\,. \tag{11c}$$

For later convenience, we also introduce $f:[0,T]\to H^1(\Omega;\mathbb{R}^3)^*$ defined by

$$\langle f(t),v\rangle_{H^1(\Omega;\mathbb{R}^3)}:=\int_\Omega f_{\mathrm{V}}(t)\cdot v\,\mathrm{d}x+\int_{\partial_N\Omega} f_{\mathrm{S}}\cdot v\,\mathrm{d}\mathcal{H}^2(x)\qquad\text{for all }v\in H^1(\Omega;\mathbb{R}^3)\,, \tag{12}$$

$\mathcal{H}^2$ denoting the 2-dimensional Hausdorff measure. It follows from (11a) that $f\in H^1(0,T;H^1(\Omega;\mathbb{R}^3)^*)$.

Observe that the requirements on g and on $f_{\mathrm{V}},\ldots,h$ could be slightly refined, cf. [2] and [3]. However, we choose to overlook this point to avoid overburdening the analysis with technicalities.

The enthalpy transformation. In view of the time-discretization procedure, it is useful to pass from the nonlinear term $c_{\mathrm{v}}(\theta)\dot\theta$ in (1c) to a linear one. This motivates the change of variables adopted in [2], by virtue of which we switch from the absolute temperature θ to the enthalpy variable w, defined via the so-called *enthalpy transformation*, viz.

$$w=\mathfrak{h}(\theta):=\int_0^\theta c_{\mathrm{v}}(s)\,\mathrm{d}s\,. \tag{13}$$

Thus, $\mathfrak{h}$ is a primitive function of c_V, normalized in such a way that $\mathfrak{h}(0) = 0$. Since c_V is strictly positive (cf. assumption (4a) above), $\mathfrak{h}$ is strictly increasing. Thus, we define

$$\Theta(w) := \begin{cases} \mathfrak{h}^{-1}(w) & \text{if } w \geq 0, \\ 0 & \text{if } w < 0, \end{cases} \qquad \mathbb{J}(z, w) := \frac{\mathbb{K}(z, \Theta(w))}{c_V(\Theta(w))}. \tag{14}$$

It follows from (4a) and (10) that

$$w_0 := \mathfrak{h}(\theta_0) \in L^1(\Omega) \quad \text{and} \quad w_0 \geq 0 \text{ a.e. in } \Omega, \tag{15}$$

and that

$$\exists\, C_\Theta^0,\, C_\Theta^1,\, C_\Theta^2 > 0\ \forall\, w \in [0, +\infty): \quad C_\Theta^0 w^{1/(\alpha_1+1)} - C_\Theta^1 \leq \Theta(w) \leq C_\Theta^2 w^{1/2}, \tag{16}$$

whereas (4b) ensures that

$$c_2 \leq |\mathbb{J}(z, w)| \leq c_3 \quad \text{for all } (z, w) \in [0, 1] \times [0, +\infty). \tag{17}$$

In view of (13), and replacing all occurrences of θ by $\Theta(w)$, we obtain the enthalpy reformulation of system (1):

$$\rho\ddot{u} - \operatorname{div}\left(\mathbb{D}(z, \Theta(w))e(\dot{u}) + \mathbb{C}(z)e(u) - \Theta(w)\,\mathbb{B}\right) = f_V \quad \text{in } (0, T) \times \Omega, \tag{18a}$$

$$\partial \mathrm{R}_1(\dot{z}) + \partial_z G(z, \nabla z) - \operatorname{div}\left(\mathrm{D}_\xi G(z, \nabla z)\right) + \tfrac{1}{2}\mathbb{C}'(z)e(u) : e(u) \ni 0 \quad \text{in } (0, T) \times \Omega, \tag{18b}$$

$$\dot{w} - \operatorname{div}\left(\mathbb{J}(z, w)\nabla w\right) = \mathrm{R}_1(\dot{z}) + \left(\mathbb{D}(z, \Theta(w))e(\dot{u}) - \Theta(w)\mathbb{B}\right) : e(\dot{u}) + H \quad \text{in } (0, T) \times \Omega. \tag{18c}$$

Energetic solutions. Let us now specify the notion of weak solution for system (18), supplemented with the boundary conditions (3). As already mentioned, along the lines of [2] the rate-independent flow rule for z is formulated by means of a semistability condition and of an energy balance, featuring the *mechanical* energy of the system

$$\mathcal{E}(t, u, z) := \int_\Omega \left(\tfrac{1}{2}\mathbb{C}(z)e(u) : e(u) + G(z, \nabla z)\right) \mathrm{d}x - \langle f(t), u \rangle_{H^1(\Omega;\mathbb{R}^3)}, \tag{19}$$

and the rate-independent dissipation potential

$$\mathcal{R}_1(\dot{z}) := \int_\Omega \mathrm{R}_1(\dot{z})\,\mathrm{d}x \tag{20}$$

with R_1 from (2). Observe that in the present case the mechanical energy equality (25) below shall reflect the nonhomogeneous boundary condition (3b). More specifically, the dynamics of the boundary loading g causes additional rate-dependent energy terms, cf. 3rd and 4th line of (25). Semistability and (mechanical) energy balance are coupled with the weak (distributional-type) formulations of the momentum and enthalpy equations. In particular, the enthalpy equation is weakly formulated with test functions in $W^{1,r'}(0, T; L^{r'}(\Omega)) \cap \mathrm{C}^0([0, T]; W^{1,r'}(\Omega))$ for every $1 \leq r < \frac{5}{4}$. This requirement is tailored to the $L^r(0, T; W^{1,r}(\Omega)) \cap \mathrm{BV}([0, T]; W^{1,r'}(\Omega)^*)$-regularity for the enthalpy variable, which results from the Boccardo-Gallouët-type estimates developed in [2] (and only briefly hinted at in Section 3.3).

The right-hand side of the weakly formulated enthalpy equation will feature the total variation measure $|\dot{z}|$ of z (i.e., the heat produced by the rate-independent dissipation), which is defined on every closed set of the form $A := [t_1, t_2] \times C \subset [0, T] \times \overline{\Omega}$ by

$$|\dot{z}|\,(A) := \int_C \mathrm{R}_1(z(t_2) - z(t_1))\,\mathrm{d}x.$$

Definition 2.1. Given initial data $(u_0, \dot{u}_0, z_0)$ satisfying (9), and θ_0 complying with (10) (so that $w_0 = \mathfrak{h}(\theta_0)$ fulfills (15)), we call a triple (u, z, w) an *energetic solution* to system (18), supplemented with the boundary conditions (3), if the functions (u, z) have the regularity

$$u \in H^1(0,T; H^1(\Omega;\mathbb{R}^3)) \cap W^{1,\infty}(0,T; L^2(\Omega;\mathbb{R}^3)), \quad \text{with } u(t) = g(t) \text{ on } \partial_{\mathrm{D}}\Omega \quad \text{for all } t \in [0,T], \tag{21a}$$

$$z \in L^\infty(0,T; W^{1,q}(\Omega)) \cap L^\infty((0,T)\times\Omega) \cap \mathrm{BV}([0,T]; L^1(\Omega)), \text{ cf. (6d)}, \quad z(t,x) \in [0,1] \text{ for a.a. } (t,x) \in (0,T)\times\Omega, \tag{21b}$$

while

$$w \in L^r(0,T; W^{1,r}(\Omega)) \cap L^\infty(0,T; L^1(\Omega)) \cap \mathrm{BV}([0,T]; W^{1,r'}(\Omega)^*) \tag{21c}$$

for every $1 \le r < \frac{5}{4}$; (u, z, w) satisfy the initial conditions

$$u(0) = u_0, \quad \dot{u}(0) = \dot{u}_0, \quad z(0) = z_0, \quad w(0) = w_0 \quad \text{a.e. in } \Omega; \tag{22}$$

the functions (u, z, w) comply with

- *unidirectionality and semistability*: for a.a. $x \in \Omega$, $z(\cdot, x)\colon [0,T] \to [0,1]$ is nonincreasing, cf. (2), and for a.a. $t \in (0,T)$ it holds

$$\forall\, \tilde{z} \in \mathcal{Z}: \quad \mathcal{E}(t, u(t), z(t)) \le \mathcal{E}(t, u(t), \tilde{z}) + \mathcal{R}_1(\tilde{z} - z(t)), \tag{23}$$

  where $\mathcal{Z}$ is defined in (7) and $\mathcal{E}(t, u, z)$ is given by (19);
- *weak formulation of the momentum equation*: for all $t \in [0,T]$

$$\begin{aligned} &\rho\int_\Omega \dot{u}(t)\cdot v(t)\,\mathrm{d}x - \rho\int_0^t\!\int_\Omega \dot{u}\cdot\dot{v}\,\mathrm{d}x\,\mathrm{d}s + \int_0^t\!\int_\Omega (\mathbb{D}(z,\Theta(w))e(\dot{u}) + \mathbb{C}(z)e(u) - \Theta(w)\,\mathbb{B}) : e(v)\,\mathrm{d}x\,\mathrm{d}s \\ &= \rho\int_\Omega \dot{u}_0\cdot v(0)\,\mathrm{d}x + \int_0^t \langle f, v\rangle_{H^1(\Omega;\mathbb{R}^3)}\,\mathrm{d}s \end{aligned} \tag{24}$$

  for all test functions $v \in L^2(0,T; H^1_{\mathrm{D}}(\Omega;\mathbb{R}^3)) \cap W^{1,1}(0,T; L^2(\Omega;\mathbb{R}^3))$;
- *mechanical energy equality*: for all $t \in [0,T]$

$$\begin{aligned} &\frac{\varrho}{2}\int_\Omega |\dot{u}(t)|^2\,\mathrm{d}x + \mathcal{E}(t, u(t), z(t)) + \int_\Omega (z_0 - z(t))\,\mathrm{d}x + \int_0^t\!\int_\Omega (\mathbb{D}(z,\Theta(w))e(\dot{u}) - \Theta(w)\,\mathbb{B}) : e(\dot{u})\,\mathrm{d}x\,\mathrm{d}s \\ &= \frac{\varrho}{2}\int_\Omega |\dot{u}_0|^2\,\mathrm{d}x + \mathcal{E}(0, u_0, z_0) + \int_0^t \partial_t\mathcal{E}(s, u(s), z(s))\,\mathrm{d}s - \int_0^t \langle f(t), \dot{g}\rangle_{H^1(\Omega;\mathbb{R}^3)}\,\mathrm{d}s \\ &\quad + \rho\left[\int_\Omega \dot{u}(t)\cdot\dot{g}(t)\,\mathrm{d}x - \int_\Omega \dot{u}_0\cdot\dot{g}(0)\,\mathrm{d}x - \int_0^t\!\int_\Omega \dot{u}\cdot\ddot{g}\,\mathrm{d}x\,\mathrm{d}s\right] \\ &\quad + \int_0^t\!\int_\Omega (\mathbb{D}(z,\Theta(w))e(\dot{u}) + \mathbb{C}(z)e(u) - \Theta(w)\,\mathbb{B}) : e(\dot{g})\,\mathrm{d}x\,\mathrm{d}s; \end{aligned} \tag{25}$$

- *weak formulation of the enthalpy equation*: for all $t \in [0,T]$

$$\begin{aligned} &\langle w(t), \eta(t)\rangle_{W^{1,r'}(\Omega)} - \int_0^t\!\int_\Omega \Theta(w)\dot{\eta}\,\mathrm{d}x\,\mathrm{d}s + \int_0^t\!\int_\Omega \mathbb{J}(z,w)\nabla w\cdot\nabla\eta\,\mathrm{d}x\,\mathrm{d}s \\ &= \int_\Omega w_0\,\eta(0)\,\mathrm{d}x + \int_0^t\!\int_\Omega \eta\,|\dot{z}|\,\mathrm{d}x\,\mathrm{d}s + \int_0^t\!\int_\Omega (\mathbb{D}(z,\Theta(w))e(\dot{u}) : e(\dot{u}) - \Theta(w)\,\mathbb{B}) : e(\dot{u})\eta\,\mathrm{d}x\,\mathrm{d}s \\ &\quad + \int_0^t\!\int_{\partial\Omega} h\eta\,\mathrm{d}\mathcal{H}^2(x)\,\mathrm{d}s + \int_0^t\!\int_\Omega H\eta\,\mathrm{d}x\,\mathrm{d}s \end{aligned} \tag{26}$$

  for all test functions $\eta \in W^{1,r'}(0,T; L^{r'}(\Omega)) \cap \mathrm{C}^0([0,T]; W^{1,r'}(\Omega))$.

Observe that, since $r < \frac{5}{4}$, its conjugate exponent r' fulfills $r' > 5$. Hence the test functions η for (26) are continuous on $[0,T] \times \overline{\Omega}$, which makes them in duality with the measure $|\dot{z}|$. Notice that, for simplicity, we write $\int_0^t \int_\Omega \eta\,|\dot{z}|\,\mathrm{d}x\,\mathrm{d}s$ instead of $\iint_{(0,t)\times\Omega} \eta\,|\dot{z}|\,(\mathrm{d}s\,\mathrm{d}x)$.

We are now in the position to state our existence result.

Theorem 2.2. *Under assumptions* (4)–(6), (8) *on the datum* g, *and* (11) *on the data* f_{V}, f_{S}, H, *and* h, *for every quadruple* $(u_0, \dot{u}_0, z_0, \theta_0)$ *fulfilling* (9) *and* (10), *with* z_0 *satisfying* (23) *at* $t = 0$, *there exists an energetic solution* (u, z, w) *to the Cauchy problem for the enthalpy-reformulated system* (18) *such that*

$$w(x,t) \geq 0 \qquad \text{for a.a. } (t,x) \in (0,T)\times\Omega\,. \tag{27}$$

3. Proof of Theorem 2.2

The proof of the existence of an energetic solution to system (18) is based on time discretization, as for many results on rate-independent processes. We consider the solutions to carefully devised incremental problems and give the time-discrete version of the energetic formulation, consisting of the semistability, the weak momentum and enthalpy equations, and the (discrete) mechanical and total energy inequalities. However, due to the time-dependent Dirichlet loading g, we will not be able to deduce from the latter inequalities the basic a priori estimates on the approximate solutions, see also the comments at the beginning of Section 3.3. Therefore, we shall have to deduce a *further* energy inequality, see (48), allowing us to establish the first set of a priori estimates and thus to pass to the time-continuous limit by compactness. Finally, the properties of the energetic solutions at the time-continuous level will be obtained by passing to the limit in the corresponding discrete properties.

In what follows, we will focus on the steps needed to decouple the time-discrete scheme and to account for time-dependent Dirichlet conditions and develop the related calculations with some detail. The other parts of the proof of Theorem 2.2 will be only sketched here and we will systematically refer to [2] and [3].

3.1. Time-discrete scheme

Given a partition

$$0 = t_n^0 < \cdots < t_n^n = T \qquad \text{with} \qquad t_n^k - t_n^{k-1} = \tfrac{T}{n} =: \tau_n\,,$$

we construct a family of discrete solutions $(u_n^k, z_n^k, w_n^k)_{k=1}^n$ by solving the time-discretization scheme (33) below, where the data f, H, and h are approximated by *local means* as follows

$$f_n^k := \tfrac{1}{\tau_n}\int_{t_n^{k-1}}^{t_n^k} f(s)\,\mathrm{d}s\,, \qquad H_n^k := \tfrac{1}{\tau_n}\int_{t_n^{k-1}}^{t_n^k} H(s)\,\mathrm{d}s\,, \qquad h_n^k := \tfrac{1}{\tau_n}\int_{t_n^{k-1}}^{t_n^k} h(s)\,\mathrm{d}s\,, \tag{28}$$

(the above integrals understood in the Bochner sense).

Let us mention in advance that the main feature of system (33) is that the three equations are decoupled one from each other and can thus be solved recursively. This simplifies the analysis of (33) in comparison with [2], where the time-discrete versions of the momentum equation, of the flow rule, and of the enthalpy equations were coupled, and it was necessary to resort to a (non-constructive) fixed point argument to prove the existence of solutions.

In [2], the coupling between the discrete enthalpy and momentum equations served to ensure the nonnegativity of the discrete enthalpy w_n^k: this resulted from the (classical) test of the discrete enthalpy equation by $-(w_n^k)^-$, and to carry out the related calculations it was essential to have the thermal expansion term on the right-hand side *implicit* (i.e. with θ_n^k in place of the term θ_n^{k-1} that now features on the right-hand side of (33c)). Accordingly, the term θ_n^k appeared also in the discrete momentum equation to guarantee the cancelations leading to the discrete (mechanical) energy inequality.

In the present setting, we will be able to prove the nonnegativity of w_n^k by a different argument from the one in [2], based on the subquadratic growth (16) of Θ, cf. Lemma 3.3 below. Thanks to this, it will be possible to keep the discrete enthalpy and momentum equations, and ultimately the *whole* scheme, decoupled.

Nonetheless, because of the quadratic terms featuring on the right-hand side of the enthalpy equation (33c), which a priori is in $L^1(\Omega)$ only, it is necessary to introduce a regularization. One option, as done in [7], is to directly truncate the quadratic terms on the right-hand side of (33c), and then pass to the limit with the truncation parameter. Alternatively, as in [2], we here add a regularizing term of the form $-\tau_n \operatorname{div}(|e(u_n^k)|^2 e(u_n^k))$ to the discrete momentum equation (33b) below. This term ensures that the right-hand side of the discrete enthalpy equation (33c) is in $L^2(\Omega)$ and thus allows us to solve it by standard arguments for elliptic equations. Clearly, $-\tau_n \operatorname{div}(|e(u_n^k)|^2 e(u_n^k))$ will pass to zero with vanishing τ_n. In view of testing the momentum equation by $u-g$, this 4-Laplacian-term requires a regularization of the (extension on $[0,T]\times\Omega$ of the) Dirichlet datum g, and of $\tilde{u}_0 := u_0 - g(0)$. More precisely, using mollifiers as in [8, p. 56, Corollary 2], we approximate $\tilde{u}_0$ by a sequence

$$\tilde{u}_n^0 \in W_0^{1,4}(\Omega;\mathbb{R}^3) \ \text{ such that } \tilde{u}_n^0 \to \tilde{u}_0 \ \text{ in } H_0^1(\Omega;\mathbb{R}^3) \text{ as } n\to\infty \tag{29}$$

and, accordingly (cf. e.g. [9, p. 189] for appropriate mollifiers in time) the datum g by a sequence

$$\mathfrak{g}_n \in W^{1,4}(0,T;W^{1,4}(\Omega;\mathbb{R}^3)) \colon \begin{cases} \sup_{n\in\mathbb{N}} \tau_n^{1/4}\|\dot{\mathfrak{g}}_n\|_{L^4(0,T;W^{1,4}(\Omega;\mathbb{R}^3))} \le C\,, \\ \mathfrak{g}_n \to g \text{ in } H^1(0,T;H^1(\Omega;\mathbb{R}^3)) \cap W^{2,2}(0,T;L^2(\Omega;\mathbb{R}^3))\,. \end{cases} \tag{30}$$

Using $\mathfrak{g}_n$, we then define the discrete data $(\mathfrak{g}_n^k)_{k=1}^n$ by setting

$$\mathfrak{g}_n^k := \mathfrak{g}_n(t_n^k) \qquad \text{for all } k=1,\ldots,n. \tag{31}$$

Finally, for each $n\in\mathbb{N}$, the initial datum is given by $u_n^0 := \tilde{u}_n^0 + \mathfrak{g}_n(0)$.

Problem 3.1. *Starting from u_n^0, $z_n^0 := z_0$, $w_n^0 := w_0$, and $u_n^{-1} := u_0 - \tau_n \dot{u}_0$, find $(u_n^k, z_n^k, w_n^k)_{k=1}^n$ such that*

$$\begin{aligned} u_n^k &\in W^{1,4}(\Omega;\mathbb{R}^3) \qquad \textit{with } u_n^k = \mathfrak{g}_n^k \textit{ on } \partial_{\mathrm{D}}\Omega\,, && (32a)\\ z_n^k &\in W^{1,q}(\Omega)\,,\ \textit{cf. (6d)}, && (32b)\\ w_n^k &\in H^1(\Omega)\,, && (32c)\end{aligned}$$

and, denoting $\theta_n^{k-1} := \Theta(w_n^{k-1})$,

$$z_n^k \in \operatorname{argmin}\left\{ \int_\Omega (z_n^{k-1} - z)\,\mathrm{d}x + \mathcal{E}(t_n^k, u_n^{k-1}, z) \colon z\in W^{1,q}(\Omega)\,,\ 0\le z\le z_n^{k-1}\le 1 \right\}, \tag{33a}$$

$$\begin{aligned} &\rho\int_\Omega \tfrac{u_n^k - 2u_n^{k-1} + u_n^{k-2}}{\tau_n^2}\cdot v\,\mathrm{d}x \\ &+ \int_\Omega \left(\mathbb{D}(z_n^{k-1},\theta_n^{k-1}) e\left(\tfrac{u_n^k - u_n^{k-1}}{\tau_n}\right) + \mathbb{C}(z_n^k) e(u_n^k) - \theta_n^{k-1}\mathbb{B} + \tau_n |e(u_n^k)|^2 e(u_n^k) \right) : e(v)\,\mathrm{d}x \quad (33b)\\ &= \left\langle f_n^k, v\right\rangle_{H^1(\Omega;\mathbb{R}^3)}, \end{aligned}$$

where the above duality pairing again is to be understood in the sense of (12),

$$\begin{aligned}&\int_\Omega \tfrac{w_n^k-w_n^{k-1}}{\tau_n}\,\eta\,\mathrm{d}x+\int_\Omega \mathbb{J}(z_n^k,w_n^{k-1})\nabla w_n^k\cdot\nabla\eta\,\mathrm{d}x\\&=\int_\Omega\Big(\mathbb{D}(z_n^{k-1},\theta_n^{k-1})e\Big(\tfrac{u_n^k-u_n^{k-1}}{\tau_n}\Big)-\theta_n^{k-1}\mathbb{B}\Big):e\Big(\tfrac{u_n^k-u_n^{k-1}}{\tau_n}\Big)\,\eta\,\mathrm{d}x\\&\quad+\int_\Omega\tfrac{z_n^{k-1}-z_n^k}{\tau_n}\,\eta\,\mathrm{d}x+\int_{\partial\Omega}h_n^k\,\eta\,\mathrm{d}\mathcal{H}^2(x)+\int_\Omega H_n^k\,\eta\,\mathrm{d}x\,,\end{aligned}\tag{33c}$$

for every $v\in W_{\mathrm{D}}^{1,4}(\Omega;\mathbb{R}^3):=\{\tilde v\in W^{1,4}(\Omega;\mathbb{R}^3),\ \tilde v=0$ *on* $\partial_{\mathrm{D}}\Omega\}$ *and every* $\eta\in H^1(\Omega)$.

In (33a) the operator "argmin" generates the argument of the minimum, i.e. $\big(\int_\Omega(z_n^{k-1}-z_n^k)\,\mathrm{d}x+\mathcal{E}(t_n^k,u_n^{k-1},z_n^k)\big)=\min_{z\in W^{1,q}(\Omega),\ 0\le z\le z_n^{k-1}\le 1}\big(\int_\Omega(z_n^{k-1}-z)\,\mathrm{d}x+\mathcal{E}(t_n^k,u_n^{k-1},z)\big)$.

We have the following existence result.

Lemma 3.2. *Assume* (4)–(6) *and* (8)–(11). *Then, for every* $n\ge 1$ *there exists a solution* $(u_n^k,z_n^k,w_n^k)_{k=1}^n$ *to Problem 3.1.*

Proof. The *existence of a minimizer for* (33a) follows from the Direct Method of the Calculus of Variations. Indeed, thanks to (6), the functional $z\mapsto\mathcal{R}_1(z-z_n^{k-1})+\mathcal{E}(t_n^k,u_n^{k-1},z)$ (with $\mathcal{R}_1$ from (20)) is coercive and (sequentially) weakly lower semicontinuous on $W^{1,q}(\Omega)$ (see the proof of [3, Prop. 3.2] for all the detailed calculations).

As for the *existence of solutions to* (33b), we observe that it is the Euler equation for the minimum problem

$$\begin{aligned}\min_{u\in\mathcal{A}_n^k}\Big\{&\tfrac{\varrho}{2}\int_\Omega\Big|\tfrac{u-2u_n^{k-1}+u_n^{k-2}}{\tau_n}\Big|^2\mathrm{d}x+\tfrac{\tau_n}{2}\int_\Omega\mathbb{D}(z_n^{k-1},\theta_n^{k-1})e\Big(\tfrac{u-u_n^{k-1}}{\tau_n}\Big):e\Big(\tfrac{u-u_n^{k-1}}{\tau_n}\Big)\,\mathrm{d}x\\&+\tfrac12\int_\Omega\mathbb{C}(z_n^k)e(u):e(u)\,\mathrm{d}x+\tfrac{\tau_n}{4}\int_\Omega|e(u_n^k)|^4\,\mathrm{d}x-\int_\Omega\theta_n^{k-1}\mathbb{B}:e(u)\,\mathrm{d}x-\Big\langle f_n^k,u\Big\rangle_{H^1(\Omega;\mathbb{R}^3)}\Big\}\end{aligned}\tag{34}$$

with $\mathcal{A}_n^k:=\{u\in W^{1,4}(\Omega;\mathbb{R}^3)\colon u=\mathfrak{g}_n^k$ on $\partial_{\mathrm{D}}\Omega\}$. The underlying functional is coercive and strictly convex on $W^{1,4}(\Omega;\mathbb{R}^3)$, hence the existence of a (unique) minimizer again ensues from the Direct Method.

Equation (33c), whose right-hand side is in $L^2(\Omega;\mathbb{R}^3)$, can be tackled by the same arguments as (33b). □

As previously mentioned, the subquadratic growth (16) of Θ is at the core of the proof of Lemma 3.3 below. Therein, the nonnegativity of the discrete enthalpy w_n^k is deduced by a *direct* argument that does not necessitate the implicit term θ_n^k on the right-hand side of (33c). Nonetheless, let us mention that a *strict positivity* result for w_n^k seems to be out of reach in the present context, while it is available with a fully implicit discrete enthalpy equation, cf. e.g. [10, Lemma 7.4] (the latter paper analyzing a temperature-dependent system for rate-independent adhesive contact).

Lemma 3.3. *Under assumptions* (4)–(6) *and* (8)–(11), *there exists* $\bar n\in\mathbb{N}$ *such that for all* $n\ge\bar n$ *we have* $w_n^k\ge 0$ *a.e. in* Ω *for every* $k=1,\ldots,n$.

Proof. We argue by induction. For all $n\in\mathbb{N}$ we have $w_n^0=w_0\ge 0$ a.e. in Ω by assumption. Consider now $n\in\mathbb{N}$ arbitrary but fixed. Suppose that $w_n^{k-1}\ge 0$. Taking $\eta=-(w_n^k)^-:=\max\{0,-w_n^k\}$ in (33c) we obtain

$$\int_\Omega\tfrac{((w_n^k)^-)^2}{\tau_n}\,\mathrm{d}x+\int_\Omega\Big(\mathbb{J}(z_n^k,w_n^{k-1})\nabla(w_n^k)^-\Big)\cdot\nabla(w_n^k)^-\,\mathrm{d}x$$

$$= -\int_\Omega \left[\tfrac{w_n^{k-1}}{\tau_n} + \left(\mathbb{D}(z_n^{k-1},\theta_n^{k-1})e\left(\tfrac{u_n^k-u_n^{k-1}}{\tau_n}\right) - \theta_n^{k-1}\mathbb{B}\right) : e\left(\tfrac{u_n^k-u_n^{k-1}}{\tau_n}\right)\right](w_n^k)^- \,\mathrm{d}x \qquad (35)$$
$$- \int_\Omega \tfrac{z_n^{k-1}-z_n^k}{\tau_n}(w_n^k)^- \,\mathrm{d}x - \int_{\partial\Omega} h_n^k (w_n^k)^- \,\mathrm{d}\mathcal{H}^2(x) - \int_\Omega H_n^k (w_n^k)^- \,\mathrm{d}x\,.$$

We now remark that the left-hand side of the previous equality is nonnegative and the right-hand side is nonpositive. Indeed, $(w_n^k)^- \geq 0$, $z_n^{k-1} - z_n^k \geq 0$, $h_n^k \geq 0$, $H_n^k \geq 0$ a.e. in Ω, and

$$\tfrac{w_n^{k-1}}{\tau_n} + \left(\mathbb{D}(z_n^{k-1},\theta_n^{k-1})e\left(\tfrac{u_n^k-u_n^{k-1}}{\tau_n}\right) - \theta_n^{k-1}\mathbb{B}\right) : e\left(\tfrac{u_n^k-u_n^{k-1}}{\tau_n}\right)$$
$$\geq \tfrac{w_n^{k-1}}{\tau_n} + C_1\left|e\left(\tfrac{u_n^k-u_n^{k-1}}{\tau_n}\right)\right|^2 - \theta_n^{k-1}|\mathbb{B}|\left|e\left(\tfrac{u_n^k-u_n^{k-1}}{\tau_n}\right)\right|$$

(recall that $\theta_n^{k-1} = \Theta(w_n^{k-1})$). The right-hand side of the last inequality is a nonnegative second-order polynomial in $\left|e\left(\tfrac{u_n^k-u_n^{k-1}}{\tau_n}\right)\right|$, since by growth condition (16)

$$(\theta_n^{k-1}|\mathbb{B}|)^2 - 4\tfrac{w_n^{k-1}}{\tau_n}C_1 \leq \left((C_\Theta^2|\mathbb{B}|)^2 - 4\tfrac{C_1}{\tau_n}\right)w_n^{k-1} \leq 0$$

for any $\tau_n = \frac{T}{n} \leq 4C_1/(C_\Theta^2|\mathbb{B}|)^2)$.

Hence, from (35) we deduce that $(w_n^k)^- = 0$ a.e. in Ω for every $n \in \mathbb{N}$ sufficiently large, whence the thesis. □

3.2. Approximate solutions and time-discrete version of the energetic formulation.
We now define the approximate solutions to (the energetic formulation of) the initial-boundary value problem for system (1) by suitably interpolating the discrete solutions $(u_n^k, z_n^k, w_n^k)_{k=1}^n$. We introduce the piecewise constant interpolants

$$\overline{u}_n(t) := u_n^k\,, \qquad \overline{w}_n(t) := w_n^k\,, \qquad \overline{z}_n(t) := z_n^k\,, \qquad (36a)$$
$$\underline{u}_n(t) := u_n^{k-1}\,, \quad \underline{w}_n(t) := w_n^{k-1}\,, \quad \underline{z}_n(t) := z_n^{k-1}\,, \qquad (36b)$$

for $t \in (t_n^{k-1}, t_n^k]$, $k = 1, \ldots, n$, and the piecewise linear interpolants

$$u_n(t) := \tfrac{t-t_n^{k-1}}{\tau_n}u_n^k + \tfrac{t_n^k-t}{\tau_n}u_n^{k-1}, \quad z_n(t) := \tfrac{t-t_n^{k-1}}{\tau_n}z_n^k + \tfrac{t_n^k-t}{\tau_n}z_n^{k-1}, \quad w_n(t) := \tfrac{t-t_n^{k-1}}{\tau_n}w_n^k + \tfrac{t_n^k-t}{\tau_n}w_n^{k-1}. \qquad (36c)$$

We set $\overline{u}_n(0) = \underline{u}_n(0) = u_n(0) := u_0$, and analogously for $\overline{z}_n, \ldots, w_n$. We will use the notation $\underline{\theta}_n$ for $\Theta(\underline{w}_n)$.

We also introduce the piecewise constant and linear interpolants of the discrete data $(f_n^k, H_n^k, h_n^k)_{k=1}^n$ in (28) by setting for $t \in (t_n^{k-1}, t_n^k]$

$$\overline{f}_n(t) := f_n^k\,, \qquad \overline{H}_n(t) := H_n^k\,, \qquad \overline{h}_n(t) := h_n^k\,,$$

and $f_n(t) := \frac{t-t_n^{k-1}}{\tau_n}f_n^k + \frac{t_n^k-t}{\tau_n}f_n^{k-1}$ with time derivative $\dot{f}_n(t) := \frac{f_n^k - f_n^{k-1}}{\tau_n}$. It follows from (11) that, as $n\to\infty$,

$$\overline{f}_n \to f \quad \text{in } L^p(0,T;H^1(\Omega;\mathbb{R}^3)^*) \ \text{ for all } 1 \leq p < \infty\,,$$
$$\overline{f}_n(t) \to f(t) \quad \text{in } H^1(\Omega;\mathbb{R}^3)^* \ \text{ for all } t \in [0,T]\,, \qquad (37a)$$
$$f_n \rightharpoonup f \quad \text{in } H^1(0,T;H^1(\Omega;\mathbb{R}^3)^*)\,, \qquad (37b)$$
$$\overline{H}_n \to H \quad \text{in } L^2(0,T;L^2(\Omega))\,, \quad \overline{h}_n \to h \quad \text{in } L^2(0,T;L^2(\partial\Omega))\,. \qquad (37c)$$

We shall denote by $\overline{\tau}_n$ the piecewise constant interpolant associated with the partition, i.e., $\overline{\tau}_n(t) := t_n^k$ for $t \in (t_n^{k-1}, t_n^k]$, with $\overline{\tau}_n(0) := 0$.

For the approximating discrete data of the Dirichlet loading $g \in H^1(0,T;H^1(\Omega;\mathbb{R}^3)) \cap W^{2,2}(0,T;L^2(\Omega;\mathbb{R}^3))$, cf. (30) and (31), observe that $\mathfrak{g}_n^k \in W^{1,4}(\Omega;\mathbb{R}^3)$ for all $k = 1,\dots,n$: hence, $u_n^k - \mathfrak{g}_n^k - (u_n^{k-1} - \mathfrak{g}_n^{k-1})$ is an admissible test function for (33b). Thus, based on (31), we introduce the piecewise constant and linear interpolants

$$\overline{g}_n(t) := \mathfrak{g}_n^k, \quad \underline{g}_n(t) := \mathfrak{g}_n^{k-1}, \quad g_n(t) := \tfrac{t-t_n^{k-1}}{\tau_n}\mathfrak{g}_n^k + \tfrac{t_n^k - t}{\tau_n}\mathfrak{g}_n^{k-1} \tag{38}$$

for $t \in (t_n^{k-1}, t_n^k]$, $k = 1,\dots,n$. Observe that, in order to make the notation more transparent we still use the letter g for the above interpolants, so that it will be clear that the functions $\overline{g}_n$, $\underline{g}_n$, g_n are approximations of g. In fact, arguing on a diagonal sequence it can be shown that

$$\begin{aligned} &\overline{g}_n(t) \to g(t) \quad \text{in } H^1(\Omega;\mathbb{R}^3) \ \text{ for all } t \in [0,T], \\ &g_n \to g \quad \text{in } H^1(0,T;H^1(\Omega;\mathbb{R}^3)), \\ &\sup_{n\in\mathbb{N}} \tau_n^{1/4} \|\dot{g}_n\|_{L^4(0,T;W^{1,4}(\Omega;\mathbb{R}^3))} \le C. \end{aligned} \tag{39}$$

We shall also work with the piecewise linear interpolant of the values $\frac{\mathfrak{g}_n^k - \mathfrak{g}_n^{k-1}}{\tau_n}$, namely with $\gamma_n : [0,T] \to W^{1,4}(\Omega;\mathbb{R}^3)$ defined by

$$\gamma_n(t) := \tfrac{t-t_n^{k-1}}{\tau_n}\tfrac{\mathfrak{g}_n^k - \mathfrak{g}_n^{k-1}}{\tau_n} + \tfrac{t_n^k - t}{\tau_n}\tfrac{\mathfrak{g}_n^{k-1} - \mathfrak{g}_n^{k-2}}{\tau_n} \quad \text{for } t \in (t_n^{k-1}, t_n^k] \text{ and } k = 2,\dots,n, \tag{40}$$

and $\gamma_n(t) := \frac{t}{\tau_n}\frac{\mathfrak{g}_n^1 - \mathfrak{g}_n^0}{\tau_n} + \frac{t_n^1 - t}{\tau_n}\frac{\mathfrak{g}_n^0 - \mathfrak{g}_n^{-1}}{\tau_n}$ with $\mathfrak{g}_n^{-1} := \mathfrak{g}_n^0 - \tau_n \dot{\mathfrak{g}}_n(0)$ for any $t \in [0, t_n^1]$. By construction we have $\dot{\gamma}_n(t) = \frac{\mathfrak{g}_n^k - 2\mathfrak{g}_n^{k-1} + \mathfrak{g}_n^{k-2}}{\tau_n^2}$ for all $t \in (t_n^{k-1}, t_n^k)$. Again, by (8), (30) and an argument along a diagonal sequence (also taking into account that $\|\gamma_n - \dot{g}_n\|_{L^\infty(0,T;L^2(\Omega))} \le \tau_n^{1/2}\|\dot{\gamma}_n\|_{L^\infty(0,T;L^2(\Omega))} \le C\tau_n^{1/2}$ to identify the limit of $(\gamma_n)_n$), one obtains

$$\gamma_n \to \dot{g} \ \text{ in } L^2(0,T;H^1(\Omega;\mathbb{R}^3)) \cap H^1(0,T;L^2(\Omega;\mathbb{R}^3)). \tag{41}$$

Proposition 3.4 below states that the approximate solutions constructed in the above lines indeed fulfill the discrete version of the energetic formulation. In the (discrete) mechanical energy inequality (44c) below, the mechanical energy $\mathcal{E}$ will be replaced by

$$\mathcal{E}_n(t,u,z) := \int_\Omega \left(\tfrac{1}{2}\mathbb{C}(z)e(u) : e(u) + \tfrac{\tau_n}{4}|e(u)|^4\right) \mathrm{d}x + \mathcal{G}(z,\nabla z) - \left\langle \overline{f}_n(t), u \right\rangle_{H^1(\Omega;\mathbb{R}^3)} \tag{42}$$

and we will understand the pointwise terms $\dot{u}_n$ and $\dot{g}_n$ as

$$\dot{u}_n(t) := \tfrac{u_n^k - u_n^{k-1}}{\tau_n}, \quad \dot{g}_n(t) := \tfrac{\mathfrak{g}_n^k - \mathfrak{g}_n^{k-1}}{\tau_n}, \quad \text{for } t \in (t_n^{k-1}, t_n^k], \quad \text{for } k = 1,\dots,n. \tag{43}$$

Proposition 3.4. *Assume (4)–(6) and (8)–(11). Then the interpolants of the time-discrete solutions $(\overline{u}_n, \underline{u}_n, u_n, \overline{z}_n, \underline{z}_n, z_n, \overline{w}_n, \underline{w}_n, w_n)$ obtained via Problem 3.1 have the following properties:*

- *unidirectionality: for a.a. $x \in \Omega$, the functions $\overline{z}_n(\cdot,x)\colon [0,T] \to [0,1]$ are nonincreasing;*
- *discrete semistability: for all $t \in [0,T]$*

$$\forall\, \tilde{z} \in \mathcal{Z}\colon \quad \mathcal{E}_n(t, \underline{u}_n(t), \overline{z}_n(t)) \le \mathcal{E}_n(t, \underline{u}_n(t), \tilde{z}) + \mathcal{R}_1(\tilde{z} - \overline{z}_n(t)); \tag{44a}$$

- *discrete formulation of the momentum equation: for all* $t \in [0,T]$ *and for every* $(n+1)$*-tuple* $(v_n^k)_{k=0}^n \subset W_{\mathrm{D}}^{1,4}(\Omega;\mathbb{R}^3)$*, setting* $\overline{v}_n(s) := v_n^k$ *and* $v_n(s) := \frac{s-t_n^{k-1}}{\tau_n}v_n^k + \frac{t_n^k-s}{\tau_n}v_n^{k-1}$ *for* $s \in (t_n^{k-1}, t_n^k]$*,*

$$
\begin{aligned}
&\rho\int_\Omega \big(\dot{u}_n(t)\cdot\overline{v}_n(t) - \dot{u}_0\cdot v_n(0)\big)\,\mathrm{d}x - \rho\int_0^{\overline{\tau}_n(t)}\int_\Omega \dot{u}_n(s-\tau_n)\cdot\dot{v}_n(s)\,\mathrm{d}x\,\mathrm{d}s\\
&\quad+\int_0^{\overline{\tau}_n(t)}\int_\Omega \Big(\mathbb{D}(\underline{z}_n,\underline{\theta}_n)e(\dot{u}_n) + \mathbb{C}(\overline{z}_n)e(\overline{u}_n) - \underline{\theta}_n\,\mathbb{B} + \tau_n|e(\overline{u}_n)|^2 e(\overline{u}_n)\Big) : e(\overline{v}_n)\,\mathrm{d}x\,\mathrm{d}s\\
&\quad= \int_0^{\overline{\tau}_n(t)} \big\langle \overline{f}_n, \overline{v}_n\big\rangle_{H^1(\Omega;\mathbb{R}^3)}\,\mathrm{d}s\,,
\end{aligned}
\tag{44b}
$$

where we have used $\underline{\theta}_n := \Theta(\underline{w}_n)$ *and extended* u_n *to* $(-\tau_n, 0]$ *by setting* $u_n(t) := u_n^0 + t\dot{u}_0$*; again, the above duality pairing has the meaning of (12);*

- *discrete mechanical energy inequality: for all* $t \in [0,T]$

$$
\begin{aligned}
&\frac{\rho}{2}\int_\Omega |\dot{u}_n(t)|^2\,\mathrm{d}x + \mathcal{E}_n(t,\overline{u}_n(t),\overline{z}_n(t)) + \int_\Omega (z_0 - \overline{z}_n(t))\,\mathrm{d}x\\
&\quad+\int_0^{\overline{\tau}_n(t)}\int_\Omega \big(\mathbb{D}(\underline{z}_n,\underline{\theta}_n)e(\dot{u}_n) - \underline{\theta}_n\,\mathbb{B}\big) : e(\dot{u}_n)\,\mathrm{d}x\,\mathrm{d}s\\
&\leq \frac{\rho}{2}\int_\Omega |\dot{u}_0|^2\,\mathrm{d}x + \mathcal{E}_n(0,u_n^0,z_0) - \int_0^{\overline{\tau}_n(t)} \big\langle \dot{f}_n, \underline{u}_n\big\rangle_{H^1(\Omega;\mathbb{R}^3)}\,\mathrm{d}s - \int_0^{\overline{\tau}_n(t)} \langle f_n, \dot{g}_n\rangle_{H^1(\Omega;\mathbb{R}^3)}\,\mathrm{d}s\\
&\quad+\rho\left[\int_\Omega \dot{u}_n(t)\cdot\dot{g}_n(t)\,\mathrm{d}x - \int_\Omega \dot{u}_0\cdot\dot{g}(0)\,\mathrm{d}x - \int_0^{\overline{\tau}_n(t)}\int_\Omega \dot{u}_n(s-\tau_n)\cdot\dot{\gamma}_n(s)\,\mathrm{d}x\,\mathrm{d}s\right]\\
&\quad+\int_0^{\overline{\tau}_n(t)}\int_\Omega \Big(\mathbb{D}(\underline{z}_n,\underline{\theta}_n)e(\dot{u}_n) + \mathbb{C}(\overline{z}_n)e(\overline{u}_n) - \underline{\theta}_n\,\mathbb{B} + \tau_n|e(\overline{u}_n)|^2 e(\overline{u}_n)\Big) : e(\dot{g}_n)\,\mathrm{d}x\,\mathrm{d}s\,;
\end{aligned}
\tag{44c}
$$

- *discrete formulation of the enthalpy equation: for all* $t \in [0,T]$ *and for every* $(n+1)$*-tuple* $(\eta_n^k)_{k=0}^n \subset H^1(\Omega)$*, setting* $\overline{\eta}_n(s) := \eta_n^k$ *and* $\eta_n(s) := \frac{s-t_n^{k-1}}{\tau_n}\eta_n^k + \frac{t_n^k-s}{\tau_n}\eta_n^{k-1}$ *for* $s \in (t_n^{k-1}, t_n^k]$*,*

$$
\begin{aligned}
&\int_\Omega \overline{w}_n(t)\overline{\eta}_n(t)\,\mathrm{d}x - \int_\Omega w_0\,\eta_n(0)\,\mathrm{d}x - \int_0^{\overline{\tau}_n(t)}\int_\Omega \underline{w}_n(s)\dot{\eta}_n(s)\,\mathrm{d}x\,\mathrm{d}s\\
&\quad+\int_0^{\overline{\tau}_n(t)}\int_\Omega \big(\mathbb{J}(\overline{z}_n,\underline{w}_n)\nabla\overline{w}_n\big)\cdot\nabla\overline{\eta}_n\,\mathrm{d}x\,\mathrm{d}s\\
&\quad= \int_0^{\overline{\tau}_n(t)}\int_\Omega \overline{\eta}_n\,|\dot{z}_n|\,\mathrm{d}x\,\mathrm{d}s \int_0^{\overline{\tau}_n(t)}\int_\Omega \big(\mathbb{D}(\underline{z}_n,\underline{\theta}_n)e(\dot{u}_n) - \underline{\theta}_n\,\mathbb{B}\big) : e(\dot{u}_n)\,\overline{\eta}_n\,\mathrm{d}x\,\mathrm{d}s\\
&\quad+\int_0^{\overline{\tau}_n(t)}\left[\int_{\partial\Omega} \overline{h}_n\,\eta_n\,\mathrm{d}\mathcal{H}^2(x) + \big\langle \overline{H}_n, \eta_n\big\rangle_{H^1(\Omega)}\right]\mathrm{d}s\,;
\end{aligned}
\tag{44d}
$$

- *discrete total energy inequality: for all* $t \in [0,T]$

$$
\begin{aligned}
&\frac{\rho}{2}\int_\Omega |\dot{u}_n(t)|^2\,\mathrm{d}x + \mathcal{E}_n(t,\overline{u}_n(t),\overline{z}_n(t)) + \int_\Omega \overline{w}_n(t)\,\mathrm{d}x\\
&\leq \frac{\rho}{2}\int_\Omega |\dot{u}_0|^2\,\mathrm{d}x + \mathcal{E}_n(0,u_n^0,z_0) + \int_\Omega w_0\,\mathrm{d}x\\
&\quad-\int_0^{\overline{\tau}_n(t)} \big\langle \dot{f}_n, \underline{u}_n\big\rangle_{H^1(\Omega;\mathbb{R}^3)}\,\mathrm{d}s - \int_0^{\overline{\tau}_n(t)} \langle f_n, \dot{g}_n\rangle_{H^1(\Omega;\mathbb{R}^3)}\,\mathrm{d}s
\end{aligned}
$$

$$
\begin{aligned}
&+\rho\left[\int_\Omega \dot{u}_n(t)\cdot\dot{g}_n(t)\,\mathrm{d}x-\int_\Omega \dot{u}_0\cdot\dot{g}(0)\,\mathrm{d}x-\int_0^{\overline{\tau}_n(t)}\int_\Omega \dot{u}_n(s-\tau_n)\cdot\dot{\gamma}_n(s)\,\mathrm{d}x\,\mathrm{d}s\right]\\
&+\int_0^{\overline{\tau}_n(t)}\int_\Omega\left(\mathbb{D}(\underline{z}_n,\underline{\theta}_n)e(\dot{u}_n)+\mathbb{C}(\overline{z}_n)e(\overline{u}_n)+\tau_n|e(\overline{u}_n)|^2e(\overline{u}_n)\right):e(\dot{g}_n)\,\mathrm{d}x\,\mathrm{d}s\\
&+\int_0^{\overline{\tau}_n(t)}\left[\int_{\partial\Omega}\overline{h}_n\,\mathrm{d}\mathcal{H}^2(x)+\int_\Omega\overline{H}_n\,\mathrm{d}x\right]\mathrm{d}s\,. \qquad (44e)
\end{aligned}
$$

The proof of Proposition 3.4 closely follows the procedure developed in [2, Lemma 4.1], cf. also [3, Prop. 3.3] for the details with regard to our particular system. Let us here just hint at the main ideas, in particular dwelling on the treatment of the Dirichlet datum:

- The discrete semistability (44a) can be directly read from the minimality of z_n^k in (33a) tested by $\tilde{z}\le z_n^{k-1}$, also using that $\mathcal{R}_1(\tilde{z}-z_n^k)=+\infty$ if $\tilde{z}>z_n^k$. This property in particular enforces $z_n^k\le z_n^{k-1}$, whence unidirectionality.
- The discrete momentum and enthalpy equations (44b) and (44d) follow from (33b) and (33c), with test functions $(v_n^k)_{k=0}^n\subset W_\mathrm{D}^{1,4}(\Omega;\mathbb{R}^3)$ and $(\eta_n^k)_{k=0}^n\subset H^1(\Omega)$, respectively, applying the following discrete integration-by-parts formula, for every $(r_k)_{k=1}^n\subset X$ and $(s_k)_{k=1}^n\subset X^*$, with X a given Banach space (and $\langle\cdot,\cdot\rangle_X$ the duality pairing between X^* and X):

$$
\sum_{k=1}^n\langle s_k,r_k-r_{k-1}\rangle_X=\langle s_n,r_n\rangle_X-\langle s_0,r_0\rangle_X-\sum_{k=1}^n\langle s_k-s_{k-1},r_{k-1}\rangle_X\,. \qquad (45)
$$

- The mechanical energy inequality (44c) results from summing (33a), tested by z_n^{k-1}, with the momentum balance (33b), tested by $v=u_n^k-\mathfrak{g}_n^k-u_n^{k-1}+\mathfrak{g}_n^{k-1}$. For the details of this calculation we refer to [3, Prop. 3.3]. Here we explain how the terms in (44c) involving the Dirichlet data (2nd and 3rd line of the RHS) emanate from (33b): Applying elementary convexity inequalities to (33b) tested by $v=u_n^k-\mathfrak{g}_n^k-u_n^{k-1}+\mathfrak{g}_n^{k-1}$ yields

$$
\rho\int_\Omega\frac{u_n^k-2u_n^{k-1}+u_n^{k-2}}{\tau_n^2}\cdot(u_n^k-u_n^{k-1})\,\mathrm{d}x\ge\rho\int_\Omega\left(\frac{1}{2}\frac{|u_n^k-u_n^{k-1}|^2}{\tau_n^2}-\frac{1}{2}\frac{|u_n^{k-1}-u_n^{k-2}|^2}{\tau_n^2}\right)\mathrm{d}x\,, \qquad (46a)
$$

$$
\int_\Omega\mathbb{C}(z_n^k)e(u_n^k):(e(u_n^k)-e(u_n^{k-1}))\,\mathrm{d}x\ge\int_\Omega\frac{1}{2}\Big(\mathbb{C}(z_n^k)e(u_n^k):e(u_n^k)-\mathbb{C}(z_n^k)e(u_n^{k-1}):e(u_n^{k-1})\Big)\,\mathrm{d}x, \qquad (46b)
$$

$$
\int_\Omega\tau_n|e(u_n^k)|^2e(u_n^k):(e(u_n^k)-e(u_n^{k-1}))\,\mathrm{d}x\ge\int_\Omega\left(\tfrac{\tau_n}{4}|e(u_n^k)|^4-\tfrac{\tau_n}{4}|e(u_n^{k-1})|^4\right)\mathrm{d}x\,. \qquad (46c)
$$

By (38), the term $-\big(\mathbb{C}(z_n^k)e(u_n^k)+\mathbb{D}(z_n^k,\theta_n^k)e(\dot{u}_n)-\mathbb{B}\theta_n^{k-1}+\tau_n|e(u_n^{k-1})|^2e(u_n^{k-1})\big):(\mathfrak{g}_n^k-\mathfrak{g}_n^{k-1})$ results in the third line on the right-hand side of (44c). Further, let $t\in(0,T]$ be fixed, and let $1\le j\le n$ be such that $t\in(t_n^{j-1},t_n^j]$. We sum (46a)–(46c) over the index $k=1,\dots,j$. Applying the integration-by-parts formula (45) we conclude that

$$
\begin{aligned}
\sum_{k=1}^j\left\langle f_n^k,u_n^k-u_n^{k-1}\right\rangle_{H^1(\Omega;\mathbb{R}^3)}&=\int_0^{\overline{\tau}_n(t)}\left\langle\overline{f}_n,\dot{u}_n\right\rangle_{H^1(\Omega;\mathbb{R}^3)}\mathrm{d}s\\
&=\left\langle\overline{f}_n(t),\overline{u}_n(t)\right\rangle_{H^1(\Omega;\mathbb{R}^3)}-\langle f(0),u_0\rangle_{H^1(\Omega;\mathbb{R}^3)}-\int_0^{\overline{\tau}_n(t)}\left\langle\dot{f}_n,\underline{u}_n\right\rangle_{H^1(\Omega;\mathbb{R}^3)}\mathrm{d}s\,.
\end{aligned} \qquad (47)
$$

Analogously, to deal with the term $\sum_{k=1}^j\rho\int_\Omega\frac{u_n^k-2u_n^{k-1}+u_n^{k-2}}{\tau_n^2}\cdot(\mathfrak{g}_n^k-\mathfrak{g}_n^{k-1})\,\mathrm{d}x$ we apply (45) with $s_k=\frac{g_n^k-g_n^{k-1}}{\tau_n}$, $r_k=\rho\frac{u_n^k-u_n^{k-1}}{\tau_n}$ and $r_{k-1}=\rho\frac{u_n^{k-1}-u_n^{k-2}}{\tau_n}$, which leads to the fifth, sixth, and seventh terms on the right-hand side of (44c).

- Finally, the discrete total energy inequality ensues from summing the discrete mechanical energy inequality (44c) with the discrete enthalpy equation (33c), tested for $\eta = \tau_n$ and added up over $k = 1, \ldots, j$. Some terms cancel, leading to (44e).

3.3. A priori estimates

Usually, the first set of a priori estimates is deduced from the (discrete versions of the) mechanical and the total energy balance by a Gronwall argument exploiting the boundedness of the initial energy and of the power of the external loadings. However observe that, due to the time-dependent Dirichlet datum g, the right-hand sides of both (44c) and (44e) contain the term $\mathbb{D}(\underline{z}_n, \underline{\theta}_n)e(\dot{u}_n) : e(\dot{g}_n)$, which cannot be estimated if the viscous tensor $\mathbb{D}$ depends on (z, θ). Note, if $\mathbb{D}$ were constant, an integration by parts in time would allow us to control that term with $\int_0^{\overline{\tau}_n(t)} \|\underline{u}_n(s)\|^2_{H^1(\Omega;\mathbb{R}^3)}\,\mathrm{d}s$, under suitable conditions on $\ddot{g}$.

Instead, we have to develop an alternative estimate that contains $\mathbb{D}(\underline{z}_n, \underline{\theta}_n)e(\dot{u}_n) : e(\dot{g}_n)$ on both sides of the inequality, such that the term on the right can be absorbed by the corresponding one on the left-hand side. Hereby, the square-root growth of the enthalpy variable, cf. (16), generated by assumption (4a) on the heat capacity, will play a crucial role. More precisely, for the above described argument we sum up (44c) with (44d) tested by $\eta = \frac{\tau_n}{2}$ and obtain the second discrete total energy inequality, namely

$$
\begin{aligned}
&\tfrac{\varrho}{2}\int_\Omega |\dot{u}_n(t)|^2\,\mathrm{d}x + \mathcal{E}_n(t, \overline{u}_n(t), \overline{z}_n(t)) + \tfrac{1}{2}\int_\Omega (z_0 - \overline{z}_n(t))\,\mathrm{d}x \\
&\quad + \tfrac{1}{2}\int_\Omega \overline{w}_n(t)\,\mathrm{d}x + \tfrac{1}{2}\int_0^{\overline{\tau}_n(t)}\int_\Omega (\mathbb{D}(\underline{z}_n, \underline{\theta}_n)e(\dot{u}_n) - \underline{\theta}_n\,\mathbb{B}) : e(\dot{u}_n)\,\mathrm{d}x\,\mathrm{d}s \\
&\le \tfrac{\varrho}{2}\int_\Omega |\dot{u}_0|^2\,\mathrm{d}x + \mathcal{E}_n(0, u_n^0, z_0) - \int_0^{\overline{\tau}_n(t)} \left\langle \dot{f}_n, \underline{u}_n \right\rangle_{H^1(\Omega;\mathbb{R}^3)} \mathrm{d}s - \int_0^{\overline{\tau}_n(t)} \langle \overline{f}_n, g_n \rangle_{H^1(\Omega;\mathbb{R}^3)}\,\mathrm{d}s \\
&\quad + \rho\left[\int_\Omega \dot{u}_n(t)\cdot\dot{g}_n(t)\,\mathrm{d}x - \int_\Omega \dot{u}_0\cdot\dot{g}(0)\,\mathrm{d}x - \int_0^{\overline{\tau}_n(t)}\int_\Omega \dot{u}_n(s-\tau_n)\cdot\dot{\gamma}_n(s)\,\mathrm{d}x\,\mathrm{d}s\right] \\
&\quad + \int_0^{\overline{\tau}_n(t)}\int_\Omega \left(\mathbb{D}(\underline{z}_n, \underline{\theta}_n)e(\dot{u}_n) + \mathbb{C}(\overline{z}_n)e(\overline{u}_n) - \underline{\theta}_n\,\mathbb{B} + \tau_n|e(\overline{u}_n)|^2 e(\overline{u}_n)\right) : e(\dot{g}_n)\,\mathrm{d}x\,\mathrm{d}s \\
&\quad + \tfrac{1}{2}\int_\Omega w_0\,\mathrm{d}x + \tfrac{1}{2}\int_0^{\overline{\tau}_n(t)}\left[\int_{\partial\Omega} \overline{h}_n\,\mathrm{d}\mathcal{H}^2(x) + \int_\Omega \overline{H}_n\,\mathrm{d}x\right]\mathrm{d}s
\end{aligned}
\tag{48}
$$

for all $t \in [0, T]$.

In the next lines we explain how to derive, starting from (48), estimates (49a)–(49g), cf. the forthcoming Proposition 3.5. For notational simplicity, we will use the symbols c, C to denote all the positive constants popping out in the following calculations, possibly varying from line to line. To control from below the left-hand side of (48) ($\doteq$ LHS(48)), first of all by Young's inequality and by the subquadratic growth (16) of Θ we get

$$
\left|\int_0^{\overline{\tau}_n(t)}\int_\Omega \underline{\theta}_n\,\mathbb{B} : e(\dot{u}_n)\,\mathrm{d}x\,\mathrm{d}s\right| \le \tfrac{1}{4\delta}\int_0^{\overline{\tau}_n(t)} \|\overline{w}_n(s)\|_{L^1(\Omega)}\,\mathrm{d}s + \delta\int_0^{\overline{\tau}_n(t)} \|e(\dot{u}_n(s))\|^2_{L^2(\Omega;\mathbb{R}^{3\times3}_{\mathrm{sym}})}\,\mathrm{d}s\,,
$$

where δ is a positive constant that shall be chosen later. Hence, by the definition (42) of the mechanical energy $\mathcal{E}_n$, and by (5), (6), and (11a) we have

$$
\begin{aligned}
\mathrm{LHS}(48) \ge -C + \tfrac{\varrho}{2}\|\dot{u}_n(t)\|^2_{L^2(\Omega;\mathbb{R}^3)} + C\left(\|\overline{u}_n(t)\|^2_{H^1(\Omega;\mathbb{R}^3)} + \tau_n\|\overline{u}_n(t)\|^4_{W^{1,4}(\Omega;\mathbb{R}^3)} + \|\overline{z}_n(t)\|^q_{W^{1,q}(\Omega)}\right) \\
+ \tfrac{1}{2}\|\overline{w}_n(t)\|_{L^1(\Omega)} + C\int_0^{\overline{\tau}_n(t)} \|e(\dot{u}_n(s))\|^2_{L^2(\Omega;\mathbb{R}^{3\times3}_{\mathrm{sym}})}\,\mathrm{d}s
\end{aligned}
$$

$$
-\tfrac{1}{4\delta}\int_0^{\overline{\tau}_n(t)} \|\overline{w}_n(s)\|_{L^1(\Omega)}\,\mathrm{d}s - \delta\int_0^{\overline{\tau}_n(t)} \|e(\dot{u}_n(s))\|^2_{L^2(\Omega;\mathbb{R}^{3\times3}_{\mathrm{sym}})}\,\mathrm{d}s\,.
$$

Here we have dropped the nonnegative term $\frac{1}{2}\int_\Omega(z_0-\overline{z}_n(t))\,\mathrm{d}x$.

On the right-hand side of (48), the terms depending only on the initial and external data are uniformly bounded thanks to (9), (15), and (37). For the third summand, we use the Cauchy inequality as follows:

$$
\left|\int_0^{\overline{\tau}_n(t)} \left\langle \dot{f}_n, \underline{u}_n\right\rangle_{H^1(\Omega;\mathbb{R}^3)}\,\mathrm{d}x\right| \le \tfrac{1}{2}\int_0^{\overline{\tau}_n(t)} \|\underline{u}_n(s)\|^2_{H^1(\Omega;\mathbb{R}^3)}\,\mathrm{d}s + \tfrac{1}{2}\int_0^{\overline{\tau}_n(t)} \|\dot{f}_n\|^2_{H^1(\Omega;\mathbb{R}^3)^*}\,\mathrm{d}s\,.
$$

Moreover,

$$
\begin{aligned}
&\int_\Omega \dot{u}_n(t)\cdot\dot{g}_n(t)\,\mathrm{d}x - \int_0^{\overline{\tau}_n(t)}\int_\Omega \dot{u}_n(s-\tau_n)\cdot\dot{\gamma}_n(s)\,\mathrm{d}x\,\mathrm{d}s\\
&\le \tfrac{1}{4\delta}\|\dot{g}_n(t)\|^2_{L^2(\Omega;\mathbb{R}^3)} + \delta\|\dot{u}_n(t)\|^2_{L^2(\Omega;\mathbb{R}^3)} + C\int_0^{\overline{\tau}_n(t)}\left(\|\dot{\gamma}_n(s)\|^2_{L^2(\Omega;\mathbb{R}^3)} + \|\dot{u}_n(s-\tau_n)\|^2_{L^2(\Omega;\mathbb{R}^3)}\right)\mathrm{d}s\,.
\end{aligned}
$$

Finally,

$$
\begin{aligned}
&\int_0^{\overline{\tau}_n(t)}\int_\Omega \left(\mathbb{D}(\underline{z}_n,\underline{\theta}_n)e(\dot{u}_n) + \mathbb{C}(\overline{z}_n)e(\overline{u}_n) - \underline{\theta}_n\,\mathbb{B} + \tau_n|e(\overline{u}_n)|^2 e(\overline{u}_n)\right) : e(\dot{g}_n)\,\mathrm{d}x\,\mathrm{d}s\\
&\le \tfrac{1}{4\delta}\int_0^{\overline{\tau}_n(t)} \|e(\dot{g}_n(s))\|^2_{L^2(\Omega;\mathbb{R}^{3\times3}_{\mathrm{sym}})}\,\mathrm{d}s + \delta\int_0^{\overline{\tau}_n(t)} \|e(\dot{u}_n(s))\|^2_{L^2(\Omega;\mathbb{R}^{3\times3}_{\mathrm{sym}})}\,\mathrm{d}s\\
&\quad + C\int_0^{\overline{\tau}_n(t)} \|e(\overline{u}_n(s))\|^2_{L^2(\Omega;\mathbb{R}^{3\times3}_{\mathrm{sym}})}\,\mathrm{d}s + C\int_0^{\overline{\tau}_n(t)} \|w_n(s)\|_{L^1(\Omega)}\,\mathrm{d}s\\
&\quad + C\tau_n\left[\int_0^{\overline{\tau}_n(t)} \|e(\dot{g}_n(s))\|^4_{L^4(\Omega;\mathbb{R}^{3\times3}_{\mathrm{sym}})}\,\mathrm{d}s + \int_0^{\overline{\tau}_n(t)} \|e(\overline{u}_n(s))\|^4_{L^4(\Omega;\mathbb{R}^{3\times3}_{\mathrm{sym}})}\,\mathrm{d}s\right],
\end{aligned}
$$

where we have used again (5) and the subquadratic growth (16) of Θ.

We then choose δ so small that the corresponding terms in the right-hand side are absorbed by larger terms on the left-hand side. Thus, taking into account the previous estimates and (39), from (48) we obtain

$$
\begin{aligned}
&c\,\|\dot{u}_n(t)\|^2_{L^2(\Omega;\mathbb{R}^3)} + c\|\overline{u}_n(t)\|^2_{H^1(\Omega;\mathbb{R}^3)} + \tau_n c\|\overline{u}_n(t)\|^4_{W^{1,4}(\Omega;\mathbb{R}^3)} + c\|\overline{z}_n(t)\|^q_{W^{1,q}(\Omega)}\\
&\quad + c\,\|\overline{w}_n(t)\|_{L^1(\Omega)} + c\int_0^{\overline{\tau}_n(t)} \|e(\dot{u}_n(s))\|^2_{L^2(\Omega;\mathbb{R}^{3\times3}_{\mathrm{sym}})}\,\mathrm{d}s\\
&\le C + C\int_0^{\overline{\tau}_n(t)} \left[\|\overline{w}_n\|_{L^1(\Omega)} + \|\underline{u}_n\|^2_{H^1(\Omega;\mathbb{R}^3)} + \|\dot{u}_n(s-\tau_n)\|^2_{L^2(\Omega;\mathbb{R}^3)} + \tau_n\|\overline{u}_n\|^4_{W^{1,4}(\Omega;\mathbb{R}^3)}\right]\mathrm{d}s\,,
\end{aligned}
$$

where we have used the assumptions on initial and external data (8), (9), (10), and (11), and the constants c and C clearly depend on δ.

All in all, by suitable application of the Cauchy inequality and of the Gronwall Lemma, we obtain the a priori estimates (49a)–(49d), the $L^\infty(0,T;W^{1,q}(\Omega))$-bound on $\overline{z}_n$ in (49f), as well as estimate (49g) below, with a constant uniform with respect to n.

Relying on (49a)–(49d) and (49g), we are in the position to deduce from the discrete mechanical energy inequality (44c) a bound on the dissipation term $\int_\Omega(z_0-\overline{z}_n(t))\,\mathrm{d}x$, which in particular ensures the $\mathrm{BV}(0,T;L^1(\Omega))$-bound on $\overline{z}_n$ in (49f). The $L^\infty((0,T)\times\Omega)$-bound on $\overline{z}_n$ is a direct consequence of the monotonicity and of the fact that $0\le z_0\le 1$.

Furthermore, estimate (49h) can be deduced using the Boccardo-Gallouët type estimates developed in [2, Prop. 4.2]. Let us just mention here, that it is based on testing the enthalpy equation (44d) by $\chi(\overline{w}_n) := 1 - 1/(1+\overline{w}_n)^\beta$ with $\beta>0$, relying on assumptions (4) in order to handle the resulting gradient term. A chain of inequalities involving Gagliardo-Nirenberg estimates ultimately leads to the $L^r(0,T;W^{1,r}(\Omega))$-estimate in (49h) and the respective regularity of w in (21c). Finally, exploiting the estimates obtained so far, the BV-estimate in (49h) results from the boundedness of $\|\overline{w}_n\|_{L^1(0,T;W^{1,r'}(\Omega)*)}$, which is, in turn, deduced by a comparison argument in (44d) using test functions $\eta\in L^\infty(0,T;W^{1,r'}(\Omega)$ with $r'>5$ by (21c) and the fact that the gradient term as well as the terms on the right-hand side of (44d) already have been proved to be bounded; we refer to [2] for the details.

Finally, estimate (49e) ensues from a comparison argument in the discrete momentum equation (44b), taking into account the previously obtained bounds (49a)–(49d), (49f)–(49h).

In total, the above arguments yield the following

Proposition 3.5 (A priori estimates). *Let the assumptions* (4)–(6) *and* (8)–(11) *hold true. Then a sequence of interpolants* $(\overline{u}_n,\underline{u}_n,u_n,\overline{z}_n,\underline{z}_n,z_n,\overline{w}_n,\underline{w}_n)_{n\in\mathbb{N}}$, *complying with the time-discrete version of the energetic formulation* (44), *satisfies*

$$\|\overline{u}_n\|_{L^\infty(0,T;H^1(\Omega;\mathbb{R}^3))} \le C, \tag{49a}$$
$$\|\dot{u}_n\|_{L^\infty(0,T;L^2(\Omega;\mathbb{R}^3))} \le C, \tag{49b}$$
$$\|\overline{u}_n\|_{L^\infty(0,T;W^{1,4}(\Omega;\mathbb{R}^3))} \le C/\sqrt[4]{\tau_n}, \tag{49c}$$
$$\|\dot{u}_n\|_{L^2(0,T;H^1(\Omega;\mathbb{R}^3))} \le C, \tag{49d}$$
$$\|\dot{u}_n\|_{\mathrm{BV}([0,T];W^{1,4}(\Omega;\mathbb{R}^3)^*)} \le C, \tag{49e}$$
$$\|\overline{z}_n\|_{L^\infty(0,T;W^{1,q}(\Omega))\cap L^\infty((0,T)\times\Omega)\cap\mathrm{BV}([0,T];L^1(\Omega))} \le C, \tag{49f}$$
$$\|\overline{w}_n\|_{L^\infty(0,T;L^1(\Omega))} \le C, \tag{49g}$$
$$\|\overline{w}_n\|_{L^r(0,T;W^{1,r}(\Omega))\cap\mathrm{BV}(0,T;W^{1,r'}(\Omega)^*)} \le C \quad \textit{for every } r<5/4. \tag{49h}$$

3.4. Limit passage from time-discrete to time-continuous

A suitable version of Helly's selection principle, cf. e.g. [11, Thm. 6.1], combined with compactness arguments based on the a priori estimates (49a)–(49f), leads to the convergence statements (50a)–(50j) and (50o) in Proposition 3.6 below. We refer to [3, Prop. 4.1] for the details of the proof, here commenting only on the further convergence (50d), which follows from the bounds (49d) and (49e) via a BV-version of the Aubin-Lions compactness theorem, cf. e.g. [9, Cor. 7.9, pag. 196]. Moreover, convergences (50k)–(50m) for the enthalpy variables can be concluded from estimates (49g) and (49h) arguing along the lines of [2], cf. also [10]. Also based on (49g) and (49h), by the aforementioned Aubin-Lions type compactness result from [9], one additionally finds the strong convergence result (50n). From this, one may extract a further, pointwise a.e. convergent subsequence in order to see that the nonnegativity of the approximate solutions deduced in Lemma 3.3 carries over to the limit w for a.e. $t\in(0,T)$.

Proposition 3.6 (Convergence of the time-discrete solutions). *Let the assumptions* (4)–(6) *and* (8)–(11) *be satisfied. Then, there exists a triple* $(u,z,w)\colon[0,T]\times\Omega\to\mathbb{R}^3\times\mathbb{R}\times[0,+\infty)$ *of regularity* (21) *such that for a.a.* $x\in\Omega$ *the function* $t\mapsto z(t,x)\in[0,1]$ *is nonincreasing, nonnegativity* (27) *of* w *holds and there exists a subsequence of the time-discrete solutions* $(\overline{u}_n,\underline{u}_n,u_n,\overline{z}_n,\underline{z}_n,\overline{w}_n,\underline{w}_n)_n$ *from* (36) *such that*

$$\overline{u}_n \overset{*}{\rightharpoonup} u \quad \textit{in } L^\infty(0,T;H^1(\Omega;\mathbb{R}^3)), \tag{50a}$$
$$u_n \rightharpoonup u \quad \textit{in } H^1(0,T;H^1(\Omega;\mathbb{R}^3)), \tag{50b}$$

$$\dot{u}_n \overset{*}{\rightharpoonup} \dot{u} \quad \textit{in } L^\infty(0,T;L^2(\Omega;\mathbb{R}^3))\,, \tag{50c}$$
$$\dot{u}_n \to u \quad \textit{in } L^2(0,T;W^{1-\varepsilon,2}(\Omega;\mathbb{R}^3))\,, \tag{50d}$$
$$\dot{u}_n(t) \rightharpoonup \dot{u}(t) \quad \textit{in } L^2(\Omega;\mathbb{R}^3) \textit{ for all } t\in[0,T]\,, \tag{50e}$$
$$\overline{z}_n\,,\,\underline{z}_n \overset{*}{\rightharpoonup} z \quad \textit{in } L^\infty(0,T;W^{1,q}(\Omega))\cap L^\infty((0,T)\times\Omega)\,, \tag{50f}$$
$$\overline{z}_n(t) \rightharpoonup z(t) \quad \textit{in } W^{1,q}(\Omega) \textit{ for all } t\in[0,T]\,, \tag{50g}$$
$$\overline{z}_n(t) \to z(t) \quad \textit{in } L^m(\Omega) \textit{ for all } m\in[1,\infty) \textit{ and for all } t\in[0,T]\,, \tag{50h}$$
$$\underline{z}_n(t) \rightharpoonup z(t) \quad \textit{in } W^{1,q}(\Omega) \textit{ for all } t\in[0,T]\backslash J\,, \tag{50i}$$
$$\underline{z}_n(t) \to z(t) \quad \textit{in } L^m(\Omega) \textit{ for all } m\in[1,\infty) \textit{ and for all } t\in[0,T]\backslash J\,, \tag{50j}$$
$$\overline{w}_n\,,\,\underline{w}_n \overset{*}{\rightharpoonup} w \quad \textit{in } L^\infty(0,T;L^1(\Omega))\,, \tag{50k}$$
$$\overline{w}_n\,,\,\underline{w}_n\,,\,w_n \rightharpoonup w \quad \textit{in } L^r(0,T;W^{1,r}(\Omega)) \quad \textit{for all } r<5/4\,, \tag{50l}$$
$$w_n(t) \overset{*}{\rightharpoonup} w(t) \quad \textit{in } W^{1,r'}(\Omega)^* \textit{ for all } t\in[0,T]\,, \tag{50m}$$
$$\overline{w}_n\,,\,\underline{w}_n\,,\,w_n \to w \quad \textit{in } L^r(0,T;W^{1-\varepsilon,r}(\Omega))\cap L^p(0,T;L^1(\Omega)) \tag{50n}$$

for all $\varepsilon\in(0,1]$ *and all* $p\in[1,\infty)$. *The set* $J\subset[0,T]$ *appearing in* (50i)–(50j) *denotes the jump set of* $z\in \mathrm{BV}([0,T];L^1(\Omega))$. *Finally,*

$$|\dot{z}_n| \to |\dot{z}| \quad \textit{in the sense of measures on } (0,T)\times\Omega\,. \tag{50o}$$

For such a limit triple (u,z,w), given by means of Proposition 3.6, it has to be verified that it solves the time-continuous energetic formulation stated in Def. 2.1. For this, one basically takes the limit $n\to\infty$ in the time-discrete energetic formulation (44). In what follows, we just outline the steps of the limit passage and comment on the main ideas and tools; for all the details the reader is referred to [3, Sect. 4], where the proof has been performed in an analogous setting.

- The *limit passage in the semistability inequality* can be carried out by verifying the *mutual recovery sequence condition*, cf. [11, 4], i.e. that for all $t\in[0,T]$, for any sequence $(v_n,\zeta_n)_{n\in\mathbb{N}}$ such that

$$\begin{aligned} &v_n\rightharpoonup v \text{ in } H^1(\Omega;\mathbb{R}^3), \quad \zeta_n\rightharpoonup\zeta \text{ in } W^{1,q}(\Omega) \text{ and}\\ &\mathcal{E}_n(t,v_n,\zeta_n)\le\mathcal{E}_n(t,v_n,\hat{\zeta}_n)+\mathcal{R}_1(\hat{\zeta}-\zeta_n)\,, \end{aligned} \tag{51}$$

  and for every $\tilde{\zeta}\in W^{1,q}(\Omega)$, there exists a mutual recovery sequence $(\tilde{\zeta}_n)_n$ such that

$$0\le\limsup_{n\to\infty}\Big(\mathcal{E}_n(t,v_n,\tilde{\zeta}_n)-\mathcal{E}_n(t,v_n,\zeta_n)+\mathcal{R}_1(\tilde{\zeta}_n-\zeta_n)\Big)\le\mathcal{E}(t,v,\tilde{\zeta})-\mathcal{E}(t,v,\zeta)+\mathcal{R}_1(\tilde{\zeta}-\zeta)\,. \tag{52}$$

  This condition is applied to the sequence $(v_n,\zeta_n)_{n\in\mathbb{N}}=(\overline{u}_n(t),\overline{z}_n(t))_n$, satisfying at every $t\in[0,T]$ the discrete semistability (44a) (whence (51)). The construction of the mutual recovery sequence $(\tilde{\zeta}_n)_{n\in\mathbb{N}}$ is developed in [3, Sect. 4.2], to which we refer for all details.
- For the *limit passage in the momentum balance* (44b) one considers test functions $v\in L^2(0,T;H^1_{\mathrm{D}}(\Omega;\mathbb{R}^3))\cap W^{1,1}(0,T;L^2(\Omega;\mathbb{R}^3))$. Because of the regularization by the 4-Laplacian, such test functions have to be regularized using mollifers of the form [8, p. 56, Corollary 2] and [9, p. 189]. These mollifed functions are discretized in time according to (31), (38), and (43), thus resulting in admissible test functions $(\bar{v}_n,v_n)$ for (44b) satisfying for all $t\in[0,T]$

$$\begin{aligned}&\bar{v}_n(t),v_n(t)\to v(t) \text{ in } H^1_{\mathrm{D}}(\Omega;\mathbb{R}^3)\,,\\ &\bar{v}_n\to v \text{ in } L^2(0,T;H^1_{\mathrm{D}}(\Omega;\mathbb{R}^3)) \text{ and } v_n\to v \text{ in } L^2(0,T;H^1_{\mathrm{D}}(\Omega;\mathbb{R}^3))\cap W^{1,1}(0,T;L^2(\Omega;\mathbb{R}^3))\,.\end{aligned}$$

Hence, in view of the weak convergences (50b)–(50e), to pass to the limit in (44b) weak-strong convergence arguments are employed. This is possible since

$$(\mathbb{D}(\underline{z}_n, \Theta(\underline{\theta}_n)) + \mathbb{C}(\underline{z}_n))e(\bar{v}_n) \to (\mathbb{D}(z, \Theta(w)) + \mathbb{C}(z))e(v) \quad \text{in } L^2(0,T, L^2(\Omega;\mathbb{R}^{3\times3})),$$

which can be deduced via dominated convergence; we refer to [3, Sect. 4.1] for the details.

- The *mechanical energy equality* (25) for the limit system is proved in two steps. The part "$\leq$" directly follows from (44c) by lower semicontinuity for the terms on the left-hand side, and by weak-strong convergence arguments for the terms on the right-hand side, making use of the convergence properties of the interpolants of the given data (29), (30), (37), (39), and (43), which, in particular, also ensure energy convergence at initial time.
- The *opposite mechanical energy inequality* "$\geq$" is deduced following the lines of [2, p. 283f]: A Riemann sum argument is applied to the already proven semistability inequality (23) of the limit (u, z, w) at times t_{i-1}, $i \in \{1, \ldots, n\}$ tested with $z(t_i)$. The result is added to the momentum balance (24) of the limit tested by $(\dot{u} - \dot{g})$. This can be done rigorously in view of the enhanced $H^1(0,T;H^1(\Omega;\mathbb{R}^3))$-regularity of $\dot{u}$, which can be gained by a comparison argument in the momentum balance, cf. [3, Rmk. 2.6] for all details. Combined with the previously deduced inequality "$\leq$", this yields the *mechanical energy balance* (25) of the limit system, see [3, Prop. 4.6] for the details. Using a lim sup-argument, i.e. starting from the discrete energy inequality (44c) and passing over to the limit in a chain of inequalities, the fact that *equality* holds for the limit system additionally allows us to conclude the *strong convergence of the viscous dissipation terms* in $L^1([0,T] \times \Omega; \mathbb{R}^{3\times3})$ by comparison.
- Finally, we carry out the *limit passage in the enthaly equation.* Starting with a function $\eta \in W^{1,r'}(0,T;L^{r'}(\Omega)) \cap \mathrm{C}^0([0,T];W^{1,r'}(\Omega))$, also in this step, time-discrete test functions $(\bar{\eta}_n, \eta_n)$ for (44d) are constructed as in (31) and (38) and thus exhibit strong convergence in $W^{1,r'}(0,T;L^{r'}(\Omega)) \cap L^{r'}(0,T;W^{1,r'}(\Omega))$. Via dominated convergence, also using convergence (50n) and the boundedness (17) of $\mathbb{J}$, it can be shown that

$$\mathbb{J}(\bar{z}_n, \underline{\theta}_n)^* \nabla \bar{\eta}_n \to \mathbb{J}(z, w)^* \nabla \eta \quad \text{strongly in } L^{r'}(0,T;L^{r'}(\Omega,\mathbb{R}^3)),$$

which is a major ingredient to obtain the *enthalpy equation* (26) of the the limit system. The strong convergence of the viscous dissipation terms in $L^1([0,T] \times \Omega; \mathbb{R}^{3\times3})$ ultimately enables us to pass to the limit in the right-hand side of (44d). We refer to [10, p. 30] for more details.

Acknowledgments

This work has been supported by the Italian Ministry of Education, University, and Research through the PRIN 2010-11 grant for the project *Calculus of Variations*, by the European Research Council through the two Advanced Grants *Quasistatic and Dynamic Evolution Problems in Plasticity and Fracture (290888)* and *Analysis of Multiscale Systems Driven by Functionals (267802)*, and by GNAMPA (Gruppo Nazionale per l'Analisi Matematica, la Probabilità e le loro Applicazioni) of INdAM (Istituto Nazionale di Alta Matematica) through the project *Modelli variariazionali per la propagazione di fratture, la delaminazione e il danneggiamento.*

References

[1] Roubíček T 2009 *Math. Methods Appl. Sci.* **32** 825–862
[2] Roubíček T 2010 *SIAM J. Math. Anal.* **40** 256–297
[3] Lazzaroni G, Rossi R, Thomas M and Toader R 2014 *WIAS-Preprint 2025 and SISSA Preprint 52/2014/MATE*
[4] Mielke A and Roubíček T 2006 *Math. Models Methods Appl. Sci.* **16** 177–209

[5] Francfort G A and Suquet P 1986 *Arch. Ration. Mech. Anal.* **96** 265–293
[6] Blanchard D and Guibé O 2000 *Advances in Differential Equations* **5** 1221–1252
[7] Rocca E and Rossi R 2014 *Math. Models Methods Appl. Sci.* **24** 1265–1341
[8] Burenkov V 1998 *Sobolev Spaces on Domains* vol 137 (B. G. Teubner)
[9] Roubíček T 2005 *Nonlinear Partial Differential Equations with Applications* (Birkhäuser)
[10] Rossi R and Roubíček T 2011 *Nonlinear Anal.* **74** 3159–3190
[11] Mielke A, Roubíček T and Stefanelli U 2008 *Calc. Var. Partial Differential Equations* **31** 387–416

Balanced-Viscosity solutions for multi-rate systems

Alexander Mielke[1,2], **Riccarda Rossi**[3], **and Giuseppe Savaré**[4]

[1] Weierstraß-Institut, Mohrenstraße 39, D–10117 Berlin, Germany.
[2] Institut für Mathematik, Humboldt-Universität zu Berlin, Rudower Chaussee 25, D–12489 Berlin (Adlershof), Germany.
[3] DICATAM – Sezione di Matematica, Università di Brescia, via Valotti 9, I–25133 Brescia, Italy.
[4] Dipartimento di Matematica "F. Casorati", Università di Pavia. Via Ferrata, I–27100 Pavia, Italy.

E-mail: mielke@wias-berlin.de, riccarda.rossi@unibs.it, giuseppe.savare@unipv.it

Abstract. Several mechanical systems are modeled by the static momentum balance for the displacement u coupled with a rate-independent flow rule for some internal variable z. We consider a class of abstract systems of ODEs which have the same structure, albeit in a finite-dimensional setting, and regularize both the static equation and the rate-independent flow rule by adding viscous dissipation terms with coefficients ε^α and ε, where $0 < \varepsilon \ll 1$ and $\alpha > 0$ is a fixed parameter. Therefore for $\alpha \neq 1$ u and z have different relaxation rates.

We address the vanishing-viscosity analysis as $\varepsilon \downarrow 0$ of the viscous system. We prove that, up to a subsequence, (reparameterized) viscous solutions converge to a parameterized curve yielding a *Balanced Viscosity* solution to the original rate-independent system, and providing an accurate description of the system behavior at jumps. We also give a reformulation of the notion of *Balanced Viscosity* solution in terms of a system of subdifferential inclusions, showing that the viscosity in u and the one in z are involved in the jump dynamics in different ways, according to whether $\alpha > 1$, $\alpha = 1$, and $\alpha \in (0, 1)$.

1. Introduction

Several mechanical systems are described by ODE or PDE systems of the type:

$$\mathrm{D}_u\mathcal{E}(t, u(t), z(t)) = 0 \qquad \text{in } \mathcal{U}^*, \qquad \text{for a.a. } t \in (0, T), \tag{1.1a}$$

$$\partial\mathcal{R}_0(z'(t)) + \mathrm{D}_z\mathcal{E}(t, u(t), z(t)) \ni 0 \qquad \text{in } \mathcal{Z}^*, \qquad \text{for a.a. } t \in (0, T), \tag{1.1b}$$

where $\mathcal{U}$, $\mathcal{Z}$ are Banach spaces, and $\mathcal{E} : [0, T] \times \mathcal{U} \times \mathcal{Z} \to \mathbb{R}$ is an energy functional. For example, within the ansatz of generalized standard materials, u is the displacement, at equilibrium, while changes in the elastic behavior due to dissipative effects are described in terms of an internal variable z in some state space $\mathcal{Z}$. In several mechanical phenomena [1], dissipation due to inertia and viscosity is negligible, and the system is governed by rate-independent evolution, which means that the (convex, nondegenerate) dissipation potential $\mathcal{R}_0 : \mathcal{Z} \to [0, \infty)$ is *positively homogeneous of degree* 1. Thus system (1.1) is invariant for time-rescalings.

It is well known that, if the map $z \mapsto \mathcal{E}(t, u, z)$ is not uniformly convex, one cannot expect the existence of absolutely continuous solutions to (1.1). This fact has motivated the development of various weak solvability concepts for (1.1), starting with the well-established notion of *energetic solution*. The latter dates back to [2] and was further developed in [3] (see [4], as well, in the context of crack growth), cf. also [1], [5] and the references therein. Despite the several good features of the energetic formulation, it is known that, in the case the energy $z \mapsto \mathcal{E}(t, u, z)$ is nonconvex, the global stability condition may lead to jumps of z as a function of time that are not motivated by, or in accord with, the mechanics of the system, cf. e.g. the discussions in [6, Ex. 6.1], [7, Ex. 6.3], and [8, Ex. 1].

Over the last years, an alternative selection criterion of mechanically feasible weak solution concepts for the rate-independent system (1.1) has been developed, moving from the finite-dimensional analysis in [9]. It is based on the interpretation of (1.1) as originating in the vanishing-viscosity limit of the *viscous* system

$$\mathrm{D}_u\mathcal{E}(t, u(t), z(t)) = 0 \qquad \text{in } \mathcal{U}^*, \qquad \text{for a.a. } t \in (0, T), \tag{1.2a}$$

$$\partial\mathcal{R}_0(z'(t)) + \varepsilon\partial\mathcal{V}_{\mathsf{z}}(z'(t)) + \mathrm{D}_z\mathcal{E}(t, u(t), z(t)) \ni 0 \qquad \text{in } \mathcal{Z}^*, \qquad \text{for a.a. } t \in (0, T), \tag{1.2b}$$

where $\mathcal{V}_{\mathsf{z}} : \mathcal{Z} \to [0, \infty)$ is a dissipation potential with *superlinear* (for instance, quadratic) growth at infinity. Observe that the existence of solutions for the *generalized gradient system* (1.2) follows from [10, 11], cf. also [12]. This vanishing-viscosity approach leads to a notion of solution featuring a *local*, rather than *global*, stability condition for the description of rate-independent evolution, thus avoiding "too early" and "too long" jumps. Furthermore, it provides an accurate description of the energetic behavior of the system at jumps, in particular highlighting how viscosity, neglected in the limit as $\varepsilon \downarrow 0$, comes back into the picture and governs the jump dynamics. This has been demonstrated in [8, 13, 14] within the frame of abstract, finite-dimensional and infinite-dimensional, rate-independent systems, and in [15] for a wide class parabolic equations with a rate-independent term. This analysis has also been developed in several applicative contexts, ranging from crack propagation [7, 16], to plasticity [17, 18, 19, 20], and to damage [21], among others.

In this note, we shall perform the vanishing-viscosity analysis of system (1.1) by considering the viscous approximation of (1.1a), in addition to the viscous approximation of (1.1b). More precisely, we will address the asymptotic analysis as $\varepsilon \downarrow 0$ of the system

$$\varepsilon^\alpha\partial\mathcal{V}_{\mathsf{u}}(u'(t)) + \mathrm{D}_u\mathcal{E}(t, u(t), z(t)) = 0 \qquad \text{in } \mathcal{U}^*, \qquad \text{for a.a. } t \in (0, T), \tag{1.3a}$$

$$\partial\mathcal{R}_0(z'(t)) + \varepsilon\partial\mathcal{V}_{\mathsf{z}}(z'(t)) + \mathrm{D}_z\mathcal{E}(t, u(t), z(t)) \ni 0 \qquad \text{in } \mathcal{Z}^*, \qquad \text{for a.a. } t \in (0, T), \tag{1.3b}$$

where $\alpha > 0$ and $\mathcal{V}_{\mathsf{u}}$ a quadratic dissipation potential for the variable u. Observe that (1.3) models systems with (possibly) *very different* relaxation times. In fact, the parameter $\alpha > 0$ sets which of the two variables u and z relaxes faster to *equilibrium* and *rate-independent* evolution, respectively.

Let us mention that the analysis developed in this paper is in the mainstream of a series of recent papers focused on the coupling between rate-independent and viscous systems. First and foremost, in [22] a wide class of rate-independent processes in viscous solids with inertia has been tackled, while the coupling with temperature has further been considered in [23]. In fact, in these systems the evolution for the internal variable z is purely rate-independent and no vanishing viscosity is added to the equation for z: viscosity and inertia only intervene in the evolution for the displacement u. For these processes, the author has proposed a notion of solution of energetic type consisting of the weakly formulated momentum equation for the displacements (and also of the weak heat equation in [23]), of an energy balance, and of a *semi-stability* condition. The latter reflects the mixed *rate dependent/independent* character of the system. In [22] and [24] a vanishing-viscosity analysis (in the momentum equation) has been performed. As discussed in [24] in the context of delamination, this approach leads to *local solutions* (cf. also [5]), describing crack initiation (i.e., delamination) in a physically feasible way. In [25], the vanishing-viscosity approach has also been developed in the context of a model for crack growth in the two-dimensional antiplane case, with a pre-assigned crack path, coupling a viscoelastic momentum equation with a viscous flow rule for the crack tip; again, this procedure leads to solutions jumping later than energetic solutions. With a rescaling technique, a vanishing-viscosity analysis both in the flow rule, and in the momentum equation, has been recently performed in [26] for perfect plasticity, recovering energetic solutions thanks to the convexity of the energy. In [27], the same analysis has led to *local solutions* for a delamination system.

With the vanishing-viscosity analysis in this paper, besides finding good *local* conditions for the limit evolution, we want to add as an additional feature a thorough description of the energetic behavior of the solutions at jumps. This shall be deduced from an *energy balance*. Moreover, in comparison to the aforementioned contributions [25, 26, 27] a greater emphasis shall be put here on how the multi-rate character of system (1.3) enters in the description of the jump dynamics. In particular, we will convey that viscosity in u and viscosity z are involved in the path followed by the system at jumps in (possibly) different ways, depending on whether the parameter α is strictly bigger than, or equal to, or strictly smaller than 1.

To focus on this and to avoid overburdening the paper with technicalities, we shall keep to a simple functional analytic setting. Namely, we shall consider the *finite-dimensional* and *smooth* case

$$\mathcal{U} = \mathbb{R}^n, \qquad \mathcal{Z} = \mathbb{R}^m, \qquad \mathcal{E} \in \mathrm{C}^1([0,T] \times \mathbb{R}^n \times \mathbb{R}^m). \tag{1.4}$$

Obviously, this considerably simplifies the analysis, since the difficulties attached to nonsmoothness of the energy and to infinite-dimensionality are completely avoided. Still, even within such a simple setting (where, however, we will allow for state-dependent dissipation potentials $\mathcal{R}_0$, $\mathcal{V}_\mathsf{z}$, and $\mathcal{V}_\mathsf{u}$), the key ideas of our vanishing-viscosity approach can be highlighted.

Let us briefly summarize our results by restricting to a simplified version of (1.3), where we assume (1.4) and use the simple choices

$$\mathcal{V}_\mathsf{u}(u') = \frac{1}{2}|u'|^2, \qquad \mathcal{V}_\mathsf{z}(z') = \frac{1}{2}|z'|^2$$

(cf. ($\mathrm{V_u}$) for the assumptions on state-dependent dissipation potentials), system (1.3) reduces to the ODE system

$$\varepsilon^\alpha u'(t) + \mathrm{D}_u\mathcal{E}(t, u(t), z(t)) = 0 \qquad \text{in } (0,T), \tag{1.5a}$$

$$\partial\mathcal{R}_0(z'(t)) + \varepsilon z'(t) + \mathrm{D}_z\mathcal{E}(t, u(t), z(t)) \ni 0 \qquad \text{in } (0,T). \tag{1.5b}$$

First of all, following [8, 13, 14], and along the lines of the *variational* approach to gradient flows by E. De Giorgi [28, 29], we will pass to the limit as $\varepsilon \downarrow 0$ in the *energy-dissipation principle* associated (and equivalent, by Fenchel-Moreau duality and the chain rule for $\mathcal{E}$) to (1.5), namely

$$\begin{aligned}
&\mathcal{E}(t, u(t), z(t)) + \int_s^t \mathcal{R}_0(z'(r)) + \frac{\varepsilon}{2}|z'(r)|^2 + \frac{\varepsilon^\alpha}{2}|u'(r)|^2 \,\mathrm{d}r \\
&\quad + \int_s^t \frac{1}{\varepsilon}\mathcal{W}_\mathsf{z}^*(-\mathrm{D}_z\mathcal{E}(r, u(r), z(r))) + \frac{1}{2\varepsilon^\alpha}|-\mathrm{D}_u\mathcal{E}(r, u(r), z(r))|^2 \,\mathrm{d}r \\
&= \mathcal{E}(s, u(s), z(s)) + \int_s^t \partial_t\mathcal{E}(r, u(r), z(r))\,\mathrm{d}r
\end{aligned} \tag{1.6}$$

for all $0 \le s \le t \le T$, where $\mathcal{W}_\mathsf{z}^*$ is the Legendre transform of $\mathcal{R}_0 + \mathcal{V}_\mathsf{z}$. As we will see in Section 4, equation (1.6) is well-suited to unveiling the role played by viscosity in the description of the energetic behavior of the system at jumps. Indeed, it reflects the competition between the tendency of the system to be governed by *viscous* dissipation both for the variable z and for the variable u (with different rates if $\alpha \neq 1$), and its tendency to be *locally stable* in z, and at equilibrium in u, cf. also the discussion in Remark 4.4.

Secondly, to develop the analysis as $\varepsilon \downarrow 0$ for a family of curves $(u_\varepsilon, z_\varepsilon)_\varepsilon \subset W^{1,2}(0,T;\mathbb{R}^n \times \mathbb{R}^m)$ fulfilling (1.6) we will adopt a by now well-established technique from [9]. Namely, to capture the viscous transition paths at jump points, we will reparameterize the curves $(u_\varepsilon, z_\varepsilon)$, for instance by their arc-length. Hence we will address the analysis as $\varepsilon \downarrow 0$ of the *parameterized curves* $(\mathsf{t}_\varepsilon, \mathsf{u}_\varepsilon, \mathsf{z}_\varepsilon)_\varepsilon$ defined on the interval $[0, S]$ with values in the extended phase space $[0,T] \times \mathbb{R}^n \times \mathbb{R}^m$, with t_ε the rescaling functions and $\mathsf{u}_\varepsilon := u_\varepsilon \circ \mathsf{t}_\varepsilon$, $\mathsf{z}_\varepsilon := z_\varepsilon \circ \mathsf{t}_\varepsilon$. Under suitable conditions it can be proved that, up to a subsequence the curves $(\mathsf{t}_\varepsilon, \mathsf{u}_\varepsilon, \mathsf{z}_\varepsilon)_\varepsilon$ converge to a triple $(\mathsf{t}, \mathsf{u}, \mathsf{z}) \in W^{1,1}([0,S];[0,T] \times \mathbb{R}^n \times \mathbb{R}^m)$. Its evolution is described by an energy-dissipation principle obtained by passing to the limit in the reparameterized version of (1.6), cf. Theorem 4.5. We will refer to $(\mathsf{t}, \mathsf{u}, \mathsf{z})$ as a *parameterized Balanced Viscosity* solution to the rate-independent system $(\mathbb{R}^n \times \mathbb{R}^m, \mathcal{E}, \mathcal{R}_0 + \varepsilon\mathcal{V}_\mathsf{z} + \varepsilon^\alpha \mathcal{V}_\mathsf{u})$.

The main result of this paper, Theorem 5.3, provides a more transparent reformulation of the energy-dissipation principle defining a parameterized Balanced Viscosity solution $(\mathsf{t}, \mathsf{u}, \mathsf{z})$. It is in terms of a system of subdifferential inclusions fulfilled by the curve $(\mathsf{t}, \mathsf{u}, \mathsf{z})$, namely

$$\begin{aligned} &\theta_\mathsf{u}(s)\mathsf{u}'(s) + (1-\theta_\mathsf{u}(s))\mathrm{D}_u\mathcal{E}(\mathsf{t}(s),\mathsf{u}(s),\mathsf{z}(s)) \ni 0 && \text{for a.a. } s \in (0,S), \\ &(1-\theta_\mathsf{z}(s))\partial\mathcal{R}_0(\mathsf{z}'(s)) + \theta_\mathsf{z}(s)\mathsf{z}'(s) + (1-\theta_\mathsf{z}(s))\mathrm{D}_z\mathcal{E}(\mathsf{t}(s),\mathsf{u}(s),\mathsf{z}(s)) \ni 0 && \text{for a.a. } s \in (0,S), \end{aligned} \tag{1.7}$$

where the Borel functions θ_u, $\theta_\mathsf{z} : [0,S] \to [0,1]$ fulfill

$$\mathsf{t}'(s)\theta_\mathsf{u}(s) = \mathsf{t}'(s)\theta_\mathsf{z}(s) = 0 \qquad \text{for a.a. } s \in (0,S), \tag{1.8}$$

plus an extra condition on the triple $(\mathsf{t}', \theta_\mathsf{u}, \theta_\mathsf{z})$ that depends explicitly on the three cases $\alpha \in (0,1)$, $\alpha = 1$, and $\alpha > 1$, see (5.13). Condition (1.8) reveals that the viscous terms $\mathsf{u}'(s)$ and $\mathsf{z}'(s)$ may contribute to (1.7) only at jumps of the system, corresponding to $\mathsf{t}'(s) = 0$ as the function t records the (slow) external time scale. In this respect, (1.7)–(1.8) is akin to the (parameterized) subdifferential inclusion

$$\begin{aligned} &\mathrm{D}_u\mathcal{E}(\mathsf{t}(s),\mathsf{u}(s),\mathsf{z}(s)) \ni 0 && \text{for a.a. } s \in (0,S), \\ &\partial\mathcal{R}_0(\mathsf{z}'(s)) + \theta(s)\mathsf{z}'(s) + \mathrm{D}_z\mathcal{E}(\mathsf{t}(s),\mathsf{u}(s),\mathsf{z}(s)) \ni 0 && \text{for a.a. } s \in (0,S), \end{aligned} \tag{1.9}$$

with the Borel function $\theta : [0,S] \to [0,\infty)$ fulfilling

$$\mathsf{t}'(s)\theta(s) = 0 \qquad \text{for a.a. } s \in (0,S). \tag{1.10}$$

Indeed, (1.9) is the subdifferential reformulation for parameterized Balanced Viscosity solutions obtained by taking the limit $\varepsilon \downarrow 0$ in (1.2), where viscosity is added only to the flow rule.

However, note that (1.7) has a much more complex structure than (1.9). In addition to the switching condition (1.8), the functions t', θ_u, and θ_z fulfill the additional condition (5.13) that differs in the three cases $\alpha > 1$, $\alpha = 1$, and $\alpha \in (0,1)$. In particular, for $\theta_\mathsf{u} > 0$ or $\theta_\mathsf{z} > 0$ the viscosity in u or z pops back into the description of the system behavior at jumps, in a way depending on whether u relaxes faster to equilibrium than z, or u and z have the same relaxation rate, or z relaxes faster to local stability than u. This explains the reason for the name (parameterized) *Balanced Viscosity solutions*, see also Definition 4.6. While this limiting system is rate-independent, it still shows a subtle balance between the different dissipative mechanisms, namely the rate-independent one given via $\mathcal{R}_0$ and the viscous ones given via $\mathcal{V}_\mathsf{u}$ and $\mathcal{V}_\mathsf{z}$.

Plan of the paper. In Section 2 we set up all the basic assumptions on the dissipation potentials $\mathcal{R}_0$, $\mathcal{V}_\mathsf{u}$, and $\mathcal{V}_\mathsf{z}$. Section 3 is devoted to the generalized gradient system driven by $\mathcal{E}$ and the "viscous" potential $\mathcal{R}_\varepsilon := \mathcal{R}_0 + \varepsilon\mathcal{V}_\mathsf{z} + \varepsilon^\alpha\mathcal{V}_\mathsf{u}$. In particular, we establish a series of estimates on the viscous solutions $(u_\varepsilon, z_\varepsilon)$ which will be at the core of the vanishing-viscosity analysis, developed in Section 4 with Theorem 4.5. In Section 5 we will prove Theorem 5.3 and explore the mechanical interpretation of parameterized Balanced Viscosity solutions. Finally, in Section 6 we will illustrate this solution notion, focusing on how it varies in the cases $\alpha > 1$, $\alpha = 1$, $\alpha \in (0,1)$, in two different examples.

Notation. In what follows, we will denote by $\langle\cdot,\cdot\rangle$ and by $|\cdot|$ the scalar product and the norm in any Euclidean space $\mathbb{R}^d$, with $d = n,\, m,\, n+m,\, \ldots$. Moreover, we will use the same symbol C to denote a positive constant depending on data, and possibly varying from line to line.

2. Setup

As mentioned in the introduction, we are going to address a more general version of system (1.5), where the 1-positively homogeneous dissipation potential $\mathcal{R}_0$, as well as the quadratic potentials $\mathcal{V}_\mathsf{u}$ and $\mathcal{V}_\mathsf{z}$ for u' and z', are also depending on the state variable

$$q := (u, z) \in \mathcal{Q} := \mathbb{R}^n \times \mathbb{R}^m.$$

Hence, the rate-independent system is

$$\partial_{q'}\mathcal{R}_0(q(t), z'(t)) + \mathrm{D}_q\mathcal{E}(t, q(t)) \ni 0 \qquad \text{in } (0,T), \tag{2.1}$$

namely

$$\begin{aligned} &\mathrm{D}_u\mathcal{E}(t, u(t), z(t)) = 0 && \text{for a.a. } t \in (0,T), && (2.2a)\\ &\partial\mathcal{R}_0(q(t), z'(t)) + \mathrm{D}_z\mathcal{E}(t, u(t), z(t)) \ni 0 && \text{for a.a. } t \in (0,T). && (2.2b) \end{aligned}$$

We approximate it with the following generalized gradient system

$$\partial_{q'}\mathcal{R}_\varepsilon(q(t), q'(t)) + \mathrm{D}_q\mathcal{E}(t, q(t)) \ni 0 \qquad \text{in } (0,T), \tag{2.3}$$

where the overall dissipation potential $\mathcal{R}_\varepsilon$ is of the form

$$\mathcal{R}_\varepsilon(q, q') = \mathcal{R}_\varepsilon(q, (u', z')) := \mathcal{R}_0(q, z') + \varepsilon\mathcal{V}_\mathsf{z}(q; z') + \varepsilon^\alpha\mathcal{V}_\mathsf{u}(q; u') \quad \text{with } \alpha > 0. \tag{2.4}$$

In what follows, let us specify our assumptions on the dissipation potentials $\mathcal{R}_0$, $\mathcal{V}_\mathsf{z}$, and $\mathcal{V}_\mathsf{u}$.

Dissipation: We require that

$$\begin{aligned} &\mathcal{R}_0 \in \mathrm{C}^0(\mathcal{Q} \times \mathbb{R}^m), \quad \forall\, q \in \mathcal{Q}:\ \mathcal{R}_0(q,\cdot) \text{ is convex, positively 1-homogeneous, and}\\ &\exists\, C_{0,R},\, C_{1,R} > 0\ \forall\,(q, z') \in \mathcal{Q} \times \mathbb{R}^m:\quad C_{0,R}|z'| \le \mathcal{R}_0(q, z') \le C_{1,R}|z'|, \end{aligned} \tag{R$_0$}$$

$$\begin{aligned} &\mathcal{V}_\mathsf{z} : \mathcal{Q} \times \mathbb{R}^m \to [0,\infty) \text{ is of the form } \mathcal{V}_\mathsf{z}(q; z') = \frac{1}{2}\langle \mathbb{V}_\mathsf{z}(q)z', z'\rangle \text{ with}\\ &\mathbb{V}_\mathsf{z} \in \mathrm{C}^0(\mathcal{Q}; \mathbb{R}^{m\times m}) \text{ and } \exists\, C_{0,V},\, C_{1,V} > 0\ \forall\, q \in \mathcal{Q}:\ C_{0,V}|z'|^2 \le \mathcal{V}_\mathsf{z}(q; z') \le C_{1,V}|z'|^2, \end{aligned} \tag{V$_\mathrm{z}$}$$

$$\begin{aligned} &\mathcal{V}_\mathsf{u} : \mathcal{Q} \times \mathbb{R}^n \to [0,\infty) \text{ is of the form } \mathcal{V}_\mathsf{u}(q; u') = \frac{1}{2}\langle \mathbb{V}_\mathsf{u}(q)u', u'\rangle \text{ with}\\ &\mathbb{V}_\mathsf{u} \in \mathrm{C}^0(\mathcal{Q}; \mathbb{R}^{n\times n}) \text{ and } \exists\, \widetilde{C}_{0,V},\, \widetilde{C}_{1,V} > 0\ \forall\, q \in \mathcal{Q}:\ \widetilde{C}_{0,V}|u'|^2 \le \mathcal{V}_\mathsf{u}(q; u') \le \widetilde{C}_{1,V}|u'|^2. \end{aligned} \tag{V$_\mathrm{u}$}$$

For later use, let us recall that, due to the 1-homogeneity of $\mathcal{R}_0(q,\cdot)$, for every $q \in \mathcal{Q}$ the convex analysis subdifferential $\partial\mathcal{R}_0(q,\cdot) : \mathbb{R}^m \rightrightarrows \mathbb{R}^m$ is characterized by

$$\zeta \in \partial\mathcal{R}_0(q,z') \quad \text{if and only if} \quad \begin{cases} \langle \zeta, w\rangle \leq \mathcal{R}_0(q,w) & \text{for all } w \in \mathbb{R}^m, \\ \langle \zeta, z'\rangle \geq \mathcal{R}_0(q,z')\,. \end{cases} \tag{2.5}$$

Furthermore, observe that $(\mathrm{V_z})$ and $(\mathrm{V_u})$ ensure that for every $q \in \mathcal{Q}$ the matrices $\mathbb{V}_\mathsf{z}(q) \in \mathbb{R}^{n\times n}$ and $\mathbb{V}_\mathsf{u}(q) \in \mathbb{R}^{m\times m}$ are positive definite, uniformly with respect to q. Furthermore, for later use we remark that the conjugate

$$\mathcal{V}_\mathsf{u}^*(q;\eta) = \sup_{v\in\mathbb{R}^n} \left(\langle \eta, v\rangle - \mathcal{V}_\mathsf{u}(q;v)\rangle\right) = \frac{1}{2}\langle \mathbb{V}_\mathsf{u}(q)^{-1}\eta,\eta\rangle$$

fulfills

$$\overline{C}_0|\eta|^2 \leq \mathcal{V}_\mathsf{u}^*(q;\eta) \leq \overline{C}_1|\eta|^2 \tag{2.6}$$

for some $\overline{C}_0$, $\overline{C}_1 > 0$. We have the analogous coercivity and growth properties for $\mathcal{V}_\mathsf{z}^*$.

Our assumptions concerning the energy functional $\mathcal{E}$, expounded below, are typical of the *variational approach* to gradient flows and generalized gradient systems. Since we are in a finite-dimensional setting, to impose *coercivity* it is sufficient to ask for boundedness of energy sublevels. The power-control condition will allow us to bound $\partial_t\mathcal{E}$ in the derivation of the basic energy estimate on system (2.3), cf. Lemma 3.1 later on. The smoothness of $\mathcal{E}$ guarantees the validity of two further, key properties, i.e. the continuity of $\mathrm{D}_q\mathcal{E}$, and the chain rule (cf. (2.10) below), which will play a crucial role for our analysis.

Later on, in Section 3, we will impose that $\mathcal{E}$ is uniformly convex with respect to u. As we will see, this condition will be at the core of the proof of an estimate for $\|u'\|_{\mathrm{L}^1(0,T;\mathbb{R}^n)}$, uniform with respect to the parameter ε. Observe that, unlike for z' such estimate does not follow from the basic energy estimate on system (2.3), since the overall dissipation potential $\mathcal{R}_\varepsilon$ is degenerate in u' as $\varepsilon \downarrow 0$. It will require additional careful calculations.

Energy: We assume that $\mathcal{E} \in \mathrm{C}^1([0,T]\times\mathcal{Q})$ and that it is bounded from below by a positive constant (indeed by adding a constant we can always reduce to this case). Furthermore, we require that

$$\begin{aligned} &\exists\, C_{0,E}, \widetilde{C}_{0,E} > 0\ \forall\,(t,q) \in [0,T]\times\mathcal{Q}: && \mathcal{E}(t,q) \geq C_{0,E}|q|^2 - \widetilde{C}_{0,E} && \textbf{(coercivity)},\\ &\exists\, C_{1,E} > 0\ \forall\,(t,q) \in [0,T]\times\mathcal{Q}: && |\partial_t\mathcal{E}(t,q)| \leq C_{1,E}\mathcal{E}(t,q) && \textbf{(power control)}. \end{aligned} \tag{E}$$

In view of (2.4), $(\mathrm{V_z})$, and $(\mathrm{V_u})$, the generalized gradient system (2.3) reads

$$\varepsilon^\alpha \mathbb{V}_\mathsf{u}(q(t))u'(t) + \mathrm{D}_u\mathcal{E}(t,u(t),z(t)) = 0 \qquad \text{in } (0,T), \tag{2.7a}$$

$$\varepsilon\mathbb{V}_\mathsf{z}(q(t))z'(t) + \partial\mathcal{R}_0(z'(t)) + \mathrm{D}_z\mathcal{E}(t,u(t),z(t)) = 0 \qquad \text{in } (0,T). \tag{2.7b}$$

Existence of solutions to the generalized gradient system (2.3). It follows from the results in [10, 12] that, under the present assumptions, for every $\varepsilon > 0$ there exists a solution $q_\varepsilon \in \mathrm{W}^{1,2}(0,T;\mathcal{Q})$ to the Cauchy problem for (2.3). Observe that q_ε also fulfills the energy-dissipation principle

$$\mathcal{E}(t,q_\varepsilon(t)) + \int_s^t \mathcal{R}_\varepsilon(q_\varepsilon(r),q_\varepsilon'(r)) + \mathcal{R}_\varepsilon^*(q_\varepsilon(r),-\mathrm{D}_q\mathcal{E}(r,q_\varepsilon(r)))\,\mathrm{d}r = \mathcal{E}(s,q_\varepsilon(s)) + \int_s^t \partial_t\mathcal{E}(r,q_\varepsilon(r))\,\mathrm{d}r. \tag{2.8}$$

In (2.8), the dual dissipation potential $\mathcal{R}_\varepsilon^* : \mathcal{Q} \times \mathbb{R}^{n+m} \to \mathbb{R}$ is the Fenchel-Moreau conjugate of $\mathcal{R}_\varepsilon$, i.e.

$$\mathcal{R}_\varepsilon^*(q,\xi) := \sup_{v\in\mathcal{Q}} \left(\langle \xi, v\rangle - \mathcal{R}_\varepsilon(q,v) \right). \tag{2.9}$$

In fact, by the Fenchel equivalence the differential inclusion (2.3) reformulates as

$$\mathcal{R}_\varepsilon(q_\varepsilon(t), q_\varepsilon'(t)) + \mathcal{R}_\varepsilon^*(q_\varepsilon(t), -\mathrm{D}_q\mathcal{E}(t,q_\varepsilon(t))) = \langle -\mathrm{D}_q\mathcal{E}(t,q_\varepsilon(t)), q_\varepsilon'(t)\rangle \qquad \text{for a.a. } t\in(0,T).$$

Combining this with the chain rule

$$\frac{\mathrm{d}}{\mathrm{d}t}\mathcal{E}(t,q(t)) = \partial_t\mathcal{E}(t,q(t)) + \langle \mathrm{D}_q\mathcal{E}(t,q(t)), q'(t)\rangle \qquad \text{for a.a. } t\in(0,T) \tag{2.10}$$

along any curve $q \in \mathrm{W}^{1,1}([0,T];\mathcal{Q})$ and integrating in time, we conclude (2.8).

The energy balance (2.8) will play a crucial role in our analysis: indeed, after deriving in Sec. 3 a series of a priori estimates, uniform with respect to the parameter $\varepsilon > 0$, we shall pass to the limit in the parameterized version of (2.8) as $\varepsilon \downarrow 0$. We will thus obtain a (parameterized) energy-dissipation principle which encodes information on the behavior of the limit system for $\varepsilon = 0$, in particular at the jumps of the limit curve q of the solutions q_ε to (2.3).

3. A priori estimates

In this section, we consider a family $(q_\varepsilon)_\varepsilon \subset \mathrm{W}^{1,2}(0,T;\mathcal{Q})$ of solutions to the Cauchy problem for (2.3), with a converging sequence of initial data $(q_\varepsilon^0)_\varepsilon$, i.e.

$$q_\varepsilon^0 \to q^0 \tag{3.1}$$

for some $q^0 \in \mathcal{Q}$. Our first result, Lemma 3.1, provides a series of basic estimates on the functions (q_ε), as well as a bound for $\|z_\varepsilon'\|_{\mathrm{L}^1(0,T;\mathbb{R}^m)}$, uniform with respect to ε. It holds under conditions (R_0), $(\mathrm{V_z})$, $(\mathrm{V_u})$, (E), as well as (3.1).

Imposing that the dissipation potential $\mathcal{V}_\mathsf{u}$ does not depend on the state variable (cf. $(\mathrm{V}_{u,1})$ below), assuming *uniform convexity* of $\mathcal{E}$ with respect to the variable u, and requiring an additional condition the initial data $(q_\varepsilon^0)_\varepsilon$ (see (3.5)), in Proposition 3.2 we will derive the following crucial estimate, uniform with respect to ε:

$$\|q_\varepsilon'\|_{\mathrm{L}^1(0,T;\mathbb{R}^{n+m})} \le C. \tag{3.2}$$

We begin with the following result, which does not require the above mentioned enhanced conditions.

Lemma 3.1. *Let* $\alpha > 0$. *Assume* (R_0), $(\mathrm{V_z})$, $(\mathrm{V_u})$, (E), *and* (3.1). *Then, there exists a constant* $C > 0$ *such that for every* $\varepsilon > 0$

$$\sup_{t\in[0,T]} \mathcal{E}(t,q_\varepsilon(t)) \le C, \tag{3.3a}$$

$$\sup_{t\in[0,T]} |q_\varepsilon(t)| \le C, \tag{3.3b}$$

$$\int_0^T |z_\varepsilon'(r)|\,\mathrm{d}r \le C. \tag{3.3c}$$

Proof. We exploit the energy identity (2.8). Observe that $\mathcal{R}_\varepsilon^*(q,\xi) \ge 0$ for all $(q,\xi) \in \mathcal{Q}\times R^{n+m}$. Therefore, we deduce from (2.8) that

$$\mathcal{E}(t,q_\varepsilon(t)) \le \mathcal{E}(0,q_\varepsilon(0)) + \int_0^t \partial_t\mathcal{E}(r,q_\varepsilon(r))\,\mathrm{d}r \le C + C_{1,E}\int_0^t \mathcal{E}(r,q_\varepsilon(r))\,\mathrm{d}r,$$

where we have used the power control from (E) and the fact that $\mathcal{E}(0, q_\varepsilon(0)) \le C$, since the $(q_\varepsilon(0))_\varepsilon$ is bounded. The Gronwall lemma then yields (3.3a), and (3.3b) ensues from the coercivity of $\mathcal{E}$. Using again the power control, we ultimately infer from (2.8) that

$$\int_0^T \mathcal{R}_\varepsilon(q_\varepsilon(r), q'_\varepsilon(r)) + \mathcal{R}^*_\varepsilon(q_\varepsilon(r), -\mathrm{D}_q\mathcal{E}(r, q_\varepsilon(r)))\,\mathrm{d}r \le C. \tag{3.4}$$

In particular, $\int_0^T \mathcal{R}_0(q_\varepsilon(r), z'_\varepsilon(r))\,\mathrm{d}r \le C$, whence (3.3c) by ($\mathrm{R}_0$). □

The derivation of the $\mathrm{L}^1(0,T;\mathbb{R}^n)$ estimate for $(u'_\varepsilon)_\varepsilon$ similar to (3.3c) clearly does not follow from (2.8), which only yields $\int_0^T \varepsilon^\alpha |u'_\varepsilon(r)|^2\,\mathrm{d}r \le C$ via (3.4) and ($\mathrm{V_u}$). It is indeed more involved, and, as already mentioned, it strongly relies on the uniform convexity of $\mathcal{E}$ with respect to u. Furthermore, we are able to obtain it only under the simplifying condition that the dissipation potential $\mathcal{V}_\mathsf{u}$ in fact *does not* depend on the state variable q, and under an additional well-preparedness condition on the data $(q^0_\varepsilon)_\varepsilon$, ensuring that the forces $\mathrm{D}_u\mathcal{E}(0, q^0_\varepsilon)$ tend to zero, as $\varepsilon \downarrow 0$, with rate ε^α.

Proposition 3.2. *Let $\alpha > 0$. Assume* (R_0), ($\mathrm{V_z}$), ($\mathrm{V_u}$), *and* (E). *In addition, suppose that*

$$\mathrm{D}_q\mathbb{V}_\mathsf{u}(q) = 0 \quad \textit{for all } q \in \mathcal{Q}, \tag{$\mathrm{V}_{u,1}$}$$

$$\begin{aligned} &\mathcal{E} \in \mathrm{C}^2([0,T]\times\mathcal{Q}) \quad \textit{and} \\ &\exists\, \mu > 0\ \forall\,(t,q) \in [0,T]\times\mathcal{Q}:\ \mathrm{D}^2_u\mathcal{E}(t,q) \ge \mu\mathbb{I}_{\mathbb{R}^{n\times n}} \quad \textbf{\textit{(uniform convexity w.r.t. }}u\textbf{\textit{)}}, \end{aligned} \tag{E_1}$$

and that the initial data $(q^0_\varepsilon)_\varepsilon$ complying with (3.1) *also fulfill*

$$|\mathrm{D}_u\mathcal{E}(0, q^0_\varepsilon)| \le C\varepsilon^\alpha. \tag{3.5}$$

Then, there exists a constant $C > 0$ such that for every $\varepsilon > 0$

$$\|u'_\varepsilon(t)\|_{\mathrm{L}^1(0,T;\mathbb{R}^n)} \le C. \tag{3.6}$$

Proof. By ($\mathrm{V}_{u,1}$) there exists a matrix $\overline{\mathbb{V}}_\mathsf{u} \in \mathbb{R}^{n\times n}$ such that $\mathbb{V}_\mathsf{u}(q) \equiv \overline{\mathbb{V}}_\mathsf{u}$ for all $q \in \mathcal{Q}$ with

$$\mathcal{V}_\mathsf{u}(q; u') = \mathcal{V}_\mathsf{u}(u') := \frac{1}{2}\langle \overline{\mathbb{V}}_\mathsf{u} u', u'\rangle. \tag{3.7}$$

Therefore (2.7a) reduces to

$$\varepsilon^\alpha \overline{\mathbb{V}}_\mathsf{u} u'_\varepsilon(t) + \mathrm{D}_u\mathcal{E}(t, u_\varepsilon(t), z_\varepsilon(t)) = 0 \qquad \text{for a.a. } t \in (0,T). \tag{3.8}$$

We differentiate (3.8) in time, and test the resulting equation by u'_ε. Thus we obtain, for almost all $t \in (0,T)$,

$$\begin{aligned} 0 = \varepsilon^\alpha \langle \overline{\mathbb{V}}_\mathsf{u} u_\varepsilon''(t), u'_\varepsilon(t)\rangle + \langle \mathrm{D}^2_u\mathcal{E}(t, u_\varepsilon(t), z_\varepsilon(t))[u'_\varepsilon(t)], u'_\varepsilon(t)\rangle \\ + \langle \mathrm{D}^2_{u,z}\mathcal{E}(t, u_\varepsilon(t), z_\varepsilon(t))[u'_\varepsilon(t)], z'_\varepsilon(t)\rangle \ \doteq\ S_1 + S_2 + S_3, \end{aligned} \tag{3.9}$$

where $\mathrm{D}^2_{u,z}$ denotes the second-order mixed derivative. Observe that

$$S_1 = \frac{\varepsilon^\alpha}{2}\frac{\mathrm{d}}{\mathrm{d}t}\mathcal{V}_\mathsf{u}(u'_\varepsilon), \qquad S_2 \ge \mu|u'^2_\varepsilon| \ge \tilde{\mu}\mathcal{V}_\mathsf{u}(u'_\varepsilon),$$
$$S_3 \ge -C|u'_\varepsilon||z'_\varepsilon| \ge -C\sqrt{\mathcal{V}_\mathsf{u}(u'_\varepsilon)}|z'_\varepsilon|.$$

Indeed, to estimate S_2 we have used the uniform convexity of $\mathcal{E}(t,\cdot,z)$, and the growth of $\mathcal{V}_{\mathsf{u}}$ from ($\mathrm{V_u}$). The estimate for S_3 follows from $\sup_{t\in(0,T)}|\mathrm{D}^2_{u,z}\mathcal{E}(t,u_\varepsilon(t),z_\varepsilon(t))|\leq C$, due to (3.3b) and the fact that $\mathrm{D}^2_{u,z}\mathcal{E}$ is continuous on $[0,T]\times\mathcal{Q}$, cf. ($\mathrm{V_u}$). We thus infer from (3.9) that

$$\frac{1}{2}\frac{\mathrm{d}}{\mathrm{d}t}\mathcal{V}_{\mathsf{u}}(u_\varepsilon'(t))+\frac{\tilde{\mu}}{\varepsilon^\alpha}\mathcal{V}_{\mathsf{u}}(u_\varepsilon'(t))\leq\frac{C}{\varepsilon^\alpha}\sqrt{\mathcal{V}_{\mathsf{u}}(u_\varepsilon'(t))}|z_\varepsilon'(t)|\qquad\text{for a.a. } t\in(0,T),$$

which rephrases as

$$\nu_\varepsilon(t)\nu_\varepsilon'(t)+\frac{\tilde{\mu}}{\varepsilon^\alpha}\nu_\varepsilon^2(t)\leq\frac{C}{\varepsilon^\alpha}\nu_\varepsilon(t)|z_\varepsilon'(t)|$$

where we have used the place-holder $\nu_\varepsilon(t):=\sqrt{\mathcal{V}_{\mathsf{u}}(u_\varepsilon'(t))}$. We now argue as in [5] and observe that, without loss of generality, we may suppose that $\nu_\varepsilon(t)>0$ (otherwise, we replace it by $\tilde{\nu}_\varepsilon=\sqrt{\nu_\varepsilon+\delta}$, which satisfies the same estimate, and then let $\delta\downarrow 0$), Hence, we deduce

$$\nu_\varepsilon'(t)+\frac{\tilde{\mu}}{\varepsilon^\alpha}\nu_\varepsilon(t)\leq\frac{C}{\varepsilon^\alpha}|z_\varepsilon'(t)|.$$

Applying the Gronwall lemma we obtain

$$\nu_\varepsilon(t)\leq C\exp\left(-\frac{\tilde{\mu}}{\varepsilon^\alpha}t\right)\nu_\varepsilon(0)+\frac{C}{\varepsilon^\alpha}\int_0^t\exp\left(-\frac{\tilde{\mu}}{\varepsilon^\alpha}(t-r)\right)|z_\varepsilon'(r)|\,\mathrm{d}r\doteq a_1^\varepsilon(t)+a_2^\varepsilon(t)\tag{3.10}$$

for all $t\in(0,T)$. We integrate the above estimate on $(0,T)$. Now, observe that (3.5) guarantees that $\nu_\varepsilon(0)=\sqrt{\mathcal{V}_{\mathsf{u}}(u_\varepsilon'(0))}\leq C|\overline{V}_{\mathsf{u}}u_\varepsilon'(0)|=C\varepsilon^{-\alpha}|\mathrm{D}\mathcal{E}(0,u_\varepsilon(0))|\leq C$. Hence, we find $\|a_1^\varepsilon\|_{\mathrm{L}^1(0,T)}\leq C\nu_\varepsilon(0)\leq C_1$. To estimate a_2^ε we use Young's convolution inequality giving

$$\begin{aligned}\|a_2^\varepsilon\|_{\mathrm{L}^1(0,T)}&=\tfrac{C}{\varepsilon^\alpha}\textstyle\int_0^T\int_0^t\exp\left(-\frac{\tilde{\mu}}{4\varepsilon^\alpha}(t-r)\right)|z_\varepsilon'(r)|\,\mathrm{d}r\,\mathrm{d}t\\&\leq\tfrac{C}{\varepsilon^\alpha}\left(\textstyle\int_0^T\exp\left(-\frac{\tilde{\mu}}{\varepsilon^\alpha}t\right)\mathrm{d}t\right)\left(\textstyle\int_0^T|z_\varepsilon'(t)|\,\mathrm{d}t\right)\;\leq\;C_2\end{aligned}$$

where we have exploited the a priori estimate (3.3c) for z_ε'. Thus, (3.10) implies (3.6), and we are done. □

4. Limit passage with vanishing viscosity

In this section, we assume that we are given a sequence $(q_\varepsilon)_\varepsilon\subset\mathrm{W}^{1,2}(0,T;\mathcal{Q})$ of solutions to (2.3), satisfying the initial conditions $q_\varepsilon(0)=q_\varepsilon^0$, such that estimate (3.2) holds. As we have shown in Proposition 3.2, the well-preparedness (3.5) of the initial data $(q_\varepsilon^0)_\varepsilon$, the condition that the dissipation potential $\mathcal{V}_{\mathsf{u}}$ does not depend on the state q, and the uniform convexity ($\mathrm{E_1}$) of $\mathcal{E}$ with respect to u guarantee the validity of (3.2). However, these conditions are not needed for the vanishing-viscosity analysis. Therefore, hereafter we will no longer impose (3.5), we will allow for a state-dependent dissipation potential $\mathcal{V}_{\mathsf{u}}=\mathcal{V}_{\mathsf{u}}(q;u')$, and we will stay with the basic conditions (E) on $\mathcal{E}$.

The energy-dissipation principle. Following the variational approach of [8, 13, 14], we will pass to the limit in (a *parameterized* version of) the energy identity (2.8).

Preliminarily, let us explicitly calculate the convex-conjugate of the dissipation potential $\mathcal{R}_\varepsilon$ (2.4).

Lemma 4.1. *Assume* ($\mathrm{R_0}$), ($\mathrm{V_z}$), *and* ($\mathrm{V_u}$). *Then, the Fenchel-Moreau conjugate* (2.9) *of* $\mathcal{R}_\varepsilon$ *is given by*

$$\mathcal{R}_\varepsilon^*(q,\xi)=\frac{1}{\varepsilon}\mathcal{W}_{\mathsf{z}}^*(q;\zeta)+\frac{1}{\varepsilon^\alpha}\mathcal{V}_{\mathsf{u}}^*(q;\eta)\qquad\textit{for all } q\in\mathcal{Q}\textit{ and }\xi=(\eta,\zeta)\in\mathbb{R}^{n+m},\tag{4.1}$$

where $\mathcal{V}_{\mathsf{u}}^*(q;\cdot)$ *is the conjugate of* $\mathcal{V}_{\mathsf{u}}(q;\cdot)$, *and*

$$\mathcal{W}_{\mathsf{z}}^*(q;\zeta) = \min_{\omega\in K(q)} \mathcal{V}_{\mathsf{z}}^*(q;\zeta-\omega) \qquad \textit{with } K(q) := \partial\mathcal{R}_0(q,0), \tag{4.2}$$

$\mathcal{V}_{\mathsf{z}}^*(q;\cdot)$ *is the conjugate of* $\mathcal{V}_{\mathsf{z}}(q;\cdot)$, *while* $\mathcal{W}_{\mathsf{z}}^*$ *is the conjugate of* $\mathcal{R}_0+\mathcal{V}_{\mathsf{z}}$.

Proof. Since $\mathcal{R}_\varepsilon(q,\cdot)$ is given by the sum of a contribution in the sole variable z' and another in the sole variable u', we have

$$\mathcal{R}_\varepsilon^*(q,\xi) = (\varepsilon^\alpha\mathcal{V}_{\mathsf{u}})^*(q,\eta) + \mathcal{W}_{\mathsf{z},\varepsilon}^*(q;\zeta) \qquad \text{for all } \xi=(\eta,\zeta)\in\mathbb{R}^{n+m}$$

where we have used the place-holder $\mathcal{W}_{\mathsf{z},\varepsilon}^*(q;\zeta) := (\mathcal{R}_0(q,\cdot)+\varepsilon\mathcal{V}_{\mathsf{z}}(q;\cdot))^*(\zeta)$. Now, taking into account that $\mathcal{V}_{\mathsf{u}}$ is quadratic, there holds

$$(\varepsilon^\alpha\mathcal{V}_{\mathsf{u}})^*(q,\eta) = \varepsilon^\alpha\mathcal{V}_{\mathsf{u}}^*\left(q,\frac{1}{\varepsilon^\alpha}\eta\right) = \frac{1}{\varepsilon^\alpha}\mathcal{V}_{\mathsf{u}}^*(q;\eta),$$

whereas the inf-sup convolution formula (see e.g. [30]) yields $\mathcal{W}_{\mathsf{z},\varepsilon}^*(q;\zeta) = \frac{1}{\varepsilon}\mathcal{W}_{\mathsf{z}}^*(q;\zeta)$ with $\mathcal{W}_{\mathsf{z}}^*(q;\cdot)$ from (4.2). □

In view of (4.1), the energy identity (2.8) rewrites as

$$\begin{aligned}\mathcal{E}(t,q_\varepsilon(t)) &+ \int_s^t \mathcal{R}_0(q_\varepsilon(r),z_\varepsilon'(r)) + \varepsilon\mathcal{V}_{\mathsf{z}}(q_\varepsilon(r);z_\varepsilon'(r)) + \varepsilon^\alpha\mathcal{V}_{\mathsf{u}}(q_\varepsilon(r);u_\varepsilon'(r))\,\mathrm{d}r \\ &+ \int_s^t \frac{1}{\varepsilon}\mathcal{W}_{\mathsf{z}}^*(q_\varepsilon(r);-\mathrm{D}_z\mathcal{E}(r,q_\varepsilon(r))) + \frac{1}{\varepsilon^\alpha}\mathcal{V}_{\mathsf{u}}^*(q_\varepsilon(r);-\mathrm{D}_u\mathcal{E}(r,q_\varepsilon(r)))\,\mathrm{d}r \\ &= \mathcal{E}(s,q_\varepsilon(s)) + \int_s^t \partial_t\mathcal{E}(r,q_\varepsilon(r))\,\mathrm{d}r.\end{aligned} \tag{4.3}$$

In fact, the two integral terms on the left-hand side of (4.3) reflect the competition between the tendency of the system to be governed by *viscous* dissipation both for the variable z and for the variable u, and its tendency to fulfill the *local stability* condition

$$\mathcal{W}_{\mathsf{z}}^*(q(t);-\mathrm{D}_z\mathcal{E}(t,q(t)))=0 \quad \text{i.e.} \quad -\mathrm{D}_z\mathcal{E}(t,q(t))\in K(q(t)) \qquad \text{for a.a. } t\in(0,T)$$

for z, and the equilibrium condition

$$\mathcal{V}_{\mathsf{u}}^*(q(t);-\mathrm{D}_u\mathcal{E}(r,q(t)))=0 \quad \text{i.e.} \quad -\mathrm{D}_u\mathcal{E}(t,q(t))=0 \qquad \text{for a.a. } t\in(0,T)$$

for u, cf. also the discussion in Remark 4.4.

The parameterized energy-dissipation principle. We now consider the parameterized curves $(\mathsf{t}_\varepsilon,\mathsf{q}_\varepsilon):[0,S_\varepsilon]\to[0,T]\times\mathcal{Q}$, where for every $\varepsilon>0$ the rescaling function $\mathsf{t}_\varepsilon:[0,S_\varepsilon]\to[0,T]$ is strictly increasing, and $\mathsf{q}_\varepsilon(s)=q_\varepsilon(\mathsf{t}_\varepsilon(s))$. We shall suppose that $\sup_{\varepsilon>0}S_\varepsilon<\infty$, and that

$$\exists\, C>0 \quad \forall\,\varepsilon>0 \quad \forall\, s\in[0,S_\varepsilon]: \qquad \mathsf{t}_\varepsilon'(s)+|\mathsf{q}_\varepsilon'(s)|\le C. \tag{4.4}$$

Remark 4.2. For instance, as in [9, 8] we might choose

$$\mathsf{t}_\varepsilon := \sigma_\varepsilon^{-1} \quad \text{with } \sigma_\varepsilon(t) := \int_0^t \left(1+|q_\varepsilon'(r)|\right)\mathrm{d}r, \tag{4.5}$$

and set $S_\varepsilon:=\sigma_\varepsilon(T)$. In fact, estimate (3.2) ensures that $\sup_\varepsilon S_\varepsilon<\infty$. With the choice (4.5) for t_ε, the functions $(\mathsf{t}_\varepsilon,\mathsf{q}_\varepsilon)$ fulfill the normalization condition

$$\mathsf{t}_\varepsilon'(s)+|\mathsf{q}_\varepsilon'(s)|=1 \qquad \text{for almost all } s\in(0,S_\varepsilon).$$

For the parameterized curves $(\mathsf{t}_\varepsilon, \mathsf{q}_\varepsilon)$, the energy-dissipation principle (4.3) reads

$$\begin{aligned} &\mathcal{E}(\mathsf{t}_\varepsilon(s_2), \mathsf{q}_\varepsilon(s_2)) + \int_{s_1}^{s_2} \mathcal{M}_\varepsilon(\mathsf{q}_\varepsilon(r), \mathsf{t}'_\varepsilon(r), \mathsf{q}'_\varepsilon(r), -\mathrm{D}_q\mathcal{E}(\mathsf{t}_\varepsilon(r), \mathsf{q}_\varepsilon(r)))\, \mathrm{d}r \\ &= \mathcal{E}(\mathsf{t}_\varepsilon(s_1), \mathsf{q}_\varepsilon(s_1)) + \int_{s_1}^{s_2} \partial_t\mathcal{E}(\mathsf{t}_\varepsilon(r), \mathsf{q}_\varepsilon(r))\mathsf{t}'_\varepsilon(r)\,\mathrm{d}r \qquad \text{for all } 0 \le s_1 \le s_2 \le S, \end{aligned} \tag{4.6}$$

where we have used the dissipation functional

$$\begin{aligned} \mathcal{M}_\varepsilon(q, \tau, q', \xi) &= \mathcal{M}_\varepsilon(q, \tau, (u', z'), (\eta, \zeta)) \\ &:= \mathcal{R}_0(q, z') + \frac{\varepsilon}{\tau}\mathcal{V}_\mathsf{z}(q; z') + \frac{\varepsilon^\alpha}{\tau}\mathcal{V}_\mathsf{u}(q; u') + \frac{\tau}{\varepsilon}\mathcal{W}^*_\mathsf{z}(q; \zeta) + \frac{\tau}{\varepsilon^\alpha}\mathcal{V}^*_\mathsf{u}(q; \eta). \end{aligned} \tag{4.7}$$

The passage from (4.3) to (4.6) follows from the change of variables $t \to \mathsf{t}_\varepsilon(r)$, whence $\mathrm{d}t \to \mathsf{t}'_\varepsilon(r)\mathrm{d}r$, while $q'_\varepsilon(t) \to \frac{1}{\mathsf{t}'_\varepsilon(r)}\mathsf{q}'_\varepsilon(r)$. In order to pass to the limit in (4.6) as $\varepsilon \downarrow 0$, it is crucial to investigate the Γ-convergence properties of the family of functionals $(\mathcal{M}_\varepsilon)_\varepsilon$. The following result reveals that the Γ-limit of $(\mathcal{M}_\varepsilon)_\varepsilon$ depends on whether the parameter α is above, equal, or below the threshold value 1. Let us point out that, for $\alpha \in (0,1)$, setting $\delta = \varepsilon^\alpha$ we rewrite $\mathcal{M}_\varepsilon$ as

$$\mathcal{M}_\varepsilon(q, \tau, (u', z'), (\eta, \zeta)) = \mathcal{R}_0(q, z') + \frac{\delta^{1/\alpha}}{\tau}\mathcal{V}_\mathsf{z}(q; z') + \frac{\delta}{\tau}\mathcal{V}_\mathsf{u}(q; u') + \frac{\tau}{\delta^{1/\alpha}}\mathcal{W}^*_\mathsf{z}(q; \zeta) + \frac{\tau}{\delta}\mathcal{V}^*_\mathsf{u}(q; \eta) \tag{4.8}$$

with $1/\alpha > 1$. It is thus natural to expect that the upcoming results will be specular in the cases $\alpha \in (0,1)$ and $\alpha > 1$.

Proposition 4.3. *Assume* (R_0), (V_z), (V_u), *and* (E). *Then, the functionals* $(\mathcal{M}_\varepsilon)_\varepsilon$ Γ*-converge as* $\varepsilon \downarrow 0$ *to* $\mathcal{M}_0 : \mathcal{Q} \times [0,\infty) \times \mathcal{Q} \times \mathbb{R}^{n+m} \to [0,\infty]$ *defined by*

$$\mathcal{M}_0(q, \tau, (u', z'), (\eta, \zeta)) := \mathcal{R}_0(q, z') + \mathcal{M}_0^{\mathrm{red}}(q, \tau, (u', z'), (\eta, \zeta)), \tag{4.9}$$

where for $\tau > 0$ *we have*

$$\mathcal{M}_0^{\mathrm{red}}(q, \tau, (u', z'), (\eta, \zeta)) = \begin{cases} 0 & \text{if } \mathcal{W}^*_\mathsf{z}(q;\zeta) = \mathcal{V}^*_\mathsf{u}(q;\eta) = 0, \\ \infty & \text{if } \mathcal{W}^*_\mathsf{z}(q;\zeta) + \mathcal{V}^*_\mathsf{u}(q;\eta) > 0, \end{cases} \tag{4.10}$$

while for $\tau = 0$ *we have the following cases:*

- *For* $\alpha > 1$

$$\mathcal{M}_0^{\mathrm{red}}(q, 0, (u', z'), (\eta, \zeta)) = \begin{cases} 2\sqrt{\mathcal{V}_\mathsf{u}(q;u')}\,\sqrt{\mathcal{V}^*_\mathsf{u}(q;\eta)} & \text{if } \mathcal{V}_\mathsf{z}(q;z') = 0, \\ 2\sqrt{\mathcal{V}_\mathsf{z}(q;z')}\,\sqrt{\mathcal{W}^*_\mathsf{z}(q;\zeta)} & \text{if } \mathcal{V}^*_\mathsf{u}(q;\eta) = 0, \\ \infty & \text{if } \mathcal{V}_\mathsf{z}(q;z')\,\mathcal{V}^*_\mathsf{u}(q;\eta) > 0, \end{cases} \tag{4.11}$$

- *For* $\alpha = 1$

$$\mathcal{M}_0^{\mathrm{red}}(q, 0, (u', z'), (\eta, \zeta)) = 2\sqrt{\mathcal{V}_\mathsf{z}(q;z') + \mathcal{V}_\mathsf{u}(q;u')}\,\sqrt{\mathcal{W}^*_\mathsf{z}(q;\zeta) + \mathcal{V}^*_\mathsf{u}(q;\eta)}, \tag{4.12}$$

- *For* $\alpha \in (0,1)$

$$\mathcal{M}_0^{\mathrm{red}}(q, 0, (u', z'), (\eta, \zeta)) = \begin{cases} 2\sqrt{\mathcal{V}_\mathsf{u}(q;u')}\,\sqrt{\mathcal{V}^*_\mathsf{u}(q;\eta)} & \text{if } \mathcal{W}^*_\mathsf{z}(q;\zeta) = 0, \\ 2\sqrt{\mathcal{V}_\mathsf{z}(q;z')}\,\sqrt{\mathcal{W}^*_\mathsf{z}(q;\zeta)} & \text{if } \mathcal{V}_\mathsf{u}(q;u') = 0, \\ \infty & \text{if } \mathcal{V}_\mathsf{u}(q;u')\,\mathcal{W}^*_\mathsf{z}(q;\zeta) > 0. \end{cases} \tag{4.13}$$

Moreover, if $(\tau_\varepsilon, q'_\varepsilon) \rightharpoonup (\tau, q')$ *in* $\mathrm{L}^1(0,S;(0,T)\times\mathcal{Q})$ *and if* $(q_\varepsilon, \xi_\varepsilon) \to (q,\xi)$ *in* $\mathrm{L}^1(0,S;\mathcal{Q}\times\mathbb{R}^{n+m})$, *then for every* $0 \le s_1 \le s_2 \le S$

$$\liminf_{\varepsilon\downarrow 0} \int_0^S \mathcal{M}_\varepsilon(q_\varepsilon(s), \tau_\varepsilon(s), q'_\varepsilon(s), \xi_\varepsilon(s))\,\mathrm{d}s \ge \int_0^S \mathcal{M}_0(q(s), \tau(s), q'(s), \xi(s))\,\mathrm{d}s\,. \tag{4.14}$$

Remark 4.4. Let us briefly comment on the expression (4.9) of the Γ-limit $\mathcal{M}_0$. To do so, we rephrase the constraints arising in the switching conditions for the reduced functional $\mathcal{M}_0^{\mathrm{red}}$, cf. (4.10), (4.11), and (4.13). Indeed, it follows from $(\mathrm{V_z})$ and $(\mathrm{V_u})$ (cf. (2.6)) that

$$\begin{aligned} &\mathcal{V}_{\mathsf{z}}(q;z') = 0 \;\Leftrightarrow\; z'=0, && \mathcal{V}_{\mathsf{u}}(q;u') = 0 \;\Leftrightarrow\; u'=0,\\ &\mathcal{V}_{\mathsf{u}}^*(q;\eta) = 0 \;\Leftrightarrow\; \eta=0, && \mathcal{W}_{\mathsf{z}}^*(q;\zeta) = 0 \;\Leftrightarrow\; \zeta\in K(q) = \partial\mathcal{R}_0(q,0). \end{aligned}$$

Therefore, from (4.10) we read that for $\tau>0$ the functional $\mathcal{M}_0^{\mathrm{red}}(q,\tau,\cdot,\cdot)$ is finite (and indeed equal to 0) only for η and ζ fulfilling

$$\eta = 0, \qquad \zeta\in K(q)\,.$$

For $\tau=0$, in the case $\alpha>1$, $\mathcal{M}_0^{\mathrm{red}}(q,0,\cdot,\cdot)$ is finite if and only if either $z'=0$ or $\eta=0$. As we will see when discussing the physical interpretation of our vanishing-viscosity result, this means that, at a jump (i.e. when $\tau=0$), either $z'=0$, i.e. z is frozen, or u fulfills the equilibrium condition $\eta = \mathrm{D}_u\mathcal{E}(t,u)=0$.

Also in view of (4.8), the switching conditions for $\alpha\in(0,1)$ are specular to the ones for $\alpha>1$ in a generalized sense. In fact, $\mathcal{M}_0^{\mathrm{red}}(q,0,\cdot,\cdot)$ is finite if and only if either u is frozen, or $\zeta = \mathrm{D}_z\mathcal{E}(t,z)\in K(q)$, meaning that z fulfills the *local stability* condition.

Proof of Proposition 4.3. Observe that

$$\mathcal{M}_\varepsilon(q,\tau,(u',z'),(\eta,\zeta)) = \mathcal{R}_0(q,z') + \mathcal{M}_\varepsilon^{\mathrm{red}}(q,\tau,(u',z'),(\eta,\zeta))$$

with $\mathcal{M}_\varepsilon^{\mathrm{red}}(q,\tau,(u',z'),(\eta,\zeta)) := \frac{\varepsilon}{\tau}\mathcal{V}_{\mathsf{z}}(q;z') + \frac{\varepsilon^\alpha}{\tau}\mathcal{V}_{\mathsf{u}}(q;u') + \frac{\tau}{\varepsilon}\mathcal{W}_{\mathsf{z}}^*(q;\zeta) + \frac{\tau}{\varepsilon^\alpha}\mathcal{V}_{\mathsf{u}}^*(q;\eta)$. Since $\mathcal{R}_0$ is continuous with respect to both variables q and z and does not depend on ε, it is clearly sufficient to prove that the functionals $\mathcal{M}_\varepsilon^{\mathrm{red}}$ Γ-converge to $\mathcal{M}_0^{\mathrm{red}}$, namely

$$\begin{aligned} &\Gamma\text{-}\liminf\ \text{estimate:}\\ &\quad (q_\varepsilon,\tau_\varepsilon,u'_\varepsilon,z'_\varepsilon,\eta_\varepsilon,\zeta_\varepsilon)\to(q,\tau,u',z',\eta,\zeta)\ \text{ for } \varepsilon\to 0\\ &\qquad \Longrightarrow\ \mathcal{M}_0^{\mathrm{red}}(q,\tau,(u',z'),(\eta,\zeta)) \le \liminf_{\varepsilon\downarrow 0}\mathcal{M}_\varepsilon^{\mathrm{red}}(q_\varepsilon,\tau_\varepsilon,(u'_\varepsilon,z'_\varepsilon),(\eta_\varepsilon,\zeta_\varepsilon)), \end{aligned} \tag{4.15}$$

$$\begin{aligned} &\Gamma\text{-}\limsup\ \text{estimate:}\\ &\quad \forall\,(q,\tau,u',z',\eta,\zeta)\ \exists\,(q_\varepsilon,\tau_\varepsilon,u'_\varepsilon,z'_\varepsilon,\eta_\varepsilon,\zeta_\varepsilon)_\varepsilon\ :\\ &\qquad \begin{cases} (q_\varepsilon,\tau_\varepsilon,u'_\varepsilon,z'_\varepsilon,\eta_\varepsilon,\zeta_\varepsilon)\to(q,\tau,u',z',\eta,\zeta) \qquad \text{and}\\ \mathcal{M}_0^{\mathrm{red}}(q,\tau,(u',z'),(\eta,\zeta)) \ge \limsup_{\varepsilon\downarrow 0}\mathcal{M}_\varepsilon^{\mathrm{red}}(q_\varepsilon,\tau_\varepsilon,(u'_\varepsilon,z'_\varepsilon),(\eta_\varepsilon,\zeta_\varepsilon)). \end{cases} \end{aligned} \tag{4.16}$$

Preliminarily, observe that minimizing with respect to τ we obtain the lower bound

$$\mathcal{M}_\varepsilon^{\mathrm{red}}(q,\tau,(u',z'),(\eta,\zeta)) \ge 2\sqrt{\varepsilon\mathcal{V}_{\mathsf{z}}(q;z') + \varepsilon^\alpha\mathcal{V}_{\mathsf{u}}(q;u')}\sqrt{\frac{1}{\varepsilon}\mathcal{W}_{\mathsf{z}}^*(q;\zeta) + \frac{1}{\varepsilon^\alpha}\mathcal{V}_{\mathsf{u}}^*(q;\eta)}. \tag{4.17}$$

In all the three cases $\alpha>1$, $\alpha=1$, and $\alpha\in(0,1)$, the expression (4.10) of $\mathcal{M}_0^{\mathrm{red}}$ for $\tau>0$ can be easily checked. Indeed, for the Γ-lim inf estimate, observe that it is trivial in the case $\mathcal{W}_{\mathsf{z}}^*(q;\zeta) = \mathcal{V}_{\mathsf{u}}^*(q;\eta) = 0$, as $\mathcal{M}_\varepsilon^{\mathrm{red}}$ takes positive values for all $\varepsilon>0$. Suppose now that $\mathcal{W}_\mathsf{z}^*(q;\zeta)+\mathcal{V}_\mathsf{u}^*(q;\eta)>0$, e.g. that $\mathcal{V}_\mathsf{u}^*(q;\eta)>0$. Now, $(q_\varepsilon,\eta_\varepsilon)\to(q,\eta)$ implies that $\mathcal{V}_\mathsf{u}^*(q_\varepsilon;\eta_\varepsilon)\geq\bar{c}>0$ for sufficiently small ε, and from (4.17) we deduce that

$$\liminf_{\varepsilon\downarrow 0}\mathcal{M}_\varepsilon^{\mathrm{red}}(q_\varepsilon,\tau_\varepsilon,(u_\varepsilon',z_\varepsilon'),(\eta_\varepsilon,\zeta_\varepsilon))=\infty=\mathcal{M}_0^{\mathrm{red}}(q,\tau,(u',z'),(\eta,\zeta))\,.$$

The Γ-lim sup estimate follows by taking the recovery sequence $(q_\varepsilon,\tau_\varepsilon,u_\varepsilon',z_\varepsilon',\eta_\varepsilon,\zeta_\varepsilon)=(q,\tau,u',z',\eta,\zeta)$. In fact, $\mathcal{W}_\mathsf{z}^*(q;\zeta)+\mathcal{V}_\mathsf{u}^*(q;\eta)>0$, then the lim sup-inequality in (4.16) is trivial. If $\mathcal{W}_\mathsf{z}^*(q;\zeta)=\mathcal{V}_\mathsf{u}^*(q;\eta)=0$, (4.16) can be checked straightforwardly.

For $\alpha=1$, in the case $\tau=0$, (4.17) clearly yields the Γ-lim inf estimate, whereas the Γ-lim sup one can be obtained by with the recovery sequence $(q_\varepsilon,\tau_\varepsilon,u_\varepsilon',z_\varepsilon',\eta_\varepsilon,\zeta_\varepsilon)=(q,\tau_\varepsilon^*,u',z',\eta,\zeta)$ with

$$\tau_\varepsilon^*=\varepsilon\frac{\sqrt{\mathcal{V}_\mathsf{z}(q;z')+\mathcal{V}_\mathsf{u}(q;u')}}{\sqrt{\mathcal{W}_\mathsf{z}^*(q;\zeta)+\mathcal{V}_\mathsf{u}^*(q;\eta)}}.$$

For $\alpha>1$ and $\tau=0$, the Γ-lim inf estimate follows taking into account that (4.17) yields

$$\mathcal{M}_\varepsilon^{\mathrm{red}}(q,\tau,(u',z'),(\eta,\zeta))\geq\frac{2}{\sqrt{\varepsilon^{\alpha-1}}}\sqrt{\mathcal{V}_\mathsf{z}(q;z')\mathcal{V}_\mathsf{u}^*(q;\eta)}. \tag{4.18}$$

Hence, if both $\mathcal{V}_\mathsf{z}(q;z')>0$ and $\mathcal{V}_\mathsf{u}^*(q;\eta)>0$, then $\liminf_{\varepsilon\downarrow 0}\mathcal{M}_\varepsilon^{\mathrm{red}}(q,\tau,(u',z'),(\eta,\zeta))=\infty$. In the case when either $\mathcal{V}_\mathsf{z}(q;z')=0$ or $\mathcal{V}_\mathsf{u}^*(q;\eta)=0$, we deduce the Γ-lim inf estimate from (4.17). For the Γ-lim sup estimate, we again take the recovery sequence $(t,q,\tau_\varepsilon^{**},u',z',\eta,\zeta)$, where now

$$\tau_\varepsilon^{**}=\varepsilon\frac{\sqrt{\mathcal{V}_\mathsf{z}(q;z')+\varepsilon^{\alpha-1}\mathcal{V}_\mathsf{u}(q;u')}}{\sqrt{\mathcal{W}_\mathsf{z}^*(q;\zeta)+\frac{1}{\varepsilon^{\alpha-1}}\mathcal{V}_\mathsf{u}^*(q;\eta)}}.$$

The discussion of the case $\alpha\in(0,1)$ is completely analogous, also in view of (4.8).

Finally, in order to prove (4.14) we use Ioffe's theorem [31]: We introduce a functional $\overline{\mathcal{M}}:[0,\infty)\times\mathcal{Q}\times[0,\infty)\times\mathcal{Q}\times\mathbb{R}^{n+m}\to[0,\infty]$ subsuming the functionals $\mathcal{M}_\varepsilon$ and $\mathcal{M}_0$, viz.

$$\overline{\mathcal{M}}(\varepsilon;q,\tau,q',\xi):=\begin{cases}\mathcal{M}_\varepsilon(q,\tau,q',\xi) & \text{if } \varepsilon>0,\\ \mathcal{M}_0(q,\tau,q',\xi) & \text{if } \varepsilon=0.\end{cases}$$

Arguing in the very same way as in the proof of [8, Lemma 3.1], it follows that the functional $\overline{\mathcal{M}}$ is lower semicontinuous on $[0,\infty)\times\mathcal{Q}\times[0,\infty)\times\mathcal{Q}\times\mathbb{R}^{n+m}$, and that $(\tau,q')\mapsto\overline{\mathcal{M}}(\varepsilon;q,\tau,q',\xi)$ is convex for all $(\varepsilon,q,\xi)\in[0,\infty)\times\mathcal{Q}\times\mathbb{R}^{n+m}$. Hence, Ioffe's theorem yields

$$\liminf_{\varepsilon\downarrow 0}\int_0^S\overline{\mathcal{M}}(\varepsilon;q_\varepsilon(s),\tau_\varepsilon(s),q_\varepsilon'(s),\xi_\varepsilon(s))\,\mathrm{d}s\geq\int_0^S\overline{\mathcal{M}}(0;q(s),\tau(s),q'(s),\xi(s))\,\mathrm{d}s,$$

whence (4.14). □

Observe that the functional $\mathcal{M}_0$ in (4.9) fulfills for all $(q,\tau)\in\mathcal{Q}\times[0,\infty)$ the estimate

$$\mathcal{M}_0(q,\tau,q',\xi)\geq\langle q',\xi\rangle=\langle u',\eta\rangle+\langle z',\zeta\rangle \text{ for } q'=(u',z')\in\mathcal{Q} \text{ and } \xi=(\eta,\zeta)\in\mathbb{R}^{n+m}. \tag{4.19}$$

Indeed, for $\tau>0$, the inequality is trivial if either $\mathcal{V}_\mathsf{u}^*(q;\eta)>0$ or $\mathcal{W}_\mathsf{z}^*(q;\zeta)>0$. When both of them equal 0, then $\eta=0$ and $\langle q',\xi\rangle=\langle\zeta,z'\rangle\leq\mathcal{R}_0(q,z')=\mathcal{M}_0(q,\tau,q',\xi)$. For $\tau=0$, e.g. in the case $\alpha>1$ we have, if $z'=0$,

$$\langle q',\xi\rangle=\langle\eta,u'\rangle\leq\sqrt{\langle\mathbb{V}_\mathsf{u}(q)u',u'\rangle}\sqrt{\langle\mathbb{V}_\mathsf{u}(q)^{-1}\eta,\eta\rangle}=\mathcal{M}_0^{\mathrm{red}}(q,\tau,q',\xi)+0=\mathcal{M}_0(q,\tau,q',\xi)$$

while, if $\eta = 0$,

$$\begin{aligned}\langle q', \xi \rangle = \langle \zeta, z' \rangle = \langle \zeta - \omega, z' \rangle + \langle \omega, z' \rangle \\ \leq \sqrt{\langle \mathbb{V}_{\mathsf{z}}(q) z', z' \rangle} \sqrt{\langle \mathbb{V}_{\mathsf{z}}(q)^{-1}(\zeta - \omega), (\zeta - \omega) \rangle} + \mathcal{R}_0(z') \;=\; \mathcal{M}_0(q, \tau, q', \xi)\end{aligned}$$

where we have chosen $w \in K(q)$ such that $\mathcal{W}_{\mathsf{z}}^*(q;\zeta) = \mathcal{V}_{\mathsf{z}}^*(q;\zeta-\omega) = \frac{1}{2}\langle \mathbb{V}_{\mathsf{z}}(q)^{-1}(\zeta-\omega), (\zeta-\omega) \rangle$, and used that $\langle \omega, z' \rangle \leq \mathcal{R}_0(z')$.

For the ensuing discussions, the set where (4.19) holds as an equality shall play a crucial role. We postpone its precise definition right before the statement of Proposition 4.8, cf. (4.30) ahead.

The vanishing-viscosity result. Theorem 4.5 below states that, up to a subsequence the parameterized solutions $(\mathsf{t}_\varepsilon, \mathsf{q}_\varepsilon)_\varepsilon$ of the (Cauchy problems for the) viscous system (2.3), converge to a parameterized curve (t, q), complying with the analog of the energy balance (4.6), with $\mathcal{M}_0$ in place of $\mathcal{M}_\varepsilon$.

We postpone after the proof of Theorem 4.5 a thorough analysis of the notion of solution to the rate-independent system (2.2) thus obtained. Let us instead mention in advance that the line of the argument for proving the limiting parameterized energy balance (4.22) is by now quite standard, cf. the proofs of [8, Thm. 3.3], [13, Thm. 5.5]. In fact, the *upper energy estimate* (i.e. the inequality $\leq$ for (4.22)) shall follow from lower semicontinuity arguments, based on the application of the Ioffe Theorem [31]. The *lower energy estimate* $\geq$ will instead ensue from the chain rule (2.10). We also point out that, for the compactness argument it is actually not necessary to start from parameterized curves for which estimate (4.4) holds, uniformly w.r.t. time. In fact, the uniform integrability of the sequence $(\mathsf{t}_\varepsilon', \mathsf{q}_\varepsilon')_\varepsilon$ is sufficient, cf. (4.20) below.

Theorem 4.5. *Assume* (R_0), (V_z), (V_u), *and* (E). *Let* $(q_\varepsilon)_\varepsilon \subset \mathrm{W}^{1,2}(0,T;\mathcal{Q})$ *be a sequence of solutions to the Cauchy problem for* (2.3). *Choose nondecreasing surjective parameterizations* $\mathsf{t}_\varepsilon : [0, S_\varepsilon] \to [0,T]$ *and set* $\mathsf{q}_\varepsilon(s) = (\mathsf{u}_\varepsilon(s), \mathsf{z}_\varepsilon(s)) := q_\varepsilon(\mathsf{t}_\varepsilon(s))$ *for* $s \in [0, S_\varepsilon]$. *Suppose that* $S_\varepsilon \to S$ *as* $\varepsilon \downarrow 0$ *up to a subsequence, and that there exist* $q_0 \in \mathcal{Q}$ *and* $m \in \mathrm{L}^1(0,S)$ *such that* $\mathsf{q}_\varepsilon(0) \to q_0$, *and*

$$m_\varepsilon := \mathsf{t}_\varepsilon' + |\mathsf{q}_\varepsilon'| \rightharpoonup m \qquad \text{in } \mathrm{L}^1(0,S) \text{ as } \varepsilon \downarrow 0. \tag{4.20}$$

Then, there exist a (not-relabeled) subsequence and a parameterized curve $(\mathsf{t},\mathsf{q}) \in \mathrm{W}^{1,1}([0,S];[0,T] \times \mathcal{Q})$ *such that as* $\varepsilon \downarrow 0$

$$(\mathsf{t}_\varepsilon, \mathsf{q}_\varepsilon) \to (\mathsf{t}, \mathsf{q}) \text{ in } \mathrm{C}^0([0,S];[0,T] \times \mathcal{Q}), \tag{4.21}$$

$\mathsf{t}' + |\mathsf{q}'| \leq m$ *a.e. in* $(0,S)$, *and* (t,q) *fulfills the (parameterized) energy identity*

$$\begin{aligned}&\mathcal{E}(\mathsf{t}(s_2), \mathsf{q}(s_2)) + \int_{s_1}^{s_2} \mathcal{M}_0(\mathsf{q}(r), \mathsf{t}'(r), \mathsf{q}'(r), -\mathrm{D}_q\mathcal{E}(\mathsf{t}(r), \mathsf{q}(r)))\,\mathrm{d}r \\ &= \mathcal{E}(\mathsf{t}(s_1), \mathsf{q}(s_1)) + \int_{s_1}^{s_2} \partial_t \mathcal{E}(\mathsf{t}(r), \mathsf{q}(r)) \mathsf{t}'(r)\,\mathrm{d}r \qquad \text{for all } 0 \leq s_1 \leq s_2 \leq S.\end{aligned} \tag{4.22}$$

Proof. Up to a reparameterization, we may suppose that the curves $(\mathsf{t}_\varepsilon, \mathsf{q}_\varepsilon)$ are defined on the fixed time interval $[0,S]$. We split the proof in three steps.

Step 1: compactness. Observe that for every $0 \leq s_1 \leq s_2 \leq S$

$$|\mathsf{q}_\varepsilon(s_1) - \mathsf{q}_\varepsilon(s_2)| \leq \int_{s_1}^{s_2} |\mathsf{q}_\varepsilon'(s)|\,\mathrm{d}s \leq \int_{s_1}^{s_2} m_\varepsilon(s)\,\mathrm{d}s\,. \tag{4.23}$$

Since $(\mathsf{q}_\varepsilon(0))_\varepsilon$ is bounded, we deduce from (4.23) that $(\mathsf{q}_\varepsilon)_\varepsilon \subset \mathrm{C}^0([0,S];\mathcal{Q})$ is bounded as well. What is more, as the family $(m_\varepsilon)_\varepsilon$ is uniformly integrable (4.20), $(\mathsf{q}_\varepsilon)_\varepsilon$ complies with the equicontinuity condition of the Ascoli-Arzelà Theorem and so does $(\mathsf{t}_\varepsilon)_\varepsilon$, by the analog of estimate (4.23). Hence, (4.21) follows. Taking into account that $\mathcal{E} \in \mathrm{C}^1([0,T]\times\mathcal{Q})$, we immediately conclude from (4.21) that

$$\mathcal{E}(\mathsf{t}_\varepsilon,\mathsf{q}_\varepsilon) \to \mathcal{E}(\mathsf{t},\mathsf{q}),\ \mathrm{D}_q\mathcal{E}(\mathsf{t}_\varepsilon,\mathsf{q}_\varepsilon) \to \mathrm{D}_q\mathcal{E}(\mathsf{t},\mathsf{q}),\ \partial_t\mathcal{E}(\mathsf{t}_\varepsilon,\mathsf{q}_\varepsilon) \to \partial_t\mathcal{E}(\mathsf{t},\mathsf{q}) \text{ uniformly on } [0,S]. \quad (4.24)$$

Furthermore, (4.20) also yields that the sequences $(\mathsf{t}_\varepsilon')_\varepsilon$ and $(\mathsf{q}_\varepsilon')_\varepsilon$ are uniformly integrable. Thus, by the Pettis Theorem, up to a further extraction we find

$$\mathsf{t}_\varepsilon' \rightharpoonup \mathsf{t}' \quad \text{in } \mathrm{L}^1(0,S), \qquad \mathsf{q}_\varepsilon' \rightharpoonup \mathsf{q}' \quad \text{in } \mathrm{L}^1(0,S;\mathcal{Q}), \tag{4.25}$$

whence $\mathsf{t}' + |\mathsf{q}'| \leq m$ a.e. in $(0,S)$.

Step 2: upper energy estimate. We now take the limit as $\varepsilon \downarrow 0$ of the (parameterized) energy-dissipation principle (4.6) for every $0 \leq s_1 \leq s_2 \leq S$:

$$\begin{aligned}
&\mathcal{E}(\mathsf{t}(s_2),\mathsf{q}(s_2)) + \int_{s_1}^{s_2} \mathcal{M}_0(\mathsf{q}(r),\mathsf{t}'(r),\mathsf{q}'(r),-\mathrm{D}_q\mathcal{E}(\mathsf{t}(r),\mathsf{q}(r)))\,\mathrm{d}r \\
&\overset{(1)}{\leq} \lim_{\varepsilon\downarrow 0}\mathcal{E}(\mathsf{t}_\varepsilon(s_2),\mathsf{q}_\varepsilon(s_2)) + \liminf_{\varepsilon\downarrow 0}\int_{s_1}^{s_2} \mathcal{M}_\varepsilon(\mathsf{q}_\varepsilon(r),\mathsf{t}_\varepsilon'(r),\mathsf{q}_\varepsilon'(r),-\mathrm{D}_q\mathcal{E}(\mathsf{t}_\varepsilon(r),\mathsf{q}_\varepsilon(r)))\,\mathrm{d}r \\
&= \lim_{\varepsilon\downarrow 0}\mathcal{E}(\mathsf{t}_\varepsilon(s_1),\mathsf{q}_\varepsilon(s_1)) + \lim_{\varepsilon\downarrow 0}\int_{s_1}^{s_2}\partial_t\mathcal{E}(\mathsf{t}_\varepsilon(r),\mathsf{q}_\varepsilon(r))\mathsf{t}_\varepsilon'(r)\,\mathrm{d}r \\
&\overset{(2)}{=} \mathcal{E}(\mathsf{t}(s_1),\mathsf{q}(s_1)) + \int_{s_1}^{s_2}\partial_t\mathcal{E}(\mathsf{t}(r),\mathsf{q}(r))\mathsf{t}'(r)\,\mathrm{d}r\,,
\end{aligned} \tag{4.26}$$

where (1) follows from the energy convergence in (4.24) and the previously proved (4.14), and (2) from (4.24), again, combined with the first of (4.25). This concludes the upper energy estimate.

Step 3: lower energy estimate. We have for all $0 \leq s_1 \leq s_2 \leq S$ that

$$\begin{aligned}
&\mathcal{E}(\mathsf{t}(s_1),\mathsf{q}(s_1)) + \int_{s_1}^{s_2}\partial_t\mathcal{E}(\mathsf{t}(r),\mathsf{q}(r))\mathsf{t}'(r)\,\mathrm{d}r \\
&\overset{(1)}{=} \mathcal{E}(\mathsf{t}(s_2),\mathsf{q}(s_2)) + \int_{s_1}^{s_2}\langle -\mathrm{D}_q\mathcal{E}(\mathsf{t}(r),\mathsf{q}(r)),\mathsf{q}'(r)\rangle\,\mathrm{d}r \\
&\overset{(2)}{\leq} \mathcal{E}(\mathsf{t}(s_2),\mathsf{q}(s_2)) + \int_{s_1}^{s_2}\mathcal{M}_0(\mathsf{q}(r),\mathsf{t}'(r),\mathsf{q}'(r),-\mathrm{D}_q\mathcal{E}(\mathsf{t}(r),\mathsf{q}(r)))\,\mathrm{d}r\,,
\end{aligned} \tag{4.27}$$

where (1) follows from the chain rule, and (2) is due to inequality (4.19). In this way, we conclude (4.22). Finally, combining (4.26) and (4.27) it is easy to deduce that

$$\lim_{\varepsilon\downarrow 0}\int_{s_1}^{s_2}\mathcal{M}_\varepsilon(\mathsf{q}_\varepsilon(r),\mathsf{t}_\varepsilon'(r),\mathsf{q}_\varepsilon'(r),-\mathrm{D}_q\mathcal{E}(\mathsf{t}_\varepsilon(r),\mathsf{q}_\varepsilon(r)))\,\mathrm{d}r = \int_{s_1}^{s_2}\mathcal{M}_0(\mathsf{q}(r),\mathsf{t}'(r),\mathsf{q}'(r),-\mathrm{D}_q\mathcal{E}(\mathsf{t}(r),\mathsf{q}(r)))\,\mathrm{d}r$$

for all $0 \leq s_1 \leq s_2 \leq S$, whence $\int_{s_1}^{s_2}\mathcal{R}_0(\mathsf{q}_\varepsilon(r),\mathsf{z}_\varepsilon'(r))\,\mathrm{d}r \to \int_{s_1}^{s_2}\mathcal{R}_0(\mathsf{q}(r),\mathsf{z}'(r))\,\mathrm{d}r$. □

Parameterized Balanced Viscosity solutions. Let us now gain further insight into the notion of solution to system (1.1) arising from the vanishing-viscosity limit. First of all, we fix its definition.

Definition 4.6. *Let* $(\mathcal{R}_0, \mathcal{V}_\mathsf{z}, \mathcal{V}_\mathsf{u}, \mathcal{E})$ *comply with* (R_0), (V_z), (V_u), *and* (E). *A curve* $(\mathsf{t}, \mathsf{q}) \in \mathrm{W}^{1,1}([0,S];[0,T]\times\mathcal{Q})$ *is called a* parameterized Balanced Viscosity *(*pBV*, for short) solution to the rate-independent system* $(\mathcal{Q}, \mathcal{E}, \mathcal{R}_0 + \varepsilon\mathcal{V}_\mathsf{z} + \varepsilon^\alpha \mathcal{V}_\mathsf{u})$ *if* $\mathsf{t} : [0,S] \to [0,T]$ *is nondecreasing, and the pair* (t, q) *complies with the energy-dissipation principle* (4.22) *for all* $0 \le s_1 \le s_2 \le S$.

Furthermore, (t, q) *is called*

- non-degenerate, *if*

$$\mathsf{t}'(s) + |\mathsf{q}'(s)| > 0 \qquad \text{for a.a. } s \in (0,S); \tag{4.28}$$

- surjective, *if* $\mathsf{t} : [0,S] \to [0,T]$ *is surjective.*

The name *parameterized Balanced Viscosity solutions* derives from the fact that the dissipation potential $\mathcal{M}_0$ (cf. (4.9)) featuring in (4.22) is defined in a nontrivial way from the three dissipation potentials $\mathcal{R}_0$, $\mathcal{V}_\mathsf{u}$, and $\mathcal{V}_\mathsf{z}$ in such a manner that the subtle balance between the different effects is obtained in the limit $\varepsilon \downarrow 0$.

Remark 4.7. Observe that, even in the case when the function m in (4.20) is a.e. strictly positive, Theorem 4.5 does not guarantee the existence of non-degenerate pBV solutions. However, any degenerate pBV solution (t, q) can be reparameterized to a non-degenerate one $(\tilde{\mathsf{t}}, \tilde{\mathsf{q}}) : [0,\tilde{S}] \to [0,T] \times \mathcal{Q}$, even fulfilling the *normalization condition*

$$\tilde{\mathsf{t}}'(\sigma) + \tilde{\mathsf{q}}'(\sigma) = 1 \qquad \text{for a.a. } \sigma \in (0,\tilde{S}). \tag{4.29}$$

Indeed, following [8, Rmk. 2], starting from a (possibly degenerate) solution (t, q), we set

$$\sigma(s) := \int_0^s \mathsf{t}'(r) + |\mathsf{q}'(r)| \,\mathrm{d}r \qquad \text{and } \tilde{S} := \sigma(S),$$

and define $(\tilde{\mathsf{t}}(\sigma), \tilde{\mathsf{q}}(\sigma)) := (\mathsf{t}(s), \mathsf{q}(s))$ if $\sigma = \sigma(s)$. Then, the very same calculations as in [8, Rmk. 2] lead to (4.29).

We conclude this section with a characterization of pBV solutions in the same spirit as [8, Prop. 2] and [13, Prop. 5.3], [14, Cor. 4.5]. We show that the energy identity (4.22) defining the concept of pBV solutions is equivalent to the corresponding energy inequality on the interval $[0,S]$, and to the energy inequality in a differential form. Finally, (4.31) below provides a further reformulation of this solution concept which involves the *contact set* (cf. [13, 14])

$$\Sigma(q) := \{(\tau, q', \xi) \in [0,\infty) \times \mathcal{Q} \times \mathbb{R}^{n+m} : \mathcal{M}_0(q, \tau, q', \xi) = \langle q', \xi \rangle\} \tag{4.30}$$

Observe that for all $q \in \mathcal{Q}$ the set $\Sigma(q)$ is closed, as the functional $\mathcal{M}_0(q, \cdot, \cdot, \cdot)$ is lower semicontinuous. In Proposition 5.1 we will provide the explicit representation of $\Sigma(q)$. This and (4.31) we will be at the core of the reformulation of pBV solutions in terms of subdifferential inclusions, which we will discuss in Sec. 5.

Proposition 4.8. *Let* $(\mathcal{R}_0, \mathcal{V}_\mathsf{z}, \mathcal{V}_\mathsf{u}, \mathcal{E})$ *comply with* (R_0), (V_z), (V_u), *and* (E). *A curve* $(\mathsf{t}, \mathsf{q}) \in \mathrm{W}^{1,1}([0,S];[0,T]\times\mathcal{Q})$, *with* t *nondecreasing, is a* pBV *solution to the rate-independent system* $(\mathcal{Q}, \mathcal{E}, \mathcal{R}_0 + \varepsilon\mathcal{V}_\mathsf{z} + \varepsilon^\alpha \mathcal{V}_\mathsf{u})$ *if and only if one of the following equivalent conditions is satisfied:*

(i) (4.22) *holds as an inequality on* $(0,S)$, *i.e.*

$$\begin{aligned} &\mathcal{E}(\mathsf{t}(S), \mathsf{q}(S)) + \int_0^S \mathcal{M}_0(\mathsf{q}(r), \mathsf{t}'(r), \mathsf{q}'(r), -\mathrm{D}_q\mathcal{E}(\mathsf{t}(r), \mathsf{q}(r))) \,\mathrm{d}r \\ &\le \mathcal{E}(\mathsf{t}(0), \mathsf{q}(0)) + \int_0^S \partial_t \mathcal{E}(\mathsf{t}(r), \mathsf{q}(r)) \mathsf{t}'(r) \,\mathrm{d}r; \end{aligned}$$

(ii) the above energy inequality holds in the differential form $\frac{\mathrm{d}}{\mathrm{d}s}\mathcal{E}(\mathsf{t},\mathsf{q})+\mathcal{M}_0(\mathsf{q},\mathsf{t}',\mathsf{q}',-\mathrm{D}_q\mathcal{E}(\mathsf{t},\mathsf{q}))\leq \partial_t\mathcal{E}(\mathsf{t},\mathsf{q})\mathsf{t}'$ *a.e. in* $(0,S)$*;*

(iii) the triple $(\mathsf{t}',\mathsf{q}',-\mathrm{D}_q\mathcal{E}(\mathsf{t},\mathsf{q}))$ *belongs to the contact set, i.e.*

$$(\mathsf{t}'(s),\mathsf{q}'(s),-\mathrm{D}_q\mathcal{E}(\mathsf{t}(s),\mathsf{q}(s)))\in\Sigma(\mathsf{q}(s)) \qquad \textit{for a.a. } s\in(0,S). \tag{4.31}$$

The proof of Proposition 4.8 is omitted: it follows by exploiting the chain rule (2.10), with arguments akin to those in the proof of Theorem 4.5, see also [8, Prop. 2] and [13, Prop. 5.3], [14, Cor. 4.5].

5. Physical interpretation

The following result provides a thorough description of the (closed) contact set $\Sigma(q)$, cf. (4.30). As we will see, the representation of $\Sigma(q)$ is substantially different in the three cases $\alpha>1$, $\alpha=1$, and $\alpha\in(0,1)$. That is why, in Proposition 5.1 below we will use the notation $\Sigma_{\alpha>1}(q)$, $\Sigma_{\alpha=1}(q)$, and $\Sigma_{\alpha\in(0,1)}(q)$. We will prove that these sets are given by the union of subsets describing the various evolution regimes for the variables u and z. The notation for these subsets will be of the form

$$\mathrm{A}_{\mathsf{r}}\mathrm{B}_{\mathsf{s}} \qquad \text{with } \mathrm{A},\mathrm{B}\in\{\mathrm{E},\mathrm{R},\mathrm{V},\mathrm{B}\} \text{ and } \mathsf{r},\mathsf{s}\in\{\mathsf{u},\mathsf{z}\}.$$

The letters E, R, V, B stand for *Equilibrated*, *Rate-independent*, *Viscous*, and *Blocked*, respectively. For instance, $\mathrm{E}_{\mathsf{u}}\mathrm{R}_{\mathsf{z}}$ is the set of (τ,q',ξ) corresponding to equilibrium for u and rate-independent evolution for z, cf. (5.2) below; we postpone more comments after the statement of Proposition 5.1. Observe that all of these sets depend on the state variable q, as does $\Sigma(q)$. However, for simplicity we will not highlight this in their notation. In their description we shall always refer to the representation $q'=(u',z')$ for the velocity variable, and $\xi=(\eta,\zeta)$ for the force variable.

Proposition 5.1. *Assume* (R_0), (V_z), (V_u), *and* (E). *Then, we have the following results:*
Case $\alpha>1$*: The contact set is given by*

$$\Sigma_{\alpha>1}(q)=\mathrm{E}_{\mathsf{u}}\mathrm{R}_{\mathsf{z}}\cup\mathrm{V}_{\mathsf{u}}\mathrm{B}_{\mathsf{z}}\cup\mathrm{E}_{\mathsf{u}}\mathrm{V}_{\mathsf{z}} \quad \textit{with} \tag{5.1}$$

$$\mathrm{E}_{\mathsf{u}}\mathrm{R}_{\mathsf{z}}:=\{(\tau,q',\xi):\tau>0,\ \ q'=(u',z'),\ \xi=(0,\zeta) \textit{ and } \partial\mathcal{R}_0(q,z')\ni\zeta\}, \tag{5.2}$$

$$\mathrm{V}_{\mathsf{u}}\mathrm{B}_{\mathsf{z}}:=\{(\tau,q',\xi):(\tau,q',\xi)=(0,(u',0),(\eta,\zeta)) \textit{ and } \exists\,\theta_{\mathsf{u}}\in[0,1]:\ \theta_{\mathsf{u}}\mathbb{V}_{\mathsf{u}}(q)u'=(1-\theta_{\mathsf{u}})\eta\}, \tag{5.3}$$

$$\begin{aligned}\mathrm{E}_{\mathsf{u}}\mathrm{V}_{\mathsf{z}}:=\{(\tau,q',\xi):\tau=0,\ \ q'=(u',z'),\ \xi=(0,\zeta) \textit{ and }&\\ \exists\,\theta_{\mathsf{z}}\in[0,1]\ :\ (1-\theta_{\mathsf{z}})\partial\mathcal{R}_0(q,z')+\theta_{\mathsf{z}}\mathbb{V}_{\mathsf{z}}(q)z'\ni(1-\theta_{\mathsf{z}})\zeta\}.&\end{aligned} \tag{5.4}$$

Case $\alpha=1$*: The contact set is given by*

$$\Sigma_{\alpha=1}(q)=\mathrm{E}_{\mathsf{u}}\mathrm{R}_{\mathsf{z}}\cup\mathrm{V}_{\mathsf{u}}\mathrm{V}_{\mathsf{z}} \quad \textit{with} \tag{5.5}$$

$$\mathrm{V}_{\mathsf{u}}\mathrm{V}_{\mathsf{z}}:=\left\{(\tau,q',\xi)\ :\ \tau=0 \textit{ and } \exists\,\theta\in[0,1]:\left\{\begin{array}{l}\theta\mathbb{V}_{\mathsf{u}}(q)u'=(1-\theta)\eta,\\(1-\theta)\partial\mathcal{R}_0(q,z')+\theta\mathbb{V}_{\mathsf{z}}(q)z'\ni(1-\theta)\zeta\end{array}\right.\right\}. \tag{5.6}$$

Case $\alpha\in(0,1)$*: The contact set is given by*

$$\Sigma_{\alpha\in(0,1)}(q)=\mathrm{E}_{\mathsf{u}}\mathrm{R}_{\mathsf{z}}\cup\mathrm{B}_{\mathsf{u}}\mathrm{V}_{\mathsf{z}}\cup\mathrm{V}_{\mathsf{u}}\mathrm{R}_{\mathsf{z}} \quad \textit{with} \tag{5.7}$$

$$\begin{aligned}\mathrm{B}_{\mathsf{u}}\mathrm{V}_{\mathsf{z}}:=\{(\tau,q',\xi):\tau=0,\ \ q'=(0,z'),\ \xi=(\eta,\zeta) \textit{ and }&\\ \exists\,\theta_{\mathsf{z}}\in[0,1]:\ (1-\theta_{\mathsf{z}})\partial\mathcal{R}_0(q,z')+\theta_{\mathsf{z}}\mathbb{V}_{\mathsf{z}}(q)z'\ni(1-\theta_{\mathsf{z}})\zeta\},&\end{aligned} \tag{5.8}$$

$$\mathrm{V}_{\mathsf{u}}\mathrm{R}_{\mathsf{z}}:=\left\{(\tau,q',\xi):\ (\tau,q',\xi)=(0,(u',z'),(\eta,\zeta)),\ \left\{\begin{array}{l}\exists\,\theta_{\mathsf{u}}\in[0,1]:\ \theta_{\mathsf{u}}\mathbb{V}_{\mathsf{u}}(q)u'=(1-\theta_{\mathsf{u}})\eta,\\ \textit{and } \partial\mathcal{R}_0(q,z')\ni\zeta\end{array}\right.\right\}. \tag{5.9}$$

As (4.31) reveals, the contact set encompasses all the relevant information on the evolution of a *parameterized Balanced Viscosity* solution. The form of the sets $\mathrm{E_u R_z}$, $\mathrm{V_u B_z}$ ... which constitute the contact set is strictly related to the mechanical interpretation of pBV solutions that shall be explored at the end of this section. Let us just explain here that

- the set $\mathrm{E_u R_z}$ corresponds to equilibrium for the variable u (as $\eta = 0$), and a *stick-slip* regime for z, which evolves rate-independently as expressed by $\partial\mathcal{R}_0(q, z') \ni \zeta$. Observe that the stationary state $u' = z' = 0$ is also encompassed.
- The set $\mathrm{V_u B_z}$ corresponds to the case in which the variable u still has to relax to an equilibrium and thus is governed by a *fast* dynamics at a jump $\tau = 0$, while z is "blocked by viscosity" and thus stays constant ($z' = 0$).
- The set $\mathrm{E_u V_z}$ corresponds to the regime in which z evolves according to viscosity at a jump $\tau = 0$, and u follows z in such a way that it is at an equilibrium ($\eta = 0$).
- The set $\mathrm{V_u V_z}$ corresponds to the case where the evolution of the system at a jump $\tau = 0$ is governed by viscosity both in u and in z.
- The set $\mathrm{B_u V_z}$ encompasses the case in which the variable z at a jump $\tau = 0$ evolves according to viscosity, while u is blocked by viscosity ($u' = 0$).
- The set $\mathrm{V_u R_z}$ describes viscous evolution for u and rate-independent evolution for z.

Remark 5.2. Let us stress once more that, as mentioned in advance, in the vanishing-viscosity limit the evolution regimes for $\alpha > 1$ and $\alpha \in (0,1)$ mirror each other. Indeed, formulae (5.1) and (5.7) are specular, up to observing that the analog of the equilibrium regime $\mathrm{E_u}$ is indeed the rate-independent regime $\mathrm{R_z}$, see also Figure 5.1.

Proof of Proposition 5.1. In all the three cases $\alpha > 1$, $\alpha = 1$, and $\alpha \in (0,1)$, for $\tau > 0$ the contact condition $\mathcal{M}_0(q, \tau, q', \xi) = \langle \xi, q' \rangle$ can hold only if the constraints $\eta = 0$ and $\zeta \in K(q)$ are satisfied. Then, $\mathcal{M}_0(q, \tau, q', \xi) = \langle \xi, q' \rangle$ reduces to $\mathcal{R}_0(q, z') = \langle \zeta, z' \rangle$. Since $\zeta \in K(q)$, this is equivalent to $\zeta \in \partial\mathcal{R}_0(q, z')$ by (2.5). This gives the set $\mathrm{E_u R_z}$, which contributes to the contact set $\Sigma(q)$ in the three cases $\alpha > 1$, $\alpha = 1$, and $\alpha \in (0,1)$.

For $\alpha = 1$, observe that in the case $\tau = 0$ the contact condition is

$$\mathcal{R}_0(q, z') + 2\sqrt{\mathcal{V}_\mathsf{z}(q; z') + \mathcal{V}_\mathsf{u}(q; u')}\sqrt{\mathcal{W}^*_\mathsf{z}(q; \zeta) + \mathcal{V}^*_\mathsf{u}(q; \eta)} = \langle \zeta, z' \rangle + \langle \eta, u' \rangle. \tag{5.10}$$

Let us first address the case in which $\sigma_1 := \sqrt{\mathcal{V}_\mathsf{z}(q; z') + \mathcal{V}_\mathsf{u}(q; u')} = 0$ or $\sigma_2 := \sqrt{\mathcal{W}^*_\mathsf{z}(q; \zeta) + \mathcal{V}^*_\mathsf{u}(q; \eta)} = 0$. The former case corresponds to the stationary state $u' = z' = 0$, which means $\theta = 1$ in (5.6). The latter to $\mathcal{W}^*_\mathsf{z}(q; \zeta) = 0$ (if and only if $\zeta \in K(q)$) and $\eta = 0$. Hence (5.10) becomes $\mathcal{R}_0(q, z') = \langle \zeta, z' \rangle$, whence $\zeta \in \partial\mathcal{R}_0(q, z')$ by (2.5), again. This corresponds to $\theta = 0$ in (5.6). If $\sigma_1\sigma_2 > 0$, then we rewrite $2\sigma_1\sigma_2$ as $\lambda\sigma_1^2 + \frac{1}{\lambda}\sigma_2^2$, with $\lambda > 0$ given by $\lambda = \frac{\sigma_2}{\sigma_1}$. With such λ, (5.10) rewrites as

$$\mathcal{R}_0(q, z') + \lambda(\mathcal{V}_\mathsf{z}(q; z') + \mathcal{V}_\mathsf{u}(q; u')) + \frac{1}{\lambda}(\mathcal{W}^*_\mathsf{z}(q; \zeta) + \mathcal{V}^*_\mathsf{u}(q; \eta)) = \langle \zeta, z' \rangle + \langle \eta, u' \rangle.$$

Upon multiplying both sides by λ, using that $\mathcal{V}_\mathsf{z}$ and $\mathcal{V}_\mathsf{u}$ are positively homogeneous of degree 2, and rearranging terms, we get

$$\mathcal{R}_0(q, \lambda z') + \mathcal{V}_\mathsf{z}(q; \lambda z') + \mathcal{W}^*_\mathsf{z}(q; \zeta) - \langle \zeta, \lambda z' \rangle = \langle \eta, \lambda u' \rangle - \mathcal{V}_\mathsf{u}(q; \lambda u') - \mathcal{V}^*_\mathsf{u}(q; \eta).$$

By the Fenchel-Moreau equivalence, this gives

$$\begin{aligned} &\mathbb{V}_\mathsf{u}(q)(\lambda u') = \eta, \\ &\partial\mathcal{R}_0(q, \lambda z') + \mathbb{V}_\mathsf{z}(q)(\lambda z') \ni \zeta \end{aligned}$$

with $\lambda > 0$. Then, (5.6) follows with $\theta \in (0,1)$ such that $\lambda = \frac{\theta}{1-\theta}$. All in all, for $\alpha = 1$ we have proved that, if $(\tau, q', \xi) \in \Sigma_{\alpha=1}(q)$, then either $(\tau, q', \xi) \in \mathrm{E_u R_z}$, or $(\tau, q', \xi) \in \mathrm{V_u V_z}$. This concludes the proof of (5.5) for $\Sigma_{\alpha=1}(q)$.

In the case $\alpha > 1$ and $\tau = 0$, $\mathcal{M}_0(q, \tau, q', \xi)$ is finite if and only if either $z' = 0$, or $\eta = 0$. In the former case, the contact condition reduces to $\sqrt{\langle \mathbb{V}_\mathsf{u}(q)u', u'\rangle}\sqrt{\langle \mathbb{V}_\mathsf{u}(q)^{-1}\eta, \eta\rangle} = \langle \eta, u'\rangle$, which is equivalent to the fact that there exists $\theta_\mathsf{u} \in [0,1]$ with $\theta_\mathsf{u}\mathbb{V}_\mathsf{u}(q)u' = (1-\theta_\mathsf{u})\eta$. This yields the set $\mathrm{V_u B_z}$. In the latter case, the contact condition rephrases as

$$\mathcal{R}_0(q, z') + \sqrt{\langle \mathbb{V}_\mathsf{z}(q)z', z'\rangle}\sqrt{\langle \mathbb{V}_\mathsf{z}(q)^{-1}(\zeta-\omega), \zeta-\omega\rangle} = \langle \zeta, z'\rangle = \langle \omega, z'\rangle + \langle \zeta-\omega, z'\rangle,$$

with $\omega \in K(q)$ such that $\mathcal{W}_\mathsf{z}^*(q;\zeta) = \frac{1}{2}\langle \mathbb{V}_\mathsf{z}(q)^{-1}(\zeta-\omega), \zeta-\omega\rangle$. It is immediate to check that the above chain of equalities implies

$$\omega \in \partial\mathcal{R}_0(q, z') \quad \text{and} \quad (1-\theta_\mathsf{z})(\zeta-\omega) = \theta_\mathsf{z}\mathbb{V}_\mathsf{z}(q)z' \ \text{ for some } \theta_\mathsf{z} \in [0,1].$$

This yields the set $\mathrm{E_u V_z}$. All in all, in the case $\alpha > 1$ we have proved that, if $(\tau, q', \xi) \in \Sigma_{\alpha>1}(q)$, then either $(\tau, q', \xi) \in \mathrm{E_u R_z}$, or $(\tau, q', \xi) \in \mathrm{V_u B_z}$, or $(\tau, q', \xi) \in \mathrm{E_u V_z}$. This concludes (5.1).

The proof of (5.7) follows the very same lines and is thus omitted. □

The **main result** of this paper is the following theorem, which is in fact a direct consequence of the characterization (4.31) of pBV solutions in terms of the contact set, and of Proposition 5.1. Observe that, we confine ourselves to *non-degenerate* pBV solutions only. This is not restrictive, in view of Remark 4.7.

Theorem 5.3 (Reformulation as a system of subdifferential inclusions). *Assume* ($\mathrm{R_0}$), ($\mathrm{V_z}$), ($\mathrm{V_u}$), *and* (E). *A curve* $(\mathsf{t},\mathsf{q}) \in \mathrm{W}^{1,1}([0,S];[0,T]\times\mathcal{Q})$ *with nondecreasing* t *is a* non-degenerate *parameterized Balanced Viscosity solution to the rate-independent system* $(\mathcal{Q}, \mathcal{E}, \mathcal{R}_0 + \varepsilon\mathcal{V}_\mathsf{z} + \varepsilon^\alpha\mathcal{V}_\mathsf{u})$ *if and only if* $\mathsf{t}' + |\mathsf{q}'| > 0$ *a.e. in* $(0,S)$ *and there exist two Borel functions* $\theta_\mathsf{u}, \theta_\mathsf{z} : [0,S] \to [0,1]$ *such that the pair* (t,q) *with* $\mathsf{q} = (\mathsf{u},\mathsf{z})$ *satisfies the system of equations for a.a.* $s \in (0,S)$*:*

$$\begin{aligned}&\theta_\mathsf{u}(s)\,\mathbb{V}_\mathsf{u}(\mathsf{q}(s))\mathsf{u}'(s) + (1-\theta_\mathsf{u}(s))\,\mathrm{D}_u\mathcal{E}(\mathsf{t}(s),\mathsf{u}(s),\mathsf{z}(s)) \ni 0,\\ &(1-\theta_\mathsf{z}(s))\,\partial\mathcal{R}_0(\mathsf{q}(s),\mathsf{z}'(s)) + \theta_\mathsf{z}(s)\,\mathbb{V}_\mathsf{z}(\mathsf{q}(s))\mathsf{z}'(s) + (1-\theta_\mathsf{z}(s))\,\mathrm{D}_z\mathcal{E}(\mathsf{t}(s),\mathsf{u}(s),\mathsf{z}(s)) \ni 0,\end{aligned} \tag{5.11}$$

with

$$\mathsf{t}'(s)\,\theta_\mathsf{u}(s) = \mathsf{t}'(s)\,\theta_\mathsf{z}(s) = 0 \tag{5.12}$$

and the following additional conditions depending on α*:*

$$\underline{\alpha > 1:} \qquad \theta_\mathsf{u}(s)\,(1-\theta_\mathsf{z}(s)) = 0; \tag{5.13a}$$

$$\underline{\alpha = 1:} \qquad \theta_\mathsf{u}(s) = \theta_\mathsf{z}(s); \tag{5.13b}$$

$$\underline{\alpha \in (0,1):} \qquad \theta_\mathsf{z}(s)\,(1-\theta_\mathsf{u}(s)) = 0. \tag{5.13c}$$

Figure 5.1 displays the structure of the allowed values for the parameters $(t', \theta_\mathsf{u}, \theta_\mathsf{z})$ depending on the value of α.

Remark 5.4. Observe that conditions (5.13a) and (5.13c) are specular (cf. Remark 5.2), revealing once more that the evolution regimes for $\alpha > 1$ and $\alpha < 1$ reflect each other. Nonetheless, a major difference occurs in that, under suitable conditions, for $\alpha > 1$ the regime $\mathrm{V_u B_z}$ only occurs at the beginning, when u relaxes fast to equilibrium, cf. Proposition 5.5.

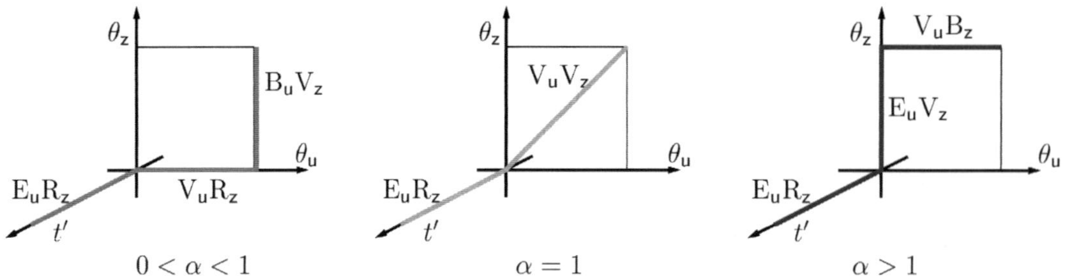

Figure 5.1. The switching between the different regimes, depending on the cases $\alpha<1$, $\alpha=1$, and $\alpha>1$, are displayed via the allowed combinations of the triples (t',θ_u,θ_z).

Finally, let us get further insight into the mechanical interpretation of system (5.11), with the constraints (5.12) and (5.13a)–(5.13c). Preliminarily, let us point out that, as in the case of parameterized solutions to the rate-independent system

$$\partial\mathcal{R}_0(z(t),z'(t))+\mathrm{D}_q\mathcal{I}(t,z(t))\ni 0 \qquad \text{in } (0,T), \tag{5.14}$$

in the *sole* variable z, we have $\mathsf{t}'(s)=0$ if and only if the system is jumping in the (slow) external time scale. Therefore, from (5.12) we gather that, in all of the three cases $\alpha>1$, $\alpha=1$, and $\alpha\in(0,1)$, when the system does not jump, then it is either in the *sticking* regime (i.e. $\mathsf{u}'=\mathsf{z}'=0$), or in the *sliding regime*, namely the evolution of z is purely rate-independent (i.e. $\partial\mathcal{R}_0(\mathsf{q},\mathsf{z}')+\mathrm{D}_z\mathcal{E}(\mathsf{t},\mathsf{q})\ni 0$), and u follows z in such a way that it is at an equilibrium (i.e. $-\mathrm{D}_u\mathcal{E}(\mathsf{t},\mathsf{q})=0$). It is the description of the system behavior at jumps that significantly differs for $\alpha>1$, $\alpha=1$, and $\alpha\in(0,1)$.

Case $\alpha>1$: fast relaxation of u. Here u relaxes faster to equilibrium than z. With (5.12) and (5.13a) we are imposing at a jump that either $\mathsf{z}'=0$ (which follows from $\theta_z=1$, i.e. V_uB_z) or u is at equilibrium (corresponding to $\theta_u=0$, i.e. E_uV_z). In fact, z cannot change until u has relaxed to equilibrium. When u has reached the equilibrium, then z may have either a *sliding jump* (i.e. $\theta_z=0$), or a *viscous jump* ($\theta_z\in(0,1)$).

Our next result shows that, in fact, under the condition that the energy $\mathcal{E}$ is uniformly convex with respect to the variable u (cf. Proposition 3.2), after an initial phase in which z is constant and u relaxes to an equilibrium evolving by viscosity (i.e. the solution is in regime V_uB_z), u never leaves the equilibrium afterwards. In that case the evolution of the system is completely described by z, which turns out to be a parameterized Balanced Viscosity solution to the rate-independent system driven by the *reduced energy functional* obtained minimizing out the variable u.

Proposition 5.5. *Assume* (R_0), (V_z), (V_u), *and* (E). *Additionally, suppose that* $\mathcal{E}$ *complies with the uniform convexity* (E_1), *and denote by* $u=M(t,z)$ *the unique solution of* $\mathrm{D}_u\mathcal{E}(t,u,z)=0$, *i.e. the minimizer of* $\mathcal{E}(t,\cdot,z)$. *Let* $(\mathsf{t},\mathsf{q})\in\mathrm{W}^{1,1}([0,S];[0,T]\times\mathcal{Q})$ *be a parameterized Balanced Viscosity solution to the rate-independent system* $(\mathcal{Q},\mathcal{E},\mathcal{R}_0+\varepsilon\mathcal{V}_z+\varepsilon^\alpha\mathcal{V}_u)$ *with* $\alpha>1$. *Set*

$$\mathfrak{S}:=\{s\in[0,S]\,:\,\mathrm{D}_u\mathcal{E}(\mathsf{t}(s),\mathsf{q}(s))=0\}. \tag{5.15}$$

Then, $\mathfrak{S}$ *is either empty or it has the form* $[s_*,S]$ *for some* $s_*\in[0,S]$.

(a) Assume $s_*>0$, *then for* $s\in[0,s_*)=[0,S]\setminus\mathfrak{S}$ *we have* $\mathsf{t}(s)=\mathsf{t}(0)$ *and* $\mathsf{z}(s)=\mathsf{z}(0)$, *whereas* u *is a solution to the reparameterized the gradient flow for* $(\mathbb{R}^n,\mathcal{E}(\mathsf{t}(0),\cdot,\mathsf{z}(0)),\mathbb{V}_u)$ *(regime* V_uB_z*), namely*

$$0=\theta_u(s)\,\mathbb{V}_u(\mathsf{u}(s),\mathsf{z}(0))\dot{\mathsf{u}}(s)+(1-\theta_u(s))\,\mathrm{D}_u\mathcal{E}(\mathsf{t}(0),\mathsf{u}(s),\mathsf{z}(0)) \text{ with } \mathsf{u}(0)\neq M(\mathsf{t}(0),\mathsf{z}(0)). \tag{5.16}$$

(b) Assume $\mathfrak{S} = [s_*, S]$ *with* $s_* < S$*, then for* $s \in [s_*, S]$ *we have* $\mathsf{u}(s) = M(\mathsf{t}(s), \mathsf{z}(s))$ *whereas the pair* (t, z) *is a parameterized Balanced Viscosity solution to the reduced rate-independent system* $(\mathbb{R}^m, \mathcal{J}, \mathcal{R}_0 + \varepsilon\mathcal{V}_\mathsf{z})$ *with the* reduced energy functional $\mathcal{J} : [0,T] \times \mathbb{R}^m \to \mathbb{R}; (t,z) \mapsto \min_{u \in \mathbb{R}^n} \mathcal{E}(t,u,z) = \mathcal{E}(t, M(t,z), z)$*, which corresponds to the regimes* $\mathrm{E}_\mathsf{u}\mathrm{V}_\mathsf{z}$ *and* $\mathrm{E}_\mathsf{u}\mathrm{R}_\mathsf{z}$.

Proof. To avoid overloaded notation we will often omit the state-dependence of the functions $\mathbb{V}_\mathsf{u}$ and $\mathbb{V}_\mathsf{z}$. For easy reference we repeat all the conditions for a BV solution (t, q) (cf. Theorem 5.3), in the case $\alpha > 1$:

$$\text{(i)}\ \ 0 = \theta_\mathsf{u}\mathbb{V}_\mathsf{u}\mathsf{u}' + (1-\theta_\mathsf{u})\mathrm{D}_u\mathcal{E}(\mathsf{t},\mathsf{u},\mathsf{z}), \quad \text{(ii)}\ \ 0 \in (1-\theta_\mathsf{z})\partial\mathcal{R}_0(\mathsf{q},\mathsf{z}') + \theta_\mathsf{z}\mathbb{V}_\mathsf{z}\mathsf{z}' + (1-\theta_\mathsf{z})\mathrm{D}_z\mathcal{E}(\mathsf{t},\mathsf{u},\mathsf{z}),$$
$$\text{(iii)}\ \ \mathsf{t}'\theta_\mathsf{u} = 0, \qquad \text{(iv)}\ \ \mathsf{t}'\theta_\mathsf{z} = 0, \qquad \text{(v)}\ \ \theta_\mathsf{u}(1-\theta_\mathsf{z}) = 0, \qquad \text{(vi)}\ \ \mathsf{t}' + |\mathsf{u}'| + |\mathsf{z}'| > 0,$$

which have to hold for a.a. $s \in (0, S)$.

Step 1: By the continuity of (t, z) and $\mathrm{D}_u\mathcal{E}$ the set $\mathfrak{S}$ is closed, hence its complement is relatively open. Consider an interval (s_1, s_2) not intersecting with $\mathfrak{S}$. Using (i) we find $\theta_\mathsf{u} > 0$ a.e. in (s_1, s_2). Hence, (iii) implies $\mathsf{t}' = 0$ a.e., and we obtain $\mathsf{t}(s) = \mathsf{t}(s_1)$ for $s \in [s_1, s_2]$. By (v) we find $\theta_\mathsf{z} = 1$ a.e. Now, (ii) implies $z' = 0$ a.e., which implies $\mathsf{z}(s) = \mathsf{z}(s_1)$ for $s \in [s_1, s_2]$. From (vi) we conclude $\mathsf{u}' \neq 0$ a.e. Thus, we summarize

$$\mathsf{t}(s) = \mathsf{t}(s_1), \quad \mathsf{z}(s) = \mathsf{z}(s_1), \quad 0 = \mathbb{V}_u(\mathsf{u}(s), \mathsf{z}(s_1))\mathsf{u}'(s) + \lambda(s)\mathrm{D}_u\mathcal{E}(\mathsf{t}(s_1), \mathsf{u}(s), \mathsf{z}(s_1)),$$

where $\lambda(s) = (1-\theta_\mathsf{u}(s))/\theta_\mathsf{u}(s) \in (0, \infty)$ a.e. In particular, u satisfies (5.16). From $u \in \mathrm{W}^{1,1}([0,S]; \mathbb{R}^m)$ and (i) we obtain $\lambda \in \mathrm{L}^1(s_1, s_2)$. Setting $\tau(s) = \int_{s_1}^s \lambda(\sigma)\,\mathrm{d}\sigma$ and defining the inverse $\hat{s}$ via $s = \hat{s}(\tau)$ we find $\hat{s}'(\tau) > 0$ and $\hat{s} \in W^{1,1}(0, \tau(s_2))$. Moreover, the function $\hat{u} : \tau \mapsto \mathsf{u}(\hat{s}(\tau))$ is a solution of the gradient flow

$$0 = \mathbb{V}_u(\hat{u}(\tau), \mathsf{z}(s_1))\hat{u}'(\tau) + \mathrm{D}_u\mathcal{E}(\mathsf{t}(s_1), \hat{u}(\tau), \mathsf{z}(s_1)). \tag{5.17}$$

Furthermore, we see that $s \mapsto \mathcal{E}(\mathsf{t}(s_1), \mathsf{u}(s), \mathsf{z}(s_1))$ is strictly decreasing on $[s_1, s_2]$, since its time derivative is given by $-\langle \mathsf{u}'(s), \mathbb{V}_u\mathsf{u}'(s)\rangle/\lambda(s)$ which is negative a.e.

Step 2: Since $\mathfrak{S}$ is closed the complement is an at most countable disjoint union of intervals of the form $(s_1, S]$, (s_2, s_3), $[0, s_4)$, or $[0, S]$, which are maximal in the sense that they cannot be extended without meeting $\mathfrak{S}$. Thus, for the "open" sides s_j this means $s_j \in \mathfrak{S}$. In the first two cases this implies $\mathsf{u}(s_j) = M(\mathsf{t}(s_j), \mathsf{z}(s_j))$, i.e. we start a gradient flow with initial condition in the global minimizer. Hence, the solution stays constant for all future times, i.e. $\mathsf{u}(s) = \mathsf{u}(s_{1,2})$ for $s \in (s_1, S]$ or (s_2, s_3), respectively. But this contradicts the fact that $s \mapsto \mathcal{E}(\mathsf{t}(s_j), \mathsf{u}(s), \mathsf{z}(s_j))$ is strictly decreasing (cf. Step 1). Hence, the first two cases cannot occur, and we conclude $\mathfrak{S} = [s_*, S]$ with $s_* = s_4$ or $\mathfrak{S} = \emptyset$. In particular, assertion (a) is established.

Step 3: To show (b) assume $s \in \mathfrak{S} = [s_*, S]$, then $\mathsf{u}(s) = M(\mathsf{t}(s), \mathsf{z}(s))$ by the definition of $\mathfrak{S}$. Observe that $\mathrm{D}_z\mathcal{J}(t,z) = \mathrm{D}_z\mathcal{E}(t, M(t,z), z) + \mathrm{D}_z M(t,z)^T \mathrm{D}_u\mathcal{E}(t, M(t,z), z) = \mathrm{D}_z\mathcal{E}(t, M(t,z), z) + 0$. Thus, (t, z) solves

$$\text{(ii)'}\ \ 0 \in (1-\theta_\mathsf{z})\partial\mathcal{R}_0(\mathsf{z},\mathsf{z}') + \theta_\mathsf{z}\mathbb{V}_\mathsf{z}\mathsf{z}' + (1-\theta_\mathsf{z})\mathrm{D}_z\mathcal{J}(\mathsf{t},\mathsf{z}), \quad \text{(iv)'}\ \ \mathsf{t}'\theta_\mathsf{z} = 0, \quad \text{(vi)'}\ \ \mathsf{t}' + |\mathsf{z}'| > 0,$$

which proves that (t, z) is a BV solution of the reduced system. For the latter relation note that $\mathsf{t}'(s) + |\mathsf{z}'(s)| = 0$ implies $\mathsf{u}'(s) = \frac{\mathrm{d}}{\mathrm{d}s}M(\mathsf{t}(s), \mathsf{z}(s)) = 0$ so that (vi)' follows from (vi). □

Our approach in Step 1 of the above proof uses the qualitative ideas from [32, 33], but our reduction to the simpler convex case makes the analysis much easier.

Case $\alpha = 1$: comparable relaxation times, Here u and z relax at the same rate. At a jump, the system may switch to the viscous regime $\mathrm{V}_\mathsf{u}\mathrm{V}_\mathsf{z}$, where *both* in the evolution of u, and in the evolution for z, viscous dissipation intervenes, modulated by the same coefficient $\theta = \theta_\mathsf{u} = \theta_\mathsf{z}$.

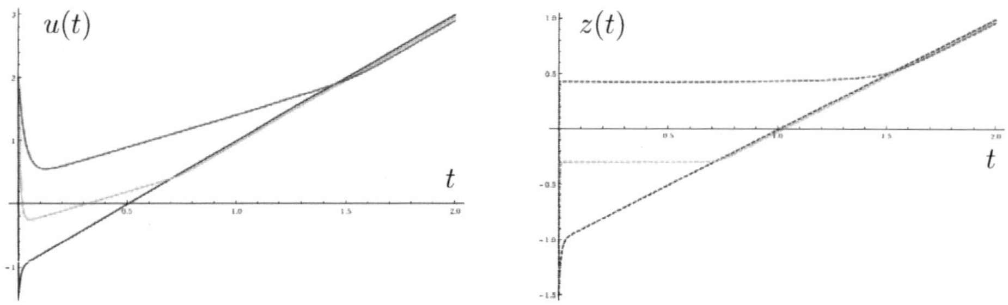

Figure 6.1. Solutions for (6.1) for the cases $\alpha = 2$ (blue), $\alpha = 1$ (green), and $\alpha = 1/2$ (red).

Case $\alpha \in (0,1)$: fast relaxation of z. Here z relaxes faster than u, and jumps in the z-component are faster than jumps in the u-component. If z jumps (possibly governed by viscous dissipation), than u stays fixed, i.e. u is blocked while z moves viscously (regime $\mathrm{B_u V_z}$). But then u has still to relax to equilibrium, and it will do it on a faster scale than the rate-independent motion of z, if z stays in locally stable states (regime $\mathrm{V_u R_z}$). Finally, full rate-independent behavior in the regime $\mathrm{E_u R_z}$ will occur, where $\mathsf{t}'(s) > 0$. Unlike in the case $\alpha > 1$, all three regimes may occur more than once in the evolution of the system, see Section 6.2 for an example.

6. Examples

To illustrate the difference between the three limit models (namely for $\alpha > 1$, $\alpha = 1$, and $\alpha \in (0,1)$), we discuss two examples. The first one treats a quadratic energy and emphasizes the different initial behavior before the solution converges to a truly rate-independent regime. In the second example we show that solutions that start in a rate-independent regime and coincide for the three different limit models may separate if viscous jumps start, leading to different rate-independent behavior afterwards.

6.1. Initial relaxation for a system with quadratic energy

We consider the energy functional $\mathcal{E}(t,u,z) = \frac{1}{2}(u-z)^2 + \frac{1}{2}z^2 - tu$ and trivial viscous energies leading to the ODE system

$$\begin{cases} 0 = \varepsilon^{\alpha}\dot{u} + u - z - t, \\ 0 \in \mathrm{Sign}(\dot{z}) + \varepsilon\dot{z} + 2z - u \end{cases} \qquad \text{with } (u(0), z(0)) = (2, -3/2). \tag{6.1}$$

We show simulations for the three cases $\alpha = 2$ (blue), $\alpha = 1$ (green), and $\alpha = 1/2$ (red) with sufficiently small ε (typically $0.001 \ldots 0.03$). The components u and z as functions of time are depicted in Figure 6.1.

However, to detect different jump behavior at $t \approx 0$ it is advantageous to look at the parameterized solutions, which are depicted in Figure 6.2, showing (t, q) for the three different cases. The parameterization was calculated using $\dot{s}(t) = \max\{0.5, |\dot{u}(t)|, \dot{z}(t)|\}$. In the parameterized form we fully see the structure of the jump for $t \approx 0$. For $\alpha = 2$ we obtain first a jump from the initial datum $(u,z) = (2,-1.5)$ to $(u,z) = (-1.5,-1.5)$ on the timescale ε^2, which is the regime $\mathrm{V_u B_z}$. Then, u is equilibrated, and a jump to $(-1,-1)$ along the diagonal $u = z$ occurs on the timescale ε, which is the regime $\mathrm{E_u V_z}$. Finally, the solution finds the rate-independent regime $\mathrm{E_u R_z}$ with $(u(t), z(t)) = q_{\mathrm{ri}}(t) := (2t-1, t-1)$.

For $\alpha = 1/2$ the solution first jumps to $(2, 0.5)$ on the time scale ε, which is the regime $\mathrm{B_u V_z}$. Next, there is a jump to $(0.5, 0.5)$ in the time scale $\varepsilon^{1/2}$, which is regime $\mathrm{V_u R_z}$. Then, the

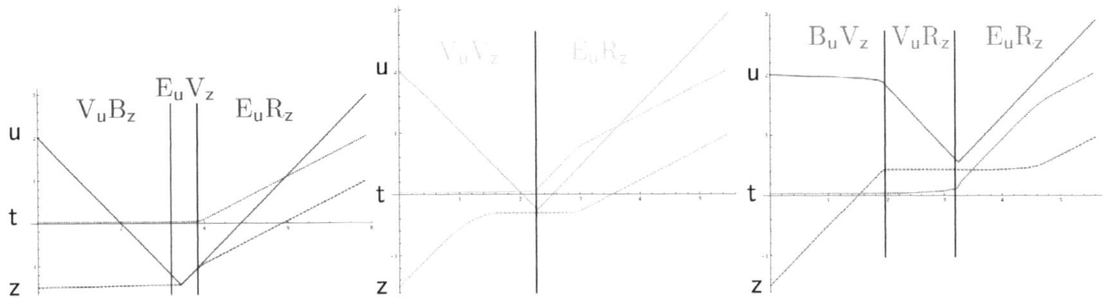

Figure 6.2. Solutions $(\mathsf{t}, \mathsf{u}, \mathsf{z})$ for (6.1) with dotted t, full u, and dashed z. Left $\alpha = 2$, middle $\alpha = 1$, right $\alpha = 1/2$.

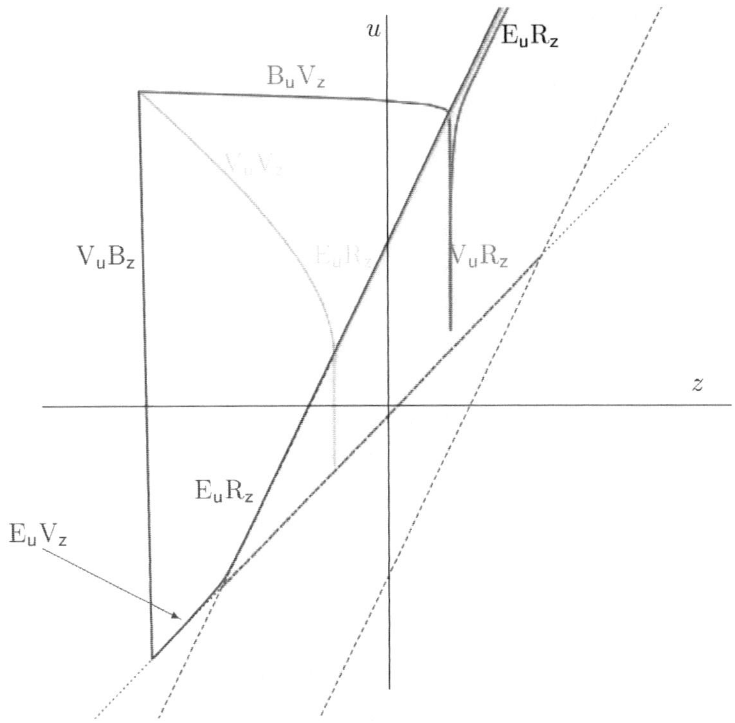

Figure 6.3. Solutions $(z(t), u(t))$ for (6.1). The dotted line is the diagonal $u = z$, while the yellow area is the locally stable region $|2z - u| \le 1$.

rate-independent regime $\mathsf{E_u R_z}$ starts, namely via $(u(t), z(t)) = (t-0.5, 0.5)$ for $t \in \,]0, 1.5]$ and $q_{\rm ri}$ for $t > 1.5$.

The behavior for $\underline{\alpha = 1}$ is intermediate: the jump occurs along a nonlinear curve in regime $\mathsf{V_u V_z}$, and $q_{\rm ri}$ is joined for $t \ge t_* \approx 0.7$, which is regime $\mathsf{E_u R_z}$.

The different behavior and the different regimes are also nicely seen by plotting the trajectories in the (u, z)-plane, see Figure 6.3, where the three different cases for α are depicted.

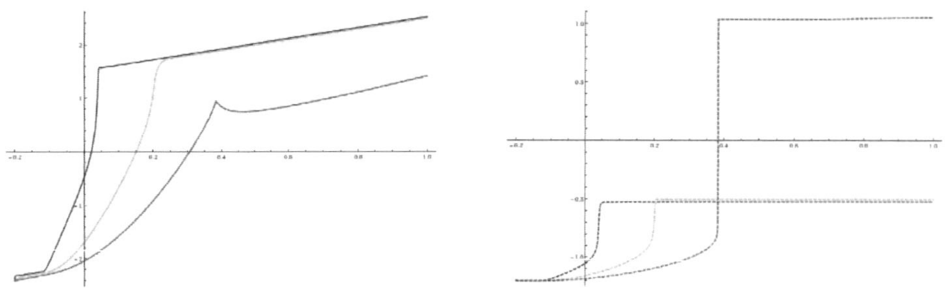

Figure 6.4. Solutions for (6.2): left $u(t)$ and right $z(t)$

6.2. Different jumps starting from the rate-independent regime

Finally we provide an example where the jumps start out of a rate-independent motion, i.e. we first have the regime $\mathrm{E_u R_z}$, and then the system becomes unstable and develops a jump. For this purpose we use the nonconvex energy

$$\mathcal{E}(t,u,z) = \frac{1}{2}(u-g(z))^2 + F(z) - tu \quad \text{with } g(z) = 4z^3 - 4z$$
$$\text{and } F'(z) = -1 + (z+1)^2\big(-40 + 10(z+1)^2 + 38\mathrm{e}^{-10(z+0.5)^2}\big).$$

Using the standard viscous potentials as above, the ODE system reads

$$\begin{cases} 0 = \varepsilon^\alpha \dot{u} + u - g(z) - t, \\ 0 \in \mathrm{Sign}(\dot{z}) + \varepsilon \dot{z} + F'(z) + g'(z)(g(z)-u) \end{cases} \quad \text{with } (u(-0.2), z(-0.2)) = (-2.4, -1.2). \quad (6.2)$$

Figure 6.4 shows simulation results of $u(t)$ and $z(t)$ for the three cases $\alpha = 2$ (blue), $\alpha = 1$ (green), and $\alpha = 1/2$ (red) with sufficiently small ε. We see that the solutions stay together for $t \in [-0.2, -0.1]$, which is exactly the time they stay in regime $\mathrm{E_u R_z}$. Then, in all three cases a jump develops, but this is quite different for different α. In Figure 6.5 we provide graphics of the same solutions, but now in the reparameterized form $(\mathsf{t}, \mathsf{u}, \mathsf{z})$ for the three α-values 2, 1, and 1/2, where again the parameterization s is chosen such that $\dot{s}(t) = \max\{0.5, |\dot{u}(t)|, \dot{z}(t)|\}$. However, for this example numerical instabilities prevented us from taking ε small enough to have a better separation of time scales. Even in the viscous regimes we still see $\mathsf{t}' > 0$ but small. Nevertheless, Figure 6.5 clearly shows the different regimes.

Figure 6.6 shows the trajectories in the (z, u)-plane.

Acknowledgments

A.M. has been partially supported by DFG via SFB 1114, subproject C5. R.R. and G.S. have been partially supported by a MIUR-PRIN'10-11 grant for the project "Calculus of Variations". R.R. also acknowledges support from the Gruppo Nazionale per l'Analisi Matematica, la Probabilità e le loro Applicazioni (GNAMPA) of the Istituto Nazionale di Alta Matematica (INdAM).

References

[1] Mielke A 2005 *Evolution in rate-independent systems.* Ch. 6 in "Handbook of Differential Equations, Evolutionary Equations, vol. 2" ed Dafermos C and Feireisl E (Amsterdam: Elsevier B.V.) pp 461–559

[2] Mielke A and Theil F 1999 *A mathematical model for rate-independent phase transformations with hysteresis.* In "Proceedings of the Workshop on Models of Continuum Mechanics in Analysis and Engineering" ed Alber H D, Balean R and Farwig R (Aachen: Shaker-Verlag) pp 117–129

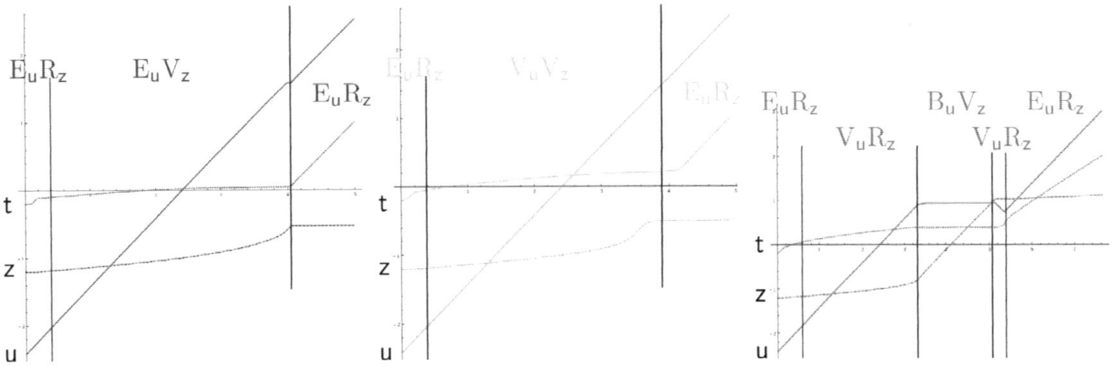

Figure 6.5. Solutions $(\mathsf{t}, \mathsf{u}, \mathsf{z})$ for (6.2) with dotted t, full u, and dashed z. Left $\alpha = 2$, middle $\alpha = 1$, right $\alpha = 1/2$.

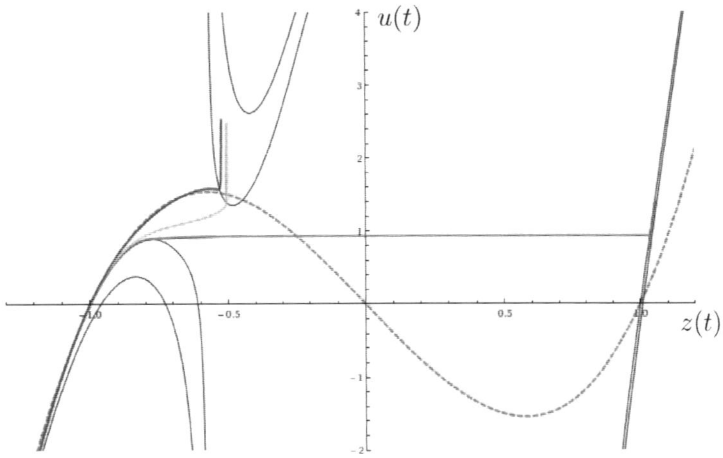

Figure 6.6. Solutions $(z(t), u(t))$ for (6.2). The dashed magenta line is $u = g(z)$, while the black curves display the boundaries of the locally stable domain $|F'(z) + g'(z)(g(z)-u)| \leq 1$.

[3] Mielke A and Theil F 2004 On rate-independent hysteresis models *Nonl. Diff. Eqns. Appl. (NoDEA)* **11** 151–189 (Accepted July 2001)

[4] Dal Maso G, Francfort G and Toader R 2005 Quasistatic crack growth in nonlinear elasticity *Arch. Rational Mech. Anal.* **176** 165–225

[5] Mielke A 2011 *Differential, energetic and metric formulations for rate-independent processes.* Ch. 3 in "Nonlinear PDEs and Applications.C.I.M.E. Summer School, Cetraro, Italy 2008" ed Ambrosio L and Savaré G (Heidelberg: Springer) pp 87–170

[6] Mielke A 2003 Energetic formulation of multiplicative elasto–plasticity using dissipation distances *Cont. Mech. Thermodynamics* **15** 351–382

[7] Knees D, Mielke A and Zanini C 2008 On the inviscid limit of a model for crack propagation *Math. Models Methods Appl. Sci.* **18** 1529–1569

[8] Mielke A, Rossi R and Savaré G 2009 Modeling solutions with jumps for rate-independent systems on metric spaces *Discrete Contin. Dyn. Syst.* **25** 585–615

[9] Efendiev M and Mielke A 2006 On the rate–independent limit of systems with dry friction and small viscosity *J. Convex Analysis* **13** 151–167

[10] Colli P and Visintin A 1990 On a class of doubly nonlinear evolution equations *Comm. Partial Differential Equations* **15** 737–756

[11] Colli P 1992 On some doubly nonlinear evolution equations in Banach spaces *Japan J. Indust. Appl. Math.* **9** 181–203

[12] Mielke A, Rossi R and Savaré G 2013 Nonsmooth analysis of doubly nonlinear equations *Calc. Var. Partial Differential Equations* **46** 253–310
[13] Mielke A, Rossi R and Savaré G 2012 Bv solutions and viscosity approximations of rate-independent systems *ESAIM Control Optim. Calc. Var.* **18** 36–80
[14] Mielke A, Rossi R and Savaré G 2014 Balanced Viscosity (BV) solutions to infinite-dimensional rate-independent systems *J. Eur. Math. Soc.* To appear. (WIAS preprint 1845. http://arxiv.org/abs/1309.6291)
[15] Mielke A and Zelik S 2014 On the vanishing viscosity limit in parabolic systems with rate-independent dissipation terms *Ann. Sc. Norm. Sup. Pisa Cl. Sci. (5)* **XIII** 67–135
[16] Toader R and Zanini C 2009 An artificial viscosity approach to quasistatic crack growth *Boll. Unione Mat. Ital. (9)* **2** 1–35
[17] Dal Maso G, DeSimone A and Solombrino F 2011 Quasistatic evolution for Cam-Clay plasticity: a weak formulation via viscoplastic regularization and time rescaling *Calc. Var. Partial Differential Equations* **40** 125–181
[18] Dal Maso G, DeSimone A and Solombrino F 2012 Quasistatic evolution for Cam-Clay plasticity: properties of the viscosity solution *Calc. Var. Partial Differential Equations* **44** 495–541
[19] Babadjian J F, Francfort G and Mora M 2012 Quasistatic evolution in non-associative plasticity - the cap model *SIAM J. Math. Anal.* **44** 245–292
[20] Francfort G A and Stefanelli U 2013 Quasi-static evolution for the Armstrong-Frederick hardening-plasticity model *Appl. Math. Res. Express. AMRX* **2013** 297–344
[21] Knees D, Rossi R and Zanini C 2013 A vanishing viscosity approach to a rate-independent damage model *Math. Models Methods Appl. Sci.* **23** 565–616
[22] Roubíček T 2009 Rate-independent processes in viscous solids at small strains *Math. Meth. Appl. Sci.* **32** 825–862
[23] Roubíček T 2010 Thermodynamics of rate-independent processes in viscous solids at small strains *SIAM J. Math. Anal.* **40** 256–297
[24] Roubíček T 2013 Adhesive contact of visco-elastic bodes and defect measures arising by vanishing viscosity *SIAM J. Math. Anal.* **45** 101–126
[25] Racca S 2012 A viscosity-driven crack evolution *Adv. Calc. Var.* **5** 433–483
[26] Dal Maso G and Scala R 2014 Quasistatic evolution in perfect plasticity as limit of dynamic processes *J. Dynam. Differential Equations* **26** 915–954
[27] Scala R 2014 Limit of viscous dynamic processes in delamination as the viscosity and inertia vanish *Preprint* Available at `http://cvgmt.sns.it/paper/2434/`
[28] Ambrosio L 1995 Minimizing movements *Rend. Accad. Naz. Sci. XL Mem. Mat. Appl. (5)* **19** 191–246
[29] Ambrosio L, Gigli N and Savaré G 2008 *Gradient flows in metric spaces and in the space of probability measures* 2nd ed Lectures in Mathematics ETH Zürich (Basel: Birkhäuser Verlag)
[30] Ioffe A D and Tihomirov V M 1979 *Theory of extremal problems* (*Studies in Mathematics and its Applications* vol 6) (Amsterdam-New York: North-Holland Publishing Co.)
[31] Ioffe A D 1977 On lower semicontinuity of integral functionals. I *SIAM J. Control Optimization* **15** 521–538
[32] Zanini C 2007 Singular perturbations of finite dimensional gradient flows *Discr. Cont. Dyanm. Syst.* **18** 657–675
[33] Agostiniani V, Rossi R and Savaré G 2014 Balanced viscosity solutions of singularly perturbed gradient flows in infinite dimension *In preparation*

Multiple Scale Reaction-Diffusion-Advection Problems with Moving Fronts

Nikolay Nefedov

Department of Mathematics, Faculty of Physics, Lomonosov Moscow State University, 119899 Moscow, Russia

E-mail: nefedov@phys.msu.ru

Abstract. In this work we discuss the further development of the general scheme of the asymptotic method of differential inequalities to investigate stability and motion of sharp internal layers (fronts) for nonlinear singularly perturbed parabolic equations, which are called in applications reaction-diffusion-advection equations. Our approach is illustrated for some new important cases of initial boundary value problems. We present results on stability and on the motion of the fronts.

1. Introduction

This work concerns nonlinear singularly perturbed parabolic equations. In applications, these problems may be interpreted as models of reaction-diffusion and reaction-diffusion-advection processes in chemical kinetics, astrophysics, biology, etc. These parabolic equations solutions often feature a narrow boundary layer region of rapid change as well as internal layers of different types (stationary internal layers - contrast structures, moving internal layers - fronts and moving spikes). It is well-known that such problems are extremely difficult for a numerical treatment as well as for asymptotic investigations and it needs new asymptotic methods to investigate them formally as well as rigorously (see the papers [1] - [4] and references therein).

In order to construct the formal asymptotics of solutions of different classes of problems with internal layers we use our recent extension of the well-known boundary layer functions method. The basic ideas of this approach have been published in the review paper [5].

Our rigorous investigation is based on the modern development of the comparison principle for elliptic and parabolic problems. The basic ideas of our approach have been presented in the papers of H. Amann and D. Sattinger (see, [6, 7]) and have got a further development in the works of P. Hess [8] and H. Amann [9]. Their results are essentially based on the Krein-Rutman theorem (see, for example, [8]) and on the results of M.A. Krasnoselskij on positive operators (see, for example, [10] and references therein).

The results of the present paper are based on the general scheme of the asymptotic method of differential inequalities, the basic ideas of which were presented in [11]. We illustrate this approach by applying it to some new important cases of initial boundary value problems

$$\varepsilon^2 \Delta u - \frac{\partial u}{\partial t} = f(u, \nabla u, x, \varepsilon), \quad x \in \mathcal{D} \subset R^N,\ t > 0. \tag{1}$$

This reaction-diffusion-advection equation and its variants are used in many applications. For these problems we state conditions which imply the existence of solutions with internal layers. Among others we discuss the following problems:

1. Existence and Ljapunov stability of stationary solutions.
2. Front formation and motion description.

Our investigation uses the so-called positivity property of the operators producing formal asymptotics and is based on some recent extensions of the Krein-Rutman theorem. The basic idea of this approach is to construct lower and upper solutions to the problem by using formal asymptotics. Applying this method, we state the existence of solutions, estimate the accuracy of the asymptotics, and investigate the asymptotic stability of the stationary solutions in the sense of Ljapunov.

Another aspect of this work is to emphasize the possibility of using this analytical treatment for numerical approaches. We have shown that the problems describing the location of fronts are significantly more simple than original problems. We use this approach in the paper [12]. To illustrate our approach, we consider the following cases of equation (1):

1. Reaction-diffusion equation

$$\varepsilon^2\frac{\partial^2 u}{\partial x^2}-\frac{\partial u}{\partial t}=f(u,x,\varepsilon),\quad x\in(0,1),\quad t>0, \tag{2}$$

2. Reaction-advection-diffusion equation

$$\varepsilon\frac{\partial^2 u}{\partial x^2}-A(u,x)\frac{\partial u}{\partial x}-\frac{\partial u}{\partial t}=f(u,x,\varepsilon),\quad x\in(0,1),\quad t>0.$$

Note that we can transform the last equation into the form of equation (1) by multiplying the equation by ε and changing the time scale. The small parameter ε models the fast reaction in case of a reaction-diffusion equation and fast reaction and advection in case of a reaction-advection-diffusion equation. For the initial boundary value problems to these equations we prove the existence of fronts and give its asymptotic approximation. The motion of fronts in the cases of reaction-diffusion equation and reaction-advection-diffusion equation is determined by the unbalanced reaction and advection term, respectively. The operators producing the asymptotics are different in these cases. Note also that in case of a reaction-advection-diffusion equation it is more complicated to analyze the stability, because the linearized operator is not selfadjoint. In the paper we show how to use comparison technique to analyze Ljapunov stability in that case. Note that our results can be applied to multidimensional problems, but we illustrate it in the one-dimensional case to escape technical difficulties.

The results of the work is a further development of the results of the papers [13] - [17]. Note that the results for equation (1) can be extended to periodic parabolic problems and to some classes for nonlocal reaction-diffusion-advection equations.

The paper is structured as follows. In Section 2 we consider the general scheme of our approach. In Section 3 we apply it to a reaction-advection-diffusion equation: in 3.1 we state the stability of the stationary solution, in 3.2 we construct the asymptotics of the moving front. Finally, in 3.3 we give some extensions of our results.

2. General scheme of the asymptotic method of differential inequalities for reaction-advection-diffusion equations

In this section we describe the general scheme of the developed approach.

We consider the initial boundary value problem

$$\begin{gathered}\varepsilon^2\Delta u-\tfrac{\partial u}{\partial t}=f(u,\nabla u,x,\varepsilon),\quad x\in\mathcal{D}\subset R^N,\, t>0,\\ Bu=h(x),\quad x\in\partial\mathcal{D},\, t>0,\\ u(x,0,\varepsilon)=u_{init}(x,\varepsilon),\quad x\in\mathcal{D},\end{gathered} \tag{3}$$

where $\varepsilon > 0$ is a small parameter, f, h, u^0 and $\partial\mathcal{D}$ are sufficiently smooth, B is the boundary operator for Dirichlet, Neumann or third order boundary conditions. We are interested in the existence and asymptotic behavior of the solution of (3) as $\varepsilon \to 0$.

We start with the description of our approach to investigate the Ljapunov stability of the stationary solutions with boundary and internal layers.

We denote by N the nonlinear operator in (3)

$$Nu \equiv \varepsilon^2 \Delta u - f(u, \nabla u, x, \varepsilon).$$

Obviously, the stationary solutions of (3) satisfy the equation $Nu = 0$.

Definition (Stationary asymptotic upper and lower solutions of order q).
We call an upper solution $\beta(x,\varepsilon)$ and a lower solution $\alpha(x,\varepsilon)$ stationary asymptotic upper and lower solutions of order $q > 0$ of problem (3) if for sufficiently small ε $(0 < \varepsilon \leq \varepsilon_0)$ they satisfy the inequalities

$$N\beta \leq -c\varepsilon^q,\ N\alpha \geq c\varepsilon^q \quad for \quad x \in \mathcal{D}$$

$$B\alpha \leq h(x) \leq B\beta \quad for \quad x \in \partial\mathcal{D},$$

where c is a positive constant.

Such asymptotic upper and lower solutions were used in [11]. A similar definition we also use in [18].

We denote by L the linear operator which we obtain from N by linearizing f at the stationary solution, and by H the following characteristic of the nonlinearity

$$H \equiv f(\beta, \nabla\beta, x, \varepsilon) - f(\alpha, \nabla\alpha, x, \varepsilon) - L_f(\beta - \alpha),$$

where L_f is the linearization of f at the stationary solution. Suppose we know how to construct the asymptotic lower and upper solutions. Note that one of the most important of our achievements is the method to construct them by using the formal asymptotic expansion. In what follows we describe this approach. Our assumption is

(A_1). *There exist a stationary asymptotic lower solution α and upper solution β of order q and a such that $\beta > \alpha$ and $|\beta - \alpha| \leq c\varepsilon^r$ for $x \in \bar{\mathcal{D}}$, $0 < \varepsilon \leq \varepsilon_0$.*

Assumption (A_1) implies the existence of a stationary solution $u(x,\varepsilon)$ of problem (3) satisfying the inequalities $\alpha \leq u(x,\varepsilon) \leq \beta$. Therefore we also have the asymptotic estimate for the solution $u(x,\epsilon)$. It differs from the upper or lower solution by a value of order $O(\varepsilon^r)$ (see Theorem 1). We also assume

(A_2). $|H| \leq c\varepsilon^p$ for $x \in \overline{\mathcal{D}}$, $0 < \varepsilon \leq \varepsilon_0$.

(A_3). $p \geq q$.

It is clear that the estimates of the assumptions (A_2), (A_3) depend on the properties of the nonlinearity f and the lower and upper solutions, and therefore on the problem under consideration.

Under the assumptions above the following theorem is valid.

Theorem 1. *Suppose the assumptions $(A_1) - (A_3)$ to be valid. Then, for sufficiently small ε there exists a solution $u(x,\varepsilon)$ of (3) which differs from the upper or lower solution by a term of order $O(\varepsilon^r)$ and is asymptotically stable in Ljapunov's sense with the local domain of stability $[\alpha,\beta]$.*

Proof. The proof of Theorem 1 is based on the revised maximum principle [9]. From $(A_2), (A_3)$ it is follows that $L(\beta - \alpha) < 0$ (see analogous calculations in [17]), $B(\beta - \alpha) > 0$ by the definition and therefore the principal eigenvalue (which exists and is real, see [9]) satisfies the estimate $\lambda_p < 0$. This implies the asymtotic stability of the stationary solution in the sense of Ljapunov.

In order to get the upper and lower solutions satisfying the assumptions of Theorem 1 we use the formal asymptotics, which can be constructed in a lot of cases by our method, proposed in [5, 13]. Under quite natural assumptions the formal asymptotics of the internal layer solution is produced by the boundary layer operators L_B, the regular expansion operators L_R and by the operators describing the location of the transition layer A^Γ. To construct the formal asymptotics we assume (or give some sufficient assumptions) that these operators are invertible. We will give some details in Section 3 (see also the papers [11, 13, 14]). For the construction of asymptotic lower and upper solutions we require that *these operators have positive inverse (they are order preserving operators) when they act in the same classes of functions in which we construct the asymptotic expansions by means of these operators.*

Using the positive solutions for the producing n-th order formal asymptotics by the operators we get asymptotic lower and upper solutions $\alpha \equiv \alpha_n$, $\beta \equiv \beta_n$ of order n of problem (3). We illustrate our approach by examples.

3. Reaction-advection-diffusion equations

In this section we illustrate our approach by considering the reaction-advection-diffusion equations. The aim of this section is to analyze the stability of the stationary solution in subsection 3.1 by applying Theorem 1. Next we consider moving fronts for this problem.

We consider the problem

$$\begin{gathered}\varepsilon\frac{\partial^2 u}{\partial x^2} - \frac{\partial u}{\partial t} = A(u,x)\frac{\partial u}{\partial x} + B(u,x), \quad x \in (0,1), \quad t \in (0,T], \\ u(0,t,\varepsilon) = u^0, \quad u(1,t,\varepsilon) = u^1, \quad t \in [0,T], \\ u(x,0,\varepsilon) = u_{init}(x,\varepsilon), \quad x \in [0,1],\end{gathered} \tag{4}$$

where A, B, u_{init} are sufficiently smooth functions, u^0 and u^1 are given constants, $\varepsilon > 0$ is a small parameter and $T > 0$. We state the existence and asymptotic stability of the stationary solution with interior layer. Moreover, we prove the existence of a moving front type solution for the non-stationary problem. We also give the asymptotic approximation of such solutions, particularly we prove that the principal term, describing the location of the moving front, is determined by the initial value problem

$$\frac{dx_0}{dt} = V(x_0), \quad x_0(0) = x_{00}, \tag{5}$$

where x_{00} is the initial location of the front, $V(x_0)$ is a known function, defined by the input data (see 3.2). Therefore, in applications, these important characteristics of the partial differential equation (4) can be determined from the ordinary differential equation (5). Problem (5) can be solved significantly simpler than the original one (4). We characterize this approach as asymptotic-numerical. It has been applied in the paper [12].

3.1. Stability of the stationary solutions

This part is an illustration of Theorem 1 in Section 2. The scheme to construct asymptotic lower and upper solution is an extension of the approaches of [11, 13]. Stationary solutions are the solutions of the following boundary value problem

$$\begin{gathered}\varepsilon\frac{\partial^2 u}{\partial x^2} = A(u,x)\frac{\partial u}{\partial x} + B(u,x), \quad x \in (0,1), \\ u(0,\varepsilon) = u^0, \quad u(1,\varepsilon) = u^1.\end{gathered} \tag{6}$$

We consider the reduced stationary equation for (6)

$$A(u,x)\frac{du}{dx} + B(u,x) = 0, \tag{7}$$

and we assume natural hypotheses:

(H_1). *On [0,1], equation (7) with the initial condition* $u(0) = u^0$ *has the solution* $u = \varphi^l(x)$; *with the initial condition* $u(1) = u^1$ *the solution* $u = \varphi^r(x)$, *moreover,*

$$\varphi^l(x) < \varphi^r(x), \quad x \in [0,1]$$

and

$$A(\varphi^l(x), x) > 0, \quad A(\varphi^r(x), x) < 0, \quad x \in [0,1].$$

We define

$$I(x) := \int_{\varphi^l(x)}^{\varphi^r(x)} A(u,x)du.$$

(H_2). *The equation*

$$I(x) = 0 \quad \textit{has a root} \quad x = x_0, \quad x_0 \in (0;1), \quad I_x(x_0) < 0,$$

and

$$\int_{\varphi^l(x_0)}^{s} A(u, x_0)du > 0 \quad \textit{for} \quad s \in (\varphi^l(x_0), \varphi^r(x_0)).$$

To characterize the location of the interior layer we introduce the point $x = x_*(\varepsilon)$ as locus of the intersection of the stationary solution $u(x,\varepsilon)$ of (4) with the curve

$$u = \frac{1}{2}\left(\varphi^l(x) + \varphi^r(x)\right) =: \varphi(x). \tag{8}$$

In what follows we construct the asymptotic expansion of $x_*(\varepsilon)$ in the form

$$x_*(\varepsilon) = x_0 + \varepsilon\, x_1 + ...,$$

where x_0 is defined by (H_2) and therefore is the leading order term to describe the layer location, and $x_k, k = 1, 2, ...$, are constants to be determined. For the following we use the notation

$$\xi := \frac{x - x_*(\varepsilon)}{\varepsilon},$$

$$\overline{\mathcal{D}}^{(l)} := \{0 \le x \le x_*(\varepsilon)\}, \quad \overline{\mathcal{D}}^{(r)} := \{x_*(\varepsilon) \le x \le 1\}.$$

First we consider in $\overline{\mathcal{D}}^{(l)}$ and in $\overline{\mathcal{D}}^{(r)}$ the boundary value problems for the stationary equation (6), where the additional boundary conditions $u(x_*, \varepsilon) = \varphi(x_*)$ is used at $x = x_*(\varepsilon)$. Formal asymptotic expansions of the solutions $U^{(l,r)}(x,\varepsilon)$ of these problems are constructed in the form

$$\begin{aligned} U^{(l,r)}(x,\varepsilon) = \bar{U}^{(l,r)}(x,\varepsilon) + Q^{(l,r)}(\xi,\varepsilon) = \\ \textstyle\sum_{i=0}^{\infty} \varepsilon^i \left(\bar{U}_i^{(l,r)}(x) + Q_i^{(l,r)}(\xi)\right), \end{aligned} \tag{9}$$

where $\bar{U}^{(l,r)}$ and $Q^{(l,r)}$ denote the regular and the interior layer parts. We get that $\bar{U}_0^{(l,r)}(x) = \varphi^{(l,r)}(x)$. By using the standard procedure of the boundary layer function method we can show that the operators $L_R^{(l,r)}$, producing regular parts of the asymptotics, have the form

$$\begin{aligned} L_R^{(l,r)} \equiv A(\varphi^{(l,r)}(x), x)\tfrac{\partial}{\partial x} + \Big(A_u(\varphi^{(l,r)}(x), x)\tfrac{\partial \varphi^{(l,r)}(x)}{\partial x} \\ -B_u(\varphi^{(l,r)}(x), x)\Big). \end{aligned} \tag{10}$$

The operators $L_R^{(l,r)}$ are positively invertible (order preserving) for a positive right hand part, in particular, the inequality $L_R^{(l,r)}\overline{u} > 0$ has a positive solution. Indeed the operators $L_R^{(l,r)}$ are first order differential operators and their positive invertibility is defined by the sign of $A(\varphi^{(l,r)}(x), x)$ (see H_1) (for reaction-diffusion equations such operators are just $f_u(\varphi^{(l,r)}(x), x)$, see e.g. [11, 13]).

By using the standard procedure of the boundary layer function method we also can construct the boundary layer parts of these expansions. The terms x_k in the expansion of x_* are determined by the C^1-matching conditions (see [13]) of the expansions $U^{(l)}$ and $U^{(r)}$

$$-\frac{\partial I}{\partial x}(x_0)x_k = h_k,\ k = 1, 2, ...$$

We see that in our case $A^\Gamma = -\frac{\partial I}{\partial x}(x_0)$ and the inequality $A^\Gamma\delta > 0$ has a positive solution according to the assumption (H_2).

We denote by $U_n^{(l,r)}$ the partial sums of order n of the expansions, where ξ is replaced by $\left(x - \sum_{i=0}^{n+1} x_i\varepsilon^i\right)/\varepsilon$ in equation (9), and we denote by $U_n(x, \varepsilon)$ these sums in the corresponding domains ($U_n(x, \varepsilon) = U_n^{(l)}(x, \varepsilon)$ on $\overline{\mathcal{D}}^{(l)}$ and $U_n(x, \varepsilon) = U_n^{(r)}(x, \varepsilon)$ on $\overline{\mathcal{D}}^{(r)}$.

In order to construct the asymptotic of $(n+1)$-th order lower and upper solution ($q = n+1$, see Section 2) we have to add positive solutions of the producing operators to the $(n+1)$-th order terms of the formal asymptotic expansion. By this way, we get the existence result and the estimate for the asymptotic expansion.

Now, we can check the assumptions (A_1) - (A_3) of the Theorem 1 from the previous section. By simple calculations (analogous calculations one can find in [13]) we get $p = 2n - 1$ (see Theorem 1). Therefore, assumption (A_3) ($p \geq q$) is satisfied for $n \geq 2$. In order to get a better estimate of the region of attraction we take $n = 2$. Thus, we have the following existence and stability theorem.

Theorem 2. *Suppose the assumptions $(H_1) - (H_2)$ to be valid. Then, for sufficiently small ε there exists a stationary solution $u(x, \varepsilon)$ of (4) which has an interior layer near the point $x_*(\varepsilon)$ and satisfies*

$$|u(x, \varepsilon) - U_n(x, \varepsilon)| \leq c\varepsilon^{n+1}, \quad x \in [0, 1],$$

where the positive constant c does not depend on ε. Moreover, this solution is asymptotically stable with a local region of attraction $[\alpha_2(x, \varepsilon), \beta_2(x, \varepsilon)]$.

3.2. Moving fronts

Now we consider the initial boundary value problem (4) assuming that

$$u(x, 0, \varepsilon) = u_{init}(x, \varepsilon)$$

is an interior layer type function, and give the asymptotic description of the solution with a moving front. We save the assumption (H_1) above. We replace the assumption (H_2) by assuming

(H_{2m}). *Let $I(x) := \int_{\varphi^l(x)}^{\varphi^r(x)} A(u, x)du > 0$ for $x \in [0, 1]$.*

We also assume

(H_{3m}). *The initial value problem*

$$\frac{dx}{dt} = \frac{I(x)}{\varphi^r(x) - \varphi^l(x)}, \quad x(0) = x_{00},$$

where x_{00} is the initial location of the front, has the solution $x_0 : [0; T] \to [0, 1]$ such that

$$x_0(t) \in (0, 1) \quad for \quad t \in [0, T].$$

The asymptotic construction is analogous to the stationary case considered above, and we only describe the main differences. The location of the interior layer is the point $x = x_*(t,\varepsilon)$ which is defined as in the previous section by (8). The asymptotic expansion of $x_*(t,\varepsilon)$ is also defined as an expansion in ε with time-dependent coefficients in the form

$$x_*(t,\varepsilon) = x_0(t) + \varepsilon\, x_1(t) + ...,$$

where the leading term $x_0(t)$ is defined by (H_{3m}) and $x_k(t), k = 1, 2, ...,$ are functions to be determined.

The structure of the asymptotic expansion is the same as for the stationary solution, but the domains $\overline{\mathcal{D}}^{(l,r)}$ are divided by the moving point $x_*(t,\varepsilon)$. The boundary layer functions $Q^{(l,r)}(\xi,\varepsilon)$, where $\xi = \frac{x - x_*(t,\varepsilon)}{\varepsilon}$, are also defined by a similar way. Particularly, zeroth order Q-functions are defined from the equations

$$\frac{\partial^2 Q_0^{l,r}}{\partial \xi^2} - (A(\varphi^{l,r}(x_*(t,\varepsilon)) + Q_0^{l,r}, x_*(t,\varepsilon)) - \frac{\partial x_*}{\partial t})\frac{\partial Q_0^{l,r}}{\partial \xi} = 0$$

with standard boundary conditions, and these problems are solvable according to assumption (H_{2m}). The terms $x_k(t)$ are determined by the C^1-matching conditions of the expansions $U^{(l)}$ and $U^{(r)}$. Particularly, the zeroth order of the C^1-matching conditions is satisfied according to assumption (H_{3m}). The next orders of these conditions yield the initial value problems to determine $x_k(t)$, $k \geq 1$:

$$\frac{dx_k}{dt} = K(x_0(t))(\varphi^l(x_0(t)) - \varphi^r(x_0(t)))^{-1}x_k + G_k(t), \quad x_k(0) = 0,$$

where $K(x_0(t))$ is a known function, $G_k(t)$ is determined on the each step. The operator producing x_k is an first order differential operator, and as in the general case, it appears as a linearization of the equation for x_0. The functions h_k are determined by known terms of the asymptotics. It is clear that this first order differential operator is an order preserving operator.

Following the results of Section 2 (see also 3.1 and for similar treatment [16]), we define $U_n(x,t,\varepsilon)$ as in Theorem 2. Finally we get the following result.

Theorem 3. *Suppose the assumptions* $(H_1), (H_{2m}), (H_{3m})$ *to be valid and* $\alpha_n(x,t,\varepsilon) \leq u_{init}(x,\varepsilon) \leq \beta_n(x,t,\varepsilon)$. *Then, for sufficiently small* ε *there exists a solution* $u(x,t,\varepsilon)$ *of (4) which is a moving front and satisfies*

$$|u(x,t,\varepsilon) - U_n(x,t,\varepsilon)| \leq c\varepsilon^{n+1} \quad \textit{for} \quad x \in [0,1] \quad \textit{and} \quad t \in [0,T].$$

3.3. Some extensions

A similar scheme can be applied for a wide class of problems for reaction-diffusion equations and systems. Particularly, an adapted version of this method was used for the boundary value problem for the equation (2) (see [16]) and nonlocal reaction-diffusion equation (see [14]). We also give an extension of our results for the reaction-diffusion system

$$\varepsilon^4\frac{\partial^2 u}{\partial x^2} - \varepsilon^2\frac{\partial u}{\partial t} = f(u,v,x,\varepsilon),$$
$$\varepsilon^2\frac{\partial^2 v}{\partial x^2} - \varepsilon^2\frac{\partial v}{\partial t} = g(u,v,x,\varepsilon), \quad x \in (0,1), \quad t > 0. \tag{11}$$

Note that in the problems above we choose an appropriate time scale (the degree of ε before the time-derivatives determines these scales). For these problems similar existence and stability theorems are valid. Moreover, it was shown that the principal term of the asymptotic approximation of the location of the internal layer is determined by the initial value problem

$$\frac{dx_0}{dt} = V(x_0), \quad x_0(0) = x_{00},$$

with known function V. We also note that the stable (in the sense of Ljapunov) stationary internal layer solutions are located near the stable stationary points of the equations for $x_0(t)$.

Note that analogies of Theorem 1 are valid for periodic-parabolic problems [15] and nonlocal reaction-advection-diffusion equations (to appear).

Acknowledgments

The author thanks the referees for important comments to improve this paper.
This work is supported by RFBR, pr. N 13-01-00200, and RFBR-DFG, pr. N14-01-91333

References

[1] O'Riordan, E. and Shishkin, G.I. 2004 *Singularly perturbed parabolic problems with non-smooth data.* Journal of Computational and Applied Mathematics, 1:, 233-245.

[2] Kumar, M. and Singh, A.K. 2014 *Singular perturbation problems in nonlinear elliptic partial differential equations: a survey.* International Journal of Nonliner Science, 17(3), 195-214.

[3] Franz, S. and Kopteva N. 2012 *Green's function estimates for a singularly perturbed convection-diffusion problem.* J. Diff. Equations, 252, 1521-1545.

[4] Kopteva, N. 2009 *Numerical analysis of a 2d singularly perturbed semilinear reaction-diffusion problem.* Lecture Notes in Comput. Sci., 5434, 80-91.

[5] Vasilieva, A.B. and Butuzov, V.F. and Nefedov, N.N. 1998 *Contrast structures in singularly perturbed problems (in russian).* Fundamentalnaja i prikladnala matematika, 4(3), 799-851.

[6] Amann, H. 1978 *Periodic Solutions of Semilinear Parabolic Equations.* Nonlinear Analysis: a Collection of Papers in Honor of Erich Rothe, 1-29. New York: Academic.

[7] Sattinger, D.H. 1972 *Monotone Methods in Elliptic and Parabolic Boundary Value Problems.* Indiana Univ. Math. J., 21(11), 979-1001.

[8] Hess, P. 1991 *Periodic-Parabolic Boundary Value Problems and Positivity.* Pitman Research Notes in Math. Series 247, Longman Scientific & Technical, Harlow.

[9] Amann, H. 2005 *Maximum priciples and principal eigenvalues.* In Ten Mahtmatical Essays in Analysis and Topology. Elsevier.

[10] Zabrejko, P.P. and Koshelev, A.I. and Krasnoseiskij, M.A et al. 1968 *Integral equations (in russian).* M.: Nauka.

[11] Nefedov, N.N. 1995 *The Method of Differential Inequalities for Some Classes of Nonlinear Singularly Perturbed Problems with Internal Layers.* Differ. Uravn., 31(7), 1142-1149.

[12] Volkov, V. and Nefedov, N. 2013 *Asymptotic-numerical investigation of generation and motion of fronts in phase transition models.* Lecture Notes in Computer Science 8236, 524-531.

[13] Vasil'eva, A.B. and Butuzov, V.F. and Nefedov, N.N. 2010 *Singularly Perturbed problems with Boundary and Internal Layers.* Proceedings of the Steklov Institute of Mathmatics, 268, 258-273.

[14] Nefedov, N.N. and Nikitin, A.G. and Petrova, M.A. and Recke, L. 2011 *Moving fronts in integro-parabolic reaction-advection-diffusion equations.* Diff. Equations 47, 1318-1332.

[15] Nefedov, N.N. and Recke, L. and Schneider, K.R. 2013 *Existence and asymptotic stability of periodic solutions with an interior layer of reaction-advection-diffusion equations.* J. Math. Anal. Appl. 405, 90-103.

[16] Bozhevol'nov, Yu.V. and Nefedov N.N. 2010 *Front motion in parabolic reaction-diffusion problem.* Computational Mathematics and Mathematical Physics 50(2), 276-285.

[17] Nefedov, N. 2013 *Comparison principle for reaction-diffusion-advection problems with boundary and internal layers.* Lecture Notes in Computer Science 8236, 62-72.

[18] Nefedov, N.N. and Recke, L. and Schneider, K.R. 2010 *Asymptotic stability via the Krein-Rutman theorem for singularly perturbed parabolic periodic Dirichlet problems.* Regular and Chaotic Dynamics, 15(2-3), 382-389.

Robust homoclinic orbits in planar systems with Preisach hysteresis operator

Alexander Pimenov
Weierstrass Institute, Mohrenstr. 39, 10117 Berlin
E-mail: alexander.pimenov@wias-berlin.de

Dmitrii Rachinskii
Department of Mathematical Sciences, The University of Texas at Dallas, TX, USA
E-mail: Dmitry.Rachinskiy@utdallas.edu

Abstract. We construct examples of robust homoclinic orbits for systems of ordinary differential equations coupled with the Preisach hysteresis operator. Existence of such orbits is demonstrated for the first time. We discuss a generic mechanism that creates robust homoclinic orbits and a method for finding them. An example of a homoclinic orbit in a population dynamics model with hysteretic response of the prey to variations of the predator is studied numerically.

1. Introduction

The theory of hysteresis operators developed in [1–3] as well as other mathematical descriptions of hysteresis (such as differential inclusions [4], differential inequalities [5] and variational descriptions of rate-independent processes [6]) allow one to analyze stability [7,8], bifurcations [9–12] and complex dynamics [13] of systems exhibiting hysteresis. Coupled systems of differential equations and hysteresis operators have been used to model mechanical [2,14,15], electro-magnetic [16,17], electro-mechanical [18,19] and engineered systems [20–22], phase transition processes [3,23], dynamics of fluids in porous materials [24,25], dynamics of populations [26–28] and economic systems [29–31]; see [32] for further examples of applications. In particular, these models include ordinary differential equations coupled with the Preisach hysteresis operator that contain the time derivative of the output of the Preisach operator. Although more complex situations are possible, a generic solution of such a system can be obtained by matching solutions of several ordinary differential systems (the switching points from one to another system include the turning points of the input of the Preisach operator as well as some values stored by the dynamic memory configuration of this operator) [33–35]. While these ordinary differential systems are different, they all have the same set of equilibrium points [36].

This picture suggests a possibility of existence of robust homoclinic trajectories in the system with hysteresis operator provided that (a) the same point is a stable equilibrium of one ordinary differential system and, simultaneously, an unstable equilibrium of the other; and, (b) a trajectory of the first system converging to this equilibrium point in forward time could be matched with a trajectory of the second system converging to the same point in backward time in such a way that the system with hysteresis would switch from the second trajectory to the first trajectory. The concatenation of the two trajectories would then form a homoclinic loop of the differential system with the hysteresis operator.

In this work, we show the existence of such homoclinic trajectories. To the best of our knowledge, homoclinic trajectories of systems with hysteresis operators have not been discussed before.

Existence of homoclinic trajectories in the systems with Preisach operator under the time derivative is closely related to the so-called *partially stable* equilibria [36], i.e. unstable equilibria, which have an open basin of attraction (in the infinite-dimensional phase space of the system). Such equilibria simultaneously attract and repel many trajectories, and they can be compared to a saddle-node singular point of an ordinary differential system, however they are robust. Systems with Preisach operator available in the literature do not demonstrate abundance of partially stable equilibria, and, moreover, it is even harder to find special kind of partially stable equilibria that would lead to the appearance of robust homoclinic orbits in such systems. We present our results in the context of mathematical ecology. Namely, we are motivated by a predator-prey system [37], where the prey switches between two modes of behaviour, safe and risky, in response to varying abundance of predator. Similar type of non-hysteretic safe-risky adaptive behaviour of the prey in response to the pressure of the predation was recently suggested as a possible mechanism leading to the so-called predator pit phenomena, where a bistability between two steady states with nonzero predator-prey densities appears [38]. When the switch between two modes of behaviour becomes hysteretic, one can observe numerically the existence of a robust homoclinic cycle [37].

We first construct a few examples of simple systems that have homoclinic orbits (Section 2). Then we give a numerical evidence of the existence of such orbits in a more complex predator-prey model with a refuge patch where we assume hysteresis in the reaction of the prey to variations of the predator abundance (Section 3). Unlike the familiar homoclinic orbits of ordinary differential systems, the homoclinic trajectories that we obtain are robust, that is they are preserved under variations of parameters and more general (sufficiently small) perturbations. Moreover, our systems typically have a continuous family of homoclinic orbits.

2. Examples of homoclinic orbits

2.1. Preisach operator

In this work we consider a coupled system of differential equations

$$\dot{u} = f(u,v) + \dot{x}h(u,v), \qquad \dot{v} = g(u,v) \tag{1}$$

and an operator equation

$$x(t) = (\mathcal{P}[\eta_0]v)(t), \tag{2}$$

where $t \geq t_0$, and $f, h, g : \mathbb{R}^2 \to C^1$.

The Preisach operator is defined by

$$(\mathcal{P}[\eta_0]v)(t) = \int_0^\infty \int_0^{\alpha_S} \mu(\alpha_R,\alpha_S)(R_{\alpha_R,\alpha_S}[\eta_0(\alpha_R,\alpha_S)]v)(t)\, d\alpha_R d\alpha_S, \tag{3}$$

where $v = v(t)$, $t \geq t_0$ is the input; the function $\eta_0 = \eta_0(\alpha_R,\alpha_S)$ which takes values 0 and 1 is the initial state function; $\mu(\alpha_R,\alpha_S)$ is an integrable density function; and R_{α_R,α_S} is the non-ideal relay operator with thresholds α_R, α_S satisfying $\alpha_R \le \alpha_S$:

$$(R_{\alpha_R,\alpha_S}[\eta_0]v)(t) = \begin{cases} 0 & \text{if} \quad v(\tau) \le \alpha_R \text{ for some } \tau \in [t_0, t] \\ & \quad\ \text{and } v(s) < \alpha_S \text{ for all } s \in [\tau, t]; \\ 1 & \text{if} \quad v(\tau) \ge \alpha_S \text{ for some } \tau \in [t_0, t] \\ & \quad\ \text{and } v(s) > \alpha_R \text{ for all } s \in [\tau, t]; \\ \eta_0(\alpha_R, \alpha_S) & \text{if} \quad \alpha_R < v(\tau) < \alpha_S \text{ for all } \tau \in [t_0, t]. \end{cases} \tag{4}$$

We represent states (4) of the relays graphically on the half-plane $\{(\alpha_R, \alpha_S)\colon \alpha_R \le \alpha_S\}$, which is divided into two parts by a staircase polyline $\Omega = \Omega(t)$ with the relays in state 1 below (to the left) of this line and in state 0 above (to the right) of this line, see Fig. 1 (left). At any moment $t \ge t_0$ the right end point of $\Omega(t)$ is the point $\alpha_R = \alpha_S = v(t)$ [1]. For simplicity, the measure $\mu(\alpha_R, \alpha_S)$ is often assumed to have triangular support $\alpha_m \le \alpha_R \le \alpha_S \le \alpha_M$, hence we are interested only in the representation of the states of the relays inside this triangle.

The staircase polyline $\Omega(t)$ changes over time in accordance with formula (4) in response to variations of continuous non-negative input $v(t)$. We illustrate the evolution of $\Omega(t)$ with time $t \ge t_0$ using the following example with $\alpha_m = 0$. Let $\Omega(t_0)$ be the horizontal segment $\alpha_S = v(t_0)$, $0 \le \alpha_R \le v(t_0)$ (see Fig. 1, right). If $v(t)$ monotonically decreases on an interval $t \in [t_0, t_1]$, then $\Omega(t)$ acquires the vertical link $\alpha_R = v(t)$, $v(t) \le \alpha_S \le v(t_0)$ as shown in Fig. 2, left. Now, suppose that the input monotonically increases for $t \in [t_1, t_3]$. Then $\Omega(t)$ acquires a new horizontal link $\alpha_S = v(t)$, $v(t_1) \le \alpha_R \le v(t) < v(t_0)$ (see Fig. 2, right). If the input reaches its initial value $v(t_0)$ at some moment t_2^*, then $\Omega(t_2^*)$ coincides with the initial horizontal segment $\Omega(t_0)$ (Fig. 1, right). For $t \in [t_2^*, t_3]$ the input further increases and $\Omega(t)$ remains horizontal until $v(t)$ reaches a maximum $v(t_3)$ (Fig. 3, left). If $v(t)$ decreases after the moment t_3, then $\Omega(t)$ acquires a vertical link, see Fig. 3, right. This example demonstrates how the staircase polyline $\Omega(t)$ can acquire and loose multiple vertical and horizontal links in response to variations of the input $v(t)$; for more details see [13]. Below $\Omega(t)$ is referred to as the polyline of the state $\eta(t)$ of the Preisach operator or, simply, the state.

In equations (1), (2), the derivative of the output of the Preisach operator is used. For the evaluation of this derivative, the most right link $\Omega_e = \Omega_e(t)$ which is attached to the right end point $\alpha_R = \alpha_S = v(t)$ of the staircase polyline $\Omega(t)$ is important, see Fig. 4, left (if Ω has infinitely many links, then $\Omega_e = \emptyset$). Denote by $(v_m, v), (v, v_M)$ the end points of the segment Ω_e, where $v_m = v$ if Ω_e is a vertical segment and $v_M = v$ if Ω_e is horizontal. If $v = v(t)$ increases, then the time derivative of the output of the Preisach operator satisfies [33]

$$\frac{d(\mathcal{P}[\eta_0]v)}{dt} = \dot{v}H(v, v_m) \quad \text{with} \quad H(v, v_m) = \int_{v_m}^{v} \mu(\alpha_R, v)\, d\alpha_R. \tag{5}$$

If v decreases, then

$$\frac{d(\mathcal{P}[\eta_0]v)}{dt} = \dot{v}V(v, v_M) \quad \text{with} \quad V(v, v_M) = \int_{v}^{v_M} \mu(v, \alpha_S)\, d\alpha_S. \tag{6}$$

2.2. Homoclinic solutions

Equilibria of the system (1), (2) can be obtained by setting all the derivatives including the derivative $\dot{x}$ of the Preisach operator to zero.

Definition 1. *We call a solution* (u^*, v^*) *of the equations* $f(u^*, v^*) = 0$, $g(u^*, v^*) = 0$ *an equilibrium of the system* (1), (2).

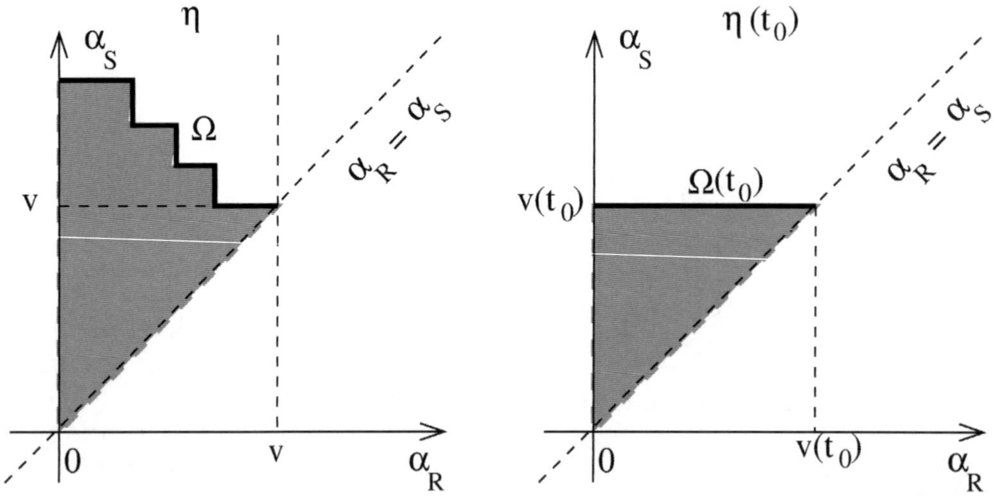

Figure 1. The domain $\{(\alpha_R, \alpha_S)\colon\ \alpha_m \le \alpha_R \le \alpha_S\}$ with $\alpha_m = 0$ is divided into two parts by a staircase polyline $\Omega = \Omega(t)$ with the relays in state 1 below (to the left) of this line (grey colour) and in state 0 above (to the right) of this line (white colour). Here all relays with $\alpha_S \le v$ are in the state 1, all relays with $\alpha_R \ge v$ are in the state 0, and the state of other relays is defined by the polyline Ω. In the right figure $\Omega(t_0)$ is the horizontal segment $\alpha_S = v(t_0)$, $0 \le \alpha_R \le v(t_0)$.

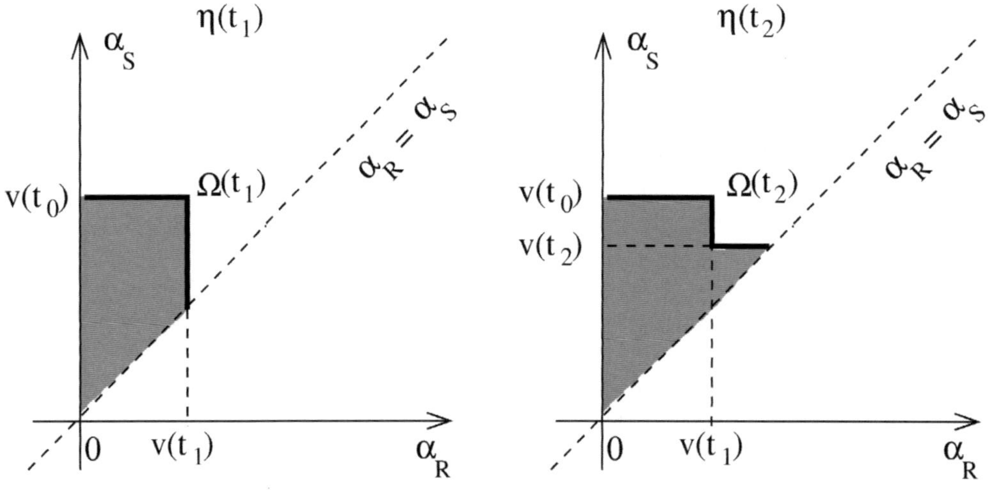

Figure 2. Evolution of the staircase polyline $\Omega = \Omega(t)$ from the initial state $\Omega(t_0)$ shown in Fig. 1 (right) in response to an input $v(t)$. For $t_0 < t \le t_1$ the input $v(t)$ monotonically decreases, and the line Ω consists of two segments (left panel, $t = t_1$): the vertical link $\alpha_R = v(t)$, $v(t) \le \alpha_S \le v(t_0)$ connects to the horizontal link, which is a part of the segment $\Omega(t_0)$ shown in Fig. 1 (right). For $t_1 < t \le t_2$ the input $v(t)$ monotonically increases, where $v(t_1) < v(t_2) < v(t_0)$, and the state Ω has three links as shown on the right panel for the moment $t = t_2$. The leftmost horizontal link and the vertical link of the line $\Omega(t_2)$ are parts of the staircase $\Omega(t_1)$ presented on the left panel, and the rightmost horizontal link is $\alpha_S = v(t_2)$, $v(t_1) \le \alpha_R \le v(t_2)$.

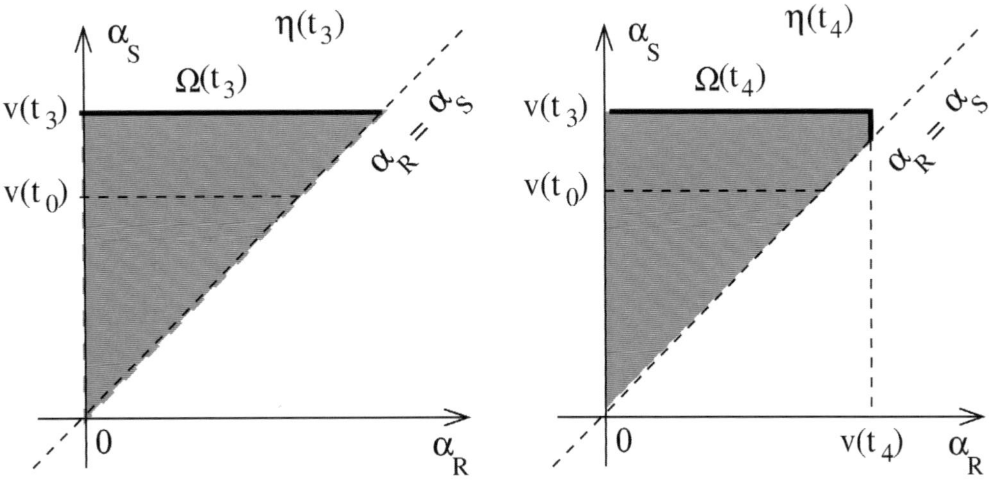

Figure 3. Evolution of $\Omega(t)$ from the state $\Omega(t_2)$ shown in Fig. 2 (right). For $t_2 \leq t < t_3$ the input $v(t)$ monotonically increases and, at the moment $t_2^* \in [t_2, t_3)$, reaches the value $v(t_2^*) = v(t_0)$. The state $\Omega(t_2^*)$ becomes the horizontal segment (which coincides with $\Omega(t_0)$) and remains the horizontal segment $\alpha_S = v(t)$, $0 \leq \alpha_R \leq v(t)$ for all $t_2^* \leq t \leq t_3$ (that is, as long as the input increases), as shown on the left panel. After the moment t_3 the input $v(t)$ decreases again, and the state $\Omega(t_4)$ presented on the right panel for a moment $t_4 > t_3$ is similar to the staircase in Fig. 2 (left).

Every equilibrium defines stationary solutions $\{u(t), v(t), \eta(t)\} \equiv \{u^*, v^*, \eta_0\}$ of system (1), (2).

Assume that system (1), (2) has an equilibrium. Without loss of generality, assume that $u^* = v^* = 0$. Since the output of the Preisach operator (2) is defined on a semiaxis $t \geq t_0$, we use the following definition of a solution to system (1), (2) on the whole axis $t \in \mathbb{R}$.

Definition 2. *A triplet* $\{u(t), v(t), \eta(t)\}$ *is a solution of* (1), (2) *if for each* t_0 *this triplet is a solution of* (1), (2) *with the initial values* $u_0 = u(t_0), v_0 = v(t_0), \eta_0 = \eta(t_0)$ *on the semiaxis* $t \geq t_0$.

Definition 3. *We call a solution* $\{u(t), v(t), \eta(t)\}$ homoclinic *if* $u(t) \to 0, v(t) \to 0$ *for* $t \to \pm\infty$, *and either* $u(t) \neq 0$ *or* $v(t) \neq 0$ *for all* $t \in \mathbb{R}$.

In this work, we focus on the piecewise monotone homoclinic solutions $\{u(t), v(t), \eta(t)\}$ with *monotone tails*.

Assumption 1. *There exist* $t_1, t_2 \in \mathbb{R}$ *such that* $t_1 \leq t_2$, *and the component* v *of the homoclinic solution* $\{u(t), v(t), \eta(t)\}$ *is monotone for* $t \leq t_1$ *and for* $t \geq t_2$.

For such trajectories we can define the limit of the state of the Preisach operator at $t \to \pm\infty$. We will write $\eta(t) \to \eta^*$ as $t \to -\infty$ (or, as $t \to \infty$) if the polyline $\Omega(t)$ of $\eta(t)$ converges to the polyline Ω^* of η^* (as a set on the plane (α_R, α_S)).

2.3. Existence of robust homoclinic orbits

To demonstrate rigorously the possibility of existence of homoclinic orbits, we consider the following example of system (1):

$$\dot{u} = au + bv - \kappa(\dot{x} - |cu + dv|v), \qquad \dot{v} = cu + dv \tag{7}$$

Journal of Physics: Conference Series **727** (2016) 012012 doi:10.1088/1742-6596/727/1/012012

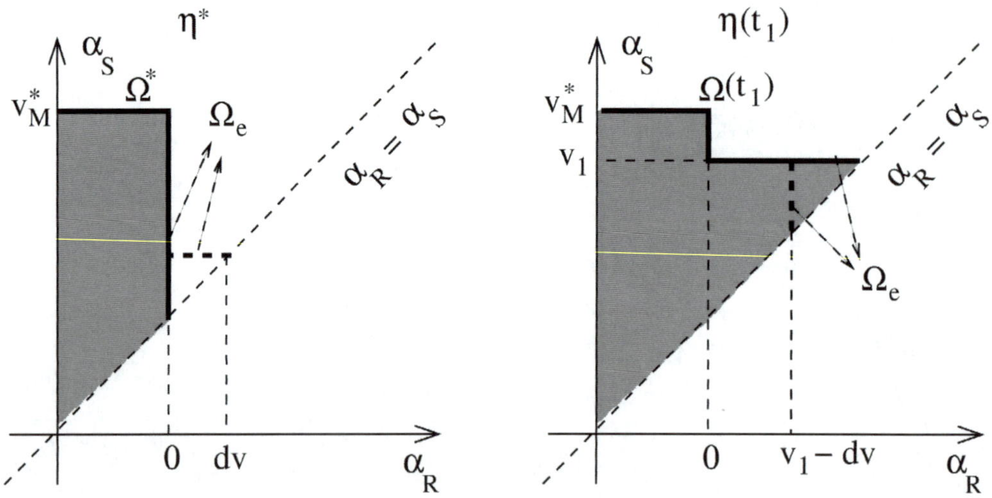

Figure 4. Evolution of the polyline $\Omega(t)$ for an input $v(t)$ that has one maximum point $v(t_1) = v_1$; $\Omega_e(t)$ is the segment of this staircase polyline with the end point $\alpha_R = \alpha_S = v(t)$ on the bisector. The limit state Ω^* (solid line on the left panel) in this example consists of a horizontal segment and the vertical segment Ω_e^*. During the time interval $(-\infty, t_1]$ when the input $v(t)$ increases, $\Omega(t)$ has three segments, the segment $\Omega_e(t)$ is horizontal (thick dashed line). The right panel shows the state $\Omega(t_1)$ (solid line) at the moment when the input achieves its maximum $v(t_1) = v_1$. After this moment, the input decreases, $\Omega(t)$ consists of four segments, the segment $\Omega_e(t)$ is vertical (thick dashed line).

and assume that the density function of the Preisach operator (2) is uniform, $\mu(\alpha_R, \alpha_S) \equiv 1$. This system has an equilibrium at zero, $u = v = 0$.

First, we prove the existence of a simple homoclinic orbit with the limit state η^* shown in Fig. 4 (left) such that the polyline Ω^* has a sufficiently long vertical segment $\Omega_e^* = \{(\alpha_R, \alpha_S) : \alpha_R = 0,\ \alpha_S \in [0, v_M]\}$. We aim to construct a homoclinic trajectory with a component $v(t)$ that increases for $t < t_1$, decreases for $t > t_1$ and has one maximum point $v(t_1)$ (cf. Assumption 1), see Fig. 5.

Proposition 1. *Assume that*

$$\Delta = ad - bc > 0,\ \tau = a + d > 0,\ D = \tau^2 - 4\Delta > 0, \tag{8}$$

$\kappa c > 0$, *and the segment* Ω_e^* *of the polyline* Ω^* *of the state* η^* *is vertical with the end points* $(\alpha_R, \alpha_S) = (0, 0)$ *and* $(\alpha_R, \alpha_S) = (0, v_M^*)$, *where* $v_M^* > \frac{a+d+2\sqrt{\Delta}}{c\kappa} =: v_{1m}$. *Then, for each* $v_1 \in (v_{1m}, v_M^*]$ *there is a homoclinic solution* $\{u(t), v(t), \eta(t)\}$ *such that* $v(t_1) = v_1$ *is a single extremum point of the component* $v(t)$, *which is a point of maximum. That is, system* (2), (7) *has infinitely many robust homoclinic trajectories satisfying* $\eta(t) \to \eta^*$ *as* $t \to \pm\infty$.

Robustness means that the continual set of homoclinic trajectories persists under perturbations of system parameters.

Proof. A homoclinic solution of system (2), (7) described in this proposition satisfies relation (5) with $v_m = 0$ for $t < t_1$ and relationship (6) with $v_M = v_1$ for $t > t_1$. Hence, for such a solution, $\dot{x} = v\dot{v}$ for $t < t_1$ and $\dot{x} = (v_1 - v)\dot{v}$ for $t > t_1$ (where we use the relationship $\mu(\alpha_R, \alpha_S) \equiv 1$ when calculating $H(v, 0)$ and $V(v, v_1)$). Substituting these expressions for $\dot{x}$ in (7), we see that the homoclinic solution satisfies the linear differential system

$$\dot{u} = au + bv, \qquad \dot{v} = cu + dv \tag{9}$$

on the semiaxis $t < t_1$; and, the linear differential system

$$\dot{u} = (a - \kappa v_1 c)u + (b - \kappa v_1 d)v, \qquad \dot{v} = cu + dv \tag{10}$$

on the semiaxis $t > t_1$; the point $(u(t_1), v(t_1)) = (u_1, v_1)$ lies on the nullcline $\dot{v} = cu + dv = 0$ of these systems, as v_1 is the maximum value of the v-component. Furthermore, to prove the proposition it suffices to show that for every $v_1 \in (v_{1m}, v_M^*] \subset (0, v_M^*]$:

(a) The solution of system (9) with the initial conditions $u(t_1) = -dv_1/c$, $v(t_1) = v_1$ satisfies $\dot{v} > 0$ for all $t < t_1$ and $v \to 0$ as $t \to -\infty$;

(b) The solution of system (10) with the same initial conditions $u(t_1) = -dv_1/c$, $v(t_1) = v_1$ satisfies $\dot{v} < 0$ for all $t > t_1$ and $v \to 0$ as $t \to \infty$.

Here (a) follows from conditions (8), which ensure that the zero equilibrium of system (9) is a proper unstable node; (b) follows from the relationship $v_1 > v_{1m}$, which implies that the zero equilibrium of system (9) is a proper asymptotically stable node. □

The case $\kappa c < 0$ can be considered similarly. In this case, the following analog of Proposition 1 is valid.

Proposition 2. *Assume that conditions (8) and $\kappa c < 0$ hold and the segment Ω_e^* of the polyline Ω^* of the state η^* is horizontal with the end points $(\alpha_R, \alpha_S) = (0, 0)$ and $(\alpha_R, \alpha_S) = (v_m^*, 0)$ with $v_m^* < -v_{1m}$, where v_{1m} is defined in Proposition 1. Then, for each $v_1 \in [v_m^*, -v_{1m})$ there is a homoclinic trajectory with $\min_{t\in\mathbb{R}} v(t) = v_1$. That is, system (2), (7) has infinitely many robust homoclinic trajectories satisfying $\eta(t) \to \eta^*$ as $t \to \pm\infty$ such that the v-component of the trajectory has a single extremum point, which is a point of minimum.*

As the last example of this section, we consider homoclinic trajectories which have more complex limit states η^* at $t \to -\infty$, see in Fig. 6, and more than two monotonicity intervals of the v-component, see Fig. 7.

Proposition 3. *Assume that conditions (8) hold, $\kappa c > 0$ and the polyline Ω^* of state η^* includes three links connecting the corner points $(0,0)$, $(0, v_M^*)$, (v_m^*, v_M^*), (v_m^*, v_{M2}^*), where $0 < v_M^* < \frac{a+d-2\sqrt{\Delta}}{c\kappa} =: v_{1M}$, $v_{M2}^* > \frac{a+d+2\sqrt{\Delta}}{c\kappa} =: v_{1m}$, and $v_m^* < -v_{1m}$. Then system (2), (7) has a robust homoclinic trajectory with the limit state $\eta(t) \to \eta^*$ as $t \to -\infty$. The v-component of this trajectory has a single maximum point and a single minimum point with $\max_{t\in\mathbb{R}} v(t) = v_M^*$ and $\min_{t\in\mathbb{R}} v(t) < v_m^*$.*

We note that the homoclinic solution described in this proposition has different limit states at $t \to \pm\infty$, that is $\lim_{t\to\infty} \eta(t) \neq \lim_{t\to-\infty} \eta(t) = \eta^*$. As a matter of fact, there are again infinitely many homoclinic trajectories with the maximum value $v_1 = \max_{t\in\mathbb{R}} v(t)$ of the v-component ranging over the interval $v_1 \in (0, v_M^*]$. We consider one of these trajectories as an example.

Proof. The proof is similar to that of Proposition 1. Here, it suffices to show that:

(a) The solution of system (9) with the initial point $u(t_1) = -dv_M^*/c$, $v(t_1) = v_M^*$ on the nullcline $\dot{v} = cu + dv = 0$ satisfies $\dot{v} > 0$ for all $t < t_1$ and $v \to 0$ as $t \to -\infty$;

(b) The solution of system (10) with $v_1 = v_M^*$ starting from the initial point $u(t_1) = -dv_M^*/c$, $v(t_1) = v_M^*$ satisfies $\dot{v} < 0$ on an interval $t \in (t_1, t_2]$ where the end point $(u(t_2), v(t_2)) =: (u_m, v_m)$ satisfies $v(t_2) = v_m^*$.

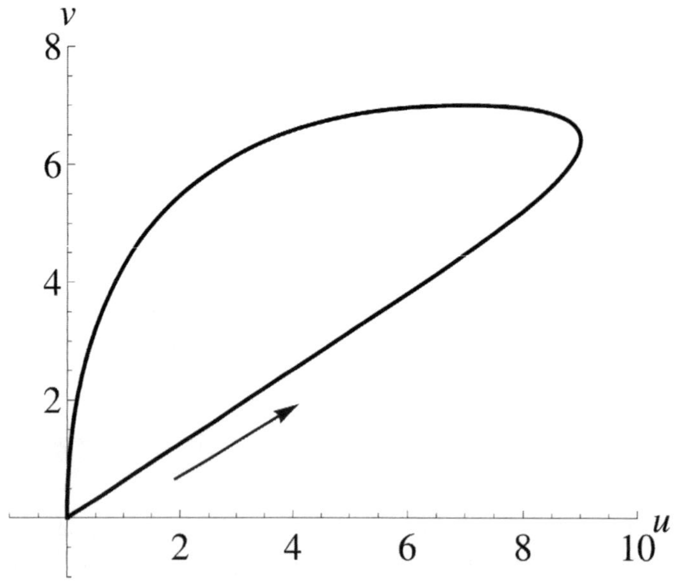

Figure 5. A homoclinic orbit of system (2), (7) satisfying the assumptions of Proposition 1 for $\kappa = 1$, $a = 5$, $b = -7$, $c = 1$, $d = -1$, $v_M^* \geq 7$. The orbit is located in the positive quadrant; the arrow shows the direction of motion; zero is the equilibrium. The maximum of the v-component is achieved at the point $u = v = 7$ of the nullcline $\dot{v} = u - v = 0$. Fig. 4 illustrates the evolution of the state $\eta(t)$ of the Preisach operator for this orbit. The solid line $\Omega = \Omega^*$ on the left panel of Fig. 4 is the polyline of the limit state η^* defined by $\eta(t) \to \eta^*$ as $t \to \pm\infty$.

(c) The solution of system (10) with $v_1 = v_{M2}^*$ starting from the initial point $(u(t_2), v(t_2)) =:$ (u_m, v_m^*) satisfies $\dot{v} < 0$ on an interval $t \in [t_2, t_3)$ where the end point $(u(t_3), v(t_3)) =:$ $(-dv_{min}/c, v_{min})$ lies on the nullcline $\dot{v} = cu + dv = 0$.

(d) The solution of the equation

$$\dot{u} = (a + \kappa v_3 c)u + (b + \kappa v_3 d)v, \qquad \dot{v} = cu + dv \tag{11}$$

with $v_3 = v_{min}$ starting from the initial point $(u(t_3), v(t_3)) = (-dv_{min}/c, v_{min})$ on the nullcline $\dot{v} = cu + dv = 0$ satisfies $\dot{v} > 0$ for all $t > t_3$ and $v \to 0$ as $t \to \infty$.

Statement (a) follows from assumptions (8), which ensure that the zero equilibrium is a proper unstable node of system (9).

The vector fields $(\dot{u}, \dot{v})$ of systems (9) and (10) coincide on the nullcline $\dot{v} = 0$. As $\dot{v} > 0$ for $t < t_1$ for the solution of (9) with the initial condition $(u(t_1), v(t_1)) = (-dv_M^*/c, v_M^*)$ on the nullcline, it follows that the solution of (10) starting from the same initial condition satisfies $\dot{v} < 0$ for sufficiently small $t - t_1 > 0$. This solution of (10) remains in the half plane $\dot{v} = cu + dv < 0$ for all $t > t_1$ and satisfies $v \to -\infty$ as $t \to \infty$, because the assumption $v_M^* < v_{1M}$ ensures that the zero equilibrium is a proper unstable node of system (10) with $v_1 = v_M^*$. As $v_M^* > 0 > v_m^*$, this solution will eventually hit the line $v = v_m^*$ at some moment t_2 at a point (u_m, v_m^*). This proves (b).

The assumption $v_{M2}^* > v_{1m}$ implies that the zero equilibrium of system (10) with $v_1 = v_{M2}^*$ is a proper asymptotically stable node. As $v_m^* < 0$ and the point (u_m, v_m^*) lies in the half plane $\dot{v} = cu + dv < 0$, it follows that the solution of this system starting at (u_m, v_m^*) will hit the nullcline $\dot{v} = cu + dv = 0$ at some point $(-dv_{min}/c, v_{min})$, that is (c) holds.

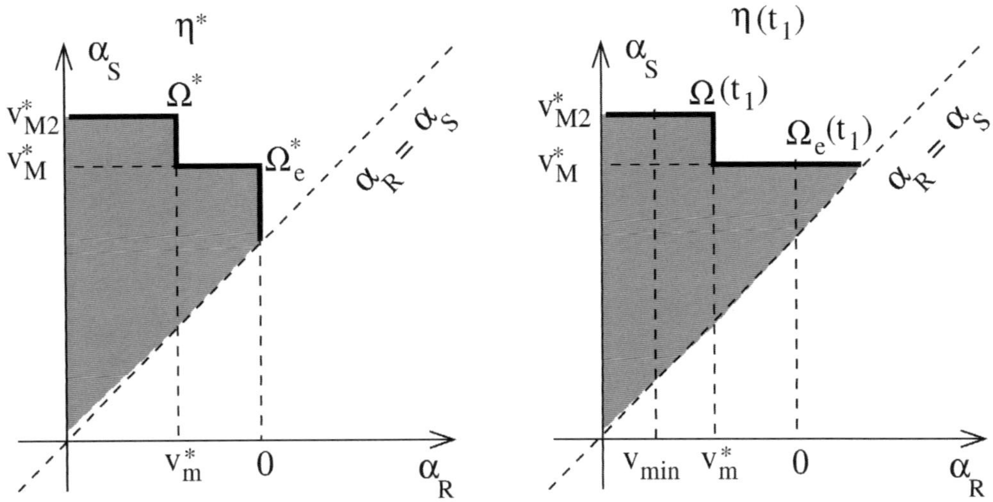

Figure 6. Evolution of the staircase polyline $\Omega(t)$ for the homoclinic trajectory described in Proposition 3. The segment $\Omega_e(t)$ of this staircase line is the link with the end point $\alpha_R = \alpha_S = v(t)$ on the bisector. The left panel shows the limit state of the homoclinic orbit at $t \to -\infty$. The right panel shows the state $\Omega(t_1)$ (solid line) at the moment when the v-component achieves its maximum value $v(t_1) = v^*_M$. After this moment, $v(t)$ decreases until it reaches its minimum value $v(t_3) = v_{min}$, then it increases and tends to zero. The polyline $\Omega(t)$ has only two links between the moment $t_2 \in (t_1, t_3)$ when the input $v(t)$ reaches the value $v(t_2) = v^*_m$ and the moment when it reaches its minimum $v(t_3) = v_{min}$.

Finally, (d) follows from the relationships $v_{min} < v^*_m < -v_{1m} < 0$, which ensure that the zero equilibrium is a proper asymptotically stable node of system (11) with $v_3 = v_{min}$. □

3. Numerical examples

3.1. Population dynamics model

In this section, we give a numerical evidence of the existence of homoclinic trajectories in a predator-prey type model [37], which has nontrivial dynamical properties such as multiple stable equilibria:

$$\dot{u} = a(u) - f(u)g(v) - h(t)u, \tag{12}$$
$$\dot{v} = \sigma f(u)g(v) - c(v), \tag{13}$$

where u is the number of prey; v is the number of predator; the term

$$a(u) = \rho u - \lambda u^2$$

describes the logistic growth of the prey with the birth rate ρ and the competition rate λ;

$$f(u) = \frac{\omega u}{\phi + u}$$

is the Holling type II functional response;

$$g(v) = \frac{v}{1 + \beta v}$$

Journal of Physics: Conference Series **727** (2016) 012012 doi:10.1088/1742-6596/727/1/012012

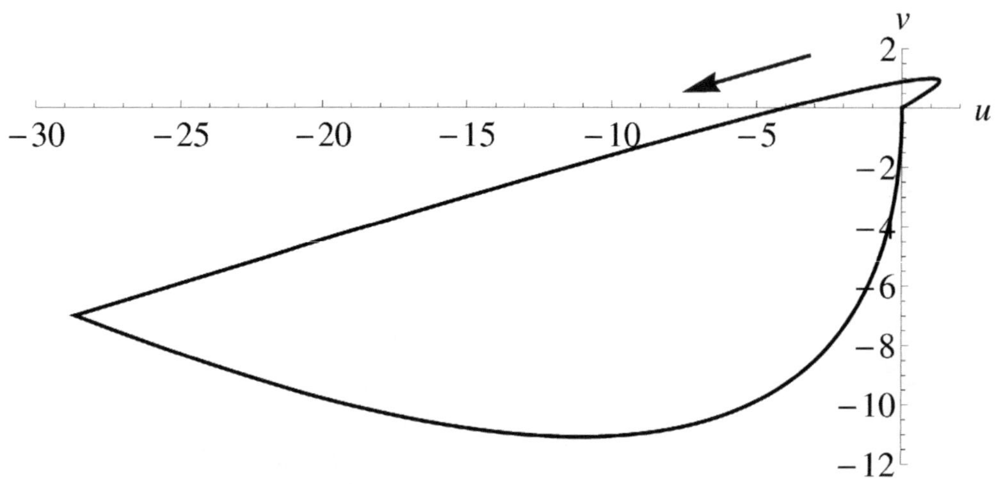

Figure 7. Homoclinic orbit of system (2), (7) satisfying the conditions of Proposition 3. The orbit converges to the zero equilibrium; the arrow shows the direction of motion. The parameters are $\kappa = 1$, $a = 5$, $b = -7$, $c = 1$, $d = -1$, $v_M^* = 1$, $v_m^* = -7$ and $v_{M2}^* = 7$. The v-component has its maximum at the point $(u(t_1), v(t_1)) = (1, 1)$ of the nullcline $\dot{v} = u - v = 0$. The trajectory makes a corner at the point $(u(t_2), v(t_2)) = (-7, -23 - 4\sqrt{2})$ of the line $v = v_m^*$ at the moment $t_2 = t_1 + \log(1 + 2\sqrt{2})$. The v-component achieves its minimum $v_{min} = -(697 + 64\sqrt{2})/71$ at the point $(u(t_3), v(t_3)) = (v_{min}, v_{min})$ of the nullcline $u - v = 0$ at the moment $t_3 = t_2 + 2\,\text{artanh}(10 + \sqrt{2})/49$.

is the predator interference; σ is the efficiency of conversion of food to growth; and, the term

$$c(v) = \gamma v$$

describes death of the predator with the death rate γ (all the parameters are positive). The term $h(t)$ describes the flow of the prey to some refuge patch from which it never returns. System (12)-(13) is analagous to the two-patch predator-prey system proposed in [37], where the safe patch is completely separated from the environment; alternatively, the term $h(t)$ can be considered as an additional predator-induced death rate.

We consider the flow rate in response to the change of predator abundance in the following form:

$$h(t) = \left(k_0 + k\frac{d}{dt}(\mathcal{P}[\eta_0]v)(t) - 2k^\dagger|\dot{v}|(v - v^\dagger)\right)^+, \tag{14}$$

where we ensure that the flow satisfies $h(t) \geq 0$ using the function $x^+ = \max\{x, 0\}$, and $\dot{v}$ can be replaced with the right hand side of equation (13). The main ingredients here are the constant flow rate k_0 and the hysteretic reaction of the prey to variations of the predator abundance, $k\frac{d}{dt}(\mathcal{P}[\eta_0]v)(t)$, where hysteresis appears due to a delay of the response of the prey to a change of the trend in predator dynamics (i.e., the change of the sign of $\dot{v}$). A specific choice of the model for this hysteretic response in the form of the Preisach operator is related to a number of phenomenological assumptions. Namely, the environment is divided into many small interconnected patches, and it assumed that the change of the rate of flow of the prey to the refuge from any given patch is delayed until the abundance of the predator drops/increases from its extremum value by a certain positive amount, which is specific to the patch. More detailed derivation of the model is presented in [37]. We assume that the measure $\mu(\alpha_R, \alpha_S)$ of the Preisach operator $\mathcal{P}[\eta_0]$ is zero outside the triangle $0 \le \alpha_R \le \alpha_S \le 1$, and $\mu(\alpha_R, \alpha_S) = 2$ inside the triangle.

The additional term $2k^\dagger|\dot{v}|(v - v^\dagger)$ in (14) strengthens the reaction of the prey to the change of the predator abundance if v is below the threshold $v^\dagger$, and weakens this reaction if v is above the threshold. In other words, the prey is attentive to the change of the predator numbers if the predator is not abundant and its attention to the predator becomes saturated for $v \gg v^\dagger$. This artificial term is similar to the term we introduced in (7) and, indeed, allows us to demonstrate numerically dynamics of (12)-(14), which are similar to dynamics of equations (2), (7). We will consider the system with ($k^\dagger > 0$) and without ($k^\dagger = 0$) this term.

In the case of increasing v, substituting formula (5) in equations (14), we obtain the following expression for the rate of flow to the refuge:

$$h(t) = \Big(k_0 + \dot{v}(kH(v, v_m) - 2k^\dagger(v - v^\dagger))\Big)^+, \tag{15}$$

where $\dot{v}$ can be replaced with the right hand side of equation (13). Similarly, when v decreases,

$$h(t) = \Big(k_0 + \dot{v}(kV(v, v_M) + 2k^\dagger(v - v^\dagger))\Big)^+. \tag{16}$$

Equilibria of system (12)-(14) can be found from the algebraic system, which is obtained by setting the derivatives of all the variables, including $d/dt(\mathcal{P}[\eta_0]v)$, to zero in (12)-(14).

Numerical examples of homoclinic orbits of system (12)-(14) presented in the next section were obtained by an algorithm, which is similar to the method used in the previous section.

3.2. Numerical results

We set $\rho = 1.35$, $\phi = 0.1$, $\beta = 1.2$, $\gamma = 0.5$, $\omega = 2$, $k_0 = 0.01$, $\lambda = 0.01$ to ensure that system (12)-(14) has three positive equilibrium points

$$(u^*, v^*) = (0.183649, 0.245752), \tag{17}$$
$$(u^\dagger, v^\dagger) = (0.340215, 0.45473), \tag{18}$$
$$(u^\ddagger, v^\ddagger) = (133.376, 0.832085). \tag{19}$$

If $k = k^\dagger = 0$ (the rate to the refuge (14) is constant, there is no hysteresis), then these equilibrium points of the ordinary differential system (12)-(14) have the eigenvalues $(0.136, 0.614)$, $(-0.089, 0.942)$, and $(-1.33, -0.25)$, respectively. That is, the first equilibrium is an unstable node, the second equilibrium is a saddle and the third equilibrium is a stable node.

When the hysteresis term is present ($k > 0$), we give a numerical evidence that equilibrium (17) can have a homoclinic orbit attached to it. We consider a constant density function of the Preisach operator (3),

$$\mu(\alpha_R, \alpha_S) = 2,$$

in the triangle $0 \le \alpha_R \le \alpha_S \le 1$ and set $\mu = 0$ outside this triangle. The integral of μ over the whole half plane $\alpha_S \ge \alpha_R$ is normalized to 1.

As the limit state of the Preisach operator, $\lim_{t \to -\infty} \eta(t) = \eta^*$, we choose the polyline $\Omega(t_0)$ which has two links: a vertical link $\Omega_e(t_0) = \{(\alpha_R, \alpha_S) : \alpha_R = v^*,\ v^* \le \alpha_S \le v_M^*\}$ and a horizontal link $\{(\alpha_R, \alpha_S) : 0 \le \alpha_R \le v^*,\ \alpha_S = v_M^*\}$, where v^* is the second component of equilibrium (17) (see Fig. 4, left, where the origin is shifted to the point $\alpha_R = \alpha_S = v^*$).

First, we consider system (12)-(14) with $k = k^\dagger = 8000$ and $v^\dagger = 0.246 > v^*$. The parameter v_M^* of the limit state for this example is set to $v_M^* = 1$. We solve the ordinary differential system (12), (13), (15) with $v_m = v^*$ numerically backward in time using as the

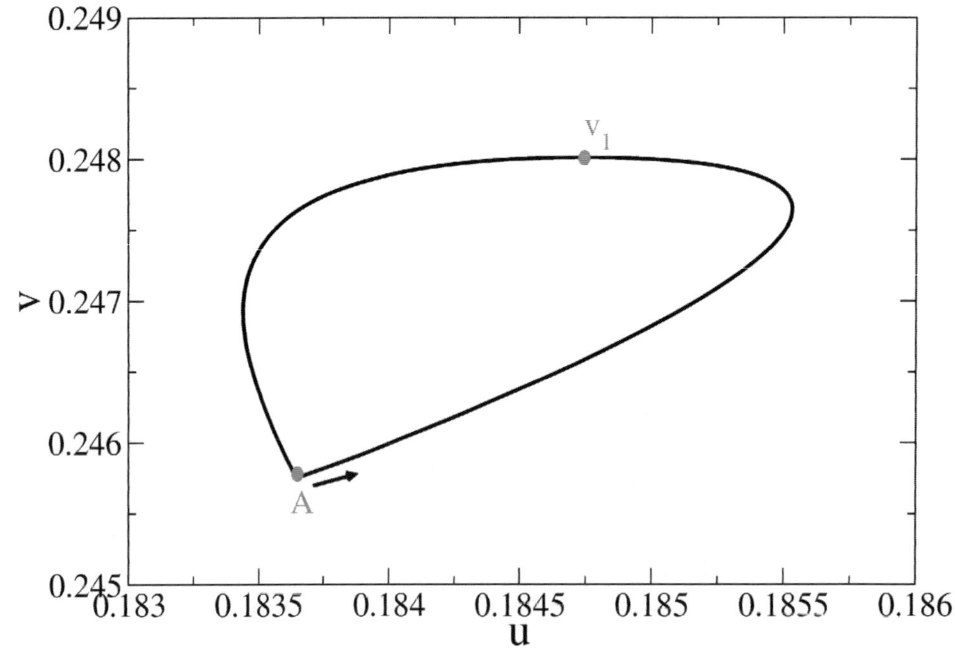

Figure 8. Homoclinic loop of system (12)-(14) with $k = k^\dagger = 8000$, $v^\dagger = 0.246$ and equilibrium (17) at the point A.

initial value a point $(u_1, v_1) = (v(t_0), v(t_0)) \approx (0.248, 0.185)$ on the nullcline $\dot{v} = 0$. The eigenvalues of the equilibrium (17) for this system are $(0.136, 0.614)$. We check that $\dot{v} > 0$ for $t < t_0$, and the solution $(u(t), v(t))$ approaches (u^*, v^*) as t decreases (see Fig. 8): $(v(t_0 - 30), u(t_0 - 30)) \approx (v^* + 3.55 \times 10^{-6}, u^* + 5.61 \times 10^{-5})$. Then we solve system (12), (13), (16) with $v_M = v_1$ numerically forward in time for $t \geq t_0$ starting from the same initial point (u_1, v_1). We observe that the solution $(u(t), v(t))$ approaches the equilibrium (u^*, v^*) and $\dot{v} < 0$ for $t > t_0$. The eigenvalues of the equilibrium (17) for this ordinary differential system are $(-5.73072, -0.0146083)$.

Next, we construct a homoclinic trajectory similar to the one shown on Fig. 7 for system (12)-(14) with $k^\dagger = 0$, $k = 700$. The parameter of the limit state is set to $v_M^* = v_1 = 0.2596$. As in the previous example, we solve system (12)-(13), (15) with $v_m = v^*$ numerically backward in time starting from the point (u_1, v_1) on the nullcline $\dot{v} = 0$ with $v(t_0) = v_1$, see Fig. 9. This solution satisfies $\dot{v} > 0$ for $t < t_0$. Then, we solve system (12)-(13), (16) with $v_M = v_1$ for $t > t_0$ starting from the same point (u_1, v_1). We observe that $v(t)$ decreases, while the trajectory spirals around the equilibrium (17) for $t_0 < t < t_{2a}$, where $t_{2a} \approx t_0 + 1.612$, because (17) is a stable focus for this ordinary differential system. However, for $t = t_{2a}$, $v(t_{2a}) \approx 0.25741$ the flow $h(t)$ in (12) becomes 0; at this point, the term $k\dot{v}V(v, v_1)$ is negative, and $|\dot{v}|$ becomes large enough to compensate for the term $k_0 > 0$ in relation (16). We observe that $k\dot{v}V(v, v_1) + k_0 < 0$ and $h(t) \equiv 0$ for $t_{2a} < t < t_{2b}$, where $t_{2b} \approx t_0 + 11.043$, $v(t_{2b}) \approx 0.0384$, and $k\dot{v}V(v, v_1) + k_0 = 0$ at $t = t_{2b}$. After this moment, $h(t)$ becomes positive again. Shortly after this point, the v-component of the solution reaches its (global) minimum $v(t_3) = v_3 \approx 0.0384$ at $t_3 \approx 11.046$ as the solution arrives at the nullcline $\dot{v} = 0$. From the nullcline, we continue the trajectory for $t > t_3$ as the solution of ordinary differential system (12)-(13), (15) with $v_m = v_3$. For this system, the equilibrium (17) has the eigenvalues $(-11.8176, -0.00708397)$ and the trajectory is attracted to the equilibrium. The component $v(t)$ increases and the term $h(t)$ remains positive

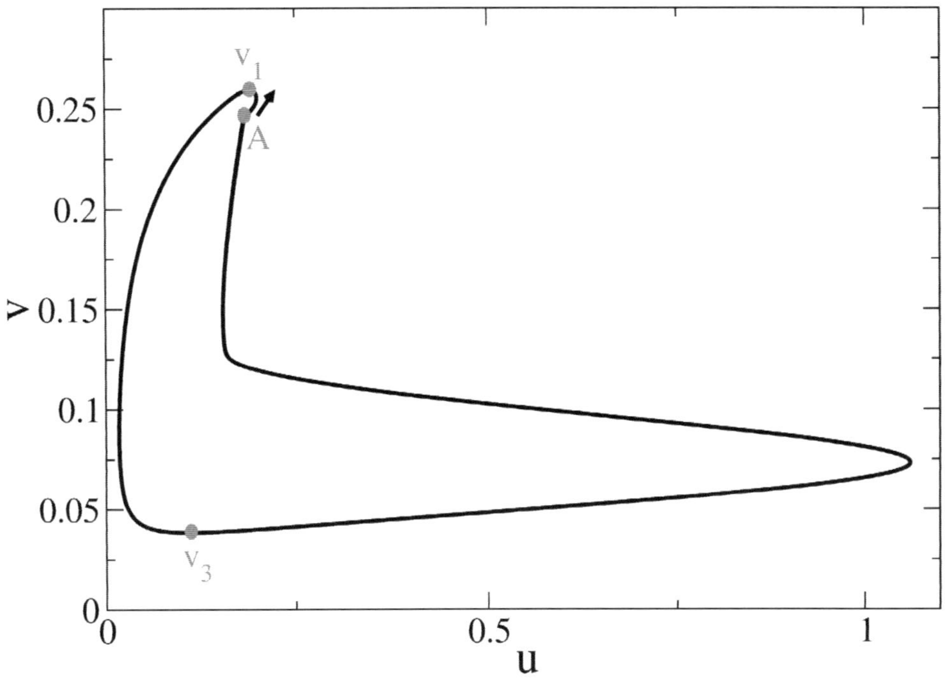

Figure 9. Homoclinic loop of system (12)-(14) with $k^{\dagger} = 0, k = 700$; the equilibrium (17) is denoted by A.

for $t > t_3$, see Fig. 9.

4. Conclusion

We have demonstrated the existence of robust families of homoclinic trajectories for operator-differential equations (1), (2) with the Preisach hysteresis operator. These systems can be considered as switching systems where switching from one planar ordinary differential equation to another occurs when a variable either passes a turning point or achieves a value stored in the memory state of the hysteresis operator. Homoclinic trajectories are possible when the basin of attraction of an equilibrium of one planar system in forward time overlaps with the basin of attraction of the equilibrium of another planar system in reversed time.

Acknowledgements

A. P. acknowledges the support of SFB 787 of the DFG. D. R. acknowledges the support of NSF through grant DMS-1413223.

References

[1] Krasnosel'skii M A and Pokrovskii A V 1989 *Systems with Hysteresis* (Springer, New York) zbl 0665.47038, MR0987431

[2] Visintin A 1994 *Differential Models of Hysteresis* (Springer, Berlin) mR1329094

[3] Brokate M and Sprekels J 1996 *Hysteresis and Phase Transitions* (Berlin: Springer) zbl 0951.74002, MR1411908

[4] Krejči P and Recupero V 2014 Bv solutions of rate independent differential inclusions *Mathematica Bohemica* **139** 607–619

[5] Krejčí P 1996 *Hysteresis, Convexity and Dissipation in Hyperbolic Equations* (Gakkotosho, Tokyo) zbl 1187.35003, MR2466538

[6] Mielke A, Rossi R and Savaré G 2012 Variational convergence of gradient flows and rate-independent evolutions in metric spaces *Milan J. Math.* **80** 381–410
[7] Brokate M, Pokrovskii A and Rachinskii D 2006 Asymptotic stability of continuum sets of periodic solutions to systems with hysteresis *Journal of Differential Equations* **319** 94–109 zbl 1111.34035, MR2217849, DOI 10.1016/j.jmaa.2006.02.060
[8] Rachinskii D 1999 Asymptotic stability of large-amplitude oscillations in systems with hysteresis *Nonlinear Differential Equations and Applications NoDEA* **6** 267–288
[9] Krasnosel'skii A and Rachinskii D 2002 On a bifurcation governed by hysteresis nonlinearity *Nonlinear Differential Equations and Applications (NoDEA)* **9** 93–115 zbl 1013.34036, MR1891697, DOI 10.1007/s00030-002-8120-2
[10] Krasnosel'skii A and Rachinskii D I 2001 On the continua of cycles in systems with hysteresis *DOKLADY MATHEMATICS C/C OF DOKLADY-AKADEMIIA NAUK* **63** 339–344 zbl 1052.34052, MR1863745
[11] Diamond P, Rachinskii D and Yumagulov M 2000 Stability of large cycles in a nonsmooth problem with Hopf bifurcation at infinity *Nonlinear Analysis: Theory, Methods & Applications* **42** 1017–1031 zbl 0963.34034, MR1780452, DOI 10.1016/S0362-546X(99)00162-5
[12] Krasnosel'skii A and Rachinskii D 2005 Bifurcation of forced periodic oscillations for equations with preisach hysteresis *Journal of Physics: Conference Series* **22** 93
[13] Brokate M, Pokrovskii A, Rachinskii D and Rasskazov O 2005 *The Science of Hysteresis* vol I ed Mayergoyz I and Bertotti G (Elsevier, Academic Press) chap II, pp 125–291 zbl 1142.34026, MR2307929
[14] Eleuteri M, Kopfova J and KrejčıP 2015 A new phase field model for material fatigue in an oscillating elastoplastic beam *Discret Cont. Dyn. S. A* **35** 2465–2495
[15] Ruderman M, Hoffmann F and Bertram T 2009 Modeling and identification of elastic robot joints with hysteresis and backlash *IEEE Tran. Ind. Electron* **56** 3840–3847
[16] Mayergoyz I 2003 *Mathematical Models of Hysteresis and Their Applications* (Elsevier) zbl 0723.73003, MR1083150
[17] Appelbe B, Rachinskii D and Zhezherun A 2008 Hopf bifurcation in a van der Pol type oscillator with magnetic hysteresis *Physica B: Condensed Matter* **403** 301–304 dOI 10.1016/j.physb.2007.08.034
[18] Kuhnen K 2003 Modeling, identification and compensation of complex hysteretic nonlinearities - a modified Prandtl-Ishlinskii approach *European Journal of Control* **9** 407–418 dOI 10.3166/ejc.9.407-418
[19] Davino D, Krejčí P and Visone C 2013 Fully coupled modeling of magneto-mechanical hysteresis through 'thermodynamic' compatibility *Smart Materials and Structures* **22** 095009 dOI 10.1088/0964-1726/22/9/095009 URL `http://stacks.iop.org/0964-1726/22/i=9/a=095009`
[20] Aström K J 1995 *Adaptive Control, Filtering and Signal Processing* ed Aström K J, Goodwin G C and Kumar P R (Springer) pp 381–410
[21] Dimial M and Andrei P 2014 *'Noise-Driven Phenomena in Hysteretic Systems* (Springer) zbl 0723.73003, MR1083150
[22] Gurevich P L, Jaeger W and Skubachevskii A L 2009 On periodicity of solutions for thermocontrol problems with hysteresis-type switches *SIAM J. Math. Anal.* **41** 733–752
[23] KrejčıP, Rocca E and Sprekels J 2007 Non-local temperature-dependent phase-field models for non-isothermal phase transitions *Journal London Mathematical Society* **76** 197–210
[24] Appelbe B, Flynn D, McNamara H, O'Kane P, Pimenov A, Pokrovskii A, Rachinskii D and Zhezherun A 2009 Rate-independent hysteresis in terrestrial hydrology *Control Systems Magazine, IEEE* **29** 44–69 dOI 10.1109/MCS.2008.930923
[25] Krejčí P, O'Kane J P, Pokrovskii A and Rachinskii D 2011 Stability results for a soil model with singular hysteretic hydrology *Journal of Physics: Conference Series* **268** 012016 dOI 10.1088/1742-6596/268/1/012016
[26] Pimenov A, Kelly T C, Korobeinikov A, O'Callaghan M J, Pokrovskii A V and Rachinskii D 2012 Memory effects in population dynamics: Spread of infectious disease as a case study *Mathematical Modelling of Natural Phenomena* **7** 204–226 zbl 06074655, MR2928740, DOI 10.1051/mmnp/20127313
[27] Chiu C, Hoppensteadt F C and Jäger W 1994 Analysis and computer simulation of accretion patterns in bacterial cultures *J. Math. Biol.* **32** 841–855
[28] Friedman G, McCarthy S and Rachinskii D 2014 Hysteresis can grant fitness in stochastically varying environment *PLOS ONE* **9** e103241
[29] Cross R, Grinfeld M, Lamba H and Seaman T 2005 A threshold model of investor psychology *Physica A: Statistical Mechanics and its Applications* **354** 463–478
[30] Cross R, McNamara H, Pokrovskii A and Rachinskii D 2008 A new paradigm for modelling hysteresis in macroeconomic flows *Physica B: Condensed Matter* **403** 231–236 dOI 10.1016/j.physb.2007.08.017
[31] KrejčıP, Lamba H, Melnik S and Rachinskii D 2014 Analytical solution for a class of network dynamics with mechanical and financial applications *Phys. Rev. E* **90** 032822

[32] Mayergoyz I and Bertotti G (eds) 2005 *The Science of Hysteresis, 3-volume set* (Elsevier, Academic Press) zbl 1142.34028, MR2307931

[33] Pimenov A and Rachinskii D 2009 Linear stability analysis of systems with Preisach memory *Discrete and Continuous Dynamical Systems - Series B* **11** 997–1018 zbl 1181.47075, MR2505656, DOI 10.3934/dcdsb.2009.11.997

[34] Pokrovskii A, Power T, Rachinskii D and Zhezherun A 2006 Differentiability of evolution operators for dynamical systems with hysteresis *Journal of Physics: Conference Series* **55** 171 dOI 10.1088/1742-6596/55/1/017

[35] Krejčí P, O'Kane J P, Pokrovskii A and Rachinskii D 2012 Properties of solutions to a class of differential models incorporating Preisach hysteresis operator *Physica D: Nonlinear Phenomena* **241** 2010–2028 mR2994340, DOI 10.1016/j.physd.2011.05.005

[36] McCarthy S and Rachinskii D 2014 Dynamics of systems with Preisach memory near equilibria *Mathematica Bohemica* **139** 39–73

[37] Pimenov A and Rachinskii D 2014 Homoclinic orbits in a two-patch predator-prey model with preisach hysteresis operator *Mathematica Bohemica* **139** 285–298

[38] Pimenov A, Kelly T C, Korobeinikov A, O'Callaghan M J A and Rachinskii D 2015 Adaptive behaviour and multiple equilibrium states in a predator-prey model *Theoretical Population Biology* **101** 24–30

Error estimates for nonlinear reaction-diffusion systems involving different diffusion length scales

Sina Reichelt
Weierstrass Institute, Mohrenstraße 39, 10117 Berlin, Germany
E-mail: sina.reichelt@wias-berlin.de

Abstract. We derive quantitative error estimates for coupled reaction-diffusion systems, whose coefficient functions are quasi-periodically oscillating modeling the microstructure of the underlying macroscopic domain. The coupling arises via nonlinear reaction terms and we allow for different diffusion length scales, i.e. whereas some species have characteristic diffusion length of order 1 other species may diffuse with the order of the characteristic microstructure-length scale. We consider an effective system, which is rigorously obtained via two-scale convergence, and we derive quantitative error estimates.

1. Introduction

Many mathematical models arising from biological, physical or engineering problems involve effects on microscopic scales, e.g. spatial inhomogeneities of the underlying material. In view of numerical simulations as well as more profound structural insight, we are interested in finding effective, or homogenized, models. From the analytical perspective, we ask for a rigorous justification of the effective model and, if available, error estimates describing the difference to the original macroscopic model.

We refer to the books [1, 2, 3, 4] for a general survey of homogenization theory. An important step in the theory of periodic homogenization was the introduction of two-scale convergence in [5, 6], which allows to rigorously treat systems involving different diffusion length scales, see e.g. [7, 8]. So far, the notion of two-scale convergence is a weak convergence. The periodic unfolding technique, introduced in [9], allows for a natural definition of strong two-scale convergence and, hence, the treatment of nonlinear problems, cf. [10, 11, 12, 13, 14, 15, 16]. Based on this strong notion of convergence, one can ask for quantitative error estimates, see e.g. [17, 18, 19, 20, 21], as well as for numerical simulations, see e.g. [22, 23, 24, 25, 26, 27]. For applications of periodic homogenization in physics and engineering, we refer to e.g. [28, 29, 30, 31] for systems of reaction-diffusion type in heterogeneous media as well as to e.g. [32, 33] for two-scale models on the evolution of damage.

The objective of this contribution are coupled reaction-diffusion systems of the following type

$$\begin{aligned} u_t^\varepsilon &= \operatorname{div}(\mathbb{D}_1(x, \tfrac{x}{\varepsilon})\nabla u^\varepsilon) &+ F_1(x, \tfrac{x}{\varepsilon}, u^\varepsilon, v^\varepsilon) \\ v_t^\varepsilon &= \operatorname{div}(\varepsilon^2\mathbb{D}_2(x, \tfrac{x}{\varepsilon})\nabla v^\varepsilon) &+ F_2(x, \tfrac{x}{\varepsilon}, u^\varepsilon, v^\varepsilon) \end{aligned} \quad \text{in } \Omega \tag{1.1}$$

supplemented with homogeneous Neumann boundary conditions and initial conditions. Here, $(u^\varepsilon, v^\varepsilon) : [0, T] \times \Omega \to \mathbb{R}^{m_1+m_2}$ denote the concentrations of m_1 "classically" diffusing species with characteristic diffusion length of order $O(1)$ and m_2 slowly diffusing species of order $O(\varepsilon)$. Moreover, $\mathbb{D}_i : \Omega \times \mathcal{Y} \to \mathbb{R}^{(m_i \times d)\times(m_i \times d)}$ denotes the diffusion coefficients and $F_i : \Omega \times \mathcal{Y} \times \mathbb{R}^{m_1+m_2} \to \mathbb{R}^{m_i}$ the nonlinear reaction terms and both, $\mathbb{D}_i$ and F_i, are assumed to be periodic in $y = x/\varepsilon$ w.r.t. a prescribed microstructure.

It was shown in [34] that the solutions $(u^\varepsilon, v^\varepsilon)$ converge for $\varepsilon \to 0$ to a limit (u, V) that decomposes into a one-scale function $u(t, x)$ and a two-scale function $V(t, x, y)$, which solve the effective system

$$\begin{aligned} u_t &= \operatorname{div}(\mathbb{D}_{\text{eff}}(x)\nabla u) + ⨍_{\mathcal{Y}} F_1(x, y, u(x), V(x, y))\,\mathrm{d}y && \text{in } \Omega, \\ V_t &= \operatorname{div}_y(\mathbb{D}_2(x, y)\nabla_y V) + F_2(x, y, u, V) && \text{in } \Omega \times \mathcal{Y}. \end{aligned} \tag{1.2}$$

In order to install the limit passage (1.1) $\to$ (1.2), we employ the technique of two-scale convergence via periodic unfolding, cf. (2.6). This involves the periodic unfolding operator $\mathcal{T}_\varepsilon : L^1(\Omega) \to L^1(\Omega \times \mathcal{Y})$, the folding operator $\mathcal{F}_\varepsilon : L^1(\Omega \times \mathcal{Y}) \to L^1(\Omega)$ and the gradient folding operators $\mathcal{G}^0_\varepsilon$ respective $\mathcal{G}^1_\varepsilon$, cf. Section 2.1. With this method, the strong two-scale convergence of the slowly diffusing species v^ε, i.e. $\max_{0\le t\le T} \| \mathcal{T}_\varepsilon v^\varepsilon(t) - V(t)\|_{L^2(\Omega\times\mathcal{Y})} \to 0$, was proved in [34], cf. Section 3.1, whereas the strong convergence $u^\varepsilon \to u$ follows immediately from the compact embedding $H^1(\Omega) \subset L^2(\Omega)$. This result was obtained under the assumption of L^∞-regularity of the coefficients and global Lipschitz continuity of the reaction terms, cf. (3.6.A1)–(3.6.A4). One major analytical difficulty to overcome is the *periodicity defect* [17] or *$\mathcal{T}_\varepsilon$-property of recovered periodicity* [34], i.e.

$$\text{for all } u^\varepsilon \in H^1(\Omega): \quad \begin{aligned} &\mathcal{T}_\varepsilon u^\varepsilon \in L^2(\Omega; H^1(Y)) \not\subseteq L^2(\Omega; H^1(\mathcal{Y})), \text{ but} \\ &\text{w-}\lim_{\varepsilon\to 0} \mathcal{T}_\varepsilon u^\varepsilon \in L^2(\Omega; H^1(\mathcal{Y})), \text{ if the limit exists.} \end{aligned} \tag{1.3.PD}$$

The aim of this paper is to derive in Theorem 3.2 the error estimate

$$\max_{0\le t\le T} \left\{ \| \mathcal{T}_\varepsilon v^\varepsilon(t) - V(t)\|_{L^2(\Omega\times\mathcal{Y})} + \|u^\varepsilon(t) - u(t)\|_{L^2(\Omega)} \right\} \le \varepsilon^{1/4} C. \tag{1.4}$$

In the interior of the domain Ω, the convergence rate in (1.4) can be improved to $\varepsilon^{1/2}$, see Theorem 3.3. We assume additional spatial regularity w.r.t. the macroscopic scale $x \in \Omega$ of the given data (3.6.A5), i.e. $\nabla_x \mathbb{D}_i$, $\nabla_x F_i \in L^\infty(\Omega \times \mathcal{Y})$, and the effective solution (u, V) (3.6.A6), i.e. $u \in H^2(\Omega)$, $V \in H^1(\Omega; H^1(\mathcal{Y}))$. We assume neither additional spatial regularity of the original solutions $(u^\varepsilon, v^\varepsilon)$ nor of the corrector functions.

In [20], a reaction-diffusion system predicting concrete corrosion is considered, but the system does not include slowly diffusing species v^ε. Nevertheless, for the classically diffusing species u^ε and its gradient ∇u^ε the convergence rate $\varepsilon^{1/2}$ and $\varepsilon^{1/4}$, respectively, is rigorously proved by the method of periodic unfolding. For systems involving slowly diffusing species v^ε, convergence rates of order $\varepsilon^{1/2}$ are derived in [24, 21] via the method of asymptotic expansion assuming continuous given data and limit solutions.

The distinctive feature of this contribution is the nonlinear coupling of the classically and slowly diffusing species combined with the periodic unfolding method, which allows to avoid any assumption of spatial continuity. Our proof to (1.4), in the first part, follows along the lines of [34] and we derive the Gronwall-type estimate

$$\tfrac{\mathrm{d}}{\mathrm{d}t} \left(\||\mathcal{T}_\varepsilon v^\varepsilon - V\||^2 + \|u^\varepsilon - u\|^2 \right) \le C \left(\||\mathcal{T}_\varepsilon v^\varepsilon - V\||^2 + \|u^\varepsilon - u\|^2 \right) + \Delta^{v^\varepsilon} + \Delta^{u^\varepsilon}, \tag{1.5}$$

where $\||\cdot\|| := \|\cdot\|_{L^2(\Omega\times\mathcal{Y})}$ and $\|\cdot\| := \|\cdot\|_{L^2(\Omega)}$ and $\Delta^{v^\varepsilon}, \Delta^{u^\varepsilon}$ comprise errors terms. In [34], it was shown that these errors vanish as $\varepsilon \to 0$. The novelty of this contribution, the second part of the proof, is the quantification of their convergence, namely $|\Delta^{u^\varepsilon} + \Delta^{v^\varepsilon}| \le \varepsilon^{1/2} C$. In order to quantify those error terms, we have to find, in particular, error estimates for the folding and unfolding operators, see the lemmas 3.5, 3.6, and 3.7 in Section 3.3, which heavily rely on the improved regularity w.r.t. $x \in \Omega$ and ideas from [17]. Moreover, we use a quantification result for the periodicity defect (1.3.PD) from [18], see Lemma 3.8.

The structure of the paper is as follows: In Section 2, we introduce basic notations, definitions, and results concerning periodic unfolding (Sec. 2.1) and two-scale convergence (Sec. 2.2). In Section 3, we consider the coupled systems (1.1)–(1.2) and derive the error estimate (1.4). Therefore, we list our assumptions and recall the existing convergence result (Sec. 3.1), state our Main Theorem (Thm. 3.2 & 3.3), explain the structure of its proof (Sec. 3.2), and we derive preparatory error estimates (Sec. 3.3). Finally, we give the proof of Theorem 3.2 (Sec. 3.4) and we discuss the obtained results (Sec. 3.5).

2. Two-scale convergence

Here, and throughout this paper, x denotes the macroscopic variable and the microscopic variable y captures periodic oscillations in x/ε. In order to describe the convergence from (1.1) to (1.2), we introduce the concept of two-scale convergence, which is designed for problems with underlying periodic microstructure. The definition of two-scale convergence (2.6), introduced in Section 2.2, is based on the periodic unfolding technique, described in Section 2.1, and with this it reduces to the notion of classical weak and strong convergence in the two-scale space $L^2(\Omega \times \mathcal{Y})$.

2.1. Periodic unfolding, folding, and gradient folding operators

Throughout this paper, let $\Omega \subset \mathbb{R}^d$ be a bounded domain with Lipschitz boundary $\partial\Omega$. Following [9, 35, 13], $Y = [0,1)^d$ denotes the *unit cell* so that $\mathbb{R}^d$ is the disjoint union of translated cells $\lambda + Y$, where $\lambda \in \mathbb{Z}^d$. Identifying opposite faces of $\overline{Y}$ gives the *periodicity cell* $\mathcal{Y}$, i.e. the torus

$$\mathcal{Y} := \mathbb{R}^d/_{\mathbb{Z}^d}.$$

But, in notation, we will not distinguish between elements of the unit cell $y \in Y$ and the ones of the periodicity cell $y \in \mathcal{Y}$. Using the mappings $[\,\cdot\,]_Y : \mathbb{R}^d \to \mathbb{Z}^d$ and $\{\,\cdot\,\}_Y : \mathbb{R}^d \to Y$, we have the unique decomposition

$$\text{for all } x \in \mathbb{R}^d : \ x = [x]_Y + \{x\}_Y, \quad \text{where } [x]_Y \in \mathbb{Z}^d \text{ and } \{x\}_Y \in Y.$$

A function $f \in L^1_{\text{loc}}(\mathbb{R}^d)$ is called *Y-periodic*, if $f(x) = f(\{x\}_Y)$ for a.a. $x \in \mathbb{R}^d$. Then, we can identify every periodic function f with a function $\widetilde{f}$ on $\mathcal{Y}$. Introducing the small length scale parameter $\varepsilon > 0$, we define the sets

$$\Lambda_\varepsilon := \{\lambda \in \mathbb{Z}^d \,|\, \varepsilon(\lambda + Y) \subset \overline{\Omega}\} \quad \text{and} \quad \widehat{\Omega}_\varepsilon := \text{int}\left(\textstyle\bigcup_{\lambda_i \in \Lambda_\varepsilon} \varepsilon(\lambda_i + Y)\right).$$

With this definition of the subset $\widehat{\Omega}_\varepsilon \subset \Omega$, we sort out microscopic cells $\varepsilon[x/\varepsilon]_Y + Y$ which overlap the boundary $\partial\Omega$. Moreover, we have $\text{vol}(\Omega \backslash \widehat{\Omega}_\varepsilon) = O(\varepsilon)$ for those cells which are only partially contained in Ω. Based in these notations, the *periodic unfolding operator* $\mathcal{T}_\varepsilon : L^1(\Omega) \to L^1(\Omega \times \mathcal{Y})$ is defined via, cf. [9, 35],

$$(\mathcal{T}_\varepsilon u)(x,y) := \begin{cases} u\left(\varepsilon\left[\frac{x}{\varepsilon}\right]_Y + \varepsilon y\right) & \text{if } (x,y) \in \widehat{\Omega}_\varepsilon \times \mathcal{Y}, \\ 0 & \text{otherwise.} \end{cases} \tag{2.1}$$

Moreover, we have the crucial properties, cf. [35],

$$\begin{array}{lll} \textit{product rule:} & \mathcal{T}_\varepsilon(uv) = (\mathcal{T}_\varepsilon u)(\mathcal{T}_\varepsilon v) & \text{for all } u, v \in L^2(\Omega), \\ \textit{unfolding criterion:} & \int_\Omega F \,\mathrm{d}x = \int_{\Omega \times \mathcal{Y}} \mathcal{T}_\varepsilon F \,\mathrm{d}x\,\mathrm{d}y + \omega_F(\varepsilon) & \text{for all } F \in L^1(\Omega), \end{array} \tag{2.2}$$

where $\omega_F(\varepsilon) = \int_{\Omega\setminus\widehat{\Omega}_\varepsilon} F\,\mathrm{d}x$. It holds $\omega_F(\varepsilon) \to 0$ as $\varepsilon \to 0$ for all $F \in L^p(\Omega)$ with $p > 1$, due to $\mathrm{vol}(\Omega\setminus\widehat{\Omega}_\varepsilon) \to 0$. The rate of $\omega(\varepsilon)$ depends on the norm of the function F.

For the reverse operation, we define the *folding operator* $\mathcal{F}_\varepsilon : L^1(\Omega \times \mathcal{Y}) \to L^1(\Omega)$ via

$$(\mathcal{F}_\varepsilon U)(x) := \fint_{\varepsilon[\frac{x}{\varepsilon}]_Y + \varepsilon Y} U(\xi, \{\tfrac{x}{\varepsilon}\}_Y)\,\mathrm{d}\xi \tag{2.3}$$

for all $x \in \widehat{\Omega}_\varepsilon$ and $(\mathcal{F}_\varepsilon U)(x) = 0$ otherwise.

Even for smooth functions $U : \Omega \times \mathcal{Y} \to \mathbb{R}$ the folded function $\mathcal{F}_\varepsilon U$ is only piecewise constant in x, hence $\nabla(\mathcal{F}_\varepsilon U)$ cannot be determined in the classical sense. Therefore, we define the so-called *gradient folding operator* $\mathcal{G}^0_\varepsilon$, respective $\mathcal{G}^1_\varepsilon$, which suitably regularizes the folded function $\mathcal{F}_\varepsilon U$. The definition of the above mentioned gradient folding operator is taken from [34, Def. 3.7], cf. also [16, Prop. 2.11], [10, Thm. 6.1], and [13, Prop. 2.10]. At first, we define the functions with zero average via

$$H^1_{\mathrm{av}}(\mathcal{Y}) := \left\{ u \in H^1(\mathcal{Y}) \mid \textstyle\int_{\mathcal{Y}} u(y)\,\mathrm{d}y = 0 \right\}.$$

Definition 2.1 (Gradient folding). $\gamma = 0$: *The gradient folding operator* $\mathcal{G}^0_\varepsilon : H^1(\Omega) \times L^2(\Omega; H^1_{\mathrm{av}}(\mathcal{Y})) \to H^1(\Omega)$ *maps a pair of functions* $(u, U) \in H^1(\Omega) \times L^2(\Omega; H^1_{\mathrm{av}}(\mathcal{Y}))$ *to* $u^\varepsilon := \mathcal{G}^0_\varepsilon(u, U)$, *where* $u^\varepsilon \in H^1(\Omega)$ *is the unique weak solution of the elliptic problem*

$$\int_\Omega (u^\varepsilon - u) \cdot \varphi + (\nabla u^\varepsilon - \mathcal{F}_\varepsilon[\nabla u + \nabla_y U]) : \nabla\varphi\,\mathrm{d}x = 0 \quad \text{for all } \varphi \in H^1(\Omega). \tag{2.4}$$

$\gamma = 1$: *The gradient folding operator* $\mathcal{G}^1_\varepsilon : L^2(\Omega; H^1(\mathcal{Y})) \to H^1(\Omega)$ *maps a two-scale function* $U \in L^2(\Omega; H^1(\mathcal{Y}))$ *to* $u^\varepsilon := \mathcal{G}^1_\varepsilon U$, *where* $u^\varepsilon \in H^1(\Omega)$ *is the unique weak solution of the elliptic problem*

$$\int_\Omega (u^\varepsilon - \mathcal{F}_\varepsilon U) \cdot \varphi + (\varepsilon\nabla u^\varepsilon - \mathcal{F}_\varepsilon(\nabla_y U)) : \varepsilon\nabla\varphi\,\mathrm{d}x = 0 \quad \text{for all } \varphi \in H^1(\Omega). \tag{2.5}$$

For $\varepsilon > 0$ fixed, the Lax-Milgram lemma yields the existence of a unique weak solution $u^\varepsilon \in H^1(\Omega)$ of (2.4)/(2.5), so that the gradient folding operators are indeed well-defined.

2.2. Weak and strong two-scale convergence

We are now in the position to give the definition of weak and strong two-scale convergence following again [9, 35, 13]. The notion of two-scale convergence was first introduced in [5] and coincides for bounded sequences with Definition (2.6a), here below. For a more detailed comparison of the different definitions see [13, Sec. 2.3].

For $(u^\varepsilon)_\varepsilon \subset L^2(\Omega)$, we say u^ε *weakly* (2.6a) respective *strongly* (2.6b) *two-scale converges* to U in $L^2(\Omega \times \mathcal{Y})$, if

$$u^\varepsilon \xrightarrow{2\mathrm{w}} U \quad \text{in } L^2(\Omega \times \mathcal{Y}) \quad :\overset{\mathrm{Def.}}{\Longleftrightarrow} \quad \mathcal{T}_\varepsilon u^\varepsilon \rightharpoonup U \quad \text{in } L^2(\Omega \times \mathcal{Y}), \tag{2.6a}$$

$$u^\varepsilon \xrightarrow{2\mathrm{s}} U \quad \text{in } L^2(\Omega \times \mathcal{Y}) \quad :\overset{\mathrm{Def.}}{\Longleftrightarrow} \quad \mathcal{T}_\varepsilon u^\varepsilon \to U \quad \text{in } L^2(\Omega \times \mathcal{Y}). \tag{2.6b}$$

The unfolding operator $\mathcal{T}_\varepsilon$ is defined for the class of Lebesgue-integrable functions, where boundary values play no role, so that in particular $L^2(\Omega \times \mathcal{Y}) = L^2(\Omega \times Y)$. In view of the *periodicity defect* (1.3.PD), we carefully distinguish the spaces $H^1(Y)$ and $H^1(\mathcal{Y}) = H^1_{\mathrm{per}}(Y)$, where the latter one is a closed subspace of $H^1(Y)$. For brevity, we set

$$X = H^1(\Omega), \quad H = L^2(\Omega), \quad \mathbb{X} = L^2(\Omega; H^1(\mathcal{Y})), \\ \mathbb{X}_{\mathrm{av}} = L^2(\Omega; H^1_{\mathrm{av}}(\mathcal{Y})), \quad \text{and} \quad \mathbb{H} = L^2(\Omega \times \mathcal{Y}). \tag{2.7}$$

We have sequential compactness w.r.t. the weak two-scale convergence and it is shown in e.g. [5], [6, Prop. 1.14], [36, Thm. 5.2, Thm. 5.4], [7, Thm. 3.4] that bounded sequences of one-scale functions $(u^\varepsilon)_\varepsilon$ admit a weakly two-scale converging subsequence, i.e.

$$
\begin{aligned}
&\text{(i)} \quad \|u^\varepsilon\|_H \le C \;\Rightarrow\; \exists U \in \mathbb{H}:\; u^{\varepsilon'} \xrightharpoonup{2\mathrm{w}} U \text{ in } \mathbb{H},\\
&\text{(ii)} \quad \|u^\varepsilon\|_H + \varepsilon\|\nabla u^\varepsilon\|_H \le C \;\Rightarrow\; \exists U \in \mathbb{X}:\; u^{\varepsilon'} \xrightharpoonup{2\mathrm{w}} U \;\&\; \varepsilon'\nabla u^{\varepsilon'} \xrightharpoonup{2\mathrm{w}} \nabla_y U \text{ in } \mathbb{H},\\
&\text{(ii)} \quad \|u^\varepsilon\|_X \le C \;\Rightarrow\; \exists (u,U) \in X \times \mathbb{X}_{\mathrm{av}}:\; u^{\varepsilon'} \rightharpoonup u \text{ in } X \text{ and } \nabla u^{\varepsilon'} \xrightharpoonup{2\mathrm{w}} \nabla u + \nabla_y U \text{ in } \mathbb{H}.
\end{aligned}
$$

Since (2.4) implies $\|\mathcal{G}_\varepsilon^1 U\|_H + \varepsilon\|\nabla(\mathcal{G}_\varepsilon^1 U)\|_H \le C$, (ii) implies the existence of a weakly two-scale convergent subsequence. However, for given $U \in \mathbb{X}$ the gradient folding operator guarantees even *strong* two-scale convergence. So, $(\mathcal{G}_\varepsilon^1 U)_\varepsilon \subset X$ recovers any function $U \in \mathbb{X}$ via strong two-scale convergence and it is shown in [16, Prop. 2.11] that

$$
\begin{aligned}
\gamma = 0:& \quad \text{for all } (u,U) \in X \times \mathbb{X}_{\mathrm{av}}:\; \mathcal{G}_\varepsilon^0(u,U) \xrightarrow{2\mathrm{s}} u \;\&\; \nabla[\mathcal{G}_\varepsilon^0(u,U)] \xrightarrow{2\mathrm{s}} \nabla u + \nabla_y U \text{ in } \mathbb{H},\\
\gamma = 1:& \quad \text{for all } U \in \mathbb{X}:\; \mathcal{G}_\varepsilon^1 U \xrightarrow{2\mathrm{s}} U \;\&\; \varepsilon\nabla[\mathcal{G}_\varepsilon^1 U] \xrightarrow{2\mathrm{s}} \nabla_y U \text{ in } \mathbb{H}.
\end{aligned}
$$

Convenient commutation relations, such as $\mathcal{F}_\varepsilon(\nabla_y U) = \varepsilon\nabla(\mathcal{F}_\varepsilon U)$ or $\mathcal{G}_\varepsilon^1(\nabla_y U) = \varepsilon\nabla(\mathcal{G}_\varepsilon^1 U)$, cannot be expected, since $\mathcal{F}_\varepsilon U \notin X$ and $\nabla_y U \notin \mathbb{X}$. Instead, we have that the different folding operators are comparable in the sense that their difference vanishes, see [34, Prop. 3.9],

$$
\begin{aligned}
\gamma = 0:& \quad \text{for all } (u,U) \in X \times \mathbb{X}_{\mathrm{av}}:\\
& \quad \|u - \mathcal{G}_\varepsilon^0(u,U)\|_H + \|\mathcal{F}_\varepsilon[\nabla u + \nabla_y U] - \nabla[\mathcal{G}_\varepsilon^0(u,U)]\|_H \to 0, \qquad (2.8)\\
\gamma = 1:& \quad \text{for all } U \in \mathbb{X}:\; \|\mathcal{F}_\varepsilon U - \mathcal{G}_\varepsilon^1 U\|_H + \|\mathcal{F}_\varepsilon(\nabla_y U) - \varepsilon\nabla(\mathcal{G}_\varepsilon^1 U)\|_H \to 0.
\end{aligned}
$$

3. Error estimates for reaction-diffusion systems

We consider a system of two coupled reaction-diffusion systems, where the coupling arises via the nonlinear reaction term $(f_1^\varepsilon, f_2^\varepsilon)$, whereas the diffusion tensor has block structure

$$
\begin{pmatrix} u_t^\varepsilon \\ v_t^\varepsilon \end{pmatrix} = \begin{pmatrix} \mathrm{div}(\mathbb{D}_1^\varepsilon \nabla u^\varepsilon) \\ \mathrm{div}(\varepsilon^2 \mathbb{D}_2^\varepsilon \nabla v^\varepsilon) \end{pmatrix} + \begin{pmatrix} f_1^\varepsilon(u^\varepsilon, v^\varepsilon) \\ f_2^\varepsilon(u^\varepsilon, v^\varepsilon) \end{pmatrix} \quad \text{in } [0,T] \times \Omega. \qquad (3.1.\mathrm{P}_\varepsilon)
$$

We supplement (3.1.P$_\varepsilon$) with homogenous Neumann boundary conditions on $\partial\Omega$ and prescribed initial values $u^\varepsilon(0) = u_0^\varepsilon$ respective $v^\varepsilon(0) = v_0^\varepsilon$. In [34] (see Theorem 3.1 below) it was proven that $(u^\varepsilon, v^\varepsilon)$ converges for $\varepsilon \to 0$ to a limit (u, V) that decomposes into a one-scale function $u(t,x)$ and a two-scale function $V(t,x,y)$ which solve the effective system

$$
\begin{pmatrix} u_t \\ V_t \end{pmatrix} = \begin{pmatrix} \mathrm{div}(\mathbb{D}_{\mathrm{eff}} \nabla u) \\ \mathrm{div}_y(\mathbb{D}_2 \nabla_y V) \end{pmatrix} + \begin{pmatrix} f_{\mathrm{eff}}(u, V) \\ F_2(u, V) \end{pmatrix} \quad \text{in } [0,T] \times \Omega \times \mathcal{Y}. \qquad (3.2.\mathrm{P}_0)
$$

Here, the effective diffusion tensor $\mathbb{D}_{\mathrm{eff}}$ and the effective u-reaction f_{eff} only depend on the macroscopic variable $x \in \Omega$, while the diffusion tensor $\mathbb{D}_2$ and the V-reaction F_2 depend on the two-scale variables $(x,y) \in \Omega \times \mathcal{Y}$, see (3.6.A1)–(3.6.A2) and (3.3)-(3.5), below. The function-to-function map $f_{\mathrm{eff}} : \Omega \times \mathbb{R}^{m_1} \times L^2(\mathcal{Y}; \mathbb{R}^{m_2}) \to \mathbb{R}^{m_1}$ is defined as

$$
f_{\mathrm{eff}}(x, u, Z) := \int_{\mathcal{Y}} F_1(x, y, u, Z(y))\, \mathrm{d}y. \qquad (3.3)
$$

The effective diffusion tensor $\mathbb{D}_{\mathrm{eff}} : \Omega \to \mathbb{R}^{(m_1 \times d) \times (m_1 \times d)}$ is given componentwise via the classical homogenization formula, see e.g. [1, 6, 37],

$$
\mathbb{D}_{\mathrm{eff}}(x)_{ijkl} := \int_{\mathcal{Y}} \mathbb{D}_1(x,y)_{ijkl} + \sum_{r=1}^{d} \mathbb{D}_1(x,y)_{ijkr} \cdot \partial_{y_r} z(y)_{kl}\, \mathrm{d}y, \qquad (3.4)
$$

for $i,k=1,...,m_1$, $j,l=1,...,d$, where the so-called correctors $z_{ij}\in H^1_{\rm av}(\mathcal{Y})$ solve the local problem in the weak sense:

$$\mathrm{div}_y\left(\mathbb{D}_1(x,y)_{ijkl}+\sum_{r=1}^d \mathbb{D}_1(x,y)_{ijkr}\cdot\partial_{y_r}z(y)_{kl}\right)=0 \quad \text{in } \mathcal{Y} \text{ for a.a. } x\in\Omega. \tag{3.5}$$

3.1. Assumptions and existing results

We recall the definition of the function spaces $(X,H,\mathbb{X},\mathbb{X}_{\rm av},\mathbb{H})$ in (2.7) and we impose the following assumptions on the given data of (3.1.P$_\varepsilon$)–(3.2.P$_0$) for $i=1,2$:

The diffusion tensor
$\mathbb{D}_i:\Omega\times\mathcal{Y}\to\mathbb{R}^{(m_i\times d)\times(m_i\times d)}$ is uniformly bounded and elliptic, i.e. (3.6.A1)
$\exists\,\mu>0:\ \mathbb{D}_i(x,y)\xi:\xi\geq\mu|\xi|^2$ for all $\xi\in\mathbb{R}^{m_i\times d}$, $(x,y)\in\Omega\times\mathcal{Y}$.

The reaction term
$F_i:\Omega\times\mathcal{Y}\times\mathbb{R}^{m_1+m_2}\to\mathbb{R}^{m_i}$ is uniformly bounded in $\Omega\times\mathcal{Y}$
as well as differentiable and globally Lipschitz continuous in $\mathbb{R}^{m_1+m_2}$, i.e. (3.6.A2)
$\exists\,L>0:\ |F_i(x,y,A_1,B_1)-F_i(x,y,A_2,B_2)|\leq L(|A_1-A_2|+|B_1-B_2|)$
for all $(A_i,B_i)\in\mathbb{R}^{m_1+m_2}$, $(x,y)\in\Omega\times\mathcal{Y}$.

The initial values
satisfy u_0, $\mathrm{div}(\mathbb{D}_{\rm eff}\nabla u_0)\in H$ and V_0, $\mathrm{div}_y(\mathbb{D}_2\nabla_y V_0)\in\mathbb{H}$. (3.6.A3)

The dependence on ε
$\mathbb{D}_i^\varepsilon:=\mathcal{F}_\varepsilon\,\mathbb{D}_i$ and $f_i^\varepsilon(\cdot,A,B):=\mathcal{F}_\varepsilon\,F_i(\cdot,\cdot,A,B)$ for all $(A,B)\in\mathbb{R}^{m_1+m_2}$, (3.6.A4)
$\exists\,c\geq0:\ \|u_0^\varepsilon\|_H+\|\,\mathrm{div}(\mathbb{D}_1^\varepsilon\nabla u_0^\varepsilon)\|_H+\|v_0^\varepsilon\|_H+\|\,\mathrm{div}(\varepsilon^2\mathbb{D}_2^\varepsilon\nabla v_0^\varepsilon)\|_H\leq c$.

Spatial Lipschitz continuity of the given data
For $(A,B)\in\mathbb{R}^{m_1+m_2}$ fixed, it holds $\nabla_x\mathbb{D}_i,\nabla_x F_i(A,B)\in L^\infty(\Omega\times\mathcal{Y})$ (3.6.A5)
and we write $C_F:=\sup_{(x,y)\in\Omega\times\mathcal{Y}}\{|F(x,y,A,B)|+|\nabla_x F(x,y,A,B)|\}$.

Improved spatial regularity of the effective solutions
$\forall\,t\in[0,T]:\ u(t)\in H^2(\Omega)$ and $V(t)\in H^1(\Omega;H^1(\mathcal{Y}))$, $V_t(t)\in H^1(\Omega;L^2(\mathcal{Y}))$. (3.6.A6)

Convergence rates for the initial values
$\exists\,c\geq0:\ \ \|\,\mathcal{T}_\varepsilon v_0^\varepsilon-V_0\|_{\mathbb{H}}+\|u_0^\varepsilon-u_0\|_H\leq\varepsilon^{1/2}c$. (3.6.A7)

We obtain the two evolution triples $X\subset H\subset X^*$ and $\mathbb{X}\subset\mathbb{H}\subset\mathbb{X}^*$. The assumptions (3.6.A1)–(3.6.A4) guarantee the existence of unique weak solutions $(u^\varepsilon,v^\varepsilon)$ of (3.1.P$_\varepsilon$) and (u,V) of (3.2.P$_0$). Further, the differentiability of the reaction terms and the additional regularity of the initial values (3.6.A4) ensure improved time-regularity of the solutions and the following a priori bounds: there exists $C_b\geq0$ independent of ε so that, cf. [34, Thm. 2.1 & Prop. 2.2],

$$\begin{aligned}\|u^\varepsilon\|_{C^1([0,T];H)}&+\|\nabla u^\varepsilon\|_{C([0,T];H)}&+\|v^\varepsilon\|_{C^1([0,T];H)}&+\varepsilon\|\nabla v^\varepsilon\|_{C([0,T];H)}\leq C_b,\\ \|u\|_{C^1([0,T];H)}&+\|\nabla u\|_{C([0,T];H)}&+\|V\|_{C^1([0,T];\mathbb{H})}&+\|\nabla_y V\|_{C([0,T];\mathbb{H})}\leq C_b.\end{aligned} \tag{3.7}$$

Moreover, we have the following convergence result.

Theorem 3.1 ([34, Thm. 5.1]). *Let the assumptions* (3.6.A1)–(3.6.A4) *as well as* $u_0^\varepsilon\to u_0$ *in* H *and* $v_0^\varepsilon\xrightarrow{2s}V_0$ *in* $\mathbb{H}$ *be satisfied. The sequence of weak solutions* $(u^\varepsilon,v^\varepsilon)$ *of* (3.1.P$_\varepsilon$) *converges* *to the weak solution* (u, V) *of* $(3.2.\mathrm{P}_0)$ *in the following sense:*

$$\begin{aligned}&\max_{0\le t\le T} \| \mathcal{T}_\varepsilon v^\varepsilon(t) - V(t)\|_{\mathbb{H}} \to 0,\ \varepsilon\nabla v^\varepsilon \xrightarrow{2\mathrm{s}} \nabla_y V \ \textit{in}\ L^2(0,T;\mathbb{H}),\ \textit{and}\\ &v_t^\varepsilon \xrightarrow{2\mathrm{w}} V_t \ \textit{in}\ L^2(0,T;\mathbb{H}),\ \textit{moreover}\ \forall\, t\in[0,T]:\ \varepsilon\nabla v^\varepsilon(t) \xrightarrow{2\mathrm{s}} \nabla_y V(t)\ \textit{in}\ \mathbb{H};\end{aligned} \tag{3.8a}$$

$$\begin{aligned}&u^\varepsilon \rightharpoonup u \ \textit{in}\ L^2(0,T;X)\ \textit{and}\ u_t^\varepsilon \rightharpoonup u_t\ \textit{in}\ H^1(0,T;X^*),\ \textit{moreover}\\ &\exists U\in L^2(0,T;\mathbb{X}_{\mathrm{av}})\ \textit{s.t.}\ \forall\, t\in[0,T]:\ \nabla u^\varepsilon(t)\xrightarrow{2\mathrm{w}} \nabla u(t)+\nabla_y U(t)\ \textit{in}\ \mathbb{H}.\end{aligned} \tag{3.8b}$$

One may drop the additional assumptions $\mathrm{div}(\mathbb{D}_1^\varepsilon\nabla u_0^\varepsilon), \mathrm{div}(\varepsilon^2\mathbb{D}_2^\varepsilon\nabla v_0^\varepsilon) \in H$ on the initial values, see [38]. Therein, it is shown that any solution with $u_0^\varepsilon, v_0^\varepsilon \in H$ can be approximated by a solution satisfying improved time-regularity as in (3.7).

3.2. Main Theorem and outline of the proof

Under the assumption of additional spatial regularity (3.6.A5)–(3.6.A7), we derive the following error estimates for the strong convergences in (3.8). We emphasize that we do not assume improved spatial regularity for the original macroscopic solutions $(u^\varepsilon, v^\varepsilon)$.

Theorem 3.2. *Let* $(u^\varepsilon, v^\varepsilon)$ *and* (u, V) *denote the solutions of* $(3.1.\mathrm{P}_\varepsilon)$ *and* $(3.2.\mathrm{P}_0)$*, respectively, and let the assumptions in* (3.6) *hold true. Then there exists a constant* $C \ge 0$ *independent of* ε *such that*

$$\max_{0\le t\le T}\{\| \mathcal{T}_\varepsilon v^\varepsilon(t) - V(t)\|_{\mathbb{H}} + \|u^\varepsilon(t)-u(t)\|_H\} \le \varepsilon^{1/4}C, \tag{3.9a}$$

$$\| \mathcal{T}_\varepsilon(\varepsilon\nabla v^\varepsilon) - \nabla_y V\|_{L^2(0,T;\mathbb{H})} + \| \mathcal{T}_\varepsilon(\nabla u^\varepsilon) - \{\nabla u + \nabla_y U\}\|_{L^2(0,T;\mathbb{H})} \le \varepsilon^{1/4} C. \tag{3.9b}$$

Moreover, we find the improved convergence rate in the interior of the domain Ω.

Theorem 3.3. *Let the assumptions* (3.6) *hold true. For all* $\delta > 0$*, let* Ω_{int} *denote an open subset of* Ω *with* $\inf_{x\in\Omega_{\mathrm{int}}} \mathit{dist}(x,\partial\Omega) > \delta$*. Then, there exists a constant* $C_\delta \ge 0$ *independent of* ε *such that for all* $\varepsilon < \delta/(4\sqrt{d})$ *it holds*

$$\begin{aligned}&\| \mathcal{T}_\varepsilon v^\varepsilon - V\|_{C([0,T];L^2(\Omega_{\mathrm{int}}\times\mathcal{Y}))} + \| \mathcal{T}_\varepsilon(\varepsilon\nabla v^\varepsilon) - \nabla_y V\|_{L^2(0,T;L^2(\Omega_{\mathrm{int}}\times\mathcal{Y}))}\\ &+ \|u^\varepsilon - u\|_{C([0,T];L^2(\Omega_{\mathrm{int}}))} + \| \mathcal{T}_\varepsilon(\nabla u^\varepsilon) - \{\nabla u+\nabla_y U\}\|_{L^2(0,T;L^2(\Omega_{\mathrm{int}}\times\mathcal{Y}))} \le \varepsilon^{1/2}C_\delta.\end{aligned}$$

Here, we focus on Theorem 3.2 and for the proof of Theorem 3.3, we refer to [38]. Therein, it shown that away from the boundary $\partial\Omega$, the error $\sqrt{\varepsilon}$ of lower order does not need to be considered, cf. Lemma 3.4, and the periodicity defect error is of improved order ε using [17, Prop. 3.3 & Thm. 3.4].

Thanks to (3.6.A5), we can equally choose $\mathbb{D}^\varepsilon(x) = \mathbb{D}(x, x/\varepsilon)$ or $\mathbb{D}^\varepsilon = \mathcal{F}_\varepsilon\mathbb{D}$ in (3.6.A4) because we can identify $W^{1,\infty}(\Omega)$ with $C^{0,1}(\Omega)$.

For $U \in L^2(0,T;\mathbb{X}_{\mathrm{av}})$ in (3.8b) we have a.e. in $[0,T]$ the representation $U_i(x,y) = \sum_{j=1}^d \frac{\partial u_i}{\partial x_j}(x) z_{ij}(y)$, where the correctors $z_{ij} \in H^1_{\mathrm{av}}(\mathcal{Y})$ solve the local problem (3.5). Since $u \in H^2(\Omega)$ by (3.6.A6), we obtain immediately $U \in H^1(\Omega; H^1_{\mathrm{av}}(\mathcal{Y}))$ and in particular we do *not* assume any improved regularity for the correctors z_{ij}. Note, (3.9b) implies the strong two-scale convergence $\nabla u^\varepsilon \xrightarrow{2\mathrm{s}} \nabla u + \nabla_y U$ in $L^2(0,T;\mathbb{H})$, which also holds in (3.8b) under the assumptions of Theorem 3.1.

Outline of the proof of Theorem 3.2: The essential idea is to derive the following Gronwall-type estimate

$$\tfrac{\mathrm{d}}{\mathrm{d}t}\left(\| \mathcal{T}_\varepsilon v^\varepsilon - V\|^2_{\mathbb{H}} + \|u^\varepsilon - u\|^2_H\right) \le C\left(\| \mathcal{T}_\varepsilon v^\varepsilon - V\|^2_{\mathbb{H}} + \|u^\varepsilon - u\|^2_H + \varepsilon^{1/2}\right). \tag{3.10}$$

Then, Gronwall's lemma yields for all $t \in [0,T]$

$$\| \mathcal{T}_\varepsilon v^\varepsilon(t) - V(t)\|^2_{\mathbb{H}} + \|u^\varepsilon(t) - u(t)\|^2_H \le C\left(\| \mathcal{T}_\varepsilon v^\varepsilon_0 - V_0\|^2_{\mathbb{H}} + \|u^\varepsilon_0 - u_0\|^2_H + \varepsilon^{1/2}\right)$$

and using assumption (3.6.A7) on the initial values gives immediately (3.9a). We derive (3.10) in separate steps, namely

$$\tfrac{\mathrm{d}}{\mathrm{d}t}\|u^\varepsilon - u\|^2_H \le C\left(\| \mathcal{T}_\varepsilon v^\varepsilon - V\|^2_{\mathbb{H}} + \|u^\varepsilon - u\|^2_H + \varepsilon^{1/2}\right) \text{ in Steps 1a–b, and} \tag{3.11}$$

$$\tfrac{\mathrm{d}}{\mathrm{d}t}\| \mathcal{T}_\varepsilon v^\varepsilon - V\|^2_{\mathbb{H}} \le C\left(\| \mathcal{T}_\varepsilon v^\varepsilon - V\|^2_{\mathbb{H}} + \|u^\varepsilon - u\|^2_H + \varepsilon^{1/2}\right) \text{ in Steps 2a–b.} \tag{3.12}$$

1a. $\tfrac{\mathrm{d}}{\mathrm{d}t}\|u^\varepsilon - u\|^2_H$*-estimate:* Following the argumentation in [34, Sect. 4.2/Proof of Thm. 4.1 (Step 2–5)], we derive the Gronwall-type estimate

$$\tfrac{\mathrm{d}}{\mathrm{d}t}\|u^\varepsilon - u\|^2_H \le C\left(\| \mathcal{T}_\varepsilon v^\varepsilon - V\|^2_{\mathbb{H}} + \|u^\varepsilon - u\|^2_H + \Delta^{u^\varepsilon}\right), \tag{3.13}$$

where $\Delta^{u^\varepsilon} = \sum_{i=1}^5 |\Delta_i^{u^\varepsilon}|$ with $\Delta_1^{u^\varepsilon}$ (folding mismatch between $\mathcal{F}_\varepsilon$ and $\mathcal{G}^0_\varepsilon$ resp. $\mathcal{F}_\varepsilon$ and $\mathcal{G}^1_\varepsilon$)
$\Delta_2^{u^\varepsilon}$ (periodicity defect of $\mathcal{T}_\varepsilon$ cf. (1.3.PD))
$\Delta_3^{u^\varepsilon}$ (approximation error $\mathbb{D}_1^\varepsilon \rightsquigarrow \mathbb{D}_{\text{eff}}$ resp. $\mathbb{D}_2^\varepsilon \rightsquigarrow \mathbb{D}_2$)
$\Delta_4^{u^\varepsilon}$ (approximation error $f_1^\varepsilon \rightsquigarrow f_{\text{eff}}$ resp. $f_2^\varepsilon \rightsquigarrow F_2$)
$\Delta_5^{u^\varepsilon}$ (unfolding error $\|V - \mathcal{T}_\varepsilon \mathcal{F}_\varepsilon V\|_{\mathbb{H}}$ resp. $\| \mathcal{T}_\varepsilon u - u\|_{\mathbb{H}}$).

Above, $u \in H$ is canonically understood as two-scale function $u \in \mathbb{H}$. The last error term $\Delta_5^{u^\varepsilon}$ (resp. $\Delta_5^{v^\varepsilon}$) does not occur in [34], but is addressed as a one-liner here. Since $\frac{1}{2}\frac{\mathrm{d}}{\mathrm{d}t}\|u^\varepsilon - u\|^2_H = \int_\Omega (u^\varepsilon_t - u_t)\cdot(u^\varepsilon - u)\,\mathrm{d}x$, we ideally subtract the weak formulations of $(3.1.\mathrm{P}_\varepsilon)_1$ and $(3.2.\mathrm{P}_0)_1$ (resp. $(3.1.\mathrm{P}_\varepsilon)_2$ and $(3.2.\mathrm{P}_0)_2$), test with the difference $u^\varepsilon - u$ (resp. $\mathcal{T}_\varepsilon v^\varepsilon - V$) and we obtain (3.13). However, due to the two-scale structure of $(3.2.\mathrm{P}_0)$, analytical difficulties arise and we cannot proceed straight forward. We modify this basic idea as follows:
In the first step, we test $(3.1.\mathrm{P}_\varepsilon)_1$ (resp. $(3.1.\mathrm{P}_\varepsilon)_2$) with $u^\varepsilon - \mathcal{G}^0_\varepsilon(u,U)$ (resp. $v^\varepsilon - \mathcal{G}^1_\varepsilon V$) and then, we *reformulate the ε-problem* into a two-scale problem using the unfolding operator $\mathcal{T}_\varepsilon$ and the folding operators $\mathcal{F}_\varepsilon$, $\mathcal{G}^0_\varepsilon$ (resp. $\mathcal{G}^1_\varepsilon$). Due to regularity issues between $\mathcal{F}_\varepsilon$ and $\mathcal{G}^0_\varepsilon$, cf. (2.8), we create the error term $\Delta_1^{u^\varepsilon}$ (resp. $\Delta_1^{v^\varepsilon}$).
In the second step, due to the periodicity defect (1.3.PD), we test $(3.2.\mathrm{P}_0)_1$ (resp. $(3.2.\mathrm{P}_0)_2$) only with (u,U) (resp. V). Afterwards, we *reformulate the limit problem* and insert the missing terms u^ε and $\mathcal{T}_\varepsilon(\nabla u^\varepsilon)$ (resp. $\mathcal{T}_\varepsilon v^\varepsilon$ and $\mathcal{T}_\varepsilon(\varepsilon\nabla v^\varepsilon)$) at the cost of creating the error term $\Delta_2^{u^\varepsilon}$ (resp. $\Delta_2^{v^\varepsilon}$).
Finally, in the third step, we add both reformulations and make further rearrangements in terms of the errors $\Delta_3^{u^\varepsilon}$–$\Delta_5^{u^\varepsilon}$ (resp. $\Delta_3^{v^\varepsilon}$–$\Delta_5^{v^\varepsilon}$) so that we end up with (3.13).

1b. *Estimation of* Δ^{u^ε} *and* (3.11)*:* We show $|\Delta^{u^\varepsilon}| \le \varepsilon^{1/2}C$. In more detail, we apply Lemma 3.7 (with $\gamma = 0$) to $\Delta_1^{u^\varepsilon}$ and we use Lemma 3.8 (with $\gamma = 0$) for $\Delta_2^{u^\varepsilon}$. The remaining error terms $\Delta_3^{u^\varepsilon}$–$\Delta_5^{u^\varepsilon}$ resolve easily with Lemma 3.5 and (3.15).

2a. $\tfrac{\mathrm{d}}{\mathrm{d}t}\| \mathcal{T}_\varepsilon v^\varepsilon - V\|^2_{\mathbb{H}}$*-estimate:* In [34, Sect. 4.2/Proof of Thm. 4.1 (Step 2–5)], the following Gronwall-type estimate is proved

$$\tfrac{\mathrm{d}}{\mathrm{d}t}\| \mathcal{T}_\varepsilon v^\varepsilon - V\|^2_{\mathbb{H}} \le C\left(\| \mathcal{T}_\varepsilon v^\varepsilon - V\|^2_{\mathbb{H}} + \|u^\varepsilon - u\|^2_H + \Delta^{v^\varepsilon}\right), \tag{3.14}$$

where $\Delta^{v^\varepsilon} = \sum_{i=1}^5 |\Delta_i^{v^\varepsilon}|$.

2b. *Estimation of* Δ^{v^ε} *and* (3.12)*:* We show $|\Delta^{v^\varepsilon}| \le \varepsilon^{1/2}C$. As in Step 1b, we use Lemma 3.7 resp. Lemma 3.8 (with $\gamma = 1$) for $\Delta_1^{v^\varepsilon}$ resp. $\Delta_2^{v^\varepsilon}$ as well as Lemma 3.5 and (3.15) for $\Delta_3^{v^\varepsilon}$–$\Delta_5^{v^\varepsilon}$.

3. *Derivation of* (3.9b)*:* We derive error estimates for the gradient terms by following the lines of [34, Proof of Thm. 4.1 (Step 7)].

3.3. Preparatory error estimates

We recall that Ω is a bounded domain with Lipschitz boundary such that we have in general $\widehat{\Omega}_\varepsilon \subsetneqq \Omega$. With this, the treatment of cells $\varepsilon(\lambda_i + Y)$ intersecting the boundary $\partial\Omega$ is crucial. Therefore, we begin with a rather classical result for the error on $\Omega\backslash\Omega_\varrho$, where $\Omega_\varrho = \{x \in \Omega \,|\, \mathrm{dist}(x, \partial\Omega) > \varrho\}$. The following lemma will be applied to the estimation of the boundary terms $\omega(\varepsilon)$ in (2.2) by choosing $\varrho = \varepsilon\sqrt{d}$.

Lemma 3.4 ([17, 18, 38])**.** *For* $u \in X$ *and* $U \in H^1(\Omega; L^2(\mathcal{Y}))$*, it holds for all* $\varrho > 0$

$$\|u\|_{L^2(\Omega\backslash\Omega_\varrho)} \le (\varrho + \sqrt{\varrho})C\|u\|_X \quad \text{and} \quad \|U\|_{L^2(\Omega\backslash\Omega_\varrho\times\mathcal{Y})} \le (\varrho + \sqrt{\varrho})C\|U\|_{H^1(\Omega;L^2(\mathcal{Y}))},$$

where the constant $C \ge 0$ *only depends on the properties of the domain* Ω.

The most important observation in deriving the error estimates (3.9a)–(3.9b) is the quantification of the well-known two-scale property, cf. [13, Prop. 2.4(e)], for every $U \in L^2(\Omega\times\mathcal{Y})$ exists a sequence $(u^\varepsilon)_\varepsilon \subset L^2(\Omega)$ such that $u^\varepsilon \xrightarrow{2s} U$ in $L^2(\Omega \times \mathcal{Y})$. For example, such a sequence is given by $u^\varepsilon = \mathcal{F}_\varepsilon U$. More precisely, based in the explicit definitions of $\mathcal{T}_\varepsilon$ and $\mathcal{F}_\varepsilon$, it holds:

Lemma 3.5. *For all* $U \in H^1(\Omega; L^2(\mathcal{Y}))$*, there exists a constant* $C \ge 0$*, only depending on* Ω *and* Y*, such that*

$$\|U - \mathcal{T}_\varepsilon \mathcal{F}_\varepsilon U\|_{\mathbb{H}} \le (\varepsilon + \varepsilon^{1/2})C\|U\|_{H^1(\Omega;L^2(\mathcal{Y}))}.$$

Proof. We use the unfolding criterion (2.2) and we apply the Poincaré-Wirtinger inequality on each cell $\mathrm{int}(\varepsilon(\lambda_i + \varepsilon Y)) \subset \widehat{\Omega}_\varepsilon$ so that

$$\begin{aligned}\|U - \mathcal{T}_\varepsilon \mathcal{F}_\varepsilon U\|_{\mathbb{H}}^2 &= \sum_{\lambda_i\in\Lambda_\varepsilon} \int_{\varepsilon(\lambda_i+Y)} \int_{\mathcal{Y}} \Big(U(x,y) - \fint_{\mathcal{N}_\varepsilon(x)+\varepsilon Y} U(\xi, y)\,\mathrm{d}\xi \Big)^2 \mathrm{d}x\,\mathrm{d}y + \omega_U(\varepsilon) \\ &\le \sum_{\lambda_i\in\Lambda_\varepsilon} C\,(\mathrm{diam}(\varepsilon(\lambda_i + Y)))^2 \,\|\nabla_x U\|^2_{L^2(\lambda_i+\varepsilon Y)} + \omega_U(\varepsilon) \le \varepsilon^2 C\|U\|^2_{H^1(\Omega;L^2(\mathcal{Y}))} + \omega_U(\varepsilon).\end{aligned}$$

Using Lemma 3.4 with $\varrho = \varepsilon\sqrt{d}$ gives

$$\omega_U(\varepsilon) = \int_{(\Omega\backslash\widehat{\Omega}_\varepsilon\times\mathcal{Y})} (U - \mathcal{T}_\varepsilon \mathcal{F}_\varepsilon U)^2\,\mathrm{d}x\,\mathrm{d}y \le \big((\varepsilon + \sqrt{\varepsilon})C\|U\|_{H^1(\Omega;L^2(\mathcal{Y}))}\big)^2.$$

Hence, we have the desired estimate. □

As a direct consequence of Lemma 3.5, we have, e.g. [17, Eq. (3.4)],

$$\text{for } u \in X: \quad \|\mathcal{T}_\varepsilon u - u\|_{\mathbb{H}} \le (\varepsilon + \sqrt{\varepsilon})C\|u\|_X. \tag{3.15}$$

For possibly discontinuous functions $U \in H^1(\Omega; L^2(\mathcal{Y}))$, the "naive folding" $x \mapsto U(x, x/\varepsilon)$ is not well-defined. But, in the proof of Lemma 3.7 below, exactly such a "naive folding" is employed. Therefore, we need a suitable regularization U_ε of U so that $\vartheta_\varepsilon(x) = U_\varepsilon(x, x/\varepsilon)$ is well-defined and the difference $\|\mathcal{F}_\varepsilon U - \vartheta_\varepsilon\|_H$ is of order $O(\varepsilon+\sqrt{\varepsilon})$. Therefore, we use in addition to $\mathcal{G}_\varepsilon^0$ respective $\mathcal{G}_\varepsilon^1$ another regularization of the folding operator $\mathcal{F}_\varepsilon$, namely, the so-called *scale-splitting operator* $\mathcal{Q}_\varepsilon$, cf. [9, 35, 17].

$$\text{For } u \in L^1(\Omega)\text{, the function } \mathcal{Q}_\varepsilon u \text{ is the } \mathcal{Q}_1\text{-Lagrangian interpolant of the discrete function } \mathcal{F}_\varepsilon u. \text{ Observe, } \mathcal{Q}_\varepsilon u \in W^{1,\infty}(\Omega) \text{ and } \mathcal{F}_\varepsilon u \in L^\infty(\Omega). \tag{3.16}$$

Note, for general functions $u \in L^\infty(\Omega)$ and $z \in L^2(\mathcal{Y})$, the composition $x \mapsto u(x)z(x/\varepsilon)$ lies in $L^2(\Omega)$, see e.g. [37, Thm.4].

Lemma 3.6. *For $w \in X$ and $z \in L^2(\mathcal{Y})$, there exists a constant $C \geq 0$, only depending on Ω and Y, such that*

$$\| (\mathcal{F}_\varepsilon w - \mathcal{Q}_\varepsilon w)\, z(\tfrac{\cdot}{\varepsilon})\|_H \leq \varepsilon^{1/2} C \|w\|_X \|z\|_{L^2(\mathcal{Y})}.$$

Proof. Based on the identity

$$\| (\mathcal{F}_\varepsilon w - \mathcal{Q}_\varepsilon w)\, z(\tfrac{\cdot}{\varepsilon})\|_H^2 = \sum_{\lambda_i \in \Lambda_\varepsilon} \int_{\lambda_i + \varepsilon Y} \left|(\mathcal{F}_\varepsilon w(x) - \mathcal{Q}_\varepsilon w(x))\, z(\tfrac{x}{\varepsilon})\right|^2 \mathrm{d}x + \omega_w(\varepsilon), \tag{3.17}$$

we consider in the following only one microscopic cell $\lambda_i + \varepsilon Y$, whereby w.l.o.g. $\lambda_i = 0$. The term $\omega_w(\varepsilon)$ comprises the boundary cells and it is treated with Lemma 3.4. By definition, we have for $x \in \varepsilon Y$ and every $\kappa = (\kappa_1, \ldots, \kappa_d) \in \{0,1\}^d$:

$$(\mathcal{Q}_\varepsilon w)(x) := \sum_{\kappa \in \{0,1\}^d} (\mathcal{F}_\varepsilon w)(\varepsilon\kappa) \cdot \bar{x}_1^{(\kappa_1)} \cdots \bar{x}_d^{(\kappa_d)}, \text{ where } \bar{x}_l^{(\kappa_l)} := \begin{cases} \frac{x_l - \mathcal{N}_\varepsilon(x)_l}{\varepsilon} & \text{if } \kappa_l = 1, \\ 1 - \frac{x_l - \mathcal{N}_\varepsilon(x)_l}{\varepsilon} & \text{if } \kappa_l = 0. \end{cases}$$

With $\bar{x}_l^{(\kappa_l)} \in [0,1]$, we obtain

$$\int_{\varepsilon Y} \left|(\mathcal{F}_\varepsilon w(0) - \mathcal{Q}_\varepsilon w(x))\, z(\tfrac{x}{\varepsilon})\right|^2 \mathrm{d}x \leq 2^d \sum_{\kappa \in \{0,1\}^d} |(\mathcal{F}_\varepsilon w(0) - \mathcal{F}_\varepsilon w(\varepsilon\kappa))|^2 \int_{\varepsilon Y} \left|z(\tfrac{x}{\varepsilon})\right|^2 \mathrm{d}x$$

$$\leq 2^d \sum_{\kappa \in \{0,1\}^d} \left| \fint_{\varepsilon Y} w(\xi) - w(\xi + \varepsilon\kappa)\, \mathrm{d}\xi \right|^2 \varepsilon^d \|z\|^2_{L^2(\mathcal{Y})} \leq 2^{2d} \varepsilon^2 d \|\nabla w\|^2_{L^2(\varepsilon Y)} \|z\|^2_{L^2(\mathcal{Y})}. \tag{3.18}$$

For the last estimate in (3.18), we use the fundamental relation $w(\xi) - w(\xi + \varepsilon\kappa) = \varepsilon \int_0^1 \nabla w(\xi + \varepsilon\kappa t) \cdot \kappa \, \mathrm{d}t$ with $|\kappa| \leq \sqrt{d}$ so that we obtain for $|\,\mathrm{d}s/\,\mathrm{d}\xi| = 1$

$$\left| \fint_{\varepsilon Y} w(\xi) - w(\xi + \varepsilon\kappa)\, \mathrm{d}\xi \right|^2 \leq \varepsilon^2 d \fint_{\varepsilon Y} \int_0^1 |\nabla w(\xi + \varepsilon\kappa t)|^2 \,\mathrm{d}t\, \mathrm{d}\xi = \frac{\varepsilon^2 d}{\varepsilon^d} \int_{\varepsilon Y} |\nabla w(s)|^2 \,\mathrm{d}s.$$

Inserting (3.18) into (3.17) and summing up over all $\lambda_i \in \Lambda_\varepsilon$ gives the desired result. □

The next Lemma quantifies the convergence (2.8) and relies on Lemma 3.6. It is applied to the estimation of the *folding mismatch* $\Delta_1^{u^\varepsilon}$ respective $\Delta_1^{v^\varepsilon}$.

Lemma 3.7. *For all $(u, U) \in H^1(\Omega) \times H^1(\Omega; H^1_{\mathrm{av}}(\mathcal{Y}))$ respective $U \in H^1(\Omega; H^1(\mathcal{Y}))$, there exists a constant $C \geq 0$ such that*

$$\gamma = 0: \quad \| \mathcal{G}_\varepsilon^0(u, U) - u\|_H + \|\nabla[\mathcal{G}_\varepsilon^0(u, U)] - \mathcal{F}_\varepsilon[\nabla u + \nabla_y U]\|_H \leq \varepsilon^{1/2} C, \tag{3.19a}$$

$$\gamma = 1: \quad \| \mathcal{G}_\varepsilon^1 U - \mathcal{F}_\varepsilon U\|_H + \|\varepsilon \nabla[\mathcal{G}_\varepsilon^1 U] - \mathcal{F}_\varepsilon[\nabla_y U]\|_H \leq \varepsilon^{1/2} C. \tag{3.19b}$$

Proof. The proof follows in principle [16, Prop. 2.1]. It is adjusted to the estimate (3.19b) and it utilizes the gradient folding operator $\mathcal{G}^1_\varepsilon$ in the case $\gamma = 1$. In the case $\gamma = 0$, i.e. (3.19a), we resort to $\mathcal{G}^0_\varepsilon$ and we only point out the differences afterwards.

The case $\gamma = 1$: By an orthogonality argument, cf. [38], we may assume that

$$U(x, y) = w(x)z(y) \quad \text{with} \quad w \in X \text{ and } z \in H^1(\mathcal{Y}).$$

Recalling $\mathcal{G}^1_\varepsilon$ in (2.5) and $\mathcal{Q}_\varepsilon$ in (3.16), we decompose $u^\varepsilon := \mathcal{G}^1_\varepsilon U \in X$ as follows

$$u^\varepsilon(x) = \vartheta_\varepsilon(x) + g_\varepsilon(x) \quad \text{with} \quad \vartheta_\varepsilon(x) = \mathcal{Q}_\varepsilon\, w(x)z(\tfrac{x}{\varepsilon}). \tag{3.20}$$

By construction, we have $\vartheta_\varepsilon \in X$ and the remainder $g_\varepsilon \in X$ is defined for each $\varepsilon > 0$ as the solution of the elliptic problem

$$\int_\Omega g_\varepsilon \cdot \varphi + \varepsilon\nabla g_\varepsilon : \varepsilon\nabla\varphi \,\mathrm{d}x = \ell_\varepsilon(\varphi) \quad \text{for all } \varphi \in X, \text{ where} \tag{3.21}$$

$$\ell_\varepsilon(\varphi) = \int_\Omega (\mathcal{F}_\varepsilon U - \vartheta_\varepsilon) \cdot \varphi + (\mathcal{F}_\varepsilon(\nabla_y U) - \varepsilon\nabla\vartheta_\varepsilon) : \varepsilon\nabla\varphi \,\mathrm{d}x.$$

The function g_ε can be estimated as follows

$$\begin{aligned} &\tfrac{1}{2}\left(\|g_\varepsilon\|_H + \|\varepsilon\nabla g_\varepsilon\|_H\right)^2 \le \|g_\varepsilon\|_H^2 + \|\varepsilon\nabla g_\varepsilon\|_H^2 = \ell_\varepsilon(g_\varepsilon) \\ &\le \left(\|\mathcal{F}_\varepsilon U - \vartheta_\varepsilon\|_H + \|\mathcal{F}_\varepsilon(\nabla_y U) - \varepsilon\nabla\vartheta_\varepsilon\|_H\right)\left(\|g_\varepsilon\|_H + \|\varepsilon\nabla g_\varepsilon\|_H\right), \end{aligned} \tag{3.22}$$

which yields $\|g_\varepsilon\|_H + \|\varepsilon\nabla g_\varepsilon\|_H \le 2\left(\|\mathcal{F}_\varepsilon U - \vartheta_\varepsilon\|_H + \|\mathcal{F}_\varepsilon(\nabla_y U) - \varepsilon\nabla\vartheta_\varepsilon\|_H\right)$. Now, we estimate the difference between u^ε and $\mathcal{F}_\varepsilon U$ by adding and subtracting ϑ_ε. Recalling $g_\varepsilon = u^\varepsilon - \vartheta_\varepsilon$ and computing $\varepsilon\nabla\vartheta_\varepsilon = \varepsilon\nabla_x\vartheta_\varepsilon + \nabla_y\vartheta_\varepsilon$, we arrive at

$$\begin{aligned} &\|u^\varepsilon - \mathcal{F}_\varepsilon U\|_H + \|\varepsilon\nabla u^\varepsilon - \mathcal{F}_\varepsilon(\nabla_y U)\|_H \\ &\le \left(\|\vartheta_\varepsilon - \mathcal{F}_\varepsilon U\|_H + \|g_\varepsilon\|_H + \|\varepsilon\nabla\vartheta_\varepsilon - \mathcal{F}_\varepsilon(\nabla_y U)\|_H + \|\varepsilon\nabla g_\varepsilon\|_H\right) \\ &\le 3\left(\|\vartheta_\varepsilon - \mathcal{F}_\varepsilon U\|_H + \|\varepsilon\nabla\vartheta_\varepsilon - \mathcal{F}_\varepsilon(\nabla_y U)\|_H\right) \\ &\le 3\left(\|\vartheta_\varepsilon - \mathcal{F}_\varepsilon U\|_H + \|\nabla_y\vartheta_\varepsilon - \mathcal{F}_\varepsilon(\nabla_y U)\|_H + \varepsilon\|\nabla_x\vartheta_\varepsilon\|_H\right). \end{aligned} \tag{3.23}$$

According to [35, Prop. 4.5] it holds $\|\mathcal{Q}_\varepsilon\, w\|_X \le C\|w\|_X$ and hence $\|\nabla_x\vartheta_\varepsilon\|_H \le C\|\nabla_x U\|_{\mathbb{H}}$. We proceed by estimating the remaining terms in (3.23) with the help of Lemma 3.6

$$\begin{aligned} &\|\vartheta_\varepsilon - \mathcal{F}_\varepsilon U\|_H + \|\nabla_y\vartheta_\varepsilon - \mathcal{F}_\varepsilon(\nabla_y U)\|_H \\ &= \|(\mathcal{Q}_\varepsilon\, w - \mathcal{F}_\varepsilon\, w)z(\cdot/\varepsilon)\|_H + \|(\mathcal{Q}_\varepsilon\, w - \mathcal{F}_\varepsilon\, w)\nabla_y z(\cdot/\varepsilon)\|_H \le \varepsilon^{1/2}C\|w\|_X\|z\|_{H^1(\mathcal{Y})} \end{aligned}$$

and thus (3.19b) is proved.

The case $\gamma = 0$: In (3.20), we set $u^\varepsilon := \mathcal{G}^0_\varepsilon(u, U)$ and decompose $u^\varepsilon = \eta_\varepsilon + g_\varepsilon$, where $\eta_\varepsilon = u + \varepsilon\vartheta_\varepsilon$ and $\vartheta_\varepsilon(x) = (\mathcal{Q}_\varepsilon\, w)(x)z(x/\varepsilon)$ for $U(x, y) = w(x)z(y)$.

In (3.21), we use $(g_\varepsilon, \varphi)_X = \ell_\varepsilon(\varphi)$ for all $\varphi \in X$ with $\ell_\varepsilon(\varphi) = \int_\Omega (u - \eta_\varepsilon) \cdot \varphi + (\mathcal{F}_\varepsilon[\nabla u + \nabla_y U] - \nabla\eta_\varepsilon) : \nabla\varphi \,\mathrm{d}x$.

As in (3.22), we have $\|g_\varepsilon\|_H + \|\nabla g_\varepsilon\|_H \le 2\left(\|u - \eta_\varepsilon\|_H + \|\mathcal{F}_\varepsilon[\nabla u + \nabla_y U] - \nabla\eta_\varepsilon\|_H\right)$.

In (3.23), we have $\nabla\eta_\varepsilon = \nabla u + \varepsilon\nabla_x\vartheta_\varepsilon + \nabla_y\vartheta_\varepsilon$ and hence $\|u^\varepsilon - u\|_H + \|\mathcal{F}_\varepsilon(\nabla u) - \nabla u\|_H + \|\nabla u^\varepsilon - \mathcal{F}_\varepsilon[\nabla u + \nabla_y U]\|_H \le 3(\varepsilon\|\vartheta_\varepsilon\|_H + \|\nabla_y\vartheta_\varepsilon - \mathcal{F}_\varepsilon(\nabla_y U)\|_H + \varepsilon\|\nabla_x\vartheta_\varepsilon\|_H)$. Again, the application of Lemma 3.5 & 3.6 and (3.15) as well as the improved regularity $(u, U) \in H^2(\Omega) \times H^1(\Omega; H^1_{\mathrm{av}}(\mathcal{Y}))$ give (3.19a). □

We use Lemma 3.8 below to estimate the *periodicity defect error* $\Delta_2^{u^\varepsilon}$ respective $\Delta_2^{v^\varepsilon}$.

Lemma 3.8 ([18, Thm. 2.2 & 2.3]). *For every $u \in X$ with $\|u\|_X \leq c$ ($\gamma = 0$) and $\|u\|_H + \varepsilon\|\nabla u\|_H \leq c$ ($\gamma = 1$), there exists a function $\Psi_\varepsilon \in \mathbb{X}$ and $\Psi_\varepsilon \in \mathbb{X}_{\mathrm{av}}$, respectively, and a constant $C \geq 0$, only depending on Ω and Y, such that*

$$\begin{aligned} \gamma = 0: &\quad \|\Psi_\varepsilon\|_{\mathbb{X}} \leq C\|u\|_X \text{ and } \|\mathcal{T}_\varepsilon(\nabla u) - \{\nabla u + \nabla_y \Psi_\varepsilon\}\|_{L^2(Y;X^*)} \leq \varepsilon^{1/2} C\|u\|_X, \\ \gamma = 1: &\quad \|\Psi_\varepsilon\|_{\mathbb{X}} \leq C\left(\|u\|_H + \varepsilon\|\nabla u\|_H\right) \text{ and} \\ &\quad \|\mathcal{T}_\varepsilon u - \Psi_\varepsilon\|_{H^1(Y;X^*)} \leq \varepsilon^{1/2} C\left(\|u\|_H + \varepsilon\|\nabla u\|_H\right). \end{aligned}$$

3.4. Proof of Theorem 3.2

Proof of Theorem 3.2. By the uniform bounds (3.7), all functions are continuous in time and thus we can restore to work with estimates pointwise for all $t \in [0, T]$.

Step 1a: $\frac{\mathrm{d}}{\mathrm{d}t}\|u^\varepsilon - u\|_H^2$-estimate. We refer to [38, Sect. 2.1] for the complete proof of the Gronwall-type estimate (3.13). For simplicity in notation we suppress the index $i = 1$ so that the error terms $\Delta^{u^\varepsilon} := \sum_{i=1}^5 |\Delta_i^{u^\varepsilon}|$ take the precise formulations

$$\begin{aligned} \Delta_1^{u^\varepsilon} &:= \int_\Omega (f^\varepsilon(u^\varepsilon, v^\varepsilon) - u_t^\varepsilon) \cdot (u - \mathcal{G}_\varepsilon^0(u, U)) \\ &\qquad - \mathbb{D}^\varepsilon \nabla u^\varepsilon : (\mathcal{F}_\varepsilon[\nabla u + \nabla_y U] - \nabla \mathcal{G}_\varepsilon^0(u, U))\,\mathrm{d}x, && (3.24) \\ \Delta_2^{u^\varepsilon} &:= \int_\Omega (f_{\mathrm{eff}}(u, V) - u_t) \cdot u^\varepsilon\,\mathrm{d}x - \int_{\Omega\times\mathcal{Y}} \mathbb{D}[\nabla u + \nabla_y U] : \mathcal{T}_\varepsilon(\nabla u^\varepsilon)\,\mathrm{d}x\,\mathrm{d}y, && (3.25) \\ \Delta_3^{u^\varepsilon} &:= \int_{\Omega\times\mathcal{Y}} (\mathbb{D} - \mathcal{T}_\varepsilon \mathbb{D}^\varepsilon)[\nabla u + \nabla_y U] : [\mathcal{T}_\varepsilon(\nabla u^\varepsilon) - \{\nabla u + \nabla_y U\}]\,\mathrm{d}x\,\mathrm{d}y, && (3.26) \\ \Delta_4^{u^\varepsilon} &:= \int_\Omega [f^\varepsilon(u, \mathcal{F}_\varepsilon V) - f_{\mathrm{eff}}(u, V)] \cdot (u^\varepsilon - u)\,\mathrm{d}x. && (3.27) \\ \Delta_5^{u^\varepsilon} &= 2L\|V - \mathcal{T}_\varepsilon \mathcal{F}_\varepsilon V\|_{\mathbb{H}}^2. && (3.28) \end{aligned}$$

Step 1b: Estimation of Δ^{u^ε} and (3.11). We derive quantitative estimates of the errors $\Delta_1^{u^\varepsilon}, \ldots, \Delta_5^{u^\varepsilon}$. We estimate the error $\Delta_1^{u^\varepsilon}$ in (3.24) with Lemma 3.7 and Lemma 3.5, viz.

$$\begin{aligned} |\Delta_1^{u^\varepsilon}| &= \left| \int_\Omega (f^\varepsilon(u^\varepsilon, v^\varepsilon) - u_t^\varepsilon) \cdot (u - \mathcal{G}_\varepsilon^0(u, U))\,\mathrm{d}x - \mathbb{D}^\varepsilon \nabla u^\varepsilon : [\mathcal{F}_\varepsilon[\nabla u + \nabla_y U] - \nabla \mathcal{G}_\varepsilon^0(u, U)]\,\mathrm{d}x \right| \\ &\leq C(C_b)\left(\|u - \mathcal{G}_\varepsilon^0(u, U)\|_H + \|\mathcal{F}_\varepsilon[\nabla u + \nabla_y U] - \nabla \mathcal{G}_\varepsilon^0(u, U)\|_H\right) \\ &\leq \varepsilon^{1/2} C, && (3.29) \end{aligned}$$

where $C = C(C_b, , \|U\|_{H^1(\Omega;H^1(\mathcal{Y}))}, \|u\|_{H^2(\Omega)})$ and we used (3.6.A2) and (3.7) to estimate the first integral.

We treat the second term $\Delta_2^{u^\varepsilon}$ in (3.25) with Lemma 3.8. There exists a two-scale function Ψ_ε so that $(u^\varepsilon, \Psi_\varepsilon) \in X \times \mathbb{X}_{\mathrm{av}}$ is an admissible test function in the weak formulation of $(3.2.\mathrm{P}_0)_1$ and hence

$$0 \equiv \int_\Omega (f_{\mathrm{eff}}(u, V) - u_t) \cdot u^\varepsilon\,\mathrm{d}x - \int_{\Omega\times\mathcal{Y}} \mathbb{D}[\nabla u + \nabla_y U] : [\nabla u^\varepsilon + \nabla_y \Psi_\varepsilon]\,\mathrm{d}x\,\mathrm{d}y. \qquad (3.30)$$

Subtracting (3.30) from (3.25) yields with Hölder's inequality and (3.6.A5)–(3.6.A6)

$$\begin{aligned} |\Delta_2^{u^\varepsilon}| &= \left| \int_{\Omega\times\mathcal{Y}} \mathbb{D}[\nabla u + \nabla_y U] : [\mathcal{T}_\varepsilon(\nabla u^\varepsilon) - \{\nabla u^\varepsilon + \nabla_y \Psi_\varepsilon\}]\,\mathrm{d}x\,\mathrm{d}y \right| \\ &\leq \|\mathbb{D}[\nabla u + \nabla_y U]\|_{H^1(\Omega;L^2(Y))} \|\mathcal{T}_\varepsilon(\nabla u^\varepsilon) - \{\nabla u^\varepsilon + \nabla_y \Psi_\varepsilon\}\|_{L^2(Y;X^*)} \\ &\leq \varepsilon^{1/2} C(C_b, \|\mathbb{D}\|_{W^{1,\infty}(\Omega;L^\infty(\mathcal{Y}))}, \|U\|_{H^1(\Omega;H^1(\mathcal{Y}))}). && (3.31) \end{aligned}$$

The third term $\Delta_3^{u^\varepsilon}$ in (3.26) is treated with Hölder's inequality and Lemma 3.5:

$$\begin{aligned}|\Delta_3^{u^\varepsilon}| &= \left|\int_{\Omega\times\mathcal{Y}} (\mathbb{D} - \mathcal{T}_\varepsilon\, \mathbb{D}^\varepsilon)[\nabla u + \nabla_y U] : [\nabla u + \nabla_y U - \mathcal{T}_\varepsilon(\nabla u^\varepsilon)]\,\mathrm{d}x\,\mathrm{d}y\right| \\ &\le C(C_b)\|(\mathbb{D} - \mathcal{T}_\varepsilon\, \mathcal{F}_\varepsilon\, \mathbb{D})[\nabla u + \nabla_y U]\|_{\mathbb{H}} \\ &\le \varepsilon^{1/2} C(C_b, \|\mathbb{D}\|_{W^{1,\infty}(\Omega;L^\infty(\mathcal{Y}))}). \end{aligned} \tag{3.32}$$

The estimation of $\Delta_4^{u^\varepsilon}$ in (3.27) is a little more involved. Applying (2.2) only to the first term in (3.27) yields

$$\Delta_4^{u^\varepsilon} = \int_{\Omega\times\mathcal{Y}} \mathcal{T}_\varepsilon\, f^\varepsilon(\mathcal{T}_\varepsilon\, u, \mathcal{T}_\varepsilon\, \mathcal{F}_\varepsilon\, V)\cdot \mathcal{T}_\varepsilon(u^\varepsilon - u) - F(u,V)\cdot(u^\varepsilon - u)\,\mathrm{d}x\,\mathrm{d}y.$$

Introducing the terms $\pm F(\mathcal{T}_\varepsilon\, u, \mathcal{T}_\varepsilon\, \mathcal{F}_\varepsilon\, V)\cdot \mathcal{T}_\varepsilon(u^\varepsilon - u)$ & $\pm F(u,v)\cdot \mathcal{T}_\varepsilon(u^\varepsilon - u)$, applying Hölder's inequality, and recalling the assumptions (3.7) & (3.6.A2) gives

$$\begin{aligned}|\Delta_4^{u^\varepsilon}| &\le \|\mathcal{T}_\varepsilon\, f^\varepsilon(\mathcal{T}_\varepsilon\, u, \mathcal{T}_\varepsilon\, \mathcal{F}_\varepsilon\, V) - F(\mathcal{T}_\varepsilon\, u, \mathcal{T}_\varepsilon\, \mathcal{F}_\varepsilon\, V)\|_{\mathbb{H}}\|\mathcal{T}_\varepsilon(u^\varepsilon - u)\|_{\mathbb{H}} \\ &\quad + \|F(\mathcal{T}_\varepsilon\, u, \mathcal{T}_\varepsilon\, \mathcal{F}_\varepsilon\, V) - F(u,V)\|_{\mathbb{H}}\|\mathcal{T}_\varepsilon(u^\varepsilon - u)\|_{\mathbb{H}} \\ &\quad + \|F(u,V)\|_{\mathbb{H}}\|\mathcal{T}_\varepsilon(u^\varepsilon - u) - (u^\varepsilon - u)\|_{\mathbb{H}} \\ &\le C(L, C_F, C_b)\,(\|\mathcal{T}_\varepsilon\, \mathcal{F}_\varepsilon\, F(\mathcal{T}_\varepsilon\, u, \mathcal{T}_\varepsilon\, \mathcal{F}_\varepsilon\, V) - F(\mathcal{T}_\varepsilon\, u, \mathcal{T}_\varepsilon\, \mathcal{F}_\varepsilon\, V)\|_{\mathbb{H}} &(3.33) \\ &\quad + \|\mathcal{T}_\varepsilon\, u - u\|_{\mathbb{H}} + \|\mathcal{T}_\varepsilon\, \mathcal{F}_\varepsilon\, V - V\|_{\mathbb{H}} + \|\mathcal{T}_\varepsilon(u^\varepsilon - u) - (u^\varepsilon - u)\|_{\mathbb{H}}). &(3.34)\end{aligned}$$

We exploit the Lipschitz continuity of F (3.6.A5) in (3.33) and we apply Lemma 3.5 resp. (3.15) in (3.34) so that we arrive at

$$|\Delta_4^{u^\varepsilon}| \le \varepsilon^{1/2} C(L, C_b, C_F, \|V\|_{H^1(\Omega;L^2(\mathcal{Y}))}). \tag{3.35}$$

For the last error term we have immediately

$$|\Delta_5^{u^\varepsilon}| = 2L\|V - \mathcal{T}_\varepsilon\, \mathcal{F}_\varepsilon\, V\|_{\mathbb{H}}^2 \le \varepsilon C(L, \|V\|_{H^1(\Omega;L^2(\mathcal{Y}))}). \tag{3.36}$$

Recalling the Gronwall-type estimate (3.13), we combine the estimates (3.29), (3.31)–(3.32), (3.35)–(3.36) and hence we obtain the quantitative estimate (3.11).

Step 2a: $\frac{\mathrm{d}}{\mathrm{d}t}\|\mathcal{T}_\varepsilon\, v^\varepsilon - V\|_{\mathbb{H}}^2$*-estimate.* For brevity we skip the index $i = 2$ in this step and the following. The proof of (3.14) can be found in [34, Proof of Thm. 4.1(Step2–5)], where the following error terms $\Delta^{v^\varepsilon} := \sum_{i=1}^5 |\Delta_i^{v^\varepsilon}|$ are derived

$$\Delta_1^{v^\varepsilon} := \int_\Omega (f^\varepsilon(u^\varepsilon, v^\varepsilon) - v_t^\varepsilon)\cdot(\mathcal{F}_\varepsilon\, V - \mathcal{G}_\varepsilon^1\, V) - \varepsilon\mathbb{D}^\varepsilon\nabla v^\varepsilon : [\mathcal{F}_\varepsilon(\nabla_y V) - \varepsilon\nabla(\mathcal{G}_\varepsilon^1\, V)]\,\mathrm{d}x, \tag{3.37}$$

$$\Delta_2^{v^\varepsilon} := \int_{\Omega\times\mathcal{Y}} [F(u,V) - V_t]\cdot \mathcal{T}_\varepsilon\, v^\varepsilon - \mathbb{D}\nabla_y V : \nabla_y(\mathcal{T}_\varepsilon\, v^\varepsilon)\,\mathrm{d}x\,\mathrm{d}y, \tag{3.38}$$

$$\Delta_3^{v^\varepsilon} := \int_{\Omega\times\mathcal{Y}} (\mathbb{D} - \mathcal{T}_\varepsilon\, \mathbb{D}^\varepsilon)\nabla_y V : \nabla_y(\mathcal{T}_\varepsilon\, v^\varepsilon - V)\,\mathrm{d}x\,\mathrm{d}y, \tag{3.39}$$

$$\Delta_4^{v^\varepsilon} := \int_{\Omega\times\mathcal{Y}} [\mathcal{T}_\varepsilon\, f^\varepsilon(\mathcal{T}_\varepsilon\, u, V) - F(u,V)]\cdot(\mathcal{T}_\varepsilon\, v^\varepsilon - V)\,\mathrm{d}x\,\mathrm{d}y, \tag{3.40}$$

$$\Delta_5^{v^\varepsilon} := 2L\|\mathcal{T}_\varepsilon\, u - u\|_{\mathbb{H}}^2. \tag{3.41}$$

Step 2b: Estimation of Δ^{v^ε} *and* (3.12). Applying Lemma 3.7 to the first error term $\Delta_1^{v^\varepsilon}$ in (3.37) yields

$$|\Delta_1^{v^\varepsilon}| \le C(C_b)\left(\|\mathcal{F}_\varepsilon\, V - \mathcal{G}_\varepsilon^1\, V\|_H + \|\mathcal{F}_\varepsilon(\nabla_y V) - \varepsilon\nabla(\mathcal{G}_\varepsilon^1\, V)\|_H\right) \le \varepsilon^{1/2} C, \tag{3.42}$$

where $C = C(C_b, \|V\|_{H^1(\Omega;H^1(\mathcal{Y}))})$.

For the estimation of $\Delta_2^{v^\varepsilon}$ in (3.38), let $\Psi_\varepsilon \in \mathbb{X}$ be as in Lemma 3.8. Then, in particular, Ψ_ε is an admissible test function for $(3.2.\mathrm{P}_0)_2$ and hence the application of Hölder's inequality and Lemma 3.8 gives

$$\begin{aligned}|\Delta_2^{v^\varepsilon}| &\leq \||\mathbb{D}\nabla_y V| + |F(u,V)| + |V_t|\|_{H^1(\Omega;L^2(Y))}\| \mathcal{T}_\varepsilon v^\varepsilon - \Psi_\varepsilon\|_{H^1(Y;X^*)} \\ &\leq \||\mathbb{D}\nabla_y V| + |F(u,V)| + |V_t|\|_{H^1(\Omega;L^2(\mathcal{Y}))}\varepsilon C(\Omega)\left(\|v^\varepsilon\|_H + \varepsilon\|\nabla v^\varepsilon\|_H\right) \\ &\leq \varepsilon^{1/2} C, \end{aligned} \tag{3.43}$$

where $C = C(C_b, C_F, \|\mathbb{D}\|_{W^{1,\infty}(\Omega;L^\infty(\mathcal{Y}))}, \|V\|_{H^1(\Omega;L^2(\mathcal{Y}))}, \|V_t\|_{H^1(\Omega;L^2(\mathcal{Y}))})$.

Recalling $\mathbb{D}^\varepsilon = \mathcal{F}_\varepsilon \mathbb{D}$ and $f^\varepsilon = \mathcal{F}_\varepsilon F$, the error terms $\Delta_3^{v^\varepsilon}$-$\Delta_5^{v^\varepsilon}$ in (3.39)–(3.41) are estimated easily by using Lemma 3.5:

$$|\Delta_3^{v^\varepsilon}| \leq 2C_b\|(\mathbb{D} - \mathcal{T}_\varepsilon \mathbb{D}^\varepsilon)\nabla_y V\|_{\mathbb{H}} \leq \varepsilon^{1/2} C(C_b, \Omega, \|\mathbb{D}\|_{W^{1,\infty}(L^\infty(\mathcal{Y}))}), \tag{3.44}$$

$$|\Delta_4^{v^\varepsilon}| \leq 2C_b\| \mathcal{T}_\varepsilon f^\varepsilon(\mathcal{T}_\varepsilon u, V) - F(u,V)\|_{\mathbb{H}} \leq \varepsilon^{1/2} C(C_b, C_F), \tag{3.45}$$

$$|\Delta_5^{v^\varepsilon}| = 2L\| \mathcal{T}_\varepsilon u - u\|_{\mathbb{H}}^2 \leq \varepsilon C(L, \|u\|_X). \tag{3.46}$$

Overall the Gronwall-type estimate (3.14) and the quantitative estimates (3.42)–(3.46) give (3.12). Hence, we finish the proof of (3.9a) by applying Gronwall's lemma to (3.10).

Step 3: Derivation of (3.9b). According to [34, Proof of Thm. 4.1 (Step 7)], we have the estimate

$$\tfrac{1}{2}\tfrac{\mathrm{d}}{\mathrm{d}t}\| \mathcal{T}_\varepsilon v^\varepsilon - V\|_{\mathbb{H}}^2 \leq -\mu\| \mathcal{T}_\varepsilon(\varepsilon\nabla u^\varepsilon) - \nabla_y V\|_{\mathbb{H}}^2 + 2L\left(\|u^\varepsilon - u\|_H^2 + \| \mathcal{T}_\varepsilon v^\varepsilon - V\|_{\mathbb{H}}^2\right) + \Delta^{v^\varepsilon}.$$

Integrating over $[0,T]$ and exploiting (3.9a) as well as the estimations for Δ^{v^ε} in Step 2b yields

$$\begin{aligned}&\mu\| \mathcal{T}_\varepsilon(\varepsilon\nabla u^\varepsilon) - \nabla_y V\|_{L^2(0,T;\mathbb{H})}^2 \\ &\leq \int_0^T -\tfrac{1}{2}\tfrac{\mathrm{d}}{\mathrm{d}t}\| \mathcal{T}_\varepsilon v^\varepsilon - V\|_H^2 + 2L\left(\|u^\varepsilon - u\|_H^2 + \| \mathcal{T}_\varepsilon v^\varepsilon - V\|_{\mathbb{H}}^2\right) + |\Delta^{v^\varepsilon}|\,\mathrm{d}t \leq T\varepsilon^{1/2}C.\end{aligned}$$

The estimation of the gradient follows analogously for the classically diffusing species u^ε. With this, we finish the proof of Theorem 3.2. □

3.5. Discussion

We close the paper with a brief comparison of the obtained convergence rates. In [20], a nonlinearly coupled system of reaction-diffusion systems is considered on a cubical domain $\Omega \subset \mathbb{R}^3$ with exactly periodic, porous microstructure. The system does not include slowly diffusing species v^ε, but rather nonlinear boundary conditions at the surface of the pores. For the classically diffusing species u^ε the convergence rate $\varepsilon^{1/2}$ is rigorously proved by the method of periodic unfolding and results from [17, 18]. We emphasize that the gradient term is squared in [20, Thm 3.6], which means $\|\nabla u + \nabla_y U - \mathcal{T}_\varepsilon(\nabla u^\varepsilon)\|_{L^2(0,T;\mathbb{H})} \leq O(\varepsilon^{1/4})$. This error estimate is comparable with the one in Theorem 3.2. The focus of this text is the convergence of the slowly diffusing species v^ε which is strongly two-scale converging, as are $\varepsilon\nabla v^\varepsilon$ and ∇u^ε. In contrast, u^ε is strongly converging in $L^2(\Omega)$ and hence, the improved rate $\|u^\varepsilon - u\|_H \leq O(\varepsilon^{1/2})$ up to the boundary is not to expect for v^ε.

In [24, 21], nonlinearly coupled systems of reaction-diffusion equations involving diffusion length scales of order $O(1)$ and $O(\varepsilon)$ are considered in a heterogeneous setting. Whereas in [24] the coefficient functions are of the form $\mathbb{D}(x, x/\varepsilon)$, in [21], the heterogeneities in the domain $\Omega \subset \mathbb{R}^2$ are not arranged in a strictly periodic manner. In both cases, the approach of formal asymptotic expansion is used and then, the convergence rate $O(\varepsilon^{1/2})$ is proved under the assumption of significantly more spatial regularity of the limit solution. In Theorem 3.3, our method reproduces the rate $O(\varepsilon^{1/2})$ as in [24, Thm. 4.5] and [21, Thm. 3.1] under significantly weaker assumptions on the given data and limit solutions.

Acknowledgement. This research was funded by Deutsche Forschungsgemeinschaft within the *Collaborative Research Center 910: Control of self-organized systems* through project *A5: Pattern formation in systems with multiple scales.* The author gratefully thanks M. Thomas, A. Mielke, O. Klein, and S. Neukamm for helpful discussions and comments.

References

[1] Bensoussan A, Lions J L and Papanicolaou G 1978 *Asymptotic analysis for periodic structures* (*Studies in Mathematics and its Applications* vol 5) (North-Holland Publishing Co., Amsterdam-New York)
[2] Jikov V V, Kozlov S M and Oleĭnik O A 1994 *Homogenization of differential operators and integral functionals* (Springer-Verlag, Berlin) translated from the Russian by G. A. Yosifian
[3] Marchenko V A and Khruslov E Y 2006 *Homogenization of partial differential equations* (*Progress in Mathematical Physics* vol 46) (Birkhäuser Boston, Inc., Boston, MA)
[4] Tartar L 2009 *The general theory of homogenization* (*Lecture Notes of the Unione Matematica Italiana* vol 7) (Springer-Verlag, Berlin; UMI, Bologna) a personalized introduction
[5] Nguetseng G 1989 *SIAM J. Math. Anal.* **20** 608–623
[6] Allaire G 1992 *SIAM J. Math. Anal.* **23** 1482–1518
[7] Peter M A and Böhm M 2008 *Math. Meth. Appl. Sci.* **24** 119–136
[8] Meier S A and Muntean A 2010 *Current advances in nonlinear analysis and related topics* (*GAKUTO Internat. Ser. Math. Sci. Appl.* vol 32) (Gakkōtosho, Tokyo) pp 443–461
[9] Cioranescu D, Damlamian A and Griso G 2002 *C. R. Math. Acad. Sci. Paris* **335** 99–104
[10] Visintin A 2004 *Atti Acad. Naz. Lincei Cl. Sci. Fis. MAt. Natur. Rend. Lincei (9) Mat. Appl.* **15** 93–107
[11] Visintin A 2006 *ESAIM Control Optim. Calc. Var.* **12** 371–397 (electronic)
[12] Visintin A 2008 *Proc. Roy. Soc. Edinburgh Sect. A* **138** 1363–1401
[13] Mielke A and Timofte A 2007 *SIAM J. Math. Anal.* **39** 642–668
[14] Neuss-Radu M and Jäger W 2007 *SIAM J. Math. Anal.* **39** 687–720
[15] Ptashnyk M and Roose T 2010 *SIAM J. Appl. Math.* **70** 2097–2118
[16] Hanke H 2011 *Math. Models Methods Appl. Sci.* **21** 1651–1684
[17] Griso G 2004 *Asymptot. Anal.* **40** 269–286
[18] Griso G 2005 *C. R. Math. Acad. Sci. Paris* **340** 251–254
[19] Onofrei D and Vernescu B 2007 *Asymptot. Anal.* **54** 103–123
[20] Fatima T, Muntean A and Ptashnyk M 2012 *Appl. Anal.* **91** 1129–1154
[21] Muntean A and van Noorden T L 2013 *European J. Appl. Math.* **24** 657–677
[22] Matache A M and Schwab C 2002 *M2AN Math. Model. Numer. Anal.* **36** 537–572
[23] E W and Engquist B 2003 *Comm. Math. Sci.* **1** 87–132
[24] Eck C 2004/05 *Multiscale Model. Simul.* **3** 1–27 (electronic)
[25] Chalupecký V, Fatima T and Muntean A 2010 *J. Math-for-Ind.* **2B** 171–181
[26] Chalupecký V and Muntean A 2012 *Jpn. J. Ind. Appl. Math.* **29** 289–316
[27] Fillep S, Orlik J, Bare Z and Steinmann P 2014 *Math. Mech. Solids* **19** 1011–1021
[28] Bedeaux D and Kapral R 1983 *J. Chem. Phys.* **79** 1783
[29] Alonso S, Bär M and Kapral R 2009 *J. Chem. Phys.* **131** 214102
[30] Alonso S, Bär M and Kapral R 2009 *Phys. Rev. Lett.* **102** 238302
[31] Keener J P 2000 *Physica D* **136** 1
[32] Kaliske M, Kolling S and Timmel M 2009 *ZAMM. Zeitschrift für Angewandte Mathematik und Mechanik* **89** 698–708
[33] Dascalu C, François B and Keita O 2010 *International Journal of Solids and Structures* **47** 493–502
[34] Mielke A, Reichelt S and Thomas M 2014 *Netw. Heterog. Media* **9** 353–382
[35] Cioranescu D, Damlamian A and Griso G 2008 *SIAM J. Math. Anal.* **40** 1585–1620
[36] Damlamian A 2005 *Math. Sci. Appl.* **24** 119–136
[37] Lukkassen D, Nguetseng G and Wall P 2002 *Int. J. Pure Appl. Math.* **2** 35–86
[38] Reichelt S 2015 *Two-scale homogenization of systems of nonlinear parabolic equations* Ph.D. thesis Humboldt-Universität zu Berlin in preparation

On damping characteristics of frictional hysteresis in pre-sliding range

Michael Ruderman* and Makoto Iwasaki**
* Department of Engineering Sciences
University of Agder, 4879 Grimstad, Norway
** Department of Computer Science and Engineering
Nagoya Institute of Technology, 466-8555 Nagoya, Japan
E-mail: michael.ruderman@uia.no

Abstract. Frictional hysteresis at relative motion in the pre-sliding range is considered. This effect is characterized by an elasto-plastic interaction, and that on the micro-scale, between two rubbing surfaces in contact that gives rise to nonlinear friction force. The pre-sliding friction force yields hysteresis in displacement. In this study, the damping characteristics of frictional hysteresis are analyzed. It is worth noting that we exclude the viscous damping mechanisms and focus on the pure hysteresis damping to be accounted in the friction modeling. The general properties of pre-sliding friction hysteresis are demonstrated and then compared with the limit case of discontinuous Coulomb friction. Further we consider two advanced dynamic friction models, LuGre and Maxwell-slip, so as demonstrate their damping properties and convergence of the motion system to equilibrium state. Experimental observations of the free motion in pre-sliding range are also shown and discussed.

1. Introducing remarks

The frictional hysteresis in pre-sliding range is one of the least understood phenomena of kinetic friction in mechanical systems. It should be stressed that besides a physics-based tribological study of friction effects (see e.g. [1]), the latter can be equally investigated from the system and control point of view, thus operating with the scalar force and velocity. The earlier works on modeling, analysis, and control methods of friction in machines (see the survey [2] and references therein) already emphasized the significance of pre-sliding friction range and related hysteresis in displacement. The convergence of (PID) controlled motion with Coulomb and static friction and possible appearance of the limit cycles has been studied in [3]. Later, the friction-induced limit cycles have been described and demonstrated in [4] for static and dynamic (LuGre) friction models. Recently the convergence of a feedback-controlled motion has been analyzed in context of pre-sliding hysteresis and evaluated experimentally in [5]. A theoretical analysis of pre-sliding frictional hysteresis acting as a feedback to the moving mass can be found in [6]. The pre-sliding behavior of several state-of-the-art dynamic friction models has been classified and compared in terms of the proposed common metrics (so-called Z-properties chart) in [7]. Further it is worth noting that the system-oriented dynamic friction modeling undergos continuous advancements and refinements which are related to the novel experimental observations on the one hand, and model applicability in control practice on the other hand, see e.g. [1, 8, 9].

In this paper, we analyze the damping characteristics of frictional hysteresis in pre-sliding range and investigate the generic force-displacement hysteresis properties in a vicinity to equilibrium of the free motion after an initial steady-state. Here it is worth noting that the end of a free motion due to nonlinear hysteresis is connatural to an unforced behavior of elastic-plastic oscillators. The earlier works in that direction have been also reported e.g. in [10, 11, 12].

2. General consideration

We start our analysis by considering the unforced one-degree-of-freedom relative motion subject to the Coulomb friction. For the sake of simplicity and without loss of generality we assume a unity mass and unity Coulomb friction coefficient, so that the relative motion in $(y, \dot{y})$-coordinates can be described by the following differential equation

$$\ddot{y} = -\text{sign}(\dot{y}) \,. \tag{1}$$

Introducing the vector field

$$f(\mathbf{x}) = \begin{pmatrix} x_2 \\ -\text{sign}(x_2) \end{pmatrix} \quad \text{with} \quad \mathbf{x} = (x_1,\, x_2)^T \in \mathbb{R}^2 \tag{2}$$

we obtain the state-space representation of system (1) as

$$\begin{pmatrix} \dot{x}_1 \\ \dot{x}_2 \end{pmatrix} = \begin{pmatrix} x_2 \\ -\text{sign}(x_2) \end{pmatrix}, \tag{3}$$

which is in relative displacement-velocity coordinates. Since the vector field does not depend on the relative displacement we consider the single-variate Lyapunov function $V(x_2) = 1/2\, x_2^2$ which constitutes kinetic energy of the moving mass. For the time derivative of Lyapunov function $d/dtV(\mathbf{x}) = \nabla V(\mathbf{x}) f(\mathbf{x})$ one easily obtains

$$\frac{d}{dt}V(x_2) = -|x_2|, \tag{4}$$

which implies that

$$\forall\, x_2 \neq 0 \quad \frac{d}{dt}V(x_2) < 0.$$

Since $V(x_2)$ is radially unbounded, the inequality above proves the global system stability in Lyapunov sense, and one can conclude that any unforced motion trajectory, i.e. starting at any point $\{\mathbf{x} : x_2 \neq 0\}$, will approach the equilibrium set $E = \{\mathbf{x} \in \mathbb{R}^2 : x_2 = 0\}$ in finite time. For equilibrium set E at zero velocity, which is a discontinuity manifold due to the sign-operator of Coulomb friction, we proceed similar to [13] by applying the invariance principle, also known as LaSalle's invariance principle (see e.g. [14] for details). The discontinuity manifold E divides the total phase plane P into two regions $P^+ = \{\mathbf{x} \in \mathbb{R}^2 : x_2 > 0\}$ and $P^- = \{\mathbf{x} \in \mathbb{R}^2 : x_2 < 0\}$. At each discontinuity point $\mathbf{x}_e \in E$ the vector field is oriented in opposite directions and is normal to the x_1-axis that follows from

$$f^+(\mathbf{x}_e) = \lim_{\mathbf{x} \to \mathbf{x}_e,\, \mathbf{x} \in P^+} f(\mathbf{x}) = \begin{pmatrix} 0 \\ -1 \end{pmatrix}, \tag{5}$$

$$f^-(\mathbf{x}_e) = \lim_{\mathbf{x} \to \mathbf{x}_e,\, \mathbf{x} \in P^-} f(\mathbf{x}) = \begin{pmatrix} 0 \\ +1 \end{pmatrix}. \tag{6}$$

Therefore, it can be concluded that once a trajectory reaches the x_1-axis it stays there forever. This is a significant issue since it shows that the Coulomb friction does not only provide a constant damping rate at unidirectional motion, i.e. according to $|\ddot{y}| = const$, but equally ensures an equilibrium state once reaching zero velocity.

Now, we are to analyze the friction hysteresis in displacement which is known to occur within a micro-displacement range, often denoted as pre-sliding range or pre-sliding region, each time the motion direction changes. Assuming the total friction force at an unidirectional steady-state motion is captured by the Coulomb friction[1] $F = \text{sign}(\dot{y})C$, where $C > 0$ is the Coulomb friction coefficient, we focus on the friction force transitions in a vicinity to motion reversals. Consider the friction trajectories in the force-displacement coordinates as depicted in Figure 1. Let us discuss in details and compare with each other three principally different reversal transitions as these are sketched in the figure.

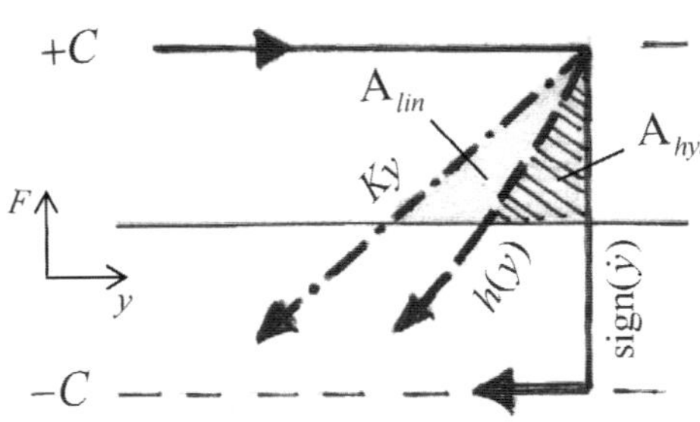

Figure 1. Friction-displacement curves after a motion reversal. The case of elastic contacts with linear stiffness is denoted by Ky, of hysteresis transition by $h(y)$, and the limit case of Coulomb friction is denoted by $\text{sign}(\dot{y})$. The shaded and scratch-through areas, A_{lin} and A_{hys} correspondingly, represent the energy recuperated at motion reversal in the linear and hysteresis case.

The first reversal transition, denoted by Ky, represents a purely elastic case with linear stiffness K of frictional contacts. It should be noted that the combination of steady-state Coulomb friction and Ky reversal transitions yields the behavior of classical Maxwell-slip element, as it will be addressed in details in Section 3.2 when discussing the Maxwell-slip friction model. Recall that in the transition characteristics the Maxwell-slip element coincides with the stop-type Prandtl-Ishlinskii operator. For more details on the latter we refer to e.g. [15], [16].

The next reversal transition, denoted by $h(y)$, is of hysteresis type. Without assuming a particular analytical form, the $h(y)$ function can be imposed on the certain (general) properties which should be fulfilled in order to map the pre-sliding hysteresis friction:

(a) the function $h(y)$ is defined on the interval $[y_r, y_r + \text{sign}(\dot{y})s]$ where y_r is the recent reversal point and $s > 0$ is a finite constant characterizing the recent pre-sliding range;

(b) the function $h(y)$ is monotonically non-decreasing and continuously differentiable (at least once);

(c) the function $dh(y)/dy$ is non-negative and monotonically non-decreasing if $\dot{y} < 0$ and non-increasing if $\dot{y} > 0$. This yields $h(y)$ convex for $\dot{y} < 0$ and concave for $\dot{y} > 0$;

(d) the function $h(y)$ is bounded by $\max|h(y)| = C$;

(e) at the domain boundary $y = y_r + \text{sign}(\dot{y})s$ the function $dh(y)/dy = 0$, i.e. $h(y)$ is saturated.

Corollary, the properties (d) and (e) imply that after a pre-sliding range the friction force reaches the Coulomb friction level and remains on that until next motion reversal.

The last reversal transition, denoted in Figure 1 by $\text{sign}(\dot{y})$, corresponds to the Coulomb friction behavior, where the friction force changes stepwise at each motion reversal. This represents a limit case for hysteresis transition when $s \to 0$. It can be seen that the $h(y)$ transition passes into the $\text{sign}(\dot{y})$ one when allowing $s = 0$.

[1] As stated in the Introduction part, the recent work is concerned only with the hysteresis damping characteristics of pre-sliding friction. Therefore, the total friction force F considered here does not include the viscous, Stribeck, and transient (dynamic) effects of kinetic friction.

We are now in a position to analyze the damping characteristics of each reversal transition as in Figure 1. For this purpose consider the unforced motion of a unity mass. The concentrated (point) mass is in contact with a homogenous fixed horizontal surface, so that the assumed one-degree-of-freedom motion is antagonized by the tangential friction force F, thus yielding

$$\ddot{y} = -F(\dot{y})\,. \tag{7}$$

The case where the total friction force is Coulomb (see eq.(1)) has been already analyzed above, turning out that the damping rate is constant until the abrupt stop, which occurs once reaching zero velocity. Henceforth, let us consider the energy balance at motion reversal for two residual transition cases, i.e. linear case denoted by Ky and hysteresis case denoted by $h(y)$. Note that the recent reversal where the sign($\dot{y}$) changes from "+" to "−" is identical to that from "−" to "+" due to the symmetry. Thus, we discuss the situation as illustrated in Figure 1 without loss of generality. Further, we denote the relative position where the corresponding force curve crosses the y-axis by y_0.

At a motion reversal, the kinetic energy is zero due to $\dot{y} = 0$. However, in case of a linear transition the potential energy $W_{lin} = 1/2\,K(y_r - y_0)^2$ is stored due to elastic contacts with stiffness K. It means that when moving back from y_r to y_0, i.e. after a motion reversal, the total amount of energy is recuperated that corresponds to the area A_{lin} indicated in Figure 1. That allows the system continuing to move into the same direction after crossing zero force, while the magnitude of counteracting friction force will increase according to $F = K(y - y_0)$. Since $y_0 = const$ for each particular reversal at y_r, we introduce the local motion variable $z = y - y_0$ with $\ddot{z} = \ddot{y}$ and obtain the system dynamics at reversal transitions as

$$\ddot{z} = -Kz\,. \tag{8}$$

It is easy to recognize that (8) constitutes an undamped harmonic oscillator for which we have an explicit time-domain solution given by

$$z(t) = Z_0 \cos(\omega t) \quad \text{with} \quad Z_0 = y_r - y_0 \quad \text{and} \quad \omega = \sqrt{K}\,. \tag{9}$$

Important to note is that once launched at Z_0, the friction force oscillator will operate between $\pm Z_0$ without passing again to the Coulomb friction state. This is due to the relative motion is unforced. The motion trajectories in the $(y,\, \dot{y})$-plane result in the closed elliptic orbits whose axes depend on the initial state Z_0 and physical parameters, i.e. stiffness and mass. Thus it can be concluded that a purely elastic reversal transition does not provide an asymptotic equilibrium state after reaching zero velocity and, instead of this, induces a stable (linear) limit cycle. This result will be reawakened in Section 3.2 when analyzing damping of the Maxwell-slip friction model.

In case of a hysteresis reversal transition, similar consideration of potential energy stored at a motion reversal can be done, however, with the total amount of recuperated energy $W_{hys} = A_{hys} < A_{lin}$, as obvious from Figure 1. Since the force transition is given by the nonlinear function $h(y)$, the energy balance is determined by $W_{hys}(y) = \int_{y_r}^{y} h(y)dy$ and that until the next motion reversal. In order to provide an analytical solution in the phase-plane we rewrite the system dynamics (7) by using $\ddot{y} = \dot{y}d\dot{y}/dy$ and obtain

$$\dot{y}d\dot{y} = -h(y)dy\,. \tag{10}$$

Integrating (10) with respect to the motion interval, i.e. from y_r to y, yields

$$\int_{y_r}^{y} h(y)dy = -\frac{1}{2}\dot{y}^2\Big|_{\dot{y}_r}^{\dot{y}}\,. \tag{11}$$

Since the velocity sign is a-priori given and the system dynamics (10) is defined between two consecutive motion reversals with $\dot{y}_r = 0$ we can calculate the velocity function as

$$\dot{y}(y) = \mathrm{sign}(\dot{y})\sqrt{2\int_{y_r}^{y} h(y)dy}\,. \tag{12}$$

Inspecting (12) and with respect to the $h(y)$-properties (a)–(e) two significant implications can be drawn.

(i) Since $\left|\int_{y_r}^{y} h(y)dy\right|$ is growing on the interval $[y_r,\, y_0)$ and reducing afterwards, the max $|\dot{y}|$ value occurs at the y_0 position, i.e. at zero-crossing of friction force. That is the $(y_0,\, \max|\dot{y}|)$ state subdivides each hysteresis transition into an acceleration and deceleration phase.

(ii) At the next motion reversal y_r^+ which occurs after the relative velocity becomes again zero $\dot{y}_r^+ = 0$, according to (12), the magnitude of friction force decreases comparing to that of previous reversal, i.e. $\left|h(y_r^+)\right| < \left|h(y_r)\right|$. This follows from the property (c) of hysteresis function $h(y)$ and implies that the potential energy, which can be recuperated after next motion reversal, is lower than that after the previous motion reversal, i.e. $W_{hys}(y_r^+) < W_{hys}(y_r)$.

From implication (ii) it can be concluded that each hysteresis transition dissipates the energy and the motion system (7) attains a stable equilibrium state $(y_r^i, \dot{y}_r^i = 0)$ at the finite number of reversals $i > 1$. As corollary, we can note that the equilibrium state lies always between the first and second reversal point, i.e. $\min(y_r^{\infty}, y_r^{\in}) \leq y_r^i \leq \max(y_r^{\infty}, y_r^{\in})$. Further, it is worth noting that at the reached equilibrium state the residual (stiction) friction force is not necessarily zero and can comply with $0 \leq |F(\dot{y}_r^i = 0)| < C$. This corresponds to adhesion mechanisms of the static (stiction) friction and is out of scope in the recent study. The single explanation, to be given here, on non-zero friction at the reached motion equilibrium is that it can be considered as a Coulomb-like friction behavior where $F(\dot{y}_r^i) = \mathrm{sign}(\dot{y})c$ with $0 < c < C$. This ensures the system remains sticking in the attained equilibrium even if $F \neq 0$, as has been analyzed at the beginning of this Section. Further we note that reaching a stable equilibrium at $i = 1$ corresponds to the Coulomb friction transition $\mathrm{sign}(\dot{y})$ which has been analyzed beforehand.

For the analysis accomplished above we are to demonstrate the unforced motion trajectories, and that for all three reversal transitions. The results are obtained by means of a numerical simulation. For the sake of simplicity we assume the unity mass, unity linear stiffness, and unity Coulomb friction coefficient. In case of hysteresis reversal transition we apply the LuGre friction model, addressed further in Section 3.1. The unforced motion trajectories with an initial state $(y, \dot{y})(0) = (0, 5)$ are shown in the phase-plane in Figure 2 for all three cases. It can be seen that the trajectories confirm the analysis made above.

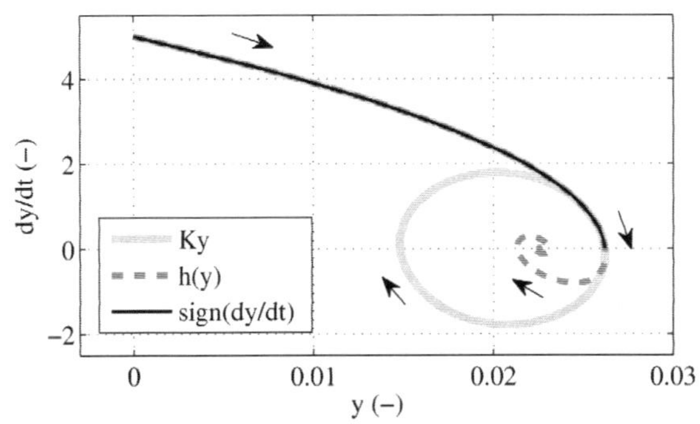

Figure 2. Unforced motion trajectories starting from an initial state $(y, \dot{y})(0) = (0, 5)$. The Coulomb reversal transition provides a stable convergence without transient oscillations. The linear reversal transition results in a stable limit cycle within pre-sliding range. The hysteresis reversal transition yields the nonlinearly-damped (oscillating) transient towards a stable equilibrium.

3. Hysteresis damping in friction models

In this Section we consider two advanced dynamic friction models, i.e. LuGre and Maxwell-slip, and analyze their damping characteristics when mapping the pre-sliding friction range. The selection of friction models for analysis is not influenced by any particular preference and solely bases on their quite different formulation and widespread use in structural mechanics and control. The *LuGre* model [17], which is most widespread in control, extends the Dahl[2] model (see [18], [19] for details) by including the damping of internal (pre-sliding) friction state and allowing for more complex, than only Coulomb, steady-state friction characteristics, i.e. the so-called Stribeck effect. The LuGre friction model incorporates also the viscous friction term which is, however, omitted in the recent study since we are only focusing on the damping characteristics of hysteresis. The *Maxwell-slip* model, or more generally Maxwell-slip structure, is suitable for describing any rate-independent hysteresis between the generalized motion and forces, see e.g. [20] for details. The approach was equally used for modeling the hysteresis behavior of materials and structures in [21], and in particular of piezoelectric stacks in [22]. It should be stressed that the parallel connection of single elasto-slip Maxwell elements renders a structure which replicates the Prandtl-Ishlinskii stop-type hysteresis model, see e.g. [15] for details. In the following Subsections, each friction model is briefly summarized for convenience of the reader. Afterwards, the analysis of damping behavior in vicinity to an equilibrium, i.e. motion stop, is performed for each model when incorporated in the system dynamics (7).

3.1. LuGre model

The LuGre friction model captures the total tangential friction force

$$F(\dot{y}) = \sigma_0 z + \sigma_1 \dot{z} + \sigma_2 \dot{y} \tag{13}$$

as a linear combination of internal friction state z, its damping, and viscous friction term which depends on the relative velocity. The positive parameters σ_0 and σ_1 can be understood as the contact stiffness and micro-damping correspondingly. The non-negative viscous friction coefficient is σ_2. Often, the internal state z is associated with the so-called 'bristle' (or also 'brush') model of frictional interaction of asperities which form the contacting surfaces, as schematically shown in Figure 3 (cf. with [23], [17]). A large number of bristles (with a rather statistical distribution and thus average lengths and distances) deflect like the springs (not necessarily linear) at a relative motion. Thus, the internal state z can be seen as an average deflection of the bristles until a steady motion sets on.

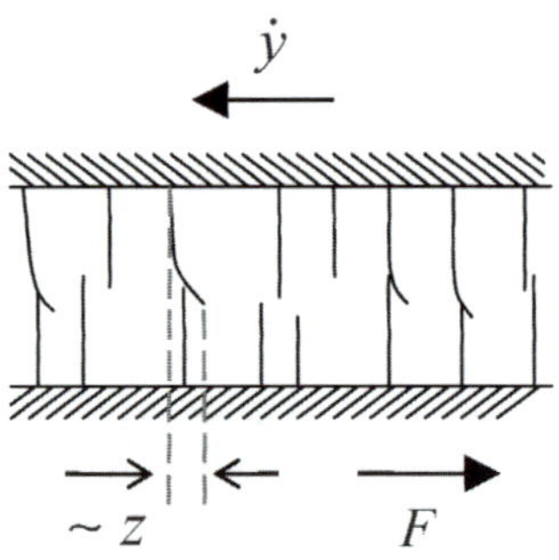

Figure 3. Schematic representation of two sliding surfaces in contact by means of so-called bristle (also brush) friction model. The tangential friction force F acts in opposite direction to the relative motion $\dot{y}$. The internal friction state z is associated with an average deflection of bristles.

[2] The Dahl model was, probably, the first trial to extend the Coulomb friction law so as to resolve the discontinuity at motion reversals. The Dahl model introduced an internal state to describe the reversal transition of friction force, thus capturing the pre-sliding hysteresis. The model was long time a standard approach in the aero-space industry to describe friction behavior in the ball-bearing type mechanisms.

The internal friction state is governed by

$$\dot{z} = \dot{y} - \sigma_0 \frac{|\dot{y}|}{g(\dot{y})} z, \tag{14}$$

where the positive nonlinear static map $g(\cdot)$ captures the Coulomb friction and Stribeck effect. In the following, however, we consider $g = C$ only so as to be in accord with the previous assumptions and hence to exclude other steady-state friction components than the Coulomb friction. For the same reason, we require $\sigma_2 = 0$ so that no viscous friction behavior will be taken into account as mentioned before.

At this stage, some significant properties and parameter conditions posed on the LuGre friction model, and known from the literature, should be briefly mentioned with respect to the dissipative nature of friction. The necessary and sufficient conditions for passivity of the LuGre friction model have been investigated in [24] and a strict, but quite simple at the same time, inequality has been derived in terms of the model parameterization. It postulates the input-output (I/O) dissipativity of an $\dot{y} \mapsto F$ map if and only if (see Property 3 in [19])

$$\sigma_2 > \sigma_1 \frac{F_s - C}{C}, \tag{15}$$

where $F_s \geq C$ corresponds to the stiction force, which is considered for mapping the Stribeck effect (in the original LuGre model). We can note that even if $(F_s - C) \to 0$ the viscous damping $\sigma_2 \neq 0$ is required for rendering the LuGre model I/O dissipative. Since no viscous damping is assumed in the recent study neither straightforward conclusions on passivity, and furthermore stability, of a frictional system can be drown based on the analysis provided in [24], [19].

With assumptions made above on the steady-state friction and viscous damping the total friction force described by the LuGre model becomes

$$F(\dot{y}) = \sigma_1 \dot{y} + \sigma_0 \Bigl(1 - \sigma_1 \frac{|\dot{y}|}{C}\Bigr) z \,. \tag{16}$$

Therefore, the unforced motion dynamics (7) can be written in the state-space form

$$[\dot{z}, \ddot{y}]^T = A(\dot{y}) [z, \dot{y}]^T \,, \tag{17}$$

with the state-dependent system matrix

$$A(\dot{y}) = \begin{pmatrix} -\sigma_0 \frac{|\dot{y}|}{C} & 1 \\ -\sigma_0 \Bigl(1 - \sigma_1 \frac{|\dot{y}|}{C}\Bigr) & -\sigma_1 \end{pmatrix}. \tag{18}$$

To be able to make a statement about the damping characteristics of system (17), (18), or generally first of all about its stability, the linearization of (18) should be made. However, this requires inspecting the total domain of velocity magnitudes which can render our analysis cumbersome. Furthermore, no particular operating points of $|\dot{y}|$ can be identified so as to perform a reasonable linearization. In fact, we are rather interested in the transient dynamics of system (17), (18), i.e. after the first motion reversal and until the final motion stop. The single particular state which is appropriate for linearization is $\dot{y} \to 0$. This is a boundary case close to the motion reversal or final motion stop. Here the system matrix becomes

$$A(0) = \lim_{\dot{y} \to 0} A(\dot{y}) = \begin{pmatrix} 0 & 1 \\ -\sigma_0 & -\sigma_1 \end{pmatrix}. \tag{19}$$

Journal of Physics: Conference Series **727** (2016) 012014 doi:10.1088/1742-6596/727/1/012014

It is evident that for any $\sigma_0 > 0$ and $\sigma_1 > 0$ the eigenvalues of $A(0)$ are in the left-half complex plane and the linearized system is asymptotically stable. This is well in accord with the results of linearization in the so-called stiction regime of LuGre friction model which have been provided in the original work [17]. The damping characteristics are fully determined by the σ_0 and σ_1 parameters. For the critically-damped solution of characteristic polynomial of matrix $A(0)$ one can show that if $\sigma_1 < 2\sqrt{\sigma_0}$ the system has an oscillating behavior. That means multiple motion reversals should occur before reaching the final equilibrium state.

In order to prove a stable convergence of system (17), (18) we apply the Aiserman's method [25] of finding a suitable Lyapunov function candidate $V(z, \dot{y})$ and evaluating its time derivative. The system matrix (18) can be split into the constant (state-independent) and variable (state-dependent) terms as

$$A(\dot{y}) = A_0 + \tilde{A}(\dot{y}) = \begin{pmatrix} 0 & 1 \\ -\sigma_0 & -\sigma_1 \end{pmatrix} + \begin{pmatrix} -f(\dot{y}) & 0 \\ \sigma_1 f(\dot{y}) & 0 \end{pmatrix}, \tag{20}$$

where $f(\dot{y}) = \sigma_0 |\dot{y}|/C > 0$ for all $\dot{y} \neq 0$. We assume the positive definite (p.d.) matrix

$$P = \begin{pmatrix} \sigma_0 & 0 \\ 0 & 1 \end{pmatrix} \tag{21}$$

that yields a p.d. Lyapunov function candidate $V = (z, \dot{y})P(z, \dot{y})^T$, which is radially unbounded since $V \to \infty$ as $\|(z, \dot{y})\| \to \infty$. Now, evaluating the Lyapunov equation for the constant part of the system matrix one obtains

$$Q_0 = -A_0^T P - P A_0 = \begin{pmatrix} 0 & 0 \\ 0 & 2\sigma_1 \end{pmatrix}. \tag{22}$$

It can be seen that for a positive micro-damping σ_1 the matrix Q_0 is positive semi-definite (p.s.d.) so that the corresponding part of derivative of the Lyapunov function candidate

$$\dot{V}_0(z, \dot{y}) = -(z, \dot{y}) Q_0 (z, \dot{y})^T \leq 0 \tag{23}$$

is negative semi-definite (n.s.d.) only. This is quite natural since for $\dot{V}_0 = -2\sigma_1 \dot{y}^2$ no impact of the z-dynamics can be taken into account and evaluated. In fact, the $\dot{V}_0 = 0$ for $\dot{y} = 0$ does not necessary imply that $z = 0$. For the state-dependent part of the system matrix the corresponding Lyapunov equation yields

$$\tilde{Q}(\dot{y}) = -\tilde{A}^T(\dot{y}) P - P\tilde{A}(\dot{y}) = \begin{pmatrix} 2f(\dot{y})\sigma_0 & -f(\dot{y})\sigma_1 \\ -f(\dot{y})\sigma_1 & 0 \end{pmatrix} \equiv \begin{pmatrix} 2\sigma_0 & -\sigma_1 \\ -\sigma_1 & 0 \end{pmatrix}. \tag{24}$$

Note that the last (equivalence) transformation in (24) is due to the positive $f(\cdot)$ for all $\dot{y} \neq 0$. The obtained (partial) solutions of both Lyapunov equations, i.e. (22) and (24), allow constructing the derivative of the total Lyapunov function candidate as

$$\dot{V} = \dot{V}_0(z, \dot{y}) - (z, \dot{y})\tilde{Q}(\dot{y})(z, \dot{y})^T = -2\sigma_1 \dot{y}^2 - 2\sigma_0 z^2 + 2\sigma_1 z\dot{y} \equiv -\dot{y}^2 - \frac{\sigma_0}{\sigma_1} z^2 + z\dot{y}. \tag{25}$$

For $\dot{V}$ to be negative definite (n.d.) the ellipsoid inequality

$$\dot{y}^2 + \frac{\sigma_0}{\sigma_1} z^2 > z\dot{y} \tag{26}$$

should hold $\forall\, z \in \mathbb{R}, y \in \mathbb{R}$. Here it is worth noting that the internal friction state z is bounded. This has been explicitly addressed and shown in [19] (see Property 1 therein). Nevertheless, for the sake of simplicity, we do not account for this auxiliary condition. It is evident that (26) holds always when $\mathrm{sign}(z) \neq \mathrm{sign}(\dot{y})$. This corresponds to all the intervals $[y_r, y_0]$, i.e. after a motion reversal and until zero-crossing of hysteresis transition curve (see Section 2). For $\mathrm{sign}(z) = \mathrm{sign}(\dot{y})$ we prove the solution of equality $\dot{y}^2 + \sigma_0/\sigma_1 z^2 = z\dot{y}$, which is

$$z = \frac{\sigma_1}{2\sigma_0}\Big(\dot{y} \pm \dot{y}\sqrt{-\frac{4\sigma_0 - \sigma_1}{\sigma_1}}\Big). \tag{27}$$

It is evident that for equation (27) obtains a real solution, the following parameter condition

$$\sigma_0 < \frac{\sigma_1}{4} \tag{28}$$

should hold. Otherwise, no real solutions of friction state z are existing and we can conclude that the inequality (26) holds everywhere in the state-space. This yields $\dot{V}$ to be n.d. and ensures a stable system convergence to the $(z, \dot{y}) = \mathbf{0}$ equilibrium. In case when however (28) holds, there are regions in the state-space where $\dot{V}$ is not n.d. and the system becomes unstable in the Lyapunov sense. Though, we will not analyze this case in more details since for real physical systems the parameter relation (28) appears as improper. Recall that σ_0 corresponds to the stiffness and σ_1 to the micro-damping of frictional contacts. Hence, in almost all physically-reasonable friction systems $\sigma_0 \gg \sigma_1$.

It is intuitively reasonable that higher micro-damping, i.e. higher σ_1 values, provides a faster convergence towards the equilibrium. This implies (indirectly) a lower number i of motion reversals until the relative motion will fully stop. However, an interesting fact occurs when allowing even zero micro-damping, i.e. $\sigma_1 = 0$. In that case one can easily show that the resulted system matrix

$$\hat{A}(\dot{y}) = \begin{pmatrix} -\sigma_0 \frac{|\dot{y}|}{C} & 1 \\ -\sigma_0 & 0 \end{pmatrix} \tag{29}$$

is n.d. for all $y \neq 0$. This yields the system asymptotically stable at hysteresis transitions and marginally stable at zero velocity, i.e. at motion reversals. Also when assuming a related Lyapunov function candidate

$$\hat{V} = \frac{1}{2}\dot{y}^2 + \frac{1}{2}\sigma_0 z^2 \tag{30}$$

one can easily obtain

$$\dot{\hat{V}} = -\sigma_0^2 \frac{|\dot{y}|}{C} z^2 \tag{31}$$

which is n.d. in the whole state-space. This yields the system asymptotically stable even if zero micro-damping is assumed.

3.2. Maxwell-slip model

The Maxwell-slip friction model can be well understood by means of the associated Maxwell-slip structure as depicted in Figure 4. The latter contains N parallel-connected elasto-plastic (or elasto-slip) Maxwell elements which are either sticking or slipping. When the j-th element is subject to the (common) input velocity $\dot{y}$ the linear spring with stiffness k_j is elongating as long as $k_j|y| \leq c_j$. Once the spring force exceeds the individual breakaway force c_j, the element begins to slip and that until the next motion reversal. It is worth noting that all the slipping blocks are assumed as massless so that no dynamic interaction occurs during their relative motion. The associated output friction force (of each block) is $F_j = k_j y$ when sticking and $F_j = c_j \mathrm{sign}(\dot{y})$

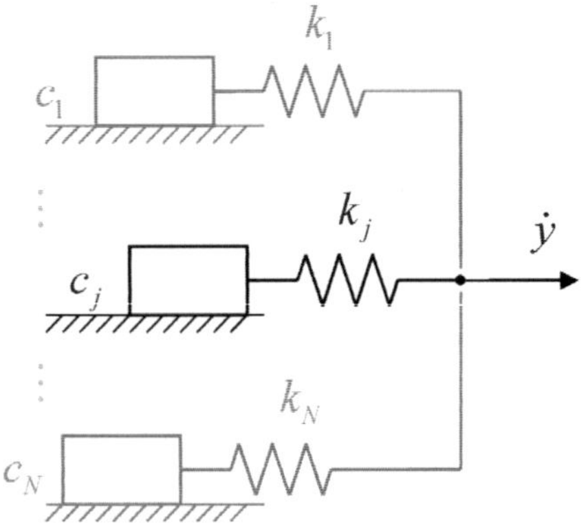

Figure 4. Maxwell-slip structure of N parallel-connected elasto-slip (Maxwell) elements. Each j-th element is characterized by ist individual stiffness $k_j > 0$ and breakaway force $c_j > 0$. The elements are driven by the common velocity input $\dot{y}$. Each element is sticking during the relative spring elongation, i.e. while $k_j|y| \leq c_j$. Afterwards the element is slipping, with a constant counteraction $c_j\mathrm{sign}(\dot{y})$, until the next motion reversal.

when slipping (compare with Ky reversal transition in Section 2). Due to a parallel connection of the blocks the total friction force is

$$F(\dot{y}) = \sum_{j=1}^{N} F_j \,. \tag{32}$$

In order to analyze the system dynamics (7) when using the Maxwell-slip friction model, we are to consider each elasto-slip Maxwell element by means of the play- and stop-type hysteresis operators (see [15], [26], [27] for details). Recall that an elasto-slip Maxwell element offers the same transfer characteristics as the stop-type hysteresis operator. To that end, the stop-type hysteresis operator can be constructed by means of his play-type counterpart. The transfer characteristics of a play-type hysteresis operator $x = \mathcal{P}(r, x_0)[y]$ are shown in Figure 5. The operator provides a multi-valued rate-independent map $y \mapsto x$ depending on the initial state x_0 and parameter r. The latter determines, as can be seen from Figure 5, the width of the play zone which equals $2r$. One can realize that each time the input direction changes, the operator is in the play zone until reaching the corresponding slope $y - \mathrm{sign}(\dot{y})r$. Within the play zone the output remains constant while keeping its previous value. The operator dynamics in differential form, of which later we will make use, is given by

$$\dot{x} = \begin{cases} \dot{y} & \text{if } x - r = y \vee x + r = y \,, \\ 0 & \text{if } x - r < y < x + r \,. \end{cases} \tag{33}$$

In the following, and that for the sake of simplicity, we will write the play-type hysteresis

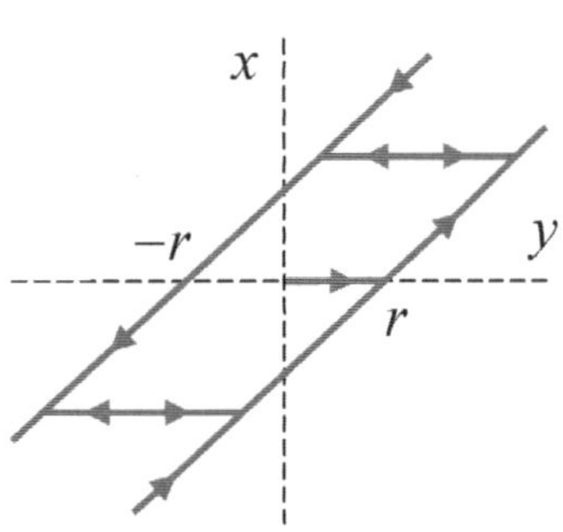

Figure 5. Transfer characteristics of rate-independent play-type hysteresis operator $\mathcal{P}$, also known as mechanical backlash. The play-type hysteresis operator $x = \mathcal{P}(r, x_0)[y]$ is parameterized by the width of the play zone $2r$. Each time the input direction changes, the operator enters the play zone in which the output remains constant while keeping its previous value.

operator by a functional notation $x(y, r)$.

The unique property of a stop-type hysteresis operator is that this can be transformed from its play-type counterpart by the simple algebraic relation $y - x(y, r)$. This allows constructing the single elasto-slip Maxwell element as

$$F_j = k_j\big(y - x(y, r_j)\big)\,, \tag{34}$$

when tacking into account the corresponding parameterization $r_j = c_j/k_j$. Recall that the closed form (34) is rate-independent and represents the hysteresis transition, in relative force-displacement coordinates, between the element sticking and slipping and vice versa.

Now, we are in a position to analyze the convergence of free motion (7), and therefore the damping characteristics of pre-sliding hysteresis, when assuming the friction force (32). For the sake of clarity we are first exploring the case $N = 1$, i.e. when a single elasto-slip Maxwell element is assumed. Afterwards, we generalize our analysis to the case $N > 1$ and discuss a limit case when $N \to \infty$. We note that the case $N = 1$ corresponds to the linear reversal transition which has been indicated by Ky and discussed before in Section 2. Here, however, we are to establish a Lyapunov-based analysis so as to make it further extendable for a general multi-element Maxwell-slip structure.

The single elasto-slip Maxwell element yields the free motion dynamics as

$$\ddot{y} = -k_1 y + k_1 x(y, r_1)\,. \tag{35}$$

Since the hysteresis operator can accumulate, or eventually release, the energy in the system, we introduce an internal state $z = x - y$, which is bounded by $-r_1 \leq z \leq r_1$, and apply the Lyapunov function candidate

$$V(\dot{y}, z) = \frac{1}{2k_1}\dot{y}^2 + \frac{1}{2}z^2. \tag{36}$$

Note that V is p.d. and radially unbounded since $V \to \infty$ as $\|(\dot{y}, z)\| \to \infty$. The time derivative of the Lyapunov function candidate results in

$$\dot{V} = \dot{y}z + \dot{z}z, \tag{37}$$

which is case-specific depending on whether the element is sticking or slipping. Here it should be noted that two cases correspond to the play zones and slopes of the play-type hysteresis operator (see Figure 5). From the internal state dynamics $\dot{z} = \dot{x} - \dot{y}$ point of view and with respect to (33) we can emphasize that $\dot{z} = 0$ when operating on one of the slopes, and $\dot{z} = -\dot{y}$ when operating within a play zone. This renders the time derivative of Lyapunov function candidate as

$$\dot{V} = \begin{cases} \dot{y}z & \text{for } \dot{x} = \dot{y} \text{ (slope)}, \\ 0 & \text{for } \dot{x} = 0 \text{ (play zone)}. \end{cases} \tag{38}$$

It can be seen from Figure 5 that the right-hand side slope, i.e. $\dot{y} > 0$, implies $z = -r$, and the left-hand side slope, i.e. $\dot{y} < 0$, implies $z = r$ correspondingly. Therefore, we can conclude that when operating on a slope the inequality $\dot{y}z < 0$ always holds. The latter renders $\dot{V}$ as n.d. and implies the system is globally asymptotically stable in the Lyapunov sense when slipping. On the contrary, within a play zone where $\dot{V} = 0$ the system yields marginally stable, but not asymptotically stable, in the Lyapunov sense. Here the system moves along the closed elliptic trajectories as has been analyzed before in Section 2.

Since the total time derivative of Lyapunov function candidate is n.s.d. only, i.e. $\dot{V} \leq 0$, we are to show that there exists an invariant set which consists of the stable linear limit cycles during the system sticking. Applying the invariant set theorem (see [14] for details), we assume for some $l = 1/2r^2 + a > 0$ the region Ω defined by $V(\dot{y}, z) < l$ which is bounded. In fact, since z is bounded by $\pm r$, and further with respect to (36), one can see that Ω is bounded on the $\pm\dot{y}$-axis by the constant $a > 0$ which is related to initial velocity. Also it is valid that $\dot{V}(\dot{y}, z) \leq 0$ for all $(\dot{y}, z) \in \Omega$. Now, we have to show that for the set $R \subset \Omega$, which is the largest invariant set with $\dot{V}(\dot{y}, z) = 0$, any trajectory with $\big(\dot{y}(0), z(0)\big) \in \Omega$ tends to R as $t \to \infty$. According to (38), zero derivative $\dot{V} = 0$ corresponds to a play zone of play-type hysteresis operator for which $\dot{x} = 0$. It should be noted that the set of all play zones is infinite and corresponds to the whole horizontal axis of relative position y. For any play zone the corresponding x-value is constant and the motion dynamics (35) becomes

$$\ddot{y} - k_1 z = 0\,. \tag{39}$$

Since $\dot{y} = -\dot{z}$ and $\ddot{y} = -\ddot{z}$ we transform the motion dynamics (39) into the relative z-coordinates and obtain the free motion of sticking system as

$$\ddot{z} + k_1 z = 0. \tag{40}$$

Note that the same result we obtained in Section 2 (see equation (8)). It is obvious that the $(\dot{y}, z)$ solutions constitute the closed elliptic trajectories of limit cycles whose union is the invariant set R. It is worth to recall that for any initial velocity $\dot{y}(0) \neq 0$, i.e. the system is in slipping, the internal state is $z = -\text{sign}(\dot{y})r$, and the play-type hysteresis operator proceeds along one of the slopes until the next motion reversal. Further we stress that for any initial state in Ω with $0 < l \leq 1/2r^2$ the trajectory $(\dot{y}, z)(t) \in R$ for all $t \geq 0$. Here the axes of elliptic limit cycles depend on the initial state $(\dot{y}, z)(0)$. For any initial state with $l > 1/2r^2$ the trajectory first proceeds towards the corresponding $z = -\text{sign}(\dot{y})r$ boundary, i.e. depending on the initial $\dot{y}(0)$ value, and afterwards slides along this boundary until the next motion reversal occurs. The motion trajectories with different initial states are shown in the $(\dot{y}, z)$ plane in Figure 6.

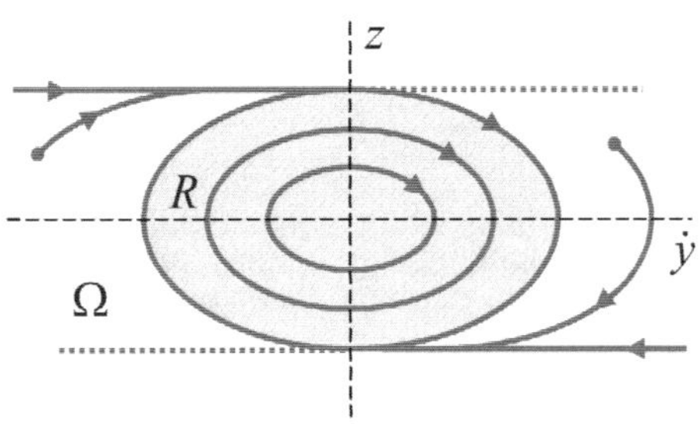

Figure 6. Trajectories of the system (35) in the $(\dot{y}, z)$ plane. The trajectories starting within R form the elliptic limit cycles. The trajectories starting in Ω outside of R first run towards the corresponding $z = -\text{sign}(\dot{y})r$ boundary and then proceed to the next motion reversal. Afterwards, the boundary (largest) limit cycles occurs.

Now, we are to generalize our analysis made above for the case when $N > 1$, therefore incorporating a multi-element Maxwell-slip structure when describing the friction force. The corresponding free motion dynamics results in

$$\ddot{y} = -\sum_{j=1}^{N} k_j y + \sum_{j=1}^{N} k_j x_j(y, r_j), \tag{41}$$

which reduces to (35) when $N = 1$. Introducing the $N \times 1$ vectors of contact stiffness $\mathbf{k} = [k_1, \ldots, k_N]^T$, play-type hysteresis operators $\mathbf{x} = [x_1, \ldots, x_N]^T$ which are parameterized by $\mathbf{r} = [r_1, \ldots, r_N]^T$, and absolute positions $\mathbf{y} = y[1, \ldots, 1]^T$ we rewrite the system dynamics (41) into the vector form

$$\ddot{y} = -\mathbf{k}^T\mathbf{y} + \mathbf{k}^T\mathbf{x}(y, \mathbf{r}). \tag{42}$$

Further we introduce, similar as in case of single Maxwell-slip element, the vector of internal states $\mathbf{z} = \mathbf{x} - \mathbf{y}$ which captures the individual transitions of play-type hysteresis operators. Defining the p.d. diagonal matrix $\tilde{\mathbf{K}} = \text{diag}(k_1, \ldots, k_N)$ we assume the Lyapunov function candidate

$$\tilde{V}(\dot{y}, \mathbf{z}) = \frac{1}{2}\dot{y}^2 + \frac{1}{2}\mathbf{z}^T\tilde{\mathbf{K}}\mathbf{z}, \tag{43}$$

which is p.d. and radially unbounded since $\tilde{V} \to \infty$ as $\|(\dot{y}, \mathbf{z})\| \to \infty$. It can be noted that the Lyapunov function candidate (43) contains the kinetic energy of the free motion and potential energy associated with each hysteresis operator, i.e. elasto-slip Maxwell element. The time derivative of the Laypunov function candidate results in

$$\dot{\tilde{V}} = \dot{y}\mathbf{k}^T\mathbf{z} + \dot{\mathbf{z}}^T\tilde{\mathbf{K}}\mathbf{z}. \tag{44}$$

With respect to the transfer characteristics of each play-type hysteresis operator (see Figure 5 and equation (38)), and herewith associated behavior of single elasto-slip Maxwell element, one can show that

$$\dot{\tilde{V}} = \begin{cases} \dot{y}\mathbf{k}^T\mathbf{z} < 0 & \text{if all } N \text{ elements are slipping,} \\ \dot{y}\mathbf{k}^T\mathbf{z} - \dot{\mathbf{y}}^T\tilde{\mathbf{K}}\mathbf{z} = 0 & \text{if all } N \text{ elements are sticking.} \end{cases} \tag{45}$$

It is easy to recognize that (45) constitutes a boundary case where all play-type hysteresis operators are either on a slope or in a play zone. However, since we are interested in analyzing the case when the system is not asymptotically stable but proceeds in a limit cycle, i.e. when $\dot{\tilde{V}} \equiv 0$, the second case of (45) provides us with the required (strict) condition. In fact, the trajectory is within invariant set R if and only if all N play-type hysteresis operators are in a play zone. In order to prove this, one can easily show that when at least one operator j is on a slope, the corresponding element has $\dot{\mathbf{z}}^T(j) = 0 \neq -\dot{\mathbf{y}}^T(j)$ and its contribution to (44), which is

$$\dot{y}\mathbf{k}^T(j)\mathbf{z}(j) + \dot{\mathbf{z}}^T(j)\tilde{\mathbf{K}}(j,j)\mathbf{z}(j) < 0\,,$$

renders $\dot{\tilde{V}}$ as n.d. Thus, one can conclude that it is sufficient that at least one element is slipping so as the overall system becomes asymptotically stable, and therefore converges to the motion equilibrium $\dot{y} = 0$.

Now, we can analyze whether and when the condition derived above satisfies the pre-sliding system behavior, in particularly in view of hysteresis reversal transitions addressed before in Section 2. While $h(y)$ should be continuously differentiable (see property (b) in Section 2), the friction force (32) constitutes only a piecewise differentiable approximation of $h(y)$. In fact, a force-displacement hysteresis branch described by (32) is a piecewise linear curve with the break points which characterize the sticking-slipping transitions of single elasto-slip Maxwell elements. Since the elasto-slip Maxwell elements are acting in parallel, each slope segment of the piecewise linear curve appears as a superimposed stiffness of all elements which are (still) sticking. At the same time, it should be stressed that immediately after each motion reversal all elements are sticking. Afterwards, they switch one by one to slipping, and that depending on their individual k_j and c_j parameters, while proceeding along the $h(y)$ curve. Therefore, one can conclude that higher numbers N approximate the $h(y)$ function with higher accuracy and, at the same time, provoke more frequent switching from sticking to slipping along the $h(y)$ curve. This implies $r_{k+1} - r_k \to 0$ as $N \to \infty$, where k and $k+1$ are two arbitrary elements whose switching occur successively. Hence, for any element $j = 1$, which switches as first after a motion reversal, $r_1 \to 0$ as $N \to \infty$. With respect to condition of the negative definiteness of $\dot{\tilde{V}}$, derived above, we can conclude that $\dot{\tilde{V}} \to$ n.d. as $N \to \infty$, and that over the whole $(\dot{y}, \mathbf{z})$ state-space. As corollary, one can show that the invariant set contains only the equilibrium point, i.e. $R \to \mathbf{0}$, as $N \to \infty$ (cf. with Figure 6).

4. Experimental observations

In this Section we show and discuss some experimental observations in favor of the analysis of pre-sliding frictional hysteresis made above. The experimental data of a measured motion response are obtained from an actuated rotary system with mechanical friction interfaces. The relative angular displacement y and velocity $\dot{y}$ are measured using a precise digital encoder mounted on the rotary shaft. The overall moving mass (inertia) is axially-symmetrical and assumed to be homogenous and lumped. The feed actuation torque u allows driving the system at certain constant (also low) velocity and can be switched off stepwise, so as to realize (afterwards) a free motion response towards the equilibrium. The principal configuration of the experimental setup in use is shown in Figure 7. Most notable is that the multiple mechanical frictional interfaces can be considered as an aggregated one due to the stiff relative motion within the same rotary degree-of-freedom y. The experimental data, shown in the following, are real-time and collected

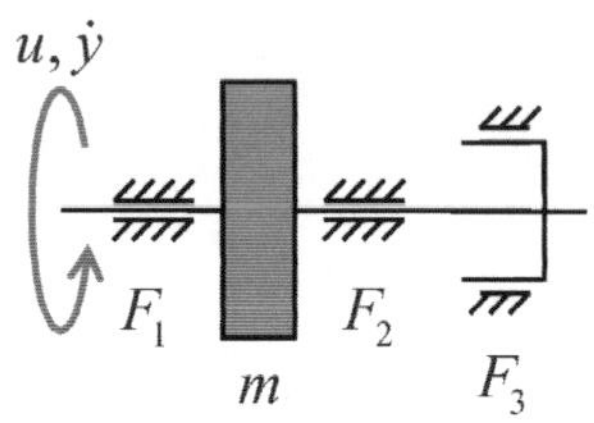

Figure 7. Principal configuration of the experimental actuated rotary system with the lumped moving mass (inertia) m. The frictional interfaces of the ball-bearing F_1, F_2 and gear teething F_3 type are considered as an aggregated friction F, since operating in the same rotary degree-of-freedom. The actuation torque u induces the angular relative displacement with velocity $\dot{y}$.

with the fixed sampling rate of 2 kHz. More details about the used experimental system can be found in [28].

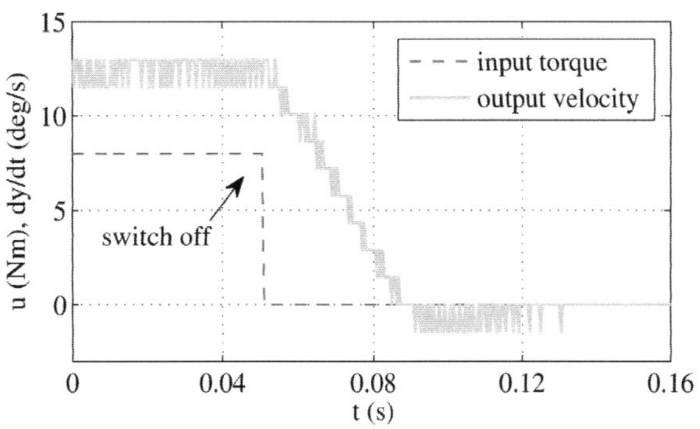

Figure 8. Experimental data of the measured motion response in pre-sliding range. The relative displacement with constant low-velocity is actuated by a constant feed force (torque) until the actuation is switched off. Afterwards, the free system response occurs until the complete motion stop. The nearly linear velocity slope indicates the predominant Coulomb friction and irrelevance of the viscous (exponential) damping.

In Figure 8, the measured motion response is shown as time series together with the corresponding input torque. At a constant torque level the nearly constant relative velocity of about 12.3 deg/s occurs until the actuation is switched off. The irregular high-frequent chattering of the measured velocity is attributed to the sampled and hold (digital) encoder signal and its discrete-time derivation. After the actuation switch off the velocity descends towards zero and the system completely stops after a low transient overshoot. The visible, nearly linear, slope of decreasing velocity indicates a predominance of Coulomb friction (cf. with Section 2) and irrelevance of the viscous (exponential) damping close to the motion stop.

The free motion trajectories are shown in Figure 9 in the $(\dot{y}, y)$ phase-plane. The measured motion response is compared with the model fit of system dynamics once using the LuGre (Figure 9(a)) and once using the Maxwell-slip (Figure 9(b)) friction model. Note that after the actuation switch off the system behavior can be captured by (7) with an appropriately set

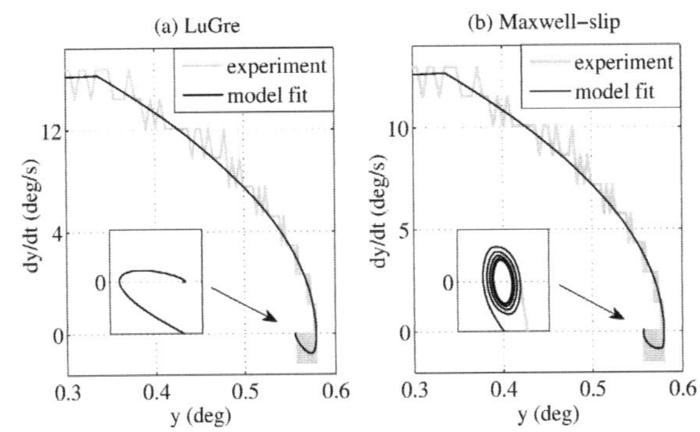

Figure 9. Free motion trajectories in the $(\dot{y}, y)$ phase-plane. Comparison of experimental measurements with the model fit of system dynamics (7) when using the LuGre friction model (a) and Maxwell-slip (with $N = 3$) friction model (b). The model fit is obtained by nonlinear least-squares minimization of the velocity error $\|\dot{\mathbf{y}} - \dot{\hat{\mathbf{y}}}\|$, where $\dot{\hat{\mathbf{y}}}$ is the model-predicted relative velocity. The zoom-in inclusions show the convergence of both models towards equilibrium.

non-zero initial velocity and initial friction $F(0) = \text{sign}(\dot{y})C$. The former is directly determined from the available measurement. The latter is inherently included into the model fit when estimating the parameters. The model fit is obtained, in case of both friction models, by the standard nonlinear least-squares minimization (Levenberg-Marquardt algorithm) of the velocity error $\|\dot{\mathbf{y}} - \dot{\hat{\mathbf{y}}}\|$, where $\dot{\hat{\mathbf{y}}}$ is the model-predicted velocity response. In case of the Maxwell-slip friction model the number of elements $N = 3$ is assumed. This renders six friction parameters to be identified and makes the model complexity, i.e. in terms of number of parameters, better comparable with that of the LuGre friction model. Note that besides the friction parameters the moving mass (inertia) is equally identified in both cases. From Figure 9, one can see that both identified friction models provide a prediction of free motion response which coincides well with experimental data. At the same time, one can see, from the zoom-in included in Figure 9, that in case of the Maxwell-slip friction model the trajectory does not converge to equilibrium but towards a stable (micro) limit cycle. This is well in accord with the analysis we made at the end of Section 3.2.

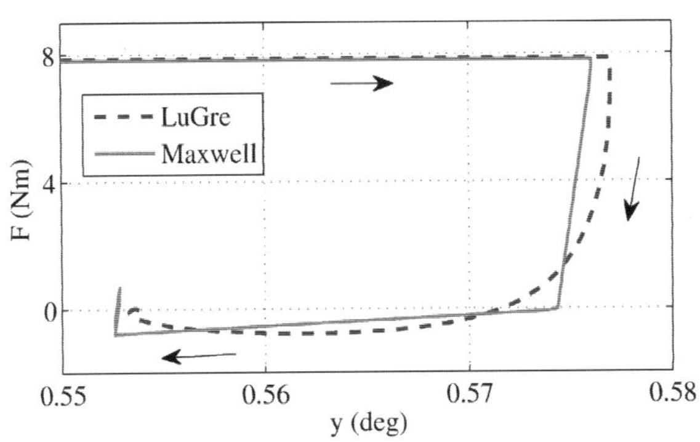

Figure 10. Friction-displacement response of the model fit of system dynamics (7) when using the LuGre and Maxwell-slip friction models. The first motion reversal (right-hand side) occurs after the system slipping with a constant Coulomb friction. The hysteresis reversal transition occurs, in case of both models, until the next motion reversal (left-hand side). The latter is followed by the attained equilibrium state. In case of Maxwell-slip friction model a micro limit cycle occurs instead of the motion stop.

Further, the pre-sliding hysteresis response of both identified models are shown in Figure 10 in the relative force-displacement coordinates. Both friction models offer nearly the same level of constant Coulomb friction until the first motion reversal (on the right-hand side). After the reversal hysteresis transitions, which are qualitatively similar but differ in the smoothness of curvature, the next motion reversal occurs (on the left-hand side) in case of both models. Close to the latter, both friction trajectories attain the final equilibrium state, while the Maxwell-slip model provides the residual (micro) limit cycles instead of a complete motion stop.

5. Conclusions

We have analyzed and demonstrated in a numerical simulation and with experiment the damping characteristics of frictional hysteresis within the so called pre-sliding range. Based on the general consideration of friction force transitions, i.e after a motion reversal, we have compared the ideal cases of linear contact stiffness and discontinuous Coulomb friction, and the generic case of hysteresis in displacement. It has been analyzed analytically and shown in a simulation how the free motion trajectories, in all three cases, proceed towards the equilibrium. Also two dynamic friction models, e.g. LuGre and Maxwell-slip, have been taken into consideration and investigated on their nonlinear dynamics in context of a free moving mass with frictional damping. Using the Lyapunov stability theory and related tools, like Aiserman method and invariant set theorem, we have derived the conditions of asymptotic and non-asymptotic stability of relative motion with both friction models. In particularly, for the LuGre friction model the impact of contact stiffness and micro-damping parameters has been emphasized and the asymptotic stability, even without viscous friction term, has been proved. For the Maxwell-slip friction model the main result consists in deriving the criterion of asymptotic stability and analysis of occurrence of the limit cycles. Some experimental observations shown and compared with the model-based motion prediction argued in favor of the analysis made before.

References

[1] Al-Bender F, De Moerlooze K and Vanherck P 2012 *Tribology Letters* **46** 23–31
[2] Armstrong B, Dupont P and De Wit C C 1994 *Automatica* **30** 1083–1138
[3] Armstrong B and Amin B 1996 *Automatica* **32** 679–692
[4] Hensen R H, Van de Molengraft M and Steinbuch M 2003 *Automatica* **39** 2131–2137
[5] Ruderman M and Iwasaki M 2014 *International Power Electronics Conference (IPEC)* pp 1665–1670
[6] Al-Bender F and Symens W 2005 *Chaos: An Interdisciplinary Journal of Nonlinear Science* **15** 013105
[7] Armstrong B and Chen Q 2008 *IEEE Control Systems Magazine* **28** 79–89
[8] Ruderman M and Bertram T 2013 *Mechanical Systems and Signal Processing* **39** 316–332
[9] Ruderman M and Iwasaki M 2015 *IEEE Transactions on Industrial Electronics* **PP** 1–1
[10] Zabreiko P, Krasnoselskii M and Lifsic E 1970 *Soviet Math. Dokl.* **11** 73–76
[11] Wielke B 1974 *Physica status solidi (a)* **23** 237–244
[12] Grossmayer R 1979 *Journal of Sound and Vibration* **65** 353–379
[13] Alvarez J, Orlov I and Acho L 2000 *Journal of Dynamic Systems, Measurement, and Control* **122** 687–690
[14] Slotine J J and Li W 1991 *Applied Nonlinear Control* 1st ed (New Jersey, USA: Prentice Hall)
[15] Krejci P 1996 *Hysteresis, Convexity and Dissipation in Hyperbolic Equations* (Tokyo: Gattötoscho)
[16] Popov V and Gray J 2012 *ZAMM - Journal of Applied Mathematics and Mechanics* **92** 683–708
[17] De Wit C C, Olsson H, Aström K J and Lischinsky P 1995 *IEEE Trans. on automatic control* **40** 419–425
[18] Dahl P R 1968 A solid friction model TOR 158(3107-18) The Aerospace Corporation, El Segundo
[19] Aström K J and De Wit C C 2008 *IEEE Control systems magazine* **28** 101–114
[20] Lazan B J 1968 *Damping of materials and members in structural mechanics* vol 214 (Pergamon press Oxford)
[21] Iwan W D 1966 *ASME Trans., Journal of Aplied Mechanics* **33** 893–900
[22] Goldfarb M and Celanovic N 1997 *IEEE Control Systems* **17** 69–79
[23] Haessig D A and Friedland B 1991 *Journal of Dynamic Systems, Measurement, and Control* **113** 354–362
[24] Barahanov N and Ortega R 2000 *IEEE Transactions on Automatic Control* **45** 830–832
[25] Aiszerman M A and Gantmakher F R 1965 *Die absolute Stabilität von Regelsystemen* (Oldenbourg)
[26] Krasnosel'skii M and Pokrovskii A 1989 *Systems with Hysteresis* (Berlin: Springer)
[27] Brokate M and Sprekels J 1996 *Hysteresis and Phase Transitions* (Berlin: Springer)
[28] Ruderman M and Iwasaki M 2014 *IEEE International Workshop on Advanced Motion Control* pp 592–597

Modelling of automotive fuel droplet heating and evaporation: mathematical tools and approximations

Sergei S. Sazhin and Mansour Al Qubeissi

Sir Harry Ricardo Laboratories, Centre for Automotive Engineering, School of Computing, Engineering and Mathematics, Faculty of Science and Engineering, University of Brighton, Brighton, BN2 4GJ, UK

E-mail: S.Sazhin@brighton.ac.uk

Abstract. New mathematical tools and approximations developed for the analysis of automotive fuel droplet heating and evaporation are summarised. The approach to modelling biodiesel fuel droplets is based on the application of the Discrete Component Model (DCM), while the approach to modelling Diesel fuel droplets is based on the application of the recently developed multi-dimensional quasi-discrete model. In both cases, the models are applied in combination with the Effective Thermal Conductivity/Effective Diffusivity model and the implementation in the numerical code of the analytical solutions to heat transfer and species diffusion equations inside droplets. It is shown that the approximation of biodiesel fuel by a single component leads to under-prediction of droplet evaporation time by up to 13% which can be acceptable as a crude approximation in some applications. The composition of Diesel fuel was simplified and reduced to only 98 components. The approximation of 98 components of Diesel fuel with 15 quasi-components/components leads to under-prediction of droplet evaporation time by about 3% which is acceptable in most engineering applications. At the same time, the approximation of Diesel fuel by a single component and 20 alkane components leads to a decrease in the evaporation time by about 19%, compared with the case of approximation of Diesel fuel with 98 components. The approximation of Diesel fuel with a single alkane quasi-component ($C_{14.763}H_{31.526}$) leads to under-prediction of the evaporation time by about 35% which is not acceptable even for qualitative analysis of the process. In the case when n-dodecane is chosen as the single alkane component, the above-mentioned under-prediction increases to about 44%.

1. Introduction

Modelling of the processes inside internal combustion engines is a challenging task. It includes modelling of coupled fluid dynamics and heat/mass transfer processes in a complex geometry. To take into account the complexity of this geometry, the internal combustion engine chambers should be split up into millions of cells. The parameters of the continuous phase in each of these cells are constant, but they change in time and from one cell to another. Moreover, the dynamics of the dispersed phase (fuel droplets) and chemical reactions in each cell need to be taken into account. The number of these droplets and chemical reactions in each cell can be many thousands in the general case [1, 2].

The above-mentioned features of the modelling of the engine processes make it impossible to perform their rigorous quantitative analysis. Two main approaches have been developed to deal with the complexities of this modelling. The first approach, most widely used in the engineering community, is based on the application of rather simplistic physical models of individual processes, but the geometry of the enclosure is approximated as accurately as possible. This approach is used in most commercial Computational Fluid Dynamics (CFD) codes. The second approach is focused on in-depth development of the physical models of individual processes, ignoring the complexities of the geometry and details of interactions between various processes. These two approaches do not contradict each other; rather, they are complementary.

In a series of our previous papers, an attempt has been made to develop the third approach to this modelling. This approach is focused on establishing a hierarchy of the processes involved (recognising multiple scales in time and space) and finding a compromise between the accuracy of the models and their CPU efficiency. This allowed us to develop models which were more advanced than those used in the conventional CFD codes, but simple enough to enable their implementation in the above-mentioned codes [3]. Various aspects of this approach were reported at the previous MURPHYS workshops and published in [4]-[8].

In the current paper, the latter approach is illustrated by our modelling of biodiesel and Diesel fuel droplet heating and evaporation, which is an integral part of the spray ignition process in Diesel engines. There will be some overlap with the previously published journal papers [9, 10]. In contrast to these papers, we focus more attention on the mathematical tools and approximations used in our analysis. Also, the models will be illustrated using examples which are different from those given in [9, 10].

2. Basic Equations

The simplest approach to modelling multi-component automotive fuel droplets, widely used in commercial CFD codes, is based on the assumptions that the thermal conductivity of droplets is infinitely large, and the diffusion of species is infinitely fast. The first assumption is commonly justified by the fact that thermal conductivity of droplets is larger than that of ambient air. In the transient processes, typical for internal combustion engines, however, the main parameter describing droplet heating is their diffusivity rather than conductivity. The former is much larger for air than for liquid, which allows us to question the above-mentioned widely used assumption about the droplet's infinitely large thermal conductivity. Regarding the second assumption, it can be questioned based on the estimate of the characteristic time of species diffusivity. This is generally much larger than the characteristic time of thermal diffusivity. Thus, the assumption that multi-component automotive fuel droplets can be approximated by single component droplets (infinitely slow species diffusivity) has more solid physical grounds than the assumption of infinitely fast species diffusivity. In what follows we will show how both these approaches can be relaxed, but in such a way that the model remains simple enough for possible implementation in CFD codes (commercial or research).

The process of heating mono- and multi-component droplets is described by the following transient one dimensional (1D) heat conduction equation for the temperature $T \equiv T(t, R)$ in the liquid phase [11, 12]:

$$\frac{\partial T}{\partial t} = \kappa \left(\frac{\partial^2 T}{\partial R^2} + \frac{2}{R} \frac{\partial T}{\partial R} \right), \tag{1}$$

where $\kappa = k_{\text{eff}}/(c_l \rho_l)$ is the effective thermal diffusivity, k_{eff}, c_l, and ρ_l are the effective thermal conductivity, specific heat capacity, and density respectively, R is the distance from the centre of the droplet (assumed to be spherical), t is time. k_{eff} is linked with the liquid thermal conductivity k_l via the following equation:

$$k_{\text{eff}} = \chi k_l, \tag{2}$$

where the coefficient χ is approximated as [13]:

$$\chi = 1.86 + 0.86 \tanh \left[2.225 \log_{10} \left(\text{Pe}_{d(l)}/30 \right) \right]. \tag{3}$$

This coefficient varies from 1 (at droplet Peclet number $\mathrm{Pe}_{d(l)} = \mathrm{Re}_{d(l)}\mathrm{Pr}_l < 10$) to 2.72 (at $\mathrm{Pe}_{d(l)} > 500$). Liquid fuel transport properties and the liquid velocity just below the droplet surface are used for calculating $\mathrm{Re}_{d(l)}$.

Equation (1) is solved for $t > 0$ and $0 \le R < R_d$, where R_d is the droplet radius, with the following boundary condition, assuming that the effects of evaporation can be ignored:

$$h(T_g - T_s) = k_{\mathrm{eff}} \left.\frac{\partial T}{\partial R}\right|_{R=R_d-0}, \tag{4}$$

and the initial condition $T(t=0) = T_{d0}(R)$, where $T_s = T_s(t)$ is the droplet's surface temperature, $T_g = T_g(t)$ is the ambient gas temperature, h is the convection heat transfer coefficient, linked with the Nusselt number Nu via the following equation $\mathrm{Nu} = 2R_d h/k_g$, k_g is the gas thermal conductivity.

To take into account the effect of droplet evaporation, gas temperature is replaced with the effective temperature defined as:

$$T_{\mathrm{eff}} = T_g + \frac{\rho_l L \dot{R}_{de}}{h}, \tag{5}$$

where L is the latent heat of evaporation, $\dot{R}_{de}$ is the rate of decrease of droplet radius due to evaporation (see Equation (13)). R_d is updated at the end of each time step Δt: $R_{d(\mathrm{new})} = R_{d(\mathrm{old})} + \dot{R}_d \Delta t$, where $\dot{R}_d = \dot{R}_{de} + \dot{R}_{ds}$, $\dot{R}_{ds}$ is the change of droplet radius due to swelling (see Equation (13)).

Equations for mass fractions of liquid species $Y_{li} \equiv Y_{li}(t, R)$ inside moving but spherically symmetric droplets are presented in the following form [14]:

$$\frac{\partial Y_{li}}{\partial t} = D_{\mathrm{eff}} \left(\frac{\partial^2 Y_{li}}{\partial R^2} + \frac{2}{R}\frac{\partial Y_{li}}{\partial R} \right), \tag{6}$$

where $i = 1, 2, 3, ...$, D_{eff} is the effective mass diffusivity, linked with the liquid diffusivity D_l via the following equation

$$D_{\mathrm{eff}} = \chi_Y D_l, \tag{7}$$

the coefficient χ_Y varies from 1 to 2.72 and can be approximated as:

$$\chi_Y = 1.86 + 0.86 \tanh\left[2.225 \log_{10}\left(\mathrm{Re}_{d(l)}\mathrm{Sc}_l/30\right)\right], \tag{8}$$

$\mathrm{Sc}_l = \nu_l/D_l$ is the liquid Schmidt number, ν_l is the liquid kinematic viscosity. As in the case of k_{eff}, liquid fuel transport properties and the liquid velocity just below the droplet surface were used for calculating $\mathrm{Re}_{d(l)}$. The model based on Equations (7) and (8) is known as the Effective Diffusivity (ED) model. The model, based on the assumption that species diffusivity is infinitely fast ($D_{\mathrm{eff}} = \infty$) is referred to as the Infinite Diffusivity (ID) model. The combined ITC/ID model is sometimes known as a well-mixed model. In the opposite limiting case, when there is no species diffusivity ($D_{\mathrm{eff}} = 0$), the multi-component fuel can be dealt with as a mono-component one.

Equation (6) is solved with the following boundary condition [14]:

$$\alpha(\epsilon_i - Y_{lis}) = -D_{\mathrm{eff}} \left.\frac{\partial Y_{li}}{\partial R}\right|_{R=R_d-0}, \tag{9}$$

and the initial condition $Y_{li}(t=0) = Y_{li0}(R)$, where $Y_{lis} = Y_{lis}(t)$ are liquid components' mass fractions at the droplet's surface,

$$\alpha = \frac{|\dot{m}_d|}{4\pi\rho_l R_d^2}, \tag{10}$$

$\dot{m}_d$ is the droplet evaporation rate. In the case of mono-component droplets, their evaporation rate is estimated from the following equation [15]:

$$\dot{m}_d = -2\pi R_d D_v \rho_{\text{total}} B_M \text{Sh}, \tag{11}$$

where D_v is the binary diffusion coefficient of vapour in air, B_M is the Spalding mass transfer number defined as:

$$B_M = \frac{\rho_{vs} - \rho_{v\infty}}{\rho_{gs}} = \frac{Y_{vs} - Y_{v\infty}}{1 - Y_{vs}}, \tag{12}$$

Y_{vs} and $Y_{v\infty}$ are the vapour mass fractions near the droplet surface and in the ambient gas, Sh is the Sherwood number (equal to 2 for an isolated stationary droplet).

When calculating the value of $\dot{R}_d$ we took into account both droplet evaporation during the time step, and the change in liquid density during this time step (see Equation (20) in [16]):

$$\dot{R}_d = \dot{R}_{de} + \dot{R}_{ds} = \frac{\dot{m}_d}{4\pi R_d^2 \rho_l} + \frac{R_d(\overline{T}_0)}{\Delta t}\left[\left(\frac{\rho(\overline{T}_0)}{\rho(\overline{T}_1)}\right)^{1/3} - 1\right], \tag{13}$$

where $\overline{T}_0$ and $\overline{T}_1$ are average droplet temperatures at the beginning $t = t_0$ and the end of the time step $t = t_1$, $\Delta t = t_1 - t_0$.

To calculate the species mass evaporation rate $\dot{m}_{di}$, we calculate first the values of Y_{vsi}. The latter depend on the partial pressure of species i in the vapour state in the immediate vicinity of the droplet surface [17]:

$$p_{vsi} = X_{lsi}\gamma_i p^*_{vsi}, \tag{14}$$

where X_{lsi} is the molar fraction of the ith species in the liquid near the droplet surface, p^*_{vsi} is the partial vapour pressure of the ith species in the case when $X_{lsi} = 1$, γ_i is the activity coefficient. In the limit when $\gamma_i = 1$, Equation (14) describes Raoult's law. This law is assumed to be valid in our analysis.

3. Approximations

The rigorous coupled numerical solutions of the equations described in the previous section would not only require substantial computer power, but would be unnecessary for practical engineering applications. In what follows we discuss the approximations which can substantially simplify the analyses of these equations, almost without sacrificing the accuracy of engineering modelling.

Firstly, we can assume that h = const. This assumption would not be acceptable for the whole period of heating of evaporating droplets, but its application for a short time step cannot be questioned except for at the very last stage of droplet evaporation. This assumption allows us to obtain the analytical solution to Equation (1), subject to the earlier mentioned boundary and initial conditions [18]:

$$T(R,t) = \frac{1}{R}\sum_{n=1}^{\infty}\left\{q_n \exp\left[-\kappa_R\lambda_n^2 t\right] - \frac{R_d^2 \sin\lambda_n}{\|v_n\|^2 \lambda_n^2}\mu_0(0)\exp\left[-\kappa_R\lambda_n^2 t\right]\right.$$

$$\left. - \frac{R_d^2 \sin\lambda_n}{\|v_n\|^2 \lambda_n^2}\int_0^t \frac{\mathrm{d}\mu_0(\tau)}{\mathrm{d}\tau}\exp\left[-\kappa_R\lambda_n^2(t-\tau)\right]\mathrm{d}\tau\right\}\sin\left[\lambda_n\left(\frac{R}{R_d}\right)\right] + T_g(t), \tag{15}$$

where λ_n are solutions to the equation:

$$\lambda\cos\lambda + h_0\sin\lambda = 0, \tag{16}$$

$$|| v_n ||^2 = \frac{R_d}{2}\left(1 - \frac{\sin 2\lambda_n}{2\lambda_n}\right) = \frac{R_d}{2}\left(1 + \frac{h_0}{h_0^2 + \lambda_n^2}\right),$$

$$q_n = \frac{1}{|| v_n ||^2}\int_0^{R_d} \tilde{T}_0(R)\sin\left[\lambda_n\left(\frac{R}{R_d}\right)\right] \mathrm{d}R,\ \kappa_R = \frac{k_{\mathrm{eff}}}{c_l\rho_l R_d^2},\ \mu_0(t) = \frac{hT_g(t)R_d}{k_{\mathrm{eff}}},$$

$h_0 = (hR_d/k_{\mathrm{eff}}) - 1$, $\tilde{T}_0(R) = RT_{d0}(R)$. The solution to Equation (16) gives a set of positive eigenvalues λ_n numbered in ascending order ($n = 1, 2, ...$). The trivial solution $\lambda = 0$ is not considered.

Solution (15) is valid for $h_0 > -1$, which is satisfied, remembering the physical background of the problem ($h > 0$). In the limit $k_{\mathrm{eff}} \to \infty$ the prediction of Expression (15) is identical to the one which follows from the model based on the assumption that $k_{\mathrm{eff}} = \infty$ [19] (Infinite Thermal Conductivity (ITC) model).

The value of Nu for an isolated moving droplet is estimated based on the following equation [13]:

$$\mathrm{Nu} = 2\frac{\ln(1 + B_T)}{B_T}\left(1 + \frac{(1 + \mathrm{Re}_d\mathrm{Pr}_d)^{1/3}\max\left[1, \mathrm{Re}_d^{0.077}\right] - 1}{2F(B_T)}\right), \tag{17}$$

where $B_T = \frac{c_{pv}(T_g - T_s)}{L_{\mathrm{eff}}}$ is the Spalding heat transfer number, $F(B_T) = (1 + B_T)^{0.7}\frac{\ln(1+B_T)}{B_T}$, $L_{\mathrm{eff}} = L + \frac{Q_L}{\dot{m}_d} = \sum_i \epsilon_i L_i + \frac{Q_L}{\sum_i \dot{m}_{di}}$, Q_L is the power spent on droplet heating, c_{pv} is the specific heat capacity of fuel vapour, $\epsilon_i = \epsilon_i(t)$ are the evaporation rates of species i, $\dot{m}_i = \epsilon_i\dot{m}_d$ ($\dot{m}_d = \sum_i \dot{m}_{di}$). The effects of the interaction between droplets are ignored.

Similarly, assuming that D_{eff} =const and the same for all species, the system of Equations (6) can be decoupled (this is implicitly assumed in the format of these equations) and each of these equations can be solved analytically as [14]:

$$Y_{li} = \epsilon_i + \frac{1}{R}\left\{\left[\exp\left[D_{\mathrm{eff}}\left(\frac{\lambda_0}{R_d}\right)^2 t\right][q_{i0} - Q_0\epsilon_i]\sinh\left(\lambda_0\frac{R}{R_d}\right)\right.\right.$$
$$\left.\left. + \sum_{n=1}^{\infty}\left[\exp\left[-D_{\mathrm{eff}}\left(\frac{\lambda_n}{R_d}\right)^2 t\right][q_{in} - Q_n\epsilon_i]\sin\left(\lambda_n\frac{R}{R_d}\right)\right]\right\}, \tag{18}$$

where λ_0 and λ_n ($n \geq 1$) are solutions to equations

$$\tanh\lambda_0 = -\frac{\lambda_0}{h_{0Y}} \quad \text{and} \quad \tan\lambda_n = -\frac{\lambda_n}{h_{0Y}} \quad (n \geq 1),$$

respectively, $h_{0Y} = -\left(1 + \frac{\alpha R_d}{D_{\mathrm{eff}}}\right)$.

$$Q_n = \begin{cases} -\frac{1}{||v_0||^2}\left(\frac{R_d}{\lambda_0}\right)^2(1 + h_{0Y})\sinh\lambda_0 & \text{when} \quad n = 0 \\ \frac{1}{||v_n||^2}\left(\frac{R_d}{\lambda_n}\right)^2(1 + h_{0Y})\sin\lambda_n & \text{when} \quad n \geq 1 \end{cases} \tag{19}$$

$$q_{in} = \frac{1}{||v_n||^2}\int_0^{R_d} RY_{li0}(R)v_n(R)\mathrm{d}R, \tag{20}$$

$n \geq 0$, where

$$v_0(R) = \sinh\left(\lambda_0\frac{R}{R_d}\right), \quad v_n(R) = \sin\left(\lambda_n\frac{R}{R_d}\right), \quad n \geq 1.$$

The validity of the assumption that D_{eff} is the same for all species can be questioned. However, if this assumption is relaxed then we would need to take into account that D_{eff} becomes the function of local mass fractions of all species and the system of Equations (6) can no longer be decoupled.

The Sherwood number for isolated droplets Sh can be approximated by the following expression [13]:

$$\mathrm{Sh} = 2\frac{\ln(1+B_M)}{B_M}\left(1+\frac{(1+\mathrm{Re}_d\mathrm{Sc}_d)^{1/3}\max\left[1,\mathrm{Re}_d^{0.077}\right]-1}{2F(B_M)}\right), \tag{21}$$

$\mathrm{Sc}_d = \frac{\nu_{\mathrm{air}}}{D_v}$ is the Schmidt number, $F(B_M) = (1+B_M)^{0.7}\frac{\ln(1+B_M)}{B_M}$. B_T and B_M are linked by the following equation [13]:

$$B_T = (1+B_M)^{\varphi} - 1, \tag{22}$$

where

$$\varphi = \left(\frac{c_{pv}}{c_{pa}}\right)\left(\frac{\mathrm{Sh}^*}{\mathrm{Nu}^*}\right)\frac{1}{\mathrm{Le}}, \tag{23}$$

$\mathrm{Le} = k_g/(c_{pa}\rho_{\mathrm{total}}D_v) = \mathrm{Sc}_d/\mathrm{Pr}_d$ is the Lewis number,

$$\mathrm{Sh}^* = 2\left(1+\frac{(1+\mathrm{Re}_d\mathrm{Sc}_d)^{1/3}\max\left[1,\mathrm{Re}_d^{0.077}\right]-1}{2F(B_M)}\right), \tag{24}$$

$$\mathrm{Nu}^* = 2\left(1+\frac{(1+\mathrm{Re}_d\mathrm{Pr}_d)^{1/3}\max\left[1,\mathrm{Re}_d^{0.077}\right]-1}{2F(B_T)}\right). \tag{25}$$

As follows from Equation (23), φ is a function of B_T. Hence, the iteration process needs to be performed to calculate B_T from Equation (22). In our previous paper [20] it was shown that, in some practically important cases, Formula (23) can be simplified assuming that $\frac{\mathrm{Sh}^*}{\mathrm{Nu}^*} = 1$. This assumption turned out to be too crude in our case.

In the case of multi-component droplets, the problem of modelling droplet evaporation is complicated by the fact that different species diffuse in air at different rates, and the evaporation rate of one of the species is affected by the evaporation rate of other species. Our analysis of $\dot{m}_d$ is based on Equation (11), assuming that the mixture of vapour species can be treated as a separate gas, similar to treating the mixture of nitrogen, oxygen and carbon dioxide as air ($Y_{vs} = \sum_i Y_{vis}$).

The approximations described above lead to the model of multi-component droplet heating and evaporation known as the Discrete Component Model (DCM). In practical engineering applications, this model is applicable to the cases when the number of components is relatively small (about a dozen or so). The most widely used approach to modelling heating and evaporation of droplets with large numbers of components is based on the probabilistic method of analysis (e.g. Continuous Thermodynamics approach [21]-[23] and the Distillation Curve Model [24]). In this approach, a number of additional simplifying assumptions were used, including the assumption that species inside droplets mix infinitely quickly. Models containing features of both approaches are described in [25, 26].

A new approach to modelling the heating and evaporation of multi-component Diesel fuel droplets, suitable for the case when a large number of components is present in the droplets, was suggested in [27]. In this approach it was assumed that the main contribution to automotive fuels (Diesel and gasoline) comes from alkanes, and the contribution of various n-alkanes was described by the distribution function $f_m(n)$ [27]:

$$f_m(n) = C_m(n_0, n_f)\frac{(M(n)-\gamma)^{\alpha-1}}{\beta^{\alpha}\Gamma(\alpha)}\exp\left[-\left(\frac{M(n)-\gamma}{\beta}\right)\right], \tag{26}$$

where $n_0 \leq n \leq n_f$, subscripts $_0$ and $_f$ stand for initial and final, $\Gamma(\alpha)$ is the Gamma function, α and β are parameters that determine the shape of the distribution, γ determines the original shift, $M(n) = 14n + 2$ is the molar mass of alkanes,

$$C_m(n_0, n_f) = \left\{ \int_{n_0}^{n_f} \frac{(M(n) - \gamma)^{\alpha-1}}{\beta^\alpha \Gamma(\alpha)} \exp\left[- \left(\frac{M(n) - \gamma}{\beta} \right) \right] \mathrm{d}n \right\}^{-1}. \tag{27}$$

This choice of C_m assures that $\int_{n0}^{n_f} f_m(n)\mathrm{d}n = 1$.

Assuming that the properties of hydrocarbons in a certain narrow range of n are close, the continuous distribution $f_m(n)$ was replaced with a discrete one, consisting of N_f quasi-components with carbon numbers

$$\overline{n}_j = \frac{\int_{n_{j-1}}^{n_j} n f_m(n)\mathrm{d}n}{\int_{n_{j-1}}^{n_j} f_m(n)\mathrm{d}n}, \tag{28}$$

(which are not integers in the general case) and the corresponding molar fractions

$$X_j = \int_{n_{j-1}}^{n_j} f_m(n)\mathrm{d}n. \tag{29}$$

where j is an integer in the range $1 \leq j \leq N_f$. Note that $\sum_{j=1}^{j=N_f} X_j = 1$. The new structures with $\overline{n}_j$ defined by Equation (28) are called quasi-components as they cannot be identified with actual components, characterised by integer n, in the general case, but can be treated as actual components in the analysis of heat and mass transfer processes in complex fuels, using the Discrete Component Model (DCM). The DCM in which the quasi-components are used alongside actual components was called the quasi-discrete model.

It was assumed that all $n_j - n_{j-1}$ are equal, i.e. all quasi-components have the same range of values of n. For the case when $N_f = 1$, this approach reduces the analysis of multi-component droplets to that of mono-component droplets. This model was generalised in [28] to take into account the dependence of density, specific heat capacity, thermal conductivity and viscosity on the values of carbon numbers. It is expected to be particularly useful when N_f is much less than the number of actual species in the hydrocarbon mixture.

There are two main problems with the application of this approach to realistic Diesel fuels. Firstly, even if we restrict our analysis only to alkanes, it does not appear to be easy to approximate their distribution with a reasonably simple distribution function $f_m(n)$, given by Expression (26). Secondly, the contributions of the other hydrocarbon groups, apart from alkanes, cannot be ignored in any realistic model of Diesel fuels. In the model suggested in [10] both of these issues were addressed.

In the new model suggested in [10] the focus is shifted from the analysis of the distribution function to the direct analysis of molar fractions of the components. These are described by the matrix X_{nm}, where n refers to the number of carbon atoms, and m refers to the groups (e.g. alkanes) or individual components (tricycloalkane, diaromatic and phenanthrene). The link between the values of m and the components is shown in Table 1.

m	Component
1	alkanes
2	cycloalkanes
3	bicycloalkanes
4	alkylbenzenes
5	indanes & tetralines
6	naphthalenes
7	tricycloalkane
8	diaromatic
9	phenanthrene

Table 1

For each m the values of $\overline{n}_{jm}$ of quasi-components can be introduced as

$$\left.\begin{array}{l}\overline{n}_{1m} = \dfrac{\sum_{n=n_{1m}}^{n=n_{(\varphi_m+1)m}} (nX_{nm})}{\sum_{n=n_{1m}}^{n=n_{(\varphi_m+1)m}} X_{nm}}, \\[2ex] \overline{n}_{2m} = \dfrac{\sum_{n=n_{(\varphi_m+2)m}}^{n=n_{(2\varphi_m+2)m}} (nX_{nm})}{\sum_{n=n_{(\varphi_m+2)m}}^{n=n_{(2\varphi_m+2)m}} X_{nm}}, \\ \ldots\ldots\ldots\ldots\ldots\ldots\ldots\ldots \\ \overline{n}_{\ell m} = \dfrac{\sum_{n=n_{((\ell-1)\varphi_m+\ell)m}}^{n=n_{k_m}} (nX_{nm})}{\sum_{n=n_{((\ell-1)\varphi_m+\ell)m}}^{n=n_{k_m}} X_{nm}}, \end{array}\right\} \tag{30}$$

where $n_{1m}{=}n_{m(\min)}$ is the minimal value of n for which $X_{nm} \neq 0$, $n_{k_m}{=}n_{m(\max)}$ is the maximal value of n for which $X_{nm} \neq 0$, $\ell = \text{integer}\ ((k_m + \varphi_m)/(\varphi_m + 1))$. Parameter φ_m is assumed to be integer; $\varphi_m + 1$ is equal to the number of components to be included into quasi-components, except possibly the last one in the group. φ_m is assumed to be the same for all quasi-components within group m. If $\varphi_m = 0$ then $\ell = k_m$ and the number of quasi-components is equal to the number of actual components. φ_m and k_m depend on m in the general case.

This approach to the generation of quasi-components is based on the selection of the number of components in each quasi-component (φ_m+1 in most cases). An alternative approach to their generation can be based on the selection of the number of quasi-components n_{qm}. In this case the number of components in each quasi-component, except possibly the last one, (n_{cm}) is taken equal to the nearest integer of the ratio k_m/n_{qm}. If k_m/n_{qm} is not an integer then the number of components in the last quasi-component (n_{lcm}) is either greater than n_{cm}, if $(k_m/n_{qm}) > n_{cm}$, or less than n_{cm}, if $(k_m/n_{qm}) \leq n_{cm}$. It was found that the second approach is more convenient for practical applications and it was used in [10]. The values of $\overline{n}_{im}$ were calculated using the same approach as in the case presented in Equation (30). Due to the additional dimensions introduced by the subscript m in Equation (30), the new model was called the multi-dimensional quasi-discrete model.

The size of quasi-components controls the number of Solutions (18) to be used in the analysis and ultimately the accuracy of calculations. Another factor affecting this accuracy is the time step used for calculation of droplet heating and evaporation (the ranges of t over which Solutions (18) are used for fixed values of input parameters). This time step is expected to be much shorter than the time steps used for calculation of droplet trajectories. The latter in their turn are much shorter than the time steps used to calculate the parameters of the carrier (gas) phase. Herein lies the multi-rate nature of spray calculations used in computational fluid dynamics (CFD) codes.

4. Results

In this section, the models described in the previous sections are applied to the analysis of heating and evaporation of biodiesel (Discrete Component Model (DCM)) and Diesel (multi-dimensional quasi-discrete model) droplets.

The application of DCM to the analysis of biodiesel fuel droplets can be justified by the presence of the relatively small number of components in these fuels. In [9], the analysis was focused on five types of biodiesel fuels: Palm Methyl Ester (PME) produced from palm oil, Hemp Methyl Esters, produced from hemp seed oil in the Ukraine (HME1) and European Union (HME2), Rapeseed oil Methyl Ester (RME), produced from rapeseed oil in the Ukraine, and Soybean Methyl Ester (SME) produced from soybean oil. These fuels contain up to 15 methyl esters and possibly small amounts of unspecified additives, which were treated as methyl esters with average characteristics. Calculations were performed: (1) taking into account the contribution of all components of biodiesel fuels; and (2) assuming that these fuels can be treated as mono-component fuels with average transport and thermodynamic coefficients.

Following [9] it is assumed that a Butterfat Fatti Acid (BFA) (see Table 5 of [29] for the composition of this fuel) fuel droplet of initial radius $R_{d0} = 12.66$ μm and initial temperature $T_0 = 300$ K moves with a velocity $U_d = 35$ m/s [30] is considered. As in [9] it is assumed that the gas temperature and pressure are equal to 880 K and 30 bars respectively. The values of droplet surface temperature (T_s) and radius (R_d) versus time (t) for this case are shown in Fig. 1. The calculations were performed: (1) taking into account the contribution of all 9 components of the BFA fuel (multi-component model) and using the Effective Thermal Conductivity/Effective Diffusivity (ETC/ED) model; and (2) replacing these components with a single component with average transport and thermodynamic properties obtained as described in Appendix C of [9] (single-component model) and using the Infinite Thermal Conductivity (ITC) model.

As follows from Fig. 1, the droplet temperatures and radii predicted by the multi- and mono-component models are reasonably close. The single component model under-predicts the evaporation time by about 13.4% and droplet surface temperature by up to about 12.0%. These differences are larger than the ones reported in [9] for other types of biodiesel fuel. This implies that biodiesel fuel droplets can be approximated by mono-component droplets with infinitely large thermal conductivities if an accuracy of about 13% can be tolerated.

Following [30], the multi-dimensional quasi-discrete model was tested for the analysis of heating and evaporation of a Diesel fuel droplet of an initial radius R_{d0} =10 μm and initial temperature $T_0 = 300$ K moving at a velocity $U_d = 35$ m/s. As in the case of biodiesel fuel droplets, it is assumed that the gas temperature and pressure are equal to 880 K and 30 bars respectively. The analysis is based on the Effective Thermal Conductivity/Effective Diffusivity (ETC/ED) model.

The plots of the droplet surface temperatures T_s and radii R_d versus time for six approximations of Diesel fuel composition are shown in Fig. 2. These are the cases shown in this figure: the contributions of all 98 components are taken into account (indicated as (98)); the contribution of 98 components is approximated by 15 quasi-components/components (indicated as (15)); the contribution of 98 components is approximated by 6 quasi-components (corresponding to the 6 groups mentioned above) and 3 components (tricycloalkane, diaromatic and phenanthrene) without taking into account the diffusion between them so that their mass fractions remain equal to the initial mass fractions and they behave like a single quasi-component (indicated as (S)), the contributions of only 20 alkane components are taken into account (standard approximation used in the original quasi-discrete model [27, 28]) (indicated as (20A)); the contributions of only 20 alkane components are taken into account and these are approximated by a single quasi-component with the average value of the carbon number ($C_{14.763}H_{31.526}$, indicated as (SA)), and the approximation of Diesel fuel by n-dodecane. The latter approximation is the one which is most widely used in engineering applications. In the cases when only the contribution of alkanes was taken into account, the mass fractions of the components were recalculated to ensure that the total mass fractions of all alkanes were equal to 1.

As can be seen from Fig. 2, the values of T_s and R_d predicted by 98 and 15 quasi-component/component approximations are reasonably close. The 15 quasi-component/component approximation under-predicts the evaporation time by about 3.2%. At the same time the single component approximation under-predicts droplet evaporation time by about 19.1% which cannot be accepted in most engineering applications. In the case when Diesel fuel is approximated with 20 alkane components the predicted droplet surface temperatures appeared to be higher and the evaporation time shorter by about 19.2% than in the case of approximation of Diesel fuel with 98 components. This means that the approximation of Diesel fuel with alkanes, a widely used assumption in the modelling of Diesel fuels (see [27, 28] and the references therein), leads to results which are less accurate, compared with the approximation of Diesel fuel by a single quasi-component. The approximation of Diesel fuel with a single alkane quasi-component ($C_{14.763}H_{31.526}$) leads to under-prediction of the evaporation time by about 35.3% which is not acceptable even for qualitative analysis of the process. In the case when Diesel fuel is approximated by n-dodecane, this error increases to about 44.2%. This leads us to question the validity of the results of numerous papers where Diesel fuel was approximated with n-dodecane (see [15, 3] for the details).

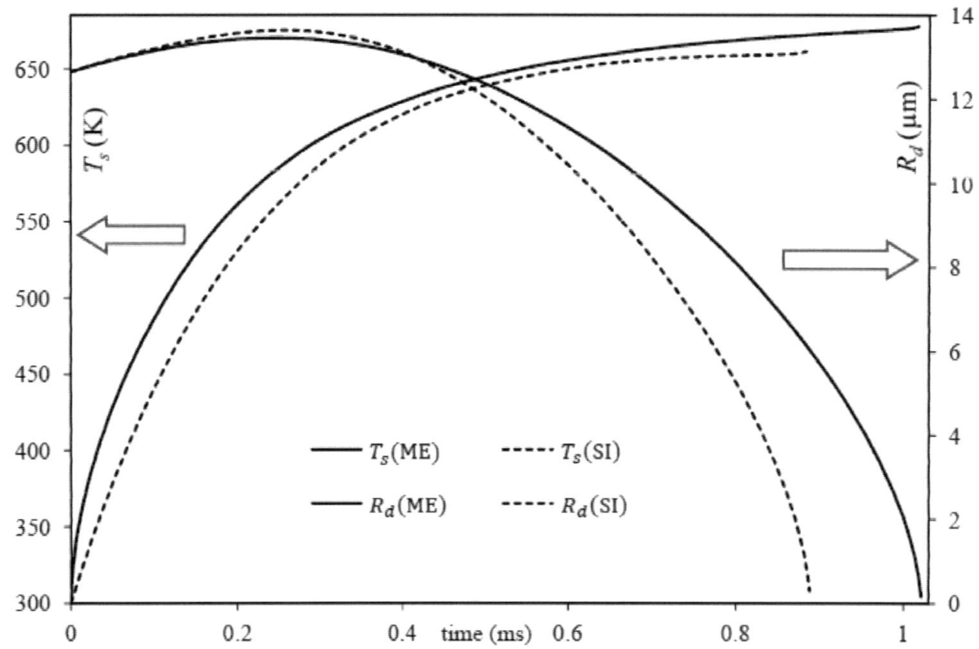

Figure 1. The plots of BFA droplet surface temperatures (T_s) and radii (R_d) versus time predicted by the multi-component (M) and single-component (S) models. Gas temperature and pressure are assumed to be equal to 880 K and 30 bar respectively. The initial droplet radius is assumed to be equal to 12.66 μm and its velocity is assumed to be constant and equal to 35 m/s. The analysis is based on the Effective Thermal Conductivity/Effective Diffusivity (ETC/ED) model (E) and Infinite Thermal Conductivity (ITC) model (I).

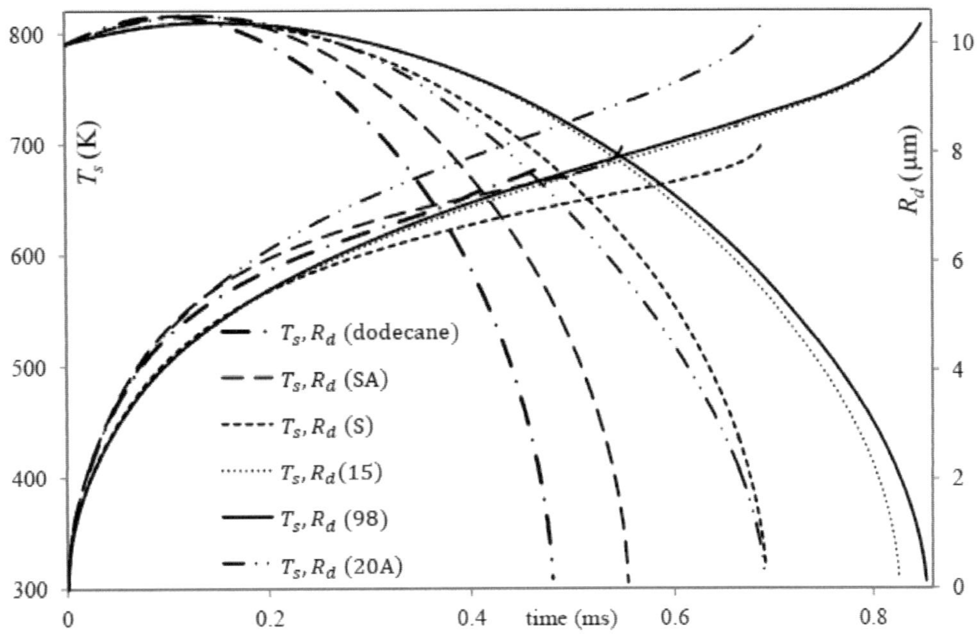

Figure 2. The plots of the droplet surface temperatures T_s and radii R_d versus time for six approximations of Diesel fuel composition, taking into account the contributions of all 98 components (98); 15 quasi-components/components (15); treating Diesel fuel as a single component (S); 20 alkane components (20A); single alkane components (SA) and n-dodecane. Gas temperature and pressure are assumed to be equal to 880 K and 30 bar respectively. The initial droplet radius is assumed to be equal to 10 μm and its velocity is assumed to be constant and equal to 35 m/s. The analysis is based on the Effective Thermal Conductivity/Effective Diffusivity (ETC/ED) model.

5. Conclusions

New mathematical tools and approximations developed for the analysis of automotive fuel droplet heating and evaporation have been summarised. The approach to modelling biodiesel fuel droplets is based on the application of the Discrete Component Model (DCM) in combination with the Effective Thermal Conductivity/Effective Diffusivity model and the implementation in a numerical code of the analytical solutions to heat transfer and species diffusion equations inside droplets. It is shown that the approximation of BFA biodiesel fuel by a single component, using the Infinite Thermal Conductivity model, leads to the under-prediction of droplet evaporation time by about 13% which can be acceptable as a crude approximation in many engineering applications.

The approach to modelling Diesel fuel droplets is based on the application of the recently developed multi-dimensional quasi-discrete model. The application of this model has been combined with the application of the Effective Thermal Conductivity/Effective Diffusivity model and the implementation of the analytical solutions to heat transfer and species diffusion equations inside droplets in the numerical code, as in the case of biodiesel fuel droplets.

The composition of Diesel fuel has been simplified and reduced to only 98 components. It has been pointed out that in contrast to biodiesel fuel droplets, the approximation of Diesel fuel by a single component leads to under-estimation of droplet evaporation time by more than 19%, which is not acceptable in many engineering applications. The approximation of

98 components of Diesel fuel by 15 quasi-components/components leads to under-prediction of this time by about 3% which is believed to be acceptable in most applications. At the same time, the approximation of Diesel fuel with 20 alkane components leads to the prediction of higher droplet surface temperatures and shorter evaporation time than in the case of approximation of Diesel fuel with 98 components. This means that the approximation of Diesel fuel with alkanes, a widely used assumption in the modelling of Diesel fuels, leads to results which are less accurate, compared with the approximation of Diesel fuel by a single quasi-component. The approximations of Diesel fuel with a single alkane quasi-component ($C_{14.763}H_{31.526}$) and n-dodecane lead to under-prediction of the evaporation time by about 35% and 44% respectively, which is not acceptable even for qualitative analysis of the process.

Acknowledgment

The authors are grateful to INTERREG IVa (Project E3C3, Reference 4274) and EPSRC (UK) (Project EP/J006793/1) for the financial support of this project.

References

[1] Flynn P F, Durrett R P, Hunter G L, zur Loye A O, Akinyemi O C, Dec J E, Westbrook C K 1999 *SAE report* 1999-01-0509.
[2] Sazhina E M, Sazhin S S, Heikal M R, Babushok V I, Johns R 2000 *Combust. Sci. Technology* **160** 317.
[3] Sazhin S S 2014 *Droplets and Sprays* (Springer).
[4] Sazhin S S 2005 J. Phys.: Conf. Series **22** 174-193.
[5] Sazhin S S 2006 J. Phys.: Conf. Series **55** 191-202.
[6] Sazhin S S, Martynov S B, Kaplanski F, Begg S 2008 J. Phys.: Conf. Series **138** paper 012034.
[7] Sazhin S S, Shishkova I N, Elwardany A, Gusev I G, Heikal M 2011 J. Phys.: Conf. Series **268** paper 012026.
[8] Sazhin S S, Elwardany A E, Heikal M R 2015 J. Phys.: Conf. Series **585** 012014.
[9] Sazhin S S, Al Qubeissi M, Kolodnytska R, Elwardany A, Nasiri R, Heikal M R 2014 *Fuel* **115** 559.
[10] Sazhin S S, Al Qubeissi M, Nasiri R, Gunko V M, Elwardany A E, Lemoine F, Grisch F, Heikal M R 2014 *Fuel* **129** 238.
[11] Carslaw H S, Jaeger J C 1986 *Conduction of Heat in Solids* (Oxford, Clarendon Press).
[12] Kartashov E M 2001 *Analytical Methods in Heat Transfer Theory in Solids* (Moscow, Vysshaya Shkola) (in Russian).
[13] Abramzon B, Sirignano W A 1989 *Int. J. Heat Mass Transfer* **32** 1605.
[14] Sazhin S S, Elwardany A, Krutitskii P A, Castanet G, Lemoine F, Sazhina E M, Heikal M R 2010 *Int. J. Heat Mass Transfer* **53** 4495.
[15] Sazhin S S. 2006 *Prog. in Energy Combustion Sci.* **32** 162.
[16] Sazhin S S, Abdelghaffar W A, Krutitskii P A, Sazhina E M, Heikal M R 2005 *Int. J. Heat Mass Transfer* **48** 4215.
[17] Atkins P, de Paula J 2002 *Atkins' Physical Chemistry*, Seventh Edition (Oxford, Oxford University Press).
[18] Sazhin S S, Krutitskii P A, Abdelghaffar W A, Mikhalovsky S V, Meikle S T, Heikal M R 2004 *Int. J. Heat Mass Transfer* **47** 3327.
[19] Sazhin S S and Krutitskii P A 2001 *Progress in Analysis Vol. II. Proceedings of the 3rd International ISAAC Congress* vol. II, ed Begehre HGW, Gilbert RP, Wong MW (Singapore: World Scientific) p 1231.
[20] Elwardany A E, Gusev I G, Castanet G, Lemoine F, Sazhin S S 2011 *Atomization and Sprays* **21** 907.
[21] Arias-Zugasti M, Rosner D F 2003 *Combustion and Flame* **135** 271.
[22] Abdel-Qader Z, Hallett W L H 2005 *Chemical Engineering Science* **60** 1629.
[23] Rivard E, Brüggemann D 2010 *Chemical Engineering Science* **65** 5137.
[24] Burger M, Schmehl R, Prommersberger K, Schäfer O, Koch R, Wittig S 2003 *Int. J. Heat and Mass Transfer* **46** 4403.
[25] Zhang L, Kong S-C 2010 *Combustion and Flame* **157** 2165.
[26] Laurent C, Lavergne G, Villedieu P 2009 *C.R. Mechanique* **337** 449.
[27] Sazhin S S, Elwardany A E, Sazhina E M, Heikal M R 2011 *Int. J. Heat and Mass Transfer* **54** 4325.
[28] Elwardany A E, Sazhin S S 2012 *Fuel* **97** 685.
[29] Yuan W, Hansen A C, Zhang Q 2005 *Fuel* **84** 943.
[30] Crua C, de Sercey G, Gold M, Heikal M R 2013 *Proceedings of ILASS-Europe, 25th conference on liquid atomization and spray systems* Chania, Greece, 1-4 September 2013.

Invariant surfaces of variable stability

Elena Shchepakina, Vladimir Sobolev

Department of Technical Cybernetics, Samara State Aerospace University (SSAU), 34, Moskovskoye shosse, Samara 443086, Russia

E-mail: shchepakina@yahoo.com, hsablem@gmail.com

Abstract. This paper presents a brief description of the theory of invariant manifolds of variable stability in the context of their connection with the theory of solutions that are bounded on the whole axis. This approach allows various generalizations both to the case of increasing of the dimension of the invariant manifolds and to the case of multiple change of their stability. The sufficient conditions for the existence of an invariant manifold of variable stability are revealed. The continuity condition for the invariant manifold yields the analytic representation of the gluing function. The theoretical developments are illustrated by several examples.

1. Introduction

In this paper we outline an approach for constructing of invariant surfaces of variable stability. As a one-dimensional example of such objects it is possible to consider a *canard*. The term "canard" (or "duck-trajectory") originally was introduced in the context of the van der Pol equation [1]–[4], but later was investigated for other types of singularly perturbed systems including the ones of higher dimensions, i.e. when the phase variables are vectors (see, for example, [5]–[10] and the references therein).

From the viewpoint of the geometrical theory of singular perturbations a canard may be considered as a result of gluing stable and unstable slow invariant manifolds at one point of the breakdown surface due to the availability of an additional scalar parameter in the differential system. This approach was first proposed in [11, 12] and was then applied in [7], [13]–[17].

Other example of invariant surfaces of variable stability may be a *canard cascade* [18, 19] which is a result of gluing stable (attractive) and unstable (repulsive) slow invariant manifolds at several points of the breakdown surface. For the possibility of this gluing an additional function of a slow variable and several additional parameters are required. If we take an additional function of a vector variable parameterizing the breakdown surface, we can glue the stable and unstable slow invariant manifolds at all points of the breakdown surface simultaneously. As a result we obtain the continuous stable/unstable invariant surface or *black swan* [8, 15, 19, 20, 21]. Such surfaces may be considered as a multidimensional analogue of the notion of a canard.

Thus, the applying of a gluing function instead of a single gluing parameter may lead to various types of the invariant surfaces of variable stability, some of which are discussed in this paper.

From the perspective of analytic representation an invariant surface of variable stability is closely connected to a bounded-on-the-whole-axis solution of the corresponding differential system. Furthermore, the continuity condition for the invariant surface yields the analytic representation of the gluing function.

In this paper we consider an approach for constructing of invariant surfaces of variable stability, which is based, on the one hand, on applying of a gluing function, and on the other hand on the analytic representation of a bounded-on-the-whole-axis solution of the corresponding differential system.

2. Theoretical Background

We briefly recall the basic definitions and notions concerning the invariant surfaces of variable stability of singularly perturbed systems. The main object of our consideration is the autonomous singularly perturbed system of the form

$$\frac{dx}{dt} = f(x, y, \alpha, \varepsilon), \tag{1}$$

$$\varepsilon \frac{dy}{dt} = g(x, y, \alpha, \varepsilon), \tag{2}$$

where x and y are vectors in Euclidean spaces R^n and R^m, respectively, ε is a small positive parameter, vector-functions f and g are sufficiently smooth and their values are comparable to unity. The slow and fast subsystems are represented by (1) and (2), respectively.

The degenerate equation $g(x, y, \alpha, 0) = 0$ describes a *slow surface* S (or *slow curve* in the case of the scalar variable x). The subset Γ of S given by

$$det \left| \frac{\partial g}{\partial y}(x, y, \alpha, 0) \right| = 0 \tag{3}$$

determines a *breakdown surface* (*breakdown, turning* or *jump points* in the scalar case) [19]. The surface Γ divides S into foliations on which

$$det \left| \frac{\partial g}{\partial y}(x, y, \alpha, 0) \right| \neq 0.$$

By the implicit function theorem, a leaf of the slow surface is determined by a well-defined vector-function: $y = \varphi(x, \alpha)$. A slow surface can consist of several foliations determined by different functions $y = \varphi_i(x, \alpha)$, domains of which can intersect, depending on the structure of the slow surface. It should be noted that the conditions of the implicit function theorem do not hold on the leaf boundary given by (3).

The leaf S^s of the slow surface is *stable* (or *attractive*) if the spectrum of the Jacobian matrix

$$\frac{\partial g}{\partial y}(x, \varphi(x, \alpha), \alpha, 0)$$

is located in the left open complex half-plane. If there is at least one eigenvalue of the Jacobian matrix with a positive real part then the leaf of S, say S^u, is *unstable* (or *repulsive*). The case of complex eigenvalues is not considered in this paper.

In an ε-neighborhood of S^s (S^u) there exists a stable (unstable) *slow invariant manifold.* The slow invariant manifold is defined as an invariant surface of slow motions.

Definition 1 *A smooth surface S_ε in $R^m \times R^n$ is called an invariant manifold of the system (1), (2) if any trajectory of the system that has at least one point in common with S_ε lies entirely on S_ε.*

The only invariant manifolds of system (1), (2) discussed here are those of dimension m (the dimension of the slow variable x) that can be represented as graphs of vector-valued functions $y = h(x, \alpha, \varepsilon) = \varphi(x, \alpha) + O(\varepsilon)$, $\varepsilon \to 0$. Here it is assumed that $h(x, \alpha, \varepsilon)$ is a sufficiently smooth

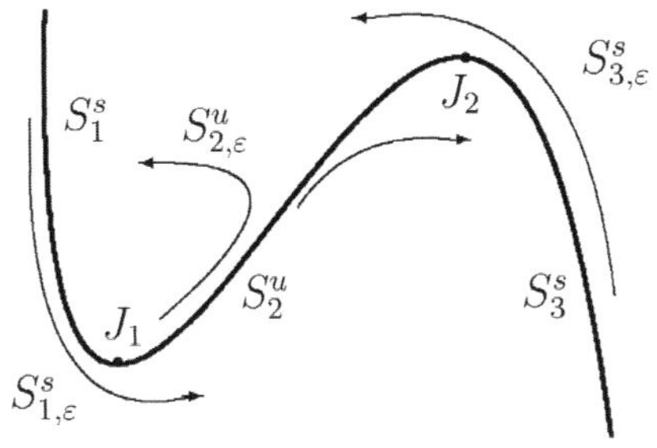

Figure 1. The slow curve (thick line) and the trajectories (thin lines) of the van der Pol's type systems. The branches S^s_1 and S^s_3 of the slow curve are stable, the branch S^u_2 is unstable, J_1 and J_2 are the jump points.

function of ε. Such integral manifolds are called manifolds of slow motions — the origin of this term lies in nonlinear mechanics.

The notions of (in)stability for slow invariant manifold become clear if we rescale the system (1), (2) via the change of variables $\tau = t/\varepsilon$. Hence we obtain the equivalent system

$$\frac{dx}{d\tau} = \varepsilon f(x, y, \alpha, \varepsilon), \tag{4}$$

$$\frac{dy}{d\tau} = g(x, y, \alpha, \varepsilon). \tag{5}$$

For simplicity we consider the case when the variables x and y are scalar. The stable S^s and unstable S^u subsets of the slow curve S for which $g_y < 0$ and $g_y > 0$, respectively, are separated by the jump point(s) at which $g_y = 0$. When $\varepsilon = 0$ S consists of equilibrium points of the system (4), (5) which are stable if $g_y < 0$ and unstable if $g_y > 0$. When $\varepsilon > 0$ and sufficiently small, the manifold of equilibria is replaced by a boundary layer close to S with slow dynamics. There is an attractive trajectory in the boundary layer close to the stable branch S^s, the *stable invariant manifold* S^s_ε, and there is a repulsive trajectory close to an unstable branch S^u, the *unstable invariant manifold* S^u_ε [22]. For example, the slow curve of the van der Pol system is S-shaped. It is divided by two jump points, say J_1 and J_2, into two stable (S^s_1 and S^s_3) and one unstable (S^u_2) branches, see Fig. 1. In the ε-neighborhood of the stable and the unstable branches of the slow curve there exist the stable ($S^s_{1,\varepsilon}$ and $S^s_{3,\varepsilon}$) and the unstable ($S^u_{2,\varepsilon}$) slow invariant manifolds, respectively. A trajectory on $S^s_{1,\varepsilon}$ must leave it when the jump point J_1 is reached. After this moment two scenarios are possible depending on the value of the control parameter α: the trajectory is either immediately repelled back to $S^s_{1,\varepsilon}$ from $S^u_{2,\varepsilon}$ (if $S^s_{1,\varepsilon}$ is situated inside of $S^u_{2,\varepsilon}$), or repelled to $S^s_{3,\varepsilon}$ and makes a large excursion from the second jump point before it can return to $S^s_{1,\varepsilon}$ (if $S^s_{1,\varepsilon}$ is situated outside of $S^u_{2,\varepsilon}$). While in the first case the limit cycles must be small, in the second case they are large and of relaxation type.

The presence of the additional scalar parameter α provides the possibility of gluing the stable and unstable invariant manifolds at the jump point to form a single trajectory, the canard. The term "canard" (or duck–trajectory) had been originally given by French mathematicians to the intermediate periodic trajectories of the van der Pol equation between the small and the large orbits due to their special shapes, see Fig. 2. However, in our work a canard is a one-dimensional slow invariant manifold of variable stability.

Definition 2 *Trajectories which at first move along the stable slow invariant manifold and then continue for a while along the unstable slow invariant manifold are called canards or duck-trajectories.*

Definition 3 *Trajectories which at first move along the unstable slow invariant manifold and then continue for a while along the stable slow invariant manifold are called false canard trajectories.*

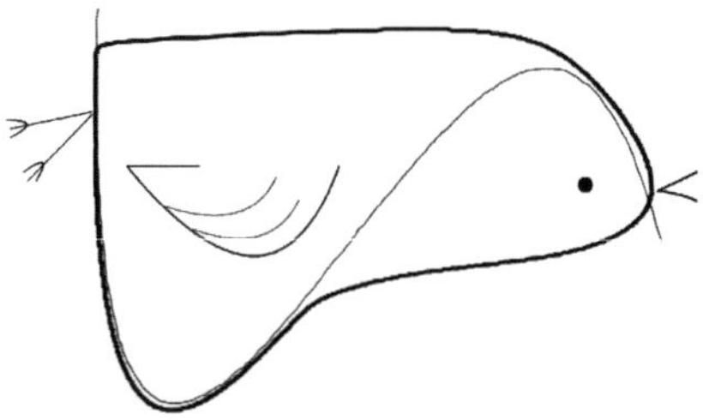

Figure 2. Canard (thick line) and the slow curve (thin line) of the van der Pol's type system.

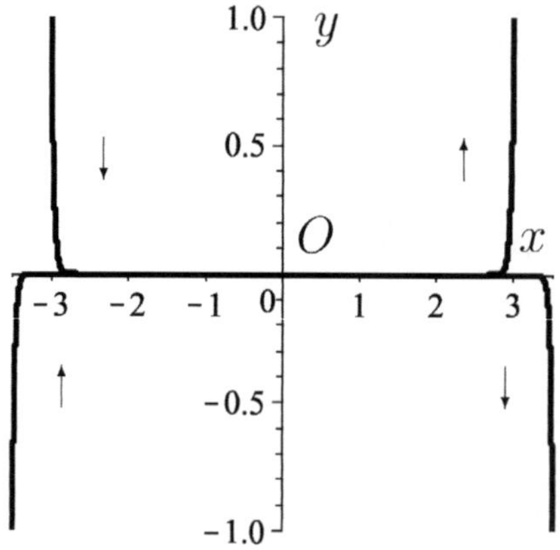

Figure 3. Global (the longest) and local canards of (6). The arrows show the direction of increasing time, O is the turning point.

Therefore it is possible to consider a straight line as a basic model of a canard.

Example 1 As the simplest system with a canard we propose

$$\dot{x} = 1, \;\; \varepsilon\dot{y} = xy. \tag{6}$$

Here $g(x, y) = xy$. It is clear that the trajectory $y = 0$ is a canard. The left part ($x < 0$) is attractive since $\partial g/\partial y < 0$ and the right part ($x > 0$) is repulsive. These two parts are divided by a turning point O, which separates stable and unstable parts of the slow curve, at $x = 0$.

Note that $y = 0$ is global (the longest) canard and the unique bounded-on-the-whole-axis solution of system (6). Others solutions of system (6) which are local canards, see Fig. 3, are infinitely increasing (decreasing) as $t \to \infty$ or $t \to -\infty$.

Example 2 Some generalization of (6) is the system

$$\dot{x} = 1, \;\; \varepsilon\dot{y} = xy + \phi(x, \varepsilon) + \alpha,$$

the unique bounded-on-the-whole-axis solution of which has the representation

$$y = \begin{cases} \dfrac{1}{\varepsilon}\displaystyle\int_{-\infty}^{x} \exp\left(\dfrac{x^2 - s^2}{2\varepsilon}\right)[\phi(s, \varepsilon) + \alpha]\, ds, & x \le 0, \\ -\dfrac{1}{\varepsilon}\displaystyle\int_{x}^{\infty} \exp\left(\dfrac{x^2 - s^2}{2\varepsilon}\right)[\phi(s, \varepsilon) + \alpha]\, ds, & x > 0. \end{cases} \tag{7}$$

The "canard value" of the parameter α can be obtained from the continuity condition for (7) at $x = 0$:

$$\varepsilon^{-1}\int_{-\infty}^{\infty} e^{-s^2/2\varepsilon}[\phi(s, \varepsilon) + \alpha]\, ds = 0$$

or

$$\alpha_\phi = -\frac{1}{\sqrt{2\pi\varepsilon}}\int_{-\infty}^{\infty} e^{-s^2/2\varepsilon}\phi(s, \varepsilon)\, ds.$$

The proof for the existence of a maximum canard is based on considering a solution to the corresponding differential system that is bounded on the whole axis. This technique canard constructing is also applicable to further generalizations.

3. Canard as a unique bounded-on-the-whole-axis solution

Will apply the technique of canard constructing described in the previous section for the planar system

$$\dot{x} = 1, \ \ \varepsilon\dot{y} = 2xy + F(x, y, \alpha, \varepsilon), \tag{8}$$

where

$$F(x, y, \alpha, \varepsilon) = f(x, y) + \varepsilon g(x, y, \varepsilon) + \varepsilon\alpha\left(1 + h(x, y, \varepsilon)\right).$$

We assume that

(i) The functions f, g and h are sufficiently smooth in the domain

$$\Omega = \{x \in R, \ \ y \in R, \ \ \alpha \in R, \ \ \varepsilon \in [0, \varepsilon_0]\}.$$

(ii) The functions f, g and h satisfy the following inequalities in the domain Ω

$$|f(x, y)| \le C|y|^2, \tag{9}$$

$$|g(x, y, \varepsilon)| \le C\mu, \tag{10}$$

$$|h(x, y, \varepsilon)| \le C\mu, \tag{11}$$

$$|f(x, y) - f(x, \bar{y})| \le \frac{L}{2}\left(|y| + |\bar{y}|\right)|y - \bar{y}|, \tag{12}$$

$$|g(x, y, \varepsilon) - g(x, \bar{y}, \varepsilon)| \le L|y - \bar{y}|, \tag{13}$$

$$|h(x, y, \varepsilon) - h(x, \bar{y}, \varepsilon)| \le L|y - \bar{y}|, \tag{14}$$

where C and L are the positive constants, μ is a sufficiently small positive constant.

Let us reduce the system (8) to the form

$$\varepsilon\frac{dy}{dx} = 2xy + F(x, y, \alpha, \varepsilon). \tag{15}$$

Due to (9) the straight line $y = 0$ is the slow curve of (15), the part of which with $x < 0$ $(x > 0)$ is stable (unstable). Our goal is to prove that for some value of the parameter α there exists a one-dimensional invariant manifold of variable stability (a canard) located in a ε-neighborhood of the slow curve $y = 0$.

We introduce the metric space Φ. Its elements are the scalar functions $\phi(x, \varepsilon)$, bounded by the value $\sqrt{\varepsilon}q\mu$, i.e.

$$|\phi(x, \varepsilon)| \le \sqrt{\varepsilon}q\mu, \tag{16}$$

and continuous in the domain

$$\Omega_1 = \{x \in R, \ \ \mu \in (0, \mu_0), \ \ \varepsilon \in (0, \varepsilon_0)\}.$$

Φ is a complete space with the metric

$$d(\phi, \bar{\phi}) = \sup_{\Omega_1}\left|\phi(x, \varepsilon) - \bar{\phi}(x, \varepsilon)\right|, \ \forall\phi \in \Phi, \ \forall\bar{\phi} \in \Phi.$$

On the space Φ we define an integral operator T by the formula

$$T(\phi)(x) = \begin{cases} \dfrac{1}{\varepsilon}\displaystyle\int_{-\infty}^{x} \exp\left(\dfrac{x^2 - s^2}{\varepsilon}\right) F(s, \phi(s, \varepsilon), \alpha_\phi, \varepsilon)ds, & x \leq 0, \\ -\dfrac{1}{\varepsilon}\displaystyle\int_{x}^{\infty} \exp\left(\dfrac{x^2 - s^2}{\varepsilon}\right) F(s, \phi(s, \varepsilon), \alpha_\phi, \varepsilon)\, ds, & x > 0. \end{cases} \tag{17}$$

It should be noted that we use a modification of the usual technique of the integral manifold theory [23, 24], and the operator in the upper (lower) line in (17) is usually used for proving the existence of stable (unstable) invariant surfaces.

The continuity condition for $T(\phi)(x)$ at $x = 0$ defines the parameter value α_ϕ corresponding to the function $\phi(x, \varepsilon)$ and implies the equation

$$\varepsilon^{-1} \int_{-\infty}^{\infty} e^{-s^2/\varepsilon} F(s, \phi(s, \varepsilon), \alpha_\phi, \varepsilon)ds = 0$$

or

$$\alpha_\phi = -\frac{I_1}{I_0},$$

where

$$I_1 = \frac{1}{\varepsilon\sqrt{\varepsilon}} \int_{-\infty}^{\infty} e^{-s^2/\varepsilon} \left[f\left(s, \phi(s, \varepsilon)\right) + \varepsilon g\left(s, \phi(s, \varepsilon), \varepsilon\right)\right] ds,$$

$$I_0 = \frac{1}{\sqrt{\varepsilon}} \int_{-\infty}^{\infty} e^{-s^2/\varepsilon} \left[1 + h\left(s, \phi(s, \varepsilon), \varepsilon\right)\right] ds.$$

From (9), (10), (16) we obtain

$$|I_1| \leq \frac{1}{\varepsilon\sqrt{\varepsilon}} \int_{-\infty}^{\infty} e^{-s^2/\varepsilon} \left|C\varepsilon\mu^2 q^2 + \varepsilon C\mu\right| ds = \sqrt{\pi}C(\mu^2 q^2 + \mu),$$

and, taking into account (11), we get

$$|I_0| \geq \frac{1}{\sqrt{\varepsilon}} \int_{-\infty}^{\infty} e^{-s^2/\varepsilon} (1 - C\mu)\, ds = \sqrt{\pi}(1 - C\mu).$$

The last two inequalities yield the estimate

$$|\alpha_\phi| \leq \frac{C\mu(\mu^2 q^2 + 1)}{1 - C\mu}. \tag{18}$$

It is clear that the operator $T(\phi)(x)$ satisfies the equation

$$\varepsilon \frac{dT(\phi)}{dx} = 2xT(\phi) + F(x, \phi(x, \varepsilon), \alpha_\phi, \varepsilon).$$

Hence, the existence problem of the canard reduces to the existence problem of the solution of the operator equation $T(\phi) = \phi$.

In accordance with (16) to check that the operator $T(\phi)$ transforms the complete metric space Φ into itself it is sufficiently to verify that

$$|T(\phi)(x)| \leq \sqrt{\varepsilon} q \mu \tag{19}$$

for all $\phi \in \Phi$. Applying (9)–(11) and (18) and taking into account the estimate

$$\int_x^\infty \exp\left(\frac{x^2 - s^2}{\varepsilon}\right) ds < \frac{\sqrt{\pi}}{2}\sqrt{\varepsilon}, \quad x > 0,$$

we get

$$|T(\phi)(x)| \leq \sqrt{\varepsilon}\frac{\sqrt{\pi}}{2} C\mu \left(\mu q^2 + 1 + \frac{1 + \mu q^2}{1 - C\mu}\right).$$

The similar inequality is obtained for $x < 0$.

The numbers μ and q will be chosen to obtain the following inequality

$$\frac{\sqrt{\pi}}{2} C \left(\mu q^2 + 1 + \frac{1 + \mu q^2}{1 - C\mu}\right) \leq q, \tag{20}$$

which means that (19) is fulfilled.

For any $\phi, \bar{\phi} \in \Phi$ from (9)–(14) and (18) we obtain

$$d\Big(T(\phi), T(\bar{\phi})\Big) \leq \frac{\sqrt{\pi}}{2}\left[L\mu q + L\sqrt{\varepsilon} + \frac{C\mu L\sqrt{\varepsilon}(1 + \mu q^2)}{1 - C\mu}\right] d(\phi, \bar{\phi}).$$

The numbers ε_0, μ and q will be chosen to obtain both (20) and the inequality

$$\frac{\sqrt{\pi}}{2}\left[L\mu q + L\sqrt{\varepsilon} + \frac{C\mu L\sqrt{\varepsilon}(1 + \mu q^2)}{1 - C\mu}\right] < 1$$

for sufficiently small ε ($0 < \varepsilon \leq \varepsilon_0$). In this case the contracting operator $T(p)$ transforms the complete metric space Φ into itself. Hence, due the Banach Contraction Principle the operator T has a unique fixed point $\phi(x, \varepsilon) \in \Phi$ corresponding to the canard of (15).

The above argument permits us to formulate the following statement.

Theorem 1 *Let the conditions (i),(ii) hold. Then there are numbers $\varepsilon_0 > 0$, q and μ such that, for all $\varepsilon \in (0, \varepsilon_0)$, there exist $\alpha = \alpha_\phi$ and $\phi(x, \varepsilon) \in \Phi$ such that $y = \phi(x, \varepsilon)$ is a canard of (8).*

Remark 1 *Usually the conditions (9)–(14) are fulfilled for $|x| \leq r_1$, $\|y\| \leq r_2$ only. In this case we deal with a local canard.*

Note that the problem of the extendability of a local canard (or local invariant manifolds of variable stability) is important and complicated but requires a separate study which is beyond the scope of this paper.

Example 3 Originally the existence of canards for the van der Pol system

$$\frac{dx}{dt} = y - \alpha, \tag{21}$$

$$\varepsilon\frac{dy}{dt} = y - y^3/3 - x \tag{22}$$

has been proven by the non-standard analysis methods [1, 2, 4]. We show that this statement also follows from the Theorem 1.

The change of variables

$$x = x_1 - \frac{2}{3}, \; y = y_1 - 1, \; \alpha = -1 + \varepsilon a,$$

transforms (21), (22) to the form

$$\dot{x}_1 = y_1 - \varepsilon a,$$

$$\varepsilon \dot{y}_1 = -x_1 + y_1^2 - \frac{1}{3} y_1^3.$$

Applying a rectifying transformation

$$\varepsilon z = -x_1 + y_1^2 - \frac{1}{3} y_1^3 + \frac{\varepsilon}{y_1 - 2}$$

and eliminating the variable t we obtain

$$\varepsilon \frac{dz}{dy_1} = \frac{y_1 (y_1 - 2)^2 z}{1 - z(y_1 - 2)} - \frac{\varepsilon}{(y_1 - 2)^2} - \frac{\varepsilon a (y_1 - 2)}{1 - z(y_1 - 2)}.$$

By the next change of the parameter $a = 1/8 + a_1$ we get

$$\varepsilon \frac{dz}{dy_1} = y_1 (y_1 - 2)^2 z \left[1 + z(y_1 - 2) + \frac{z^2 (y_1 - 2)^2}{1 - z(y_1 - 2)} \right]$$

$$- \varepsilon \left[\frac{1}{(y_1 - 2)^2} + \frac{y_1 - 2}{8 - 8z(y_1 - 2)} \right] - \varepsilon a_1 \frac{y_1 - 2}{1 - z(y_1 - 2)},$$

or, finally,

$$\varepsilon \frac{dz}{dy_1} = 4 y_1 z + 4 y_1^2 \left(\frac{y_1}{4} - 1 \right) z + y_1 (y_1 - 2)^3 \frac{z^2}{1 - z(y_1 - 2)}$$

$$- \varepsilon \left[\frac{1}{(y_1 - 2)^2} + \frac{y_1 - 2}{8 - 8z(y_1 - 2)} \right] - \varepsilon a_1 \frac{y_1 - 2}{1 - z(y_1 - 2)}. \qquad (23)$$

It easy to see that (23) has the form of equation (8) and the conditions of the Theorem 1 are fulfilled in some neighborhood of $y_1 = 0$, outside of which the functions f, g, h should be extended with preservation of the properties (9)–(14).

The existence theorem for the canards can be extended to the case of vector variables x and y [7, 25] and can be obtained as a corollary of the existence theorem for the black swans, see Subsection 4.3.

4. Further generalizations

The described method for constructing of the invariant surfaces of variable stability in the form of bounded-on-the-whole-axis solutions of corresponding differential systems assumes various generalizations, some of which are discussed in this section. A distinctive feature in these cases lies in replacing a single gluing parameter with a gluing function to ensure continuity of the invariant surface.

4.1. Case of a multiple turning point

In this Subsection we deal with a multiple turning point.

Example 4 Consider a modification of the Example 1 given by

$$\dot{x} = 1, \ \ \varepsilon\dot{y} = x^k y,$$

or

$$\varepsilon\frac{dy}{dx} = x^k y. \tag{24}$$

For $k = 3, 5, \ldots$ the straight line $y = 0$ is a canard.

Example 5 Some generalization of (24) is the equation

$$\varepsilon\frac{dy}{dx} = x^k y + f(x, y, \alpha).$$

In this case the gluing function can be found in the form

$$\alpha = \alpha(x, \varepsilon) = \alpha_1(\varepsilon)x^{k-1} + \ldots + \alpha_{k-1}(\varepsilon)x + \alpha_k(\varepsilon).$$

The other approach for investigation of canards in the case of a multiple turning point has been proposed in [26].

4.2. Canard cascades

If it is necessary to glue stable and unstable slow invariant manifolds at several breakdown points, we need several additional parameters and as a result we obtain a slow invariant manifold with multiple change of stability (a canard cascade).

Definition 4 *The continuous one-dimensional slow invariant manifold of (1), (2) which contains at least two canards or false canards is called a canard cascade.*

In the case of a planar system, if we take an additional function whose arguments are a vector parameter and a slow variable, we can glue the stable (attractive) and unstable (repulsive) slow invariant manifolds at all breakdown points at the same time. The outcome of this procedure is a canard cascade.

Example 6 The differential system

$$\dot{x} = 1, \ \ \varepsilon\dot{y} = x(x-1)y$$

gives a simplest canard cascade $y = 0$ which consists of two repulsive parts ($x < 0$ and $x > 1$ since $\partial g/\partial y = x(x-1) > 0$) and one attractive part ($0 < x < 1$ since $\partial g/\partial y = x(x-1) < 0$) with two breakdown points $x = 0$ and $x = 1$.

Example 7 An example of a system with an infinite number of breakdown points is

$$\dot{x} = 1, \ \ \varepsilon\dot{y} = y\sin x.$$

Example 8 The slow curve of the van der Pol system (21), (22) has two jump points. To construct a canard cascade we need to have two independent parameters. We can consider α as a gluing function $\alpha = \alpha(x, \lambda)$ depending on the slow variable x and the vector $\lambda = (\lambda_1, \lambda_2)$. The vector λ is a function of ε: $\lambda = \lambda(\varepsilon)$. The consideration of a variant where the function α is a polynomial in x, i.e., $\alpha = \lambda_1 x + \lambda_2$ seems quite natural. Of course, we can use a

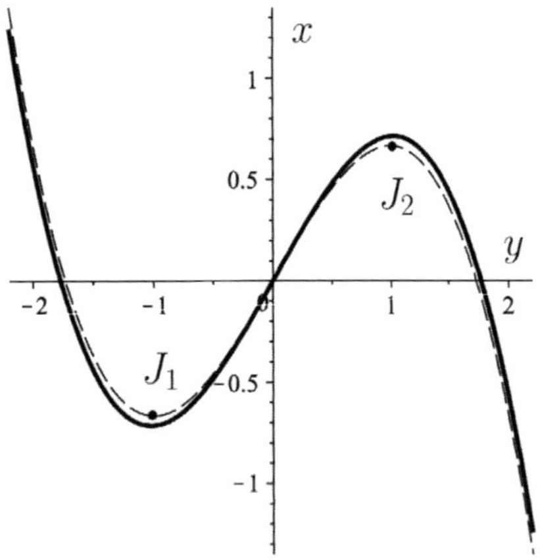

Figure 4. Slow curve (thin dashed line) and canard cascade (thick full line) in the case of the van der Pol equation. J_1 and J_2 are jump points where stable and unstable slow invariant manifolds are glued.

trigonometric polynomial or the linear combination of any linearly independent functions in x with the coefficients λ_1, λ_2. We will search the canard cascade in the form of a polynomial

$$x = y - y^3/3 + \varepsilon q_m(y, \varepsilon), \tag{25}$$

where $q_m(y, \varepsilon)$ is an mth-degree polynomial in y. From (21), (22) and (25) we have

$$\frac{\partial}{\partial y}\left(y - y^3/3 + \varepsilon q_m(y, \varepsilon)\right)(-q_m(y, \varepsilon)) = y - \alpha.$$

Balancing the degrees of polynomials on both sides of this equality implies that $m = 1$ since the highest power of y in x is y^3, and therefore, $q_m(y, \varepsilon) = ay + b$. After some calculations we obtain [18, 19] the representation for gluing function

$$\alpha = \alpha(x, \varepsilon) = 3\mu(\varepsilon)x,$$

where

$$\mu(\varepsilon) = (\sqrt{1 + 2\varepsilon} - 1)/2\varepsilon,$$

and a canard cascade

$$x = y - y^3/3 + \varepsilon\mu(\varepsilon)y.$$

Figure 4 shows that the canard cascade (the one-dimensional slow invariant manifold) passes near the slow curve of the system. We glue the stable and the unstable slow invariant manifolds at two jump points J_1 and J_2. In this case we have used only one additional parameter to obtain a canard cascade.

Note that the change of variable

$$x = -\varepsilon z + y - y^3/3 + \varepsilon\mu(\varepsilon)y$$

transforms (21), (22) to the system

$$\dot{y} = z - \mu y, \quad \varepsilon\dot{z} = (y - 1)(y + 1)z - 4\varepsilon\mu z$$

with the canard cascade $z = 0$ which plays the role of a bounded-on-the-whole-axis solution.

In the general case, when the slow curve of a system has k jump points, it is necessary to use k additional parameters to construct a canard cascade.

4.3. Black swans

We return now to the more general case and consider the multidimensional invariant surfaces of variable stability or black swans [20, 21]. These surfaces are considered as natural generalizations of the notion of a canard.

We consider a non-autonomous system obtained from an original autonomous system of type (1), (2) with a scalar fast variable by eliminating the variable t. It is assumed that this system can be represented as:

$$\frac{dy}{dx} = Y(x, y, z, \varepsilon), \qquad y \in R^n, \qquad x \in R; \tag{26}$$

$$\varepsilon \frac{dz}{dx} = 2xz + a + Z(x, y, z, a, \varepsilon), \qquad |z| \le r, \tag{27}$$

where r and a_0 are positive constants. It is supposed that the functions Y, Z are continuous and satisfy the following inequalities for $x \in R$, $y \in R^n$, $|z| \le r$, $|a| \le a_0$, $\varepsilon \in [0, \varepsilon_0]$:

$$\|Y(x, y, z, \varepsilon)\| \le k, \quad |Z(x, y, z, a, \varepsilon)| \le M\left(\varepsilon^2 + \varepsilon|z| + |z|^2\right), \tag{28}$$

$$\|Y(x, y, z, \varepsilon) - Y(x, \bar{y}, \bar{z}, \varepsilon)\| \le M(\|y - \bar{y}\| + |z - \bar{z}|), \tag{29}$$

$$|Z(x, y, z, a, \varepsilon) - Z(x, \bar{y}, \bar{z}, \bar{a}, \varepsilon)| \le M\{(\varepsilon + |\tilde{z}|)|z - \bar{z}|$$
$$+(\varepsilon^2 + \varepsilon|\tilde{z}| + |\tilde{z}|^2)\|y - \bar{y}\| + \varepsilon|a - \bar{a}|\Big\}, \quad |\tilde{z}| = \max\{|z|, |\bar{z}|\}, \tag{30}$$

where $\|\cdot\|$ denotes the usual norm in R^n and $|\cdot|$ denotes the absolute value of a scalar, k and M are positive constants.

We take a as a function: $a = a(y, \varepsilon)$.

Let F be the complete metric space of functions $a(y, \varepsilon)$ continuous with respect to y and satisfying

$$|a(y, \varepsilon)| \le \varepsilon^2 K, \quad |a(y, \varepsilon) - a(\bar{y}, \varepsilon)| \le \varepsilon^2 L\|y - \bar{y}\|,$$

for $\varepsilon \in (0, \varepsilon_0]$, where K and L are positive constants, with the metric defined by

$$\rho(a, \bar{a}) = \sup_{y \in R^n} |a(y, \varepsilon) - \bar{a}(y, \varepsilon)|.$$

Let H be the complete metric space of functions $h(x, y, \varepsilon)$ mapping $R \times R^n$ to R continuous with respect to x, y and satisfying

$$|h(x, y, \varepsilon)| \le \varepsilon^{\frac{3}{2}} q,$$

$$|h(x, y, \varepsilon) - h(x, \bar{y}, \varepsilon)| \le \varepsilon^{\frac{3}{2}} \delta \|y - \bar{y}\|,$$

for $\varepsilon \in (0, \varepsilon_0]$, where q and δ are positive constants, with the metric

$$\rho(h, \bar{h}) = \sup_{x \in R, y \in R^n} |h(x, y, \varepsilon) - \bar{h}(x, y, \varepsilon)|.$$

On the space H we define an operator T by the formula

$$Th(x, y, \varepsilon) = \begin{cases} \dfrac{1}{\varepsilon} \displaystyle\int_{-\infty}^{x} \exp\left(\frac{x^2 - s^2}{\varepsilon}\right) [Z(\cdot) + a(\varphi(s, x), \varepsilon)]\, ds, & x \le 0, \\ -\dfrac{1}{\varepsilon} \displaystyle\int_{x}^{\infty} \exp\left(\frac{x^2 - s^2}{\varepsilon}\right) [Z(\cdot) + a(\varphi(s, x), \varepsilon)]\, ds, & x > 0, \end{cases}$$

where $Z(\cdot) = Z(s, \varphi(s,x), h(s, \varphi(s,x), \varepsilon), a(\varphi(s,x), \varepsilon), \varepsilon)$, and $\varphi(s,x)$ is defined as follows. For any element $h \in H$ the initial value problem

$$\frac{d\varphi}{ds} = Y(s, \varphi, h(s, \varphi, \varepsilon), \varepsilon),$$

$$\varphi(x) = y,$$

is considered. This problem is obtained from (26) if we replace z by an element $h \in H$. The solution of this problem is denoted by $\Phi(s, x, y, \varepsilon | h) = \varphi(s,x)$. When the operator T possesses a fixed point $h(x, y, \varepsilon)$ in H then the surface $z = h(x, y, \varepsilon)$ is a slow stable/unstable invariant manifold (a black swan). It should be noted that the operator T defines a manifold of the solutions which are bounded with respect to z.

For the detailed proof, which is multidimensional modification of the proof of the Theorem 1, the readers are referred to [20, 21, 25]. The following theorem gives sufficient conditions for the system (26), (27) to have a black swan.

Theorem 2 *Let the conditions (28)–(30) be satisfied. Then there are numbers $\varepsilon_0 > 0$ and K, L, q, δ such that, for all $\varepsilon \in (0, \varepsilon_0)$, there exist functions $a(y, \varepsilon) \in F$ and $h(x, y, \varepsilon) \in H$ such that $z = h(x, y, \varepsilon)$ is a slow integral manifold (the black swan).*

Remark 2 *A original singularly perturbed differential system can be reduced to the form (26), (27) by an multidimensional analogue of the rectifying transformation described in the Example 3 (see, for instance, [7, 25]).*

Remark 3 *Usually the conditions (28)–(30) are fulfilled for $|x| \le r_1$, $\|y\| \le r_2$ only. Integral manifolds in this case are local.*

Note that Theorem 2 can be generalized to the case when z is a vector variable.

Let us consider now the system

$$\frac{dy}{dx} = Y(x, y, z_1, z_2, \varepsilon), \; y \in R^n, \; x \in R, \tag{31}$$

$$\varepsilon \frac{dz_1}{dx} = 2xz_1 + a(y, \varepsilon) + Z_1(x, y, z_1, z_2, \varepsilon), \; z_1 \in R, \tag{32}$$

$$\varepsilon \frac{dz_2}{dx} = A(x)z_2 + a(y, \varepsilon)B + Z_2(x, y, z_1, z_2, \varepsilon), \; z_2 \in R^m, \tag{33}$$

where $A(x)$ is a bounded matrix, satisfying a Lipschitz condition, with eigenvalues $\lambda_i(x)$:

$$Re\lambda_i(x) \le -2\beta < 0 \quad (i = 1, 2, \dots, m),$$

B is a constant vector, the continuous functions Y, Z_1, Z_2, a satisfy the inequalities

$$\|Y(x, y, z_1, z_2, \varepsilon)\| \le k, \tag{34}$$

$$|Z_1(x, y, z_1, z_2, \varepsilon)| \le M\left(\varepsilon^2 + \varepsilon\|z\| + \|z\|^2\right), \tag{35}$$

$$\|Z_2(x, y, z_1, z_2, \varepsilon)\| \le M\left(\varepsilon^2 + \varepsilon\|z\| + \|z\|^2\right), \tag{36}$$

$$\|Y(x, y, z_1, z_2, \varepsilon) - Y(x, \bar{y}, \bar{z}_1, \bar{z}_2, \varepsilon)\| \le M(\|y - \bar{y}\| + \|z - \bar{z}\|), \tag{37}$$

$$|Z_1(x, y, z_1, z_2, \varepsilon) - Z_1(x, \bar{y}, \bar{z}_1, \bar{z}_2, \varepsilon)| \le M\left\{(\varepsilon + \|\tilde{z}\|)\|z - \bar{z}\| \right.$$
$$\left. + (\varepsilon^2 + \varepsilon\|\tilde{z}\| + \|\tilde{z}\|^2)\|y - \bar{y}\|\right\}, \tag{38}$$

$$\|Z_2(x,y,z_1,z_2,\varepsilon)-Z_2(x,\bar y,\bar z_1,\bar z_2,\varepsilon)\|\le M\left\{(\varepsilon+\|\tilde z\|)\|z-\bar z\|\right.$$
$$\left.+(\varepsilon^2+\varepsilon\|\tilde z\|+\|\tilde z\|^2)\|y-\bar y\|\right\}, \tag{39}$$

where

$$z=\begin{pmatrix} z_1\\ z_2\end{pmatrix},\quad \bar z=\begin{pmatrix} \bar z_1\\ \bar z_2\end{pmatrix},\quad \|\tilde z\|=\max\{\|z\|,\|\bar z\|\}, k>0, M>0.$$

We consider $a(y,\varepsilon)$ to be a continuous function satisfying the inequalities

$$|a(y,\varepsilon)|\le\varepsilon^2K,\ |a(y,\varepsilon)-a(\bar y,\varepsilon)|\le\varepsilon^2L\|y-\bar y\|, \tag{40}$$

for $\varepsilon\in(0,\varepsilon_0]$ with some positive constants K and L.

Theorem 3 *Let the conditions (34)–(39) be satisfied. Then there are numbers $\varepsilon_0>0$ and K,L,q,δ such that, for all $\varepsilon\in(0,\varepsilon_0)$, there exist functions $a(y,\varepsilon)$ satisfying (40) and a continuous function $h(x,y,\varepsilon)$ satisfying*

$$|h_1(x,y,\varepsilon)|\le\varepsilon^{3/2}q,\ |h_1(x,y,\varepsilon)-h_1(x,\bar y,\varepsilon)|\le\varepsilon^{3/2}\delta\|y-\bar y\|, \tag{41}$$

$$\|h_2(x,y,\varepsilon)\|\le\varepsilon^2q,\ \|h_2(x,y,\varepsilon)-h_2(x,\bar y,\varepsilon)\|\le\varepsilon^2\delta\|y-\bar y\|, \tag{42}$$

$$h(x,y,\varepsilon)=\begin{pmatrix} h_1(x,y,\varepsilon)\\ h_2(x,y,\varepsilon)\end{pmatrix},\quad h_1\in R,\quad h_2\in R^m,$$

such that $z=h(x,y,\varepsilon)$ is a black swan of the system (31)–(33).

The proof of Theorem 3 follows the lines of the proof of Theorem 2, but now H is the complete metric space of functions $h(x,y,\varepsilon)$ mapping $R\times R^n$ to R^{m+1}, continuous with respect to x,y and satisfying (41), (42) for $\varepsilon\in(0,\varepsilon_0]$. The operator T is defined by the formula $T=\begin{pmatrix} T_1\\ T_2\end{pmatrix}$:

$$T_1h(x,y)=\begin{cases} -\varepsilon^{-1}\int\limits_x^{\infty} e^{(x^2-s^2)/\varepsilon}[Z_1(\cdot)+a(\varphi(s,x),\varepsilon)]ds, & x\ge 0,\\ \varepsilon^{-1}\int\limits_{-\infty}^{x} e^{(x^2-s^2)/\varepsilon}[Z_1(\cdot)+a(\varphi(s,x),\varepsilon)]ds, & x<0,\end{cases}$$

$$T_2h(x,y)=\varepsilon^{-1}\int\limits_{-\infty}^{x} W(x,s,\varepsilon)[Z_2(\cdot)+Ba(\varphi(s,x),\varepsilon)]ds,$$

where

$$Z_{1,2}(\cdot)=Z_{1,2}(s,\varphi(s,x),h_1(s,\varphi(s,x),\varepsilon),h_2(s,\varphi(s,x),\varepsilon),\varepsilon)$$

and $W(x,s,\varepsilon)$, $W(s,s,\varepsilon)=I$, is the fundamental matrix of the equation

$$\varepsilon\frac{dz_2}{dx}=A(x)z_2.$$

Remark 4 *The proof of the Theorem 3 can be extended to the case when the system (31)–(33) has the form*

$$\frac{dy}{dx}=Y(x,y,z_1,z_2,\varepsilon),\ y\in R^n,\ x\in R, \tag{43}$$

$$\varepsilon\frac{dz_1}{dx}=2xb(x,y)z_1+a(y,\varepsilon)+Z_1(x,y,z_1,z_2,\varepsilon),\ z_1\in R, \tag{44}$$

$$\varepsilon\frac{dz_2}{dx}=A(x,y)z_2+a(y,\varepsilon)B(x,y,\varepsilon)+Z_2(x,y,z_1,z_2,\varepsilon),\ z_2\in R^m. \tag{45}$$

Here the matrix $A(x,y)$ and the functions Y, Z_1, Z_2, and a satisfy the conditions of the Theorem 3. Moreover, the matrix $A(x,y)$, the scalar function $b(x,y)$ $(0<c_1\le b(x,y)\le c_2)$ and the vector function $B(x,y,\varepsilon)$ are continuous and bounded, and satisfy a Lipschitz condition.

Remark 5 *The existence of canards for the systems (31)–(33) and (43)–(45) follows from the Theorem 3. Indeed, the black swan* $z = h(x, y, \varepsilon)$ *can be considered as a surface consisting entirely of the canards, each of which passes through a certain point of the breakdown surface. If we take the gluing function has to be a constant,* $a(y, \varepsilon) = a(y_0, \varepsilon)$*, the stable and unstable invariant manifolds can be glued at one point* $y = y_0$ *of the breakdown surface only. The canard passes only through this point.*

5. Summary

In this paper we have looked from unified positions at the invariant surfaces of variable stability both from a geometric viewpoint, when dealing with gluing the stable and the unstable slow invariant manifolds, and in terms of their analytic representations as the unique bounded-on-the-whole-axis solutions. The invariant surfaces of variable stability are interesting not only from a mathematical viewpoint, they are successfully used for modeling of critical phenomena of diverse nature. It should be noted that the considered cases do not exhaust the possible generalizations.

Acknowledgments

This work is supported in part by the Russian Foundation for Basic Research (grants 13-01-97002-p, 14-01-97018-p) and the Ministry of education and science of the Russian Federation in the framework of the implementation of Program of increasing the competitiveness of SSAU for 2013–2020 years.

References

[1] Diener M 1979 Nessie et Les Canards (Strasbourg: Publication IRMA)
[2] Benoit E, Callot J L, Diener F and Diener M 1981–1982 Chasse au canard *Collect. Math.* **31–32** 37–119
[3] Eckhaus W 1983 *Relaxation oscillations including a standart chase on French ducks Asymptotic analysis II* (Lecture Notes in Math. vol 985) ed F Verhulst (Berlin: Springer) pp 449–494
[4] Zvonkin A K and Shubin M A 1984 Non-standard analysis and singular perturbations of ordinary differential equations *Russian Math. Surveys* **39** 69–131
[5] Benoit E 1983 Systèmes lents-rapides dans R^3 et leurs canards *Société Mathématique de France, Astérisque* **109–110** 159–191
[6] Mishchenko E F, Kolesov Yu S, Kolesov A Yu and Rozov N Kh 1995 *Asymptotic Methods in Singularly Perturbed Systems* (New York: Plenum Press)
[7] Sobolev V A and Shchepakina E A 1996 Duck trajectories in a problem of combustion theory *Differential Equations* **32** 1177–1186
[8] Shchepakina E and Sobolev V 2005 *Black Swans and Canards in Laser and Combustion Models* (Singular Perturbation and Hysteresis) ed M P Mortell *et al* (Philadelphia: SIAM) pp 207–255
[9] Marino F, Marin F, Balle S and Piro O 2007 Chaotically spiking canards in an excitable system with 2D inertial fast manifolds *Phys. Rev. Lett.* **98** 074104
[10] Desroches M, Krauskopf B and Osinga H M 2010 Numerical continuation of canard orbits in slow-fast dynamical systems *Nonlinearity* **23** 739–765
[11] Gorelov G N and Sobolev V A 1991 Mathematical modelling of critical phenomena in thermal explosion theory *Combust. Flame* **87** 203–210
[12] Gorelov G N and Sobolev V A 1992 Duck–trajectories in a thermal explosion problem *Appl. Math. Lett.* **5** 3–6
[13] Sobolev V A and Shchepakina E A 1993 Self–ignition of laden medium *J. Combustion, Explosion and Shock Waves* **29** 378–381
[14] Gol'dshtein V, Zinoviev A, Sobolev V and Shchepakina E 1996 Criterion for thermal explosion with reactant consumption in a dusty gas *Proc. London Roy. Soc. Ser. A.* **452** 2103–2119
[15] Shchepakina E 2003 Black swans and canards in self–ignition problem *Nonlinear Analysis: Real Word Applications* **4** 45–50
[16] Schneider K, Shchepakina E and Sobolev V 2003 A new type of travelling wave *Mathematical Methods in the Applied Sciences* **26** 1349–1361
[17] Gorelov G N, Sobolev V A and Shchepakina E A 2006 Canards and critical behaviour in autocatalytic combustion models *Journal of Engineering Mathematics* **56** 143–160

[18] Sobolev V 2013 Canard Cascades *Discr. and Cont. Dynam. Syst. B* **18** 513–521

[19] Shchepakina E, Sobolev V and Mortell M P 2014 *Singular Perturbations. Introduction to system order reduction methods with applications* (Lect. Notes in Math. vol 2114) (Berlin–Heidelber–London: Springer)

[20] Shchepakina E and Sobolev V 2000 Attracting/repelling invariant manifolds *Stab. Control Theory Appl.* **3** 263–274

[21] Shchepakina E and Sobolev V 2001 Integral manifolds, canards and black swans *Nonlinear Analysis* A **44** 897–908

[22] Strygin V V and Sobolev V A 1976 Effect of geometric and kinetic parameters and energy dissipation on orientation stability of satellites with double spin *Cosmic Research* **14** 331–335

[23] Daletskii Yu L and Krein M G 1974*Stability of Solutions of Differential Equations in Banach Space* (New York: AMS)

[24] Mitropol'skii Yu A and Lykova O B 1975 *Integral Manifolds in Nonlinear Mechanics* (in Russian) (Moscow: Nauka)

[25] Sobolev V A and Shchepakina E A 2010 *Model Reduction and Critical Phenomena in Macrokinetics* (in Russian) (Moscow: Fizmatlit)

[26] Forget T 2008 Asymptotic study of planar canard solutions *Bull. Belg. Math. Soc. Simon Stevin* **15** 809-824

Slow Integral Manifolds and Control Problems in Critical and Twice Critical Cases

Vladimir Sobolev

Department of Technical Cybernetics, Samara State Aerospace University (SSAU), 34, Moskovskoye shosse, Samara 443086, Russia

E-mail: `hsablem@gmail.com`

Abstract. We consider singularly perturbed differential systems in cases where the standard theory to establish a slow integral manifold existence does not work. The theory has traditionally dealt only with perturbation problems near normally hyperbolic manifold of singularities and this manifold is supposed to isolated. Applying transformations we reduce the original singularly perturbed problem to a regularized one such that the existence of slow integral manifolds can be established by means of the standard theory. We illustrate our approach by several examples.

1. Introduction

Singularly perturbed differential systems of the type

$$\begin{aligned} \frac{dx}{dt} &= f(x,y,t,\varepsilon), \\ \varepsilon\,\frac{dy}{dt} &= g(x,y,t,\varepsilon) \end{aligned} \tag{1}$$

play an important role as mathematical models of numerous nonlinear phenomena in biology, chemistry, control theory, and in other fields (see e.g. [1, 2, 3, 4, 6, 7]). A usual approach in the qualitative study of (1) is to consider first the degenerate system

$$\begin{aligned} \frac{dx}{dt} &= f(x,y,t,0), \\ 0 &= g(x,y,t,0) \end{aligned} \tag{2}$$

and then to draw conclusions for the qualitative behavior of the full system (1) for sufficiently small ε. A special case of this approach is the quasi–steady state assumption. A mathematical justification of that method can be given by means of the theory of integral manifolds for singularly perturbed systems (1) (see e.g. [3, 4, 8, 9]).

In order to recall a basic result of the geometric theory of singularly perturbed systems we introduce the following notation and assumptions.

Let I_i be the interval $I_i := \{\varepsilon \in R : 0 < \varepsilon < \varepsilon_i\}$, where $0 < \varepsilon_i \ll 1, i = 0, 1, \dots$.

(A_1). $f : R^m \times R^n \times R \times \overline{I_0} \to R^m$, $g : R^m \times R^n \times R \times \overline{I_0} \to R^n$ are sufficiently smooth and uniformly bounded together with all their derivatives.

(A_2). There are some region $\mathcal{G} \in R^m$ and a map $h : \mathcal{G} \times R \to R^m$ of the same smoothness as g such that

$$g(x, h(x,t), t, 0) \equiv 0 \quad \forall (x,t) \in \mathcal{G} \times R.$$

(A_3). The spectrum of the Jacobian matrix $B(x,t) = g_y(x, h(x,t), t, 0)$ is uniformly separated from the imaginary axis for all $(x,t) \in \mathcal{G} \times R$, i.e. the eigenvalues $\lambda_i(x,t)(i = 1, \dots, n)$ of the matrix $B(x,t)$ satisfy the inequality

$$|Re\lambda_i(x,t)| \geq \gamma \tag{3}$$

for some positive number γ.

Then the following result is valid (see e.g. [8, 9]):

Proposition 1.1. *Under the assumptions $(A_1) - (A_3)$ there is a sufficiently small positive ε_1, $\varepsilon_1 \leq \varepsilon_0$, such that for $\varepsilon \in \overline{I_1}$ system (1) has a smooth integral manifold $\mathcal{M}_\varepsilon$ (slow integral manifold) with the representation*

$$\mathcal{M}_\varepsilon := \{(x, y, t) \in R^{n+m+1} : y = \psi(x,t,\varepsilon),\ (x,t) \in \mathcal{G} \times R\}$$

and with the asymptotic expansion

$$\psi(x,t,\varepsilon) = h(x,t) + \varepsilon\psi_1(x,t) + \dots\ .$$

The motion on this manifold is described by the *slow* differential equation

$$\dot{x} = f(x, \psi(x,t,\varepsilon), t, \varepsilon). \tag{4}$$

Remark 1.1. *The global boundedness assumption in (A_1) with respect to (x,y) can be relaxed by modifying f and g outside some bounded region of $R^n \times R^m$.*

Remark 1.2. *In applications it is usually assumed that the spectrum of the Jacobian matrix $g_y(x, h(x,t), t, 0)$ is located in the left half plane. Under this additional hypothesis the manifold $\mathcal{M}_\varepsilon$ is exponentially attracting for $\varepsilon \in I_1$.*

The case that assumption (A_3) is violated is called critical. We distinguish three subcases:

(i) The Jacobian matrix $g_y(x,y,t,0)$ is singular on some subspace of $R^m \times R^n \times R$. In that case, system (1) is referred to as a singular singularly perturbed system [10]. This subcase has been treated in [3, 4, 7, 10, 11].

(ii) The Jacobian matrix $g_y(x,y,t,0)$ has eigenvalues on the imaginary axis with nonvanishing imaginary parts. A similar case has been investigated in [3, 4, 12].

(iii) The Jacobian matrix $g_y(x,y,t,0)$ is singular on the set $\mathcal{M}_0 := \{(x,y,t) \in R^m \times R^n \times R : y = h(x,t), (x,t) \in \mathcal{G} \times R\}$. In that case, $y = h(x,t)$ is generically an isolated root of $g = 0$ but not a simple one.

The critical case (i) is considered in Section 2 as applied to the high-gain control problem, the case (ii) is considered in Section 3 as applied to the manipulator control, the case (iii) is considered in Section 4 as applied to the partially cheap control problem. Two critical cases, (i) and (ii) are combined in the the optimal control problem which is analyzed in Section 5, and therefore it is possible to say that this Section is devoted to the consideration of the twice critical case. It is not inconceivable that combinations of other pairs of critical cases and even triple critical case are of interest as well and possibly they will be considered later.

2. Singular Singularly Perturbed Systems

Consider the system

$$\varepsilon \dot{z} = Z(z, t, \varepsilon), \qquad z \in R^{m+n}, \quad t \in R, \tag{5}$$

where $0 \le \varepsilon \ll 1$, and the vector-function Z is sufficiently smooth.

Formally, this system may be considered as system (1) in the case where x is absent and $z = y$. From the other hand, system (1) can be reduced to the form (5) with $z = \begin{pmatrix} x \\ y \end{pmatrix}$ by the presentation of the first equation in (1) of form $\varepsilon \frac{dx}{dt} = \varepsilon f(x, y, t, \varepsilon)$.

Suppose that the limit system $Z(z, t, 0) = 0$ $(\varepsilon = 0)$ has a family of solutions

$$z = \psi(v, t), \qquad v \in R^m, \quad t \in R, \tag{6}$$

with a sufficiently smooth vector-function ψ . We try to find a slow integral manifold

$$z = P(v, t, \varepsilon), \tag{7}$$

with a flow described by the equation

$$\dot{v} = Q(v, t, \varepsilon). \tag{8}$$

Suppose that the rank of the matrix $\psi_v(v, t)$ is equal to m, the rank of the matrix $A(v, t) = Z_z(\psi(v, t), t, 0)$ is equal to n where $Z_z = \left(\frac{\partial Z_i}{\partial z_j}\right)$ $(i, j = 1, \ldots, n + m)$, and the matrix $A(v, t)$ has an m–fold zero eigenvalue and n other eigenvalues $\lambda_i(v, t)$ have have strictly negative real parts, i.e. they satisfy the inequality

$$Re\lambda_i(v, t) \le -\gamma < 0, \qquad t \in R, \qquad v \in R^m. \tag{9}$$

We shall restrict our consideration to smooth integral surfaces situated in the ε-neighborhood of the slow surface $z = \psi(v, t)$, i. e.

$$P(v, t, 0) = \psi(v, t),$$

a motion on which is described by differential equations of form (8) with a smooth right hand side.

The equation (5) describes motions with speeds of order $O(\varepsilon^{-1})$, while (8) describes motions with speeds of order $O(1)$. Thus, the integral manifold (7) is a manifold of slow motions or a slow integral manifold. The matrix $A(v, t)$ is a singular one and therefore the system (5) is called as singular singularly perturbed [10].

2.1. High-gain control

Consider the control system

$$\dot{x} = \zeta(x) + B(x)u, \;\; x(0) = x_0,$$

where $x \in R^n$, $u \in R^r$ and $t \ge 0$. The vector function ζ and the matrix function B are taken to be sufficiently smooth and bounded. The control vector u is to be selected in such a way as to transfer the vector x from $x = x_0$ to a sufficiently small neighborhood of a smooth m–dimensional surface $S(x) = 0$. A commonly employed feedback control is

$$u = -\frac{1}{\varepsilon} KS(x), \tag{10}$$

where K is a constant $r \times m$–matrix and ε is a small positive parameter, see [6] and references therein.

Suppose that we can choose the matrix K in such a way that the matrix $-N(x,t) = -GBK$ is stable[1] and its inverse matrix is bounded, and introduce the additional variable $y = S(x)$, then x and y satisfy the system

$$\varepsilon\dot{x} = \varepsilon\zeta(x) - B(x)Ky, \quad x(0) = x_0,$$

$$\varepsilon\dot{y} = \varepsilon G(x)\zeta(x) - G(x)B(x)Ky, \quad y(0) = y_0 = S(x_0),$$

where $G(x) = \partial S/\partial x$. The reduced $(\varepsilon = 0)$ algebraic problem possesses an n–parameter family of solutions $x = v, \quad y = 0$. The role of A is played by the singular matrix

$$\begin{pmatrix} 0 & -BK \\ 0 & -N \end{pmatrix}.$$

The latter singular singularly perturbed differential system possesses an n–dimensional slow integral manifold

$$x = v, \qquad y = \varepsilon N^{-1}(v,t)G(v)\zeta(v) + O(\varepsilon^2).$$

The flow on the manifold is governed by

$$\dot{v} = [I - B(v)KN^{-1}(v)G(v)]\zeta(v) + O(\varepsilon).$$

Introduce the new variable

$$y = z + \varepsilon N^{-1}(x)G(x)\zeta(x).$$

Then for z we obtain the equations

$$\varepsilon\dot{z} = -N(x)z + O(\varepsilon).$$

It is now clear that the representations

$$y = N^{-1}G\zeta + O(\varepsilon)$$

are valid for for all $t > 0$. Thus, under the control law (10) the trajectory very quickly attains the ε–neighborhood of $S(x) = 0$.

Let us introduce the modified control

$$u = -\frac{1}{\varepsilon}K\left[S(x) - \varepsilon N^{-1}(x)G(x)\zeta(x)\right], \tag{11}$$

with the stable matrix $-N(x) = -GBK$. Under this control for the variable x we obtain the equation

$$\varepsilon\dot{x} = \varepsilon\left[I - B(x)K(GBK)^{-1}G(x)\right]\zeta(x) - B(x)KS(x),$$

and for the variable $y = S(x)$ we obtain the equation

$$\varepsilon\dot{y} = -N(x)y,$$

$$y = O\left(e^{-\nu\varepsilon^{-1}t}\right), \quad \nu > 0, \ t > 0, \ \varepsilon \to 0$$

for some positive ν. Note that the modified control law turns $y \equiv 0$ invariant and exponentially attractive. This means that the modified control law (11) is more preferable than the usually used law (10). Surprisingly, there is no need for using of asymptotic expansion to design the control law since the modified control law gives the exact result.

[1] A stable matrix is one whose eigenvalues all have strictly negative real parts.

Journal of Physics: Conference Series **727** (2016) 012017 doi:10.1088/1742-6596/727/1/012017

3. Weakly Attractive Integral Manifolds

In this section we consider the system (1) when the matrix $B = g_y(x, \phi(x,t), t, 0)$ has eigenvalues on the imaginary axis with nonvanishing imaginary parts. If the eigenvalues at $\varepsilon = 0$ are pure imaginary but after taking into account the perturbations of higher order they move to the complex left half-plane, then the system under consideration has stable slow integral manifolds. It seems reasonable to say that this kind of problem for gyroscopic systems was previously investigated by the integral manifolds method, see, for example, [12]. Note that in the eigenvalues of the matrix of the linearized fast subsystem in such problems have the form $-\alpha \pm i\beta/\varepsilon$ with positive α and the existence problem for slow integral manifolds has not any connections with the bifurcation problems when the real parts of the eigenvalues change their sign. Some problems of the mechanics of manipulators with high-frequency and weakly damped transient regimes are now discussed in this context. More results along this line can be found in [3, 4, 12].

3.1. Control of a One Rigid-Link Flexible-Joint Manipulator

Consider a simple model of a rigid-link flexible joint manipulator [13, 14], where J_m is the motor inertia, J_1 is the link inertia, M is the link mass, l is the link length, c is the damping coefficient, k is the stiffness. The model is described by the equations:

$$J_1\ddot{q}_1 + Mgl\sin q_1 + c(\dot{q}_1 - \dot{q}_m) + k(q_1 - q_m) = 0,$$

$$J_m\ddot{q}_m - c(\dot{q}_1 - \dot{q}_m) - k(q_1 - q_m) = u.$$

Here q_1 is the link angle, q_m is the rotor angle, and u is the torque input which is the controller.

The control problem under consideration consists of a tracking problem in which it is desired that the link coordinate q_1 follows a time-varying smooth and bounded desired trajectory $q_d(t)$ so that $|q_d(t) - q_1(t)| \to 0$ as $t \to \infty$ [13, 14].

If we rewrite the original system in the form

$$J_1\ddot{q}_1 + J_m\ddot{q}_m + Mgl\sin q_1 = u,$$

$$\ddot{q}_1 - \ddot{q}_m + \frac{Mgl}{J_1}\sin q_1 + k\left(\frac{1}{J_1} + \frac{1}{J_m}\right)(q_1 - q_m) + c\left(\frac{1}{J_1} + \frac{1}{J_m}\right)(\dot{q}_1 - \dot{q}_m) = -\frac{u}{J_m},$$

then the use of the small parameter $\varepsilon = 1/\sqrt{k}$ and new variables

$$x_1 = (J_1q_1 + J_mq_m)/(J_1 + J_m), \quad x_2 = \dot{x}_1, \quad y_1 = q_1 - q_m, \quad y_2 = \varepsilon\dot{y}_1, \tag{12}$$

yield the system

$$\dot{x}_1 = x_2, \quad \dot{x}_2 = -\frac{Mgl}{J_1 + J_m}\sin\left(x_1 + \frac{J_m}{J_1 + J_m}y_1\right) + \frac{u}{J_1 + J_m}, \tag{13}$$

$$\varepsilon\dot{y}_1 = y_2, \quad \varepsilon\dot{y}_2 = -\left(\frac{1}{J_1} + \frac{1}{J_m}\right)y_1 - \varepsilon c\left(\frac{1}{J_1} + \frac{1}{J_m}\right)y_2 - \tag{14}$$

$$\varepsilon^2\frac{Mgl}{J_1}\sin\left(x_1 + \frac{J_m}{J_1 + J_m}y_1\right) - \varepsilon^2\frac{u}{J_m}.$$

Note that neglecting all terms of order $O(\varepsilon^2)$ in the r.h.s. of the last equation we obtain the independent subsystem

$$\dot{\varepsilon y_1} = y_2,$$

$$\varepsilon\dot{y}_2 = -\left(\frac{1}{J_1} + \frac{1}{J_m}\right)y_1 - \varepsilon c\left(\frac{1}{J_1} + \frac{1}{J_m}\right)y_2,$$

solutions of which are characterized by high frequency $\approx \sqrt{(1/J_1 + 1/J_m)}/\varepsilon$ and relatively slow decay $c(1/J_1 + 1/J_m)/2$, since this differential system has the characteristic polynomial

$$\varepsilon^2\lambda^2 + c\left(\frac{1}{J_1} + \frac{1}{J_m}\right)\lambda + \left(\frac{1}{J_1} + \frac{1}{J_m}\right)$$

which possesses complex zeros

$$\lambda_{1,2} = -\frac{c}{2}\left(\frac{1}{J_1} + \frac{1}{J_m}\right) \pm \frac{i}{\varepsilon}\sqrt{\left(\frac{1}{J_1} + \frac{1}{J_m}\right) - \varepsilon^2\frac{c^2}{4}\left(\frac{1}{J_1} + \frac{1}{J_m}\right)^2}.$$

Since the real part of these numbers is negative, for the analysis of the manipulator model under consideration it is possible to use the slow invariant manifold noting that the reducibility principle holds for this manifold (the exact statement may be found in [3]). The terms of $O(\varepsilon^2)$ of the subsystem (14) lead us to conclude that the slow invariant manifold may be found in the form $y_1 = \varepsilon^2 Y + O(\varepsilon^3)$ and $y_2 = O(\varepsilon^3)$, where

$$Y = -\left[\frac{Mgl}{J_1}\sin(x_1) + \frac{u_0}{J_m}\right]\left(\frac{1}{J_1} + \frac{1}{J_m}\right)^{-1}. \tag{15}$$

Here we used the representation $u = u_0 + \varepsilon^2 u_1 + O(\varepsilon^3)$. Thus, the flow on this manifold is described by equations

$$\dot{x}_1 = x_2, \quad \dot{x}_2 = -\frac{Mgl}{J_1 + J_m}\sin\left(x_1 + \varepsilon^2\frac{J_m}{J_1 + J_m}Y\right) + \frac{u_0 + \varepsilon^2 u_1}{J_1 + J_m} + O(\varepsilon^3). \tag{16}$$

It is important to emphasize that due to (12) $q_1 = x_1 + \frac{J_m}{J_1+J_m}y_1$, where $y_1 = \varepsilon^2 Y + O(\varepsilon^3)$, and on the slow invariant manifold we obtain the representation

$$q_1 = x_1 + \varepsilon^2\frac{J_m}{J_1 + J_m}Y + O(\varepsilon^3). \tag{17}$$

This allows us to rewrite the system (16) on the slow invariant manifold using the original variable q_1 instead x_1 in the form

$$\ddot{q}_1 - \varepsilon^2\frac{J_m}{J_1 + J_m}\ddot{Y} = -\frac{Mgl}{J_1 + J_m}\sin(q_1) + \frac{u_0 + \varepsilon^2 u_1}{J_1 + J_m} + O(\varepsilon^3). \tag{18}$$

The function $\ddot{Y}$ will be calculated below.

Let q_d be the desired trajectory, i.e., the goal of the controlled motion is $q_1 \to q_d$ as $t \to \infty$ [13, 14]. Unlike [13, 14] we do not use a fast term added to the control input to make the fast dynamics asymptotically stable to guarantee the fast decay of fast variables y_1 and y_2, but we use the slow component of the control function u which is written as a sum $u_0 = (J_1 + J_m)u_d + Mgl\sin q_1$, where $u_d = \ddot{q}_d - a_1(x_1 - q_d) - a_2(\dot{x}_1 - \dot{q}_d)$ [13, 14]. The goal of this control law is to obtain an equation with decaying solutions for the difference between q_1 and q_d.

Setting $\varepsilon = 0$, using (18) and the definitions of u_0 and u_d we obtain, to an accuracy of order $O(\varepsilon^2)$,

$$\ddot{q}_1 - \ddot{q}_d + a_2(\dot{q}_1 - \dot{q}_d) + a_1(q_1 - q_d) = 0$$

for the difference $q_1 - q_d$, since $q_1 = x_1 + O(\varepsilon^2)$ on the slow invariant manifold by (17). Note that any solution of this second order linear homogeneous differential equation tends to zero as time increase for any positive constant coefficients a_1 and a_2. This differential equation allows us to choose the coefficients a_1 and a_2 in the control function u_d in such a way that the corresponding control function u gives the possibility of realizing a desired motion. Let, for example [4, 13, 14], $M = 1,\ \ k = 100,\ \ l = 1,\ \ J_1 = 1,\ \ J_m = 1,\ g = 9.8$ and $c = 2$. Setting $a_1 = 3,\ a_2 = 4$ for the desired trajectory $q_d = \sin t$ we obtain the following control law for the original variables

$$u = (J_1 + J_m)u_d + Mgl \sin x_1 = (J_1 + J_m)u_d + Mgl \sin q_1 = 2u_d + 9.8 \sin q_1 =$$

$$2[-\sin t - 4(\dot{q}_1 - \cos t) - 3(q_1 - \sin t)] + 9.8 \sin q_1.$$

Note that the trajectory of q_1 tends to the desired trajectory $\sin t$ as t increases.

To take into account the terms of order $O(\varepsilon^2)$ we set $u_1 = -J_m\ddot{Y}$ to obtain the same equation $\ddot{q}_1 - \ddot{q}_d + a_2(\dot{q}_1 - \dot{q}_d) + a_1(q_1 - q_d) = 0$ from (18). To calculate Y we use (15) with $u_0 = (J_1 + J_m)u_d + Mgl \sin q_1$, where $u_d = \ddot{q}_d - a_1(x_1 - q_d) - a_2(\dot{x}_1 - \dot{q}_d)$, and obtain

$$Y = -Mgl \sin q_1 - J_1 u_d.$$

It is easy now to obtain $\dot{Y}$

$$\dot{Y} = -Mgl \cos q_1 \frac{dq_1}{dt} - J_1 \frac{du_d}{dt}$$

$$= -Mgl \cos q_1 \frac{dq_1}{dt} - J_1 \left[\frac{d^3 q_d}{dt^3} - a_1 \left(\frac{dq_1}{dt} - \frac{dq_d}{d}\right) - a_2 \left(\frac{d^2 q_1}{dt^2} - \frac{d^2 q_d}{dt^2}\right)\right] =$$

$$-Mgl \cos q_1 \frac{dq_1}{dt} - J_1 \left[\frac{d^3 q_d}{dt^3} + a_1 a_2 (q_1 - q_d) + (a_2^2 - a_1)\left(\frac{dq_1}{dt} - \frac{dq_d}{dt}\right)\right],$$

because $\ddot{q}_1 - \ddot{q}_d = -a_1(q_1 - q_d) - a_2(\dot{q}_1 - \dot{q}_d)$. Similarly we obtain

$$\ddot{Y} = Mgl \sin q_1 \left(\frac{dq_1}{dt}\right)^2 - Mgl \cos q_1 \frac{dq_1}{dt}$$

$$-J_1 \left[\frac{d^4 q_d}{dt^4} - a_1(a_2^2 - a_1)(q_1 - q_d) - a_2(a_2^2 - 2a_1)\left(\frac{dq_1}{dt} - \frac{dq_d}{dt}\right)\right],$$

and, finally, with $\varepsilon^2 = 1/k$,

$$u = (J_1 + J_m)u_d + Mgl \sin q_1 - J_m \ddot{Y}/k.$$

4. The case of multiple root of the degenerate equation

We consider system (1) under the assumptions (A_1) and (A_2). Instead of hypothesis (A_3) we suppose

$$det\ g_y(x, h(x,t), t, 0) \equiv 0 \quad \forall (x,t) \in \mathcal{G} \times R, \tag{19}$$

that is, $y = h(x,t)$ is not a simple root of the degenerate equation

$$g(x, y, t, 0) = 0. \tag{20}$$

Under this assumption we cannot apply Proposition 1.1 to system (1) in order to establish the existence of a slow integral manifold near $\mathcal{M}_0$ for small ε. Our goal is to derive conditions which imply that for sufficiently small ε system (1) has at least one integral manifold $\mathcal{M}_\varepsilon$ with the representation

$$y = \psi_i(x, t, \varepsilon) = h(x,t) + \varepsilon^{q_i} h_{1,i}(x,t) + \varepsilon^{2q_i} h_{2,i}(x,t) + \dots \ .$$

where $q_i, 0 < q_i < 1$, is a rational number.
The key idea to solve this problem consists in looking for scalings and transformations of the type

$$\varepsilon = \mu^r,\ y = \tilde{y}(\mu, z, x, t),\ t = \tilde{t}(\mu, \tau)$$

such that system (1) can be reduced to a system

$$\frac{dx}{d\tau} = f(x, z, \tau, \mu),\quad \mu\,\frac{dz}{d\tau} = g(x, z, \tau, \mu) \tag{21}$$

to which Proposition 1.1 can be applied.

4.1. Partially cheap linear-quadratic optimal control problem
Consider the linear-quadratic optimal control problem

$$\dot{x} = A(t,\varepsilon)x + B_1(t,\varepsilon)u_1 + B_2(t,\varepsilon)u_2;$$

$$J = \frac{1}{2}x^T(1)Fx(1) + \frac{1}{2}\int_0^1 [x^T(t)Q(t,\varepsilon)x(t) + \varepsilon^2 u_1^T(t)R_1(t,\varepsilon)u_1(t) + u_2^T(t)R_2(t,\varepsilon)u_2(t)]dt,$$

where $Q = Q^T \geq 0$, $F = F^T \geq 0$, $R_1 = R_1^T > 0$, $R_2 = R_2^T > 0$, $t \in [0,1]$, ε is a small positive parameter. If the matrices B_1 and R_1 are zero matrices the control problem under consideration is known linear-quadratic optimal control problem. If the case B_2 and R_2 are zero matrices is considered, such a problem is called a *cheap control* problem because there is a small parameter multiplied by a control function in the cost functional [15]. The solution of this problem is given by the formula

$$u = u_1 = -\varepsilon^{-2}R_1^{-1}B_1^T K_1 x,$$

where K_1 is the solution of differential matrix Riccati equation

$$\varepsilon^2(\dot{K}_1 + A^T K_1 + K_1 A + Q) = K_1 S_1 K_1,\quad S_1 = B_1 R_1^{-1} B_1^T;\quad K_1(1) = F.$$

The asymptotic solution of this equation which contains fractional exponents of ε can be found in [15]. Under the condition $\varepsilon = 0$, the degenerate equation has the multiple solution $K_1 = 0$ and the branching of slow integral manifolds takes place for this Riccati equation [17]. In this sense we obtain a critical case in the problem under consideration.

In the case of partially cheap control the control law is given by the formula

$$u = \begin{pmatrix} u_1 \\ u_2 \end{pmatrix} = -\begin{pmatrix} \varepsilon^{-2}R_1^{-1}B_1^T \\ R_2^{-1}B_2^T \end{pmatrix} Kx,$$

i.e., $u_1 = -\varepsilon^{-2}R_1^{-1}B_1^T Kx$ and $u_2 = -R_2^{-1}B_2^T Kx$.

The method of regularization [17] of the differential matrix Riccati equation is adaptable to the Riccati equation in this case :

$$\varepsilon^2(\dot{K} + A^T K + KA + Q - KS_2K) = KS_1K,\quad S_1 = B_1R_1^{-1}B_1^T,\ S_2 = B_2R_2^{-1}B_2^T;\quad K(1) = F,$$

but below we shall restrict the consideration of the partially cheap linear-quadratic optimal control problem by a simple example.

4.2. Example 1

We investigate the optimal control problem

$$\dot{x}_1 = u_1, \quad \dot{x}_2 = x_1 + x_2 + u_2$$

with the cost functional

$$J = \frac{1}{2}\int_0^T [x_1^2(t) + \varepsilon x_2^2(t) + u_1^2(t) + \varepsilon^2 u_2^2(t)]dt \to \ \min .$$

This problem is *a partially cheap control problem* because one of the control terms in the cost functional is multiplied by a small parameter. The optimal control in this problem is given by the formula

$$\begin{pmatrix} u_1 \\ u_2 \end{pmatrix} = -\begin{pmatrix} 1 & 0 \\ 0 & \varepsilon \end{pmatrix}^{-1} K \begin{pmatrix} x_1 \\ x_2 \end{pmatrix},$$

where the matrix K is a nonnegative solution of matrix Riccati equation

$$\frac{dK}{dt} \ = \ -K\begin{pmatrix} 0 & 0 \\ 1 & 1 \end{pmatrix} - \begin{pmatrix} 0 & 1 \\ 0 & 1 \end{pmatrix} K + K\begin{pmatrix} 1 & 0 \\ 0 & \varepsilon^2 \end{pmatrix}^{-1} K - \begin{pmatrix} 1 & 0 \\ 0 & \varepsilon \end{pmatrix}$$

satisfying the condition $K(T) = 0$. If we put

$$K = \begin{pmatrix} k_1 & \varepsilon k_2 \\ \varepsilon k_2 & \varepsilon k_3 \end{pmatrix}$$

we obtain the differential system

$$\begin{aligned} \frac{dk_1}{dt} &= k_1^2 + k_2^2 - 2\varepsilon k_2 - 1, \\ \varepsilon\,\frac{dk_2}{dt} &= \varepsilon k_1 k_2 + k_2 k_3 - \varepsilon(k_2 + k_3), \\ \varepsilon\,\frac{dk_3}{dt} &= k_3^2 + \varepsilon^2 k_2^2 - 2\varepsilon k_3 - \varepsilon. \end{aligned} \tag{22}$$

In this example the method of the Newton diagram or blow-up transformations may be useful but we prefer to use more simple gauge function method [16].

The corresponding degenerate system

$$k_2 k_3 = 0, \ k_3^2 = 0$$

has the solution $k_2 = k_3 = 0$, but this solution is not simple. In order to get simple roots we apply the scaling

$$k_3 = \mu\ \kappa_3, \ \varepsilon = \mu^2. \tag{23}$$

Substituting (23) into (22) we obtain

$$\begin{aligned} \frac{dk_1}{dt} &= k_1^2 + k_2^2 - 2\mu^2 k_2 - 1, \\ \mu\,\frac{dk_2}{dt} &= \mu k_1 k_2 + k_2 \kappa_3 - \mu(k_2 + \mu\kappa_3), \\ \mu\,\frac{d\kappa_3}{dt} &= \kappa_3^2 - 1 + \mu^2 k_2^2 - 2\mu\kappa_3. \end{aligned} \tag{24}$$

The corresponding degenerate system

$$k_2\kappa_3 = 0,\ \kappa_3^2 = 1$$

has the simple nonnegative solution $k_2 = 0, \kappa_3 = 1$. Applying Proposition 1.1 we get that the original system (22) has the invariant manifold

$$k_2 = \mu\psi_1(k_1,\mu),\ k_3 = \mu + \mu^2\psi_2(k_1,\mu). \tag{25}$$

Substituting (25) into (22) leads to the initial value problem

$$\frac{dk_1}{dt} = k_1^2 - 1 + O(\mu^2),\ k_1(T) = 0.$$

By this way, the elements of the matrix K can be determined approximately.

5. Twice Critical Case

Consider the control system

$$\varepsilon\dot{x} = A(t,\varepsilon)x + B(t,\varepsilon)u,\ x \in R^{n+m},\ x(0) = x_0 \tag{26}$$

with the cost functional

$$J = \frac{1}{2}x^T(1)Fx(1) + \frac{1}{2}\int_0^1 (x^T(t)Q(t)x(t) + u^T(t)R(t)u(t))dt. \tag{27}$$

The solution of this problem is the optimal linear feedback control law

$$u = -\varepsilon^{-1}R^{-1}B^TP(t,\varepsilon)x,$$

where P satisfies the differential matrix Riccati equation

$$\varepsilon\dot{P} = -PA - A^TP + PSP - Q,\quad P(1,\varepsilon) = F.$$

Let the matrices in (26) (27) may be represented as

$$A = \begin{pmatrix} \varepsilon A_1 & \varepsilon A_2 \\ A_3 & A_4 \end{pmatrix},\ B = \begin{pmatrix} \varepsilon B_1 \\ B_2 \end{pmatrix},$$

$$F = \begin{pmatrix} F_1 & F_2 \\ F_2^T & F_3 \end{pmatrix},\ Q = \begin{pmatrix} Q_1 & Q_2 \\ Q_2^T & Q_3 \end{pmatrix},$$

where $A_1,\ F_1,\ Q_1$ are $(m\times m)$-matrices, $A_2,\ F_2,\ Q_2$ are $(m\times n)$-matrices, A_4, F_3, Q_3 are $(n\times n)$-matrices, B_1 is $(m\times r)$-matrix, B_2 is $(n\times r)$-matrix, and R is $(r\times r)$-matrix. Suppose that all these matrices have the following asymptotic presentations with respect to ε:

$A_i(t,\varepsilon) = \sum\limits_{j\geq 0}\varepsilon^j A_{ij}(t),\ i = \overline{1,4},\ B_i(t,\varepsilon) = \sum\limits_{j\geq 0}\varepsilon^j B_{ij}(t),\ i = 1,2,\ Q_i(t,\varepsilon) = \sum\limits_{j\geq 0}\varepsilon^j Q_{ij}(t),\ i = \overline{1,3},$
$R(t,\varepsilon) = \sum\limits_{j\geq 0}\varepsilon^j R_j(t),$
$F_i(\varepsilon) = \sum\limits_{j\geq 0}\varepsilon^j F_{ij},\ i = \overline{1,3}$
with smooth on t matrix coefficients, $t \in [0,1]$. The first m components of vector x are slow

variables, and the other n components are fast variables. The control law can be found in the block form

$$u = -R^{-1}(B_1^T \ \varepsilon^{-1}B_2^T)\begin{pmatrix} P_1 & \varepsilon P_2 \\ \varepsilon P_2^T & \varepsilon P_3 \end{pmatrix} x, \tag{28}$$

or

$$u = -R^{-1}(B_1^T P_1 + B_2^T P_2^T \ \ \varepsilon B_1^T P_2 + B_2^T P_3)x.$$

For P_1, P_2, P_3 we obtain the following differential system

$$\begin{aligned}
\dot{P}_1 &= -P_1A_1 - A_1^TP_1 - P_2A_3 - A_3^TP_2^T + P_1S_1P_1 + \\
&\quad + P_1S_2P_2^T + P_2S_2^TP_1 + P_2S_3P_2^T - Q_1, \\
\varepsilon\dot{P}_2 &= -P_1A_2 - P_2A_4 - \varepsilon A_1^TP_2 - A_3^TP_3 + \\
&\quad + P_1S_2P_3 + P_2S_3P_3 + \varepsilon(P_1S_1P_2 + P_2S_2^TP_2) - Q_2, \\
\varepsilon\dot{P}_3 &= \ \ P_3A_4 - A_4^TP_3 + P_3S_3P_3 + \varepsilon(-P_2^TA_2 - A_2^TP_2 + \\
&\quad + \varepsilon P_2^TS_1P_2 + P_2^TS_2P_3 + P_3S_2^TP_2) - Q_3,
\end{aligned} \tag{29}$$

$S_1 = B_1R^{-1}B_1^T, \ S_2 = B_1R^{-1}B_2^T, \ S_3 = B_2R^{-1}B_2^T$
with additional conditions

$$P_1(1,\varepsilon) = F_1, \ \ P_2(1,\varepsilon) = \varepsilon^{-1}F_2, \ \ P_3(1,\varepsilon) = \varepsilon^{-1}F_3.$$

Setting $\varepsilon = 0$ we obtain from (29) the matrix algebraic equation

$$-MA_{40} - A_{40}^TM + MS_{30}M - Q_{30} = 0,$$

where $S_{30} = B_{20}R_0^{-1}B_{20}^T$. If the last equation has a positive definite solution $M(t)$ and eigenvalues of $D_{40} = A_{40} - S_{30}M$ have negative real parts for $t \in [0,1]$ then it is possible to use the boundary functions method [1, 2] or the integral manifolds method [6, 9, 18] setting $K_2 = h_2(K_1,t,\varepsilon) = h_{20}(K_1,t) + \varepsilon\ldots, K_3 = h_3(k_1,t,\varepsilon) = M(t) + \varepsilon\ldots$. However for systems with low energy dissipation the matrices S_{30} and Q_{30} are equal to zero and the main role plays the linear operator

$$\mathbf{L}X = XA_{40} + A_{40}^TX.$$

For this class of systems the eigenvalues of A_{40} are pure imaginary and the spectrum of the linear operator $\mathbf{L}$ has a nontrivial kernel, since sums $(\lambda_i(t) + \lambda_j(t))$, $i,j = 1,\ldots,n$, form its spectrum. This means that the third equation of (29) is singular singularly perturbed. Thus, the dimension of the slow integral manifold of (29) is greater than the dimension of the matrix P_1 and the problem under consideration is critical in this sense. Moreover, under taking into account that all other, nonzero eigenvalues of $\mathbf{L}$, are pure imaginary, it is possible to say that this problem is twice critical.

5.1. Example 2

Let

$$A = \begin{pmatrix} -\varepsilon & 1 \\ -1 & 0 \end{pmatrix}, B = \begin{pmatrix} 0 \\ \varepsilon \end{pmatrix}, R = (1), Q = \begin{pmatrix} 2 & 0 \\ 0 & 1 \end{pmatrix}, S = \begin{pmatrix} 0 & 0 \\ 0 & 1 \end{pmatrix}, P = \begin{pmatrix} p_1 & p_2 \\ p_2 & p_3 \end{pmatrix}.$$

Consider the corresponding differential system

$$\begin{aligned} \varepsilon\dot{p}_1 &= 2p_2 + 2\varepsilon p_1 + \varepsilon p_2^2 - 2\varepsilon, \\ \varepsilon\dot{p}_2 &= \varepsilon p_2 - p_1 + p_3 + \varepsilon p_2 p_3, \\ \varepsilon\dot{p}_3 &= -2p_2 + \varepsilon p_3^2 - \varepsilon. \end{aligned} \tag{30}$$

First, we need to separate it into a slow and a fast subsystem. At first glance, all three equations are singularly perturbed. However, we should consider the matrix of leading terms on the right hand side of the system, which has the form

$$\begin{pmatrix} 0 & 2 & 0 \\ -1 & 0 & 1 \\ 0 & -2 & 0 \end{pmatrix}$$

Obviously, this matrix has a zero eigenvalue and two pure imaginary eigenvalues, i.e. the problem under consideration is twice critical.

Introducing the new variables

$$q_1 = p_1 + p_3 - 2, \quad q_2 = p_2, \quad q_3 = p_3 - 1,$$

we obtain the differential system

$$\begin{aligned} \dot{q}_1 &= 2q_1 + q_2^2 + q_3^2, \\ \varepsilon\dot{q}_2 &= 2\varepsilon q_2 + 2q_3 - q_1 + \varepsilon q_2 q_3, \\ \varepsilon\dot{q}_3 &= -2q_2 + 2\varepsilon q_3 + \varepsilon q_3^2 \end{aligned} \tag{31}$$

with the slow variable q_1 and two fast variables q_2, q_3.

The last system possesses one-dimensional slow invariant manifold which is weakly attractive with respect to argument $1 - t$ because the main matrix of the fast subsystem is

$$\begin{pmatrix} 2\varepsilon & 2 \\ -2 & 2\varepsilon \end{pmatrix}.$$

Thus, the dimension of the system of Riccati differential equations can be reduced from three to one. Let us construct the slow integral manifold using the fact that it can be asymptotically expanded in powers of the small parameter. Setting

$$q_2 = \varphi(q_1, \varepsilon) = \varepsilon\varphi_1(q_1) + \varepsilon^2 \dots ,$$

$$q_3 = \psi(q_1, \varepsilon) = \psi_0(q_1) + \varepsilon\psi_1(q_1) + \varepsilon^2 \dots ,$$

we obtain

$$\psi_0(q_1) = q_1/2, \quad \varphi_1(q_1) = q_1^2/16, \quad \psi_1(q_1) = 0.$$

Thus we obtain the slow invariant manifold

$$q_2 = \varepsilon q_1^2/16 + O(\varepsilon^2), \quad q_3 = q_1/2 + O(\varepsilon^2),$$

with the equation on the integral manifold

$$\dot{q}_1 = 2q_1 + q_1^2/4 + O(\varepsilon^2).$$

6. Conclusion

Critical cases for singularly perturbed differential systems are studied in the paper. We have considered singularly perturbed control problems as applications. It has been shown that the reduction of dimensions of these problems can be done by means of the integral manifold method. The slow integral manifolds for the matrix Riccati equation of linear-quadratic control problem are constructed and it is shown that the method of integral manifolds allows us to reduce the dimension of control problems. This approach was used for the investigation of optimal filtering problems in [19, 20].

Acknowledgments

The author would like to thank the anonymous referees and the handling editor for their valuable comments and suggestions which have led to a significant improvement to the paper.

This work is supported in part by the Russian Foundation for Basic Research (grant 13-01-97002-p) and the Ministry of education and science of the Russian Federation in the framework of the implementation of Program of increasing the competitiveness of SSAU for 2013–2020 years.

References

[1] Vasilieva A B, Butuzov V F and Kalachev L V 1995 *The Boundary Function Method for Singular Perturbation Problems* (Philadelphia: SIAM)
[2] O'Malley R E 1991 *Singular Perturbation Methods for Ordinary Differential Equations* Appl. Math. Sci. **89** (New-York: Springer–Verlag)
[3] Mortell M P, O'Malley R E, Pokrovskii A and Sobolev V A 2005 *Singular Perturbation and Hysteresis* (Philadelphia: SIAM)
[4] Shchepakina E, Sobolev V and Mortell M P 2014 *Singular Perturbations. Introduction to System Order Reduction Methods with Applications* (Lect Notes in Math **2114**) (Berlin–Heidelber–London: Springer)
[5] O'Malley R E 1977 *High gain feedback systems as singular singular–perturbation problems* In: Prepr. techn. pap., 18th Joint Automat. Control Conf., New York 1278–1281
[6] Kokotović P V, Khalil K H and O'Reilly J 1986 *Singular Perturbation Methods in Control: Analysis and Design* (Philadelphia: SIAM)
[7] Vasilieva A B and Butuzov V F 1980 *Singularly Perturbed Equations in the Critical Case* Tech. Report MRC–TSR 2039 (Madison: University of Visconsin)
[8] Strygin V V and Sobolev V A 1976 Effect of geometric and kinetic parameters and energy dissipation on orientation stability of satellites with double spin *Cosmic Research* **14** 331–335
[9] Sobolev V A 1984 Integral manifolds and decomposition of singularly perturbed systems *System and Control Letters* **5** 169–179
[10] Gu Z, Nefedov N N and O'Malley R E 1989 On singular singularly perturbed initial values problems *SIAM J Appl Math* **49** 1–25
[11] Kononenko L I and Sobolev V A 1994 Asymptotic expansion of slow integral manifolds *Siberian Math J* **35** 1119–1132
[12] Sobolev V A and Strygin V V 1978 Permissibility of changing over to precession equations of gyroscopic systems *Mechanics of Solids* **5** 7–13
[13] Spong M W, Khorasani K and Kokotovic P V 1987 An integral manifold approach to the feedback control of flexible joint robots *IEEE Journal of Robotics and Automation* **3:4** 291–300
[14] Ghorbel F and Spong M W 2000 Integral manifold of singularly perturbed systems with application to rigid-link flexible-joint multibody system *International Journal of Non-linear Mechanics* **35** 133–155
[15] O'Malley R E, Jameson A 1975-77 Singular perturbations and singular arcs I, II. *Trans Autom Control* AC-20 218–226 and AC-22 328–337
[16] Murdock J A 1991 *Perturbations. Theory and Methods*, John Wiley & Sons Inc.
[17] Smetannikova E , Sobolev V 2005 Regularization of Cheap Periodic Control Problems. *Autom Remote Control* 66:6, 903–916 (2005)
[18] Sobolev V A 1991 Singular perturbations in linearly quadratic optimal control problems *Autom Remote Control* 52:2 180–189
[19] Osintsev M S and Sobolev V A 2013 Reduction of dimensionality of optimal estimation and control problems for systems of low dissipativity solid bodies *Autom Remote Control* 74:8 1334–1347

[20] Osintsev M S and Sobolev V A 2014 Reduction of dimension of optimal estimation problems for dynamical systems with singular perturbations *Computational Mathematics and Mathematical Physics* 54:1 45-58

MURPHYS-HSFS-2014 IOP Publishing
Journal of Physics: Conference Series **727** (2016) 012018 doi:10.1088/1742-6596/727/1/012018

Amplitude chimeras and chimera death in dynamical networks

Anna Zakharova, Marie Kapeller, and Eckehard Schöll*

Institut für Theoretische Physik, Technische Universität Berlin, Hardenbergstraße 36, 10623 Berlin, Germany

E-mail: schoell@physik.tu-berlin.de

Abstract. We find chimera states with respect to amplitude dynamics in a network of Stuart-Landau oscillators. These partially coherent and partially incoherent spatio-temporal patterns appear due to the interplay of nonlocal network topology and symmetry-breaking coupling. As the coupling range is increased, the oscillations are quenched, amplitude chimeras disappear and the network enters a symmetry-breaking stationary state. This particular regime is a novel pattern which we call chimera death. It is characterized by the coexistence of spatially coherent and incoherent inhomogeneous steady states and therefore combines the features of chimera state and oscillation death. Additionally, we show two different transition scenarios from amplitude chimera to chimera death. Moreover, for amplitude chimeras we uncover the mechanism of transition towards in-phase synchronized regime and discuss the role of initial conditions.

1. Introduction

Chimera states represent an intriguing phenomenon in a dynamical network of coupled elements, which spontaneously separates into two coexisting domains with dramatically different behavior [1]. They occur, surprisingly, in networks of identical units and symmetric coupling schemes [2, 3]. In spite of the fact that these symmetry-breaking states have been the subject of many investigations, e.g., [4, 5, 6, 7, 8, 9, 10, 11, 12, 13, 14, 15], there still remain many issues to be addressed. A recent review on chimera states [16] discusses a number of significant open questions concerning, for instance, the necessary conditions for their existence or generalizations of chimera states. A generalization to amplitude dynamics and to steady state coherence-incoherence patterns has been provided in [1]. In particular, in [1] it has been shown that chimeras of steady states appear due to a special symmetry-breaking coupling which originates from oscillation death investigations [17, 18, 19, 20]. Oscillation death, i.e., the stabilization of inhomogeneous steady states in coupled oscillator networks, appears when a homogeneous steady state splits into at least two distinct symmetry-breaking branches - upper and lower [21]. In comparison with amplitude death, which is another type of oscillation suppression, i.e., the stabilization of an unstable homogeneous steady state, oscillation death is much less investigated [22].

Initially found for phase oscillators [2], chimera states can be treated as long chaotic transients towards the in-phase synchronized regime. In this context the network size plays an essential role. In particular, it has been shown theoretically [11] and confirmed experimentally [23] that the chimera lifetime grows exponentially with the system size. Therefore, chimera states are stable for infinitely large networks and transient for finite-size ensembles. Recently, in [24] the question of what is the smallest possible number of oscillators allowing for chimera states has been addressed. Another important problem, also considered as a necessary condition for the existence of chimera states, is the choice of initial conditions. Random initial conditions do not always guarantee chimera behavior. This is due to the fact that classical chimera states typically coexist with the completely synchronized regime. The coexistence of two stable solutions is a signature of bistability which is typically associated with hysteresis. The basin of attraction for chimera states can be relatively small in comparison with that of the synchronized state. For that reason chimera states remained for a long time undetected. Indeed, the first experimental studies [25, 26, 27, 28, 29, 30, 31] on chimera states were provided only twelve years after their theoretical discovery. Consequently, it is reasonable to use specially prepared initial conditions to ensure chimera patterns. However, it is worth noting that a chimera state is not just a temporary trace of initial conditions which disappears in time, but a persisting pattern with a long lifetime.

In the present work we investigate the connection between two symmetry-breaking effects: chimera states and oscillation death. We show that the interplay of nonlocal coupling topology and symmetry-breaking coupling leads to a number of novel partially coherent inhomogeneous spatial patterns, for instance amplitude chimeras and chimera death. Additionally, we analyze the transition scenarios between these two patterns. Moreover, the transient amplitude chimeras and the transition mechanism from amplitude chimera to a synchronized regime is investigated. Further, the newly found amplitude chimeras are analyzed with respect to initial conditions and transients. In particular, we aim to understand the mechanism of transition from amplitude chimera to in-phase synchronization. Finally, we address the question of how the lifetime of amplitude chimeras depends on initial conditions and what are the optimal initial distributions.

2. Model

We analyze the paradigmatic model of Stuart-Landau oscillators [2, 32, 33, 34, 35, 36, 37]:

$$\dot{z} = f(z) \equiv (\lambda + i\omega - |z|^2)z, \tag{1}$$

where $z = re^{i\phi} = x + iy \in \mathbb{C}$, $\lambda, \omega \in \mathbb{R}$. For $\lambda > 0$, the uncoupled system exhibits self-sustained limit cycle oscillations with radius $r_0 = \sqrt{\lambda}$ and frequency ω. Therefore, the Stuart-Landau system represents a generic model for nonlinear oscillators close to a Hopf bifurcation. We investigate a ring of N nonlocally coupled Stuart-Landau oscillators [1]:

$$\dot{z}_j = f(z_j) + \frac{\sigma}{2P} \sum_{k=j-P}^{j+P} (\mathrm{Re} z_k - \mathrm{Re} z_j), \tag{2}$$

where $j = 1, 2, ..., N$. The coupling parameters, which are identical for all links, are the coupling strength $\sigma \in \mathbb{R}$ and the coupling range P/N, where P corresponds to the number of nearest neighbors in each direction on a ring. Here we consider coupling only in the real parts, since this breaks the rotational S^1 symmetry of the system which is a necessary condition for the existence of nontrivial steady states $z_j \neq 0$ and thus for oscillation death [21]. With respect to applications this means that the oscillators are coupled only through a single real variable x. If compared with slow-fast systems such as, for example, the FitzHugh-Nagumo oscillator, which demonstrates two separated time scales, the Stuart-Landau oscillator is a generic model for sinusoidal oscillations which have only one characteristic time scale defined by the frequency ω.

3. Amplitude chimeras and chimera death states

While tuning the coupling range P for a fixed value of the coupling strength σ we uncover a variety of dynamical regimes in equation (2) which are shown as space-time plots, color-coded by the variable y, and snapshots in Fig. 1. In particular, we find chimera behavior with respect

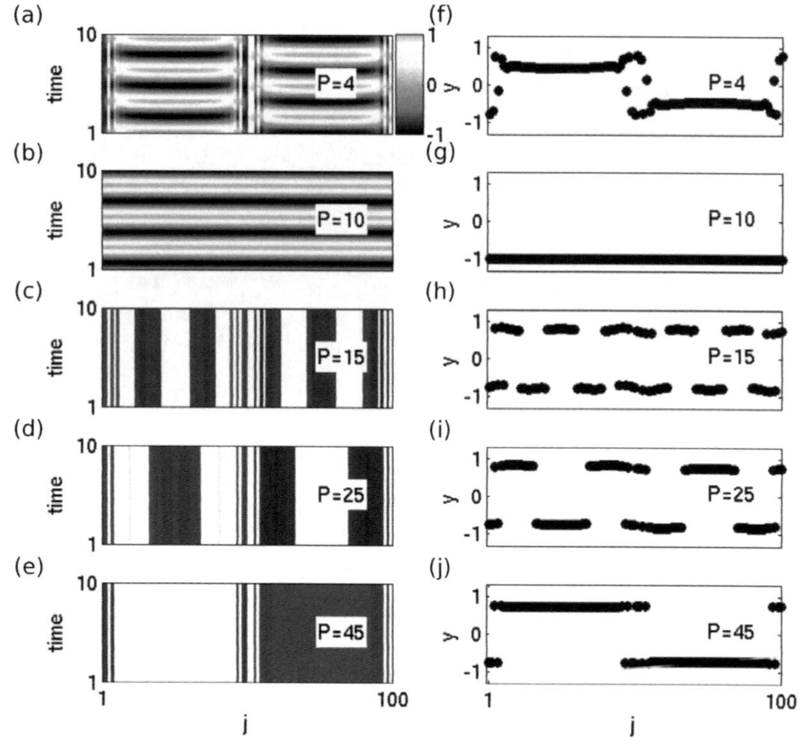

Figure 1. Space-time plots (left panel) and snapshots for the variable $y_j(t)$ (right panel) in a network of Stuart-Landau oscillators for coupling strength $\sigma = 10$ and varying nearest neighbors number P. (a),(f) $P = 4$: amplitude chimera; (b),(g) $P = 10$: in-phase synchronized oscillations; (c),(h) $P = 15$: multi-cluster (> 3) chimera death; (d),(i) $P = 25$: 3-cluster chimera death; (e),(j) $P = 45$: 1-cluster chimera death. Other parameters: $N = 100$, $\lambda = 1$, $\omega = 2$.

to amplitude dynamics, i.e., amplitude chimeras [1], where one part of the network is oscillating with spatially coherent amplitude, while the other displays oscillations with spatially incoherent amplitudes (Fig. 1a,f). It is important to note that within the incoherent domain of the amplitude chimera the center of mass of each oscillator $z_{c.m.} = \int_0^T z_j(t)dt/T$, where $T = 2\pi/\omega$ is the oscillation period, is shifted away from the origin, while the elements from the coherent subgroup oscillate around the origin. To illustrate this fact we calculate the distance between the center of mass of each oscillator and the origin $r_{c.m.}$ (Fig. 2a). Surprisingly, the shape of this profile is very similar to that of the mean phase velocity profile for the classical phase chimeras. However, for amplitude chimeras the mean phase velocity profile $\bar{\omega}_j$ is flat (Fig. 2b). This fact discloses the crucial feature of amplitude chimera: averaged phase velocities remain the same for every element of the network, since the phases are correlated even within the incoherent domain. In more detail, Fig. 2c shows the phase portraits of all oscillators in the complex z plane clearly demonstrating the limit cycles with different amplitudes and centers of mass. The nodes from the coherent part oscillating with larger amplitudes around the origin perform fast motions in the complex $z = x + iy$ phase plane, while the elements of the incoherent domain with smaller amplitudes are slowed down. Therefore, the angular frequencies of all the nodes on average remain the same. Consequently, we observe pure amplitude chimera - chimera behavior exclusively with respect to amplitude dynamics rather than the phase, in contrast to amplitude-mediated chimeras, for which both phase and amplitude are in a chimera state [29, 38].

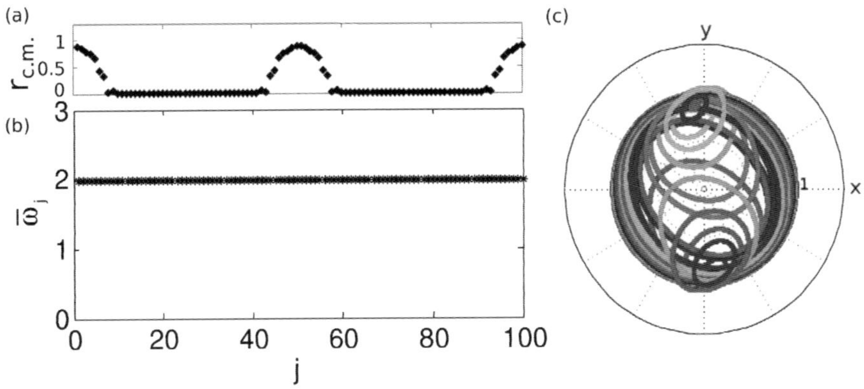

Figure 2. Amplitude chimera: (a) snapshot at $t = 1000$ for $r_{c.m.}$ (the distance between the center of mass for every oscillator and the origin); (b) mean phase velocity profile; (c) phase portraits of all oscillators in the complex $z = x+iy$ plane. Parameters: $N = 100$, $P = 4$, $\sigma = 14$, $\lambda = 1$, $\omega = 2$.

The increase of the coupling range P/N induces a transition from amplitude chimeras to an in-phase synchronized state (Fig. 1b,g). By increasing the coupling range even further we detect a novel pattern which provides bridging between chimera states and oscillation death. Therefore, we call it *chimera death* [1] (Fig. 1c-e, h-j). In this regime the oscillations die out in a peculiar way. The population of identical oscillators breaks up into two domains: (i) spatially coherent oscillation death, where the neighboring elements of the network are correlated forming a regular inhomogeneous steady state, and (ii) spatially incoherent oscillation death, where the sequence of populated branches of the inhomogeneous steady state of neighboring nodes is completely random. It is important to note that the term "coherent/incoherent" refers to the coherence/incoherence in space, i.e., spatial correlation, which should be distinguished from temporal coherence which refers to correlations in time of the dynamics. Because of symmetry reasons, for a node on the upper branch $y^{*1} \approx +1$ of the inhomogeneous steady state in the left half of the system, there always exists a mirror state shifted by phase π (anti-phase) in the complex plane, i.e., located on the lower branch $y^{*2} \approx -1$, in the right half of the system. Interestingly, the increase of the coupling range for fixed coupling strength also induces structural changes of the chimera death pattern: the coherent spatial domain may consist of one or several clusters where the nodes are on the same branch of the inhomogeneous steady state. With increasing coupling range the number of clusters in the coherent spatial domain is decreased, see the scenario in Fig. 1, from (c,h) (5 clusters) via (d,i) (3 clusters) to (e,j) (1 cluster). The illustration of the chimera death pattern with the maximum number of clusters in the coherent part is provided by Fig. 3a-c. The phase portrait typical for chimera death is shown in Fig. 3c, it consists of two fixed points shifted by a phase π, corresponding to the two branches y^{*1}, y^{*2} of the inhomogeneous steady state. It should be noted that these two fixed points are very close to the centers of mass of those oscillations in Fig. 2c with the smallest amplitude, i.e., in the center of the incoherent domain of the amplitude chimera.

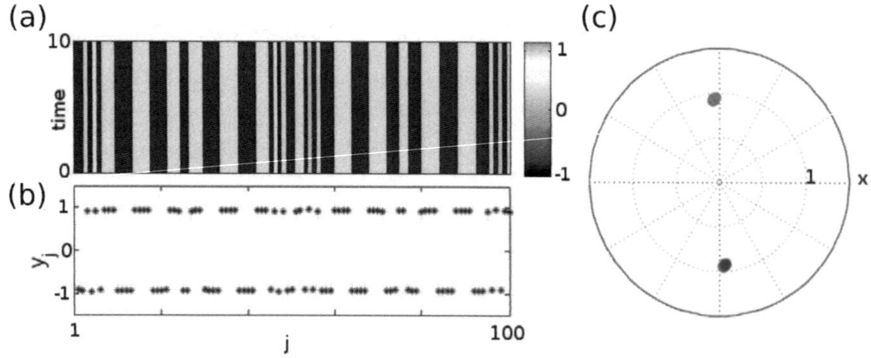

Figure 3. Multi-cluster (> 3) chimera death: (a) space-time plot for the variable y_j; (b) snapshot for the variable y_j; (c) phase portrait of all oscillators in the complex $z = x + iy$ plane. Parameters: $N = 100$, $P = 5$, $\sigma = 26$, $\lambda = 1$, $\omega = 2$.

To provide an overall view on the network behavior for the wide range of coupling parameters we plot the map of regimes in the plane of coupling range and coupling strength. The oscillatory behavior of the network is represented by amplitude chimeras (blue region in Fig. 4), which are observed for small coupling range, and by in-phase synchronized oscillations (light green region in Fig. 4). The steady state solutions occur for larger values of coupling parameters and are manifested by chimera death (red regions in Fig. 4 with different hatching). The region of chimera death is divided into several regimes depending on the number of coherent clusters. Chimera death may consist of two coherent domains, which are formed by one cluster on the

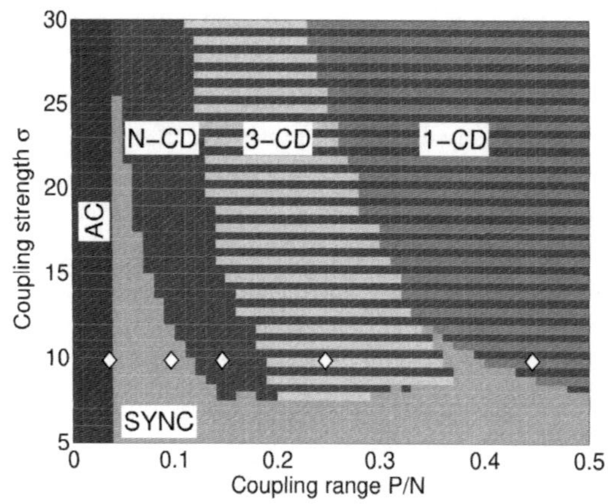

Figure 4. Map of dynamical regimes for $N = 100$, $\lambda = 1$, $\omega = 2$ in the plane of coupling range P/N and coupling strength σ for specially prepared initial conditions, showing 1-cluster chimera death (1-CD); 3-cluster chimera death (3-CD); multi-cluster (> 3) chimera death (N-CD); amplitude chimera (AC); in-phase synchronized oscillations (SYNC). Diamonds mark the parameter values chosen in Fig. 1. Other parameters: $N = 100$, $\lambda = 1$, $\omega = 2$. The detailed explanation of initial conditions is given in Section 5.

upper and one on the lower branch of the steady state (dark and light red stripes in Fig. 4).

Journal of Physics: Conference Series **727** (2016) 012018 doi:10.1088/1742-6596/727/1/012018

Moreover, the coherent regions may split into three clusters each (dark red and yellow stripes in Fig. 4) or more (dark red in Fig. 4). The typical space-time patterns and snapshots in panels (c,d,e) and (h,i,j) of Fig. 1 provide illustrations of the different multi-cluster chimera death states, and the corresponding coupling parameters are indicated in the phase diagram (Fig. 4) by empty diamonds. It is important to note that the chimera death regime is characterized by high multistability. Therefore, chimera death patterns with different number of clusters in the coherent part coexist. The map of regimes strongly depends on initial conditions, in particular, the borders between chimera death patterns with different number of clusters. However, the boundary separating the oscillatory regime from the steady state regimes (with multiple cluster numbers) appears to be less sensitive to initial conditions and remains almost the same for different realizations of initial conditions. The existence of two distinct transition scenarios from the oscillatory to the steady state regime becomes evident from Fig. 4. For a small value of coupling strength $\sigma = 10$ the amplitude chimera gives way to the chimera death state after passing through in-phase synchronized oscillations when the coupling range is increased (diamonds in Fig. 4). In contrast, for a large value of the coupling strength, for example, $\sigma = 26$ a slight increase of the coupling range from $P/N = 0.04$ to $P/N = 0.05$ destroys amplitude chimeras and directly leads to chimera death.

4. Transient behavior of amplitude chimeras

An important feature of classical chimera states is their transient character and strong dependence on initial conditions. Our numerical analysis shows that amplitude chimeras although manifesting a long lifetime (up to 10^4 dimensionless time units) for the optimal choice of initial conditions, can rapidly transform into in-phase synchronized regime for completely random initial conditions.

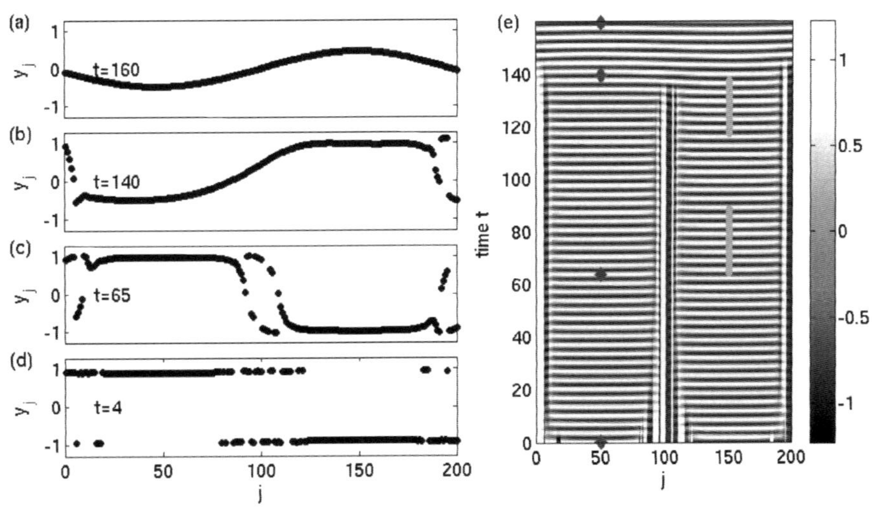

Figure 5. Transient amplitude chimera for $N = 200$, $\lambda = 1$, $\omega = 2$, $P = 5$ and $\sigma = 20$. Snapshots for the variable y_j (left panel): (a) $t = 160$: in-phase synchronized oscillations; (b) $t = 140$: deformed amplitude chimera at the transition point; (c) $t = 65$: amplitude chimera; (d) $t = 4$: trace of initial condition. The right panel shows the space-time plot for the variable $y_j(t)$. Blue diamonds mark the time values chosen for the snapshots in the left panel. The green lines mark the time ranges chosen for the phase portraits in Fig. 6.

To gain more insight into the transient behavior of amplitude chimeras we trace the system's evolution in time having fixed all other parameters. Figure 5a-d shows corresponding snapshots for different time values. As initial condition here we take chimera death ($t = 0$). The pattern at an early stage ($t = 4$) shown in Fig. 5d is reminiscent of the initial conditions, while the regime of amplitude chimeras (Fig. 5c) is reached later ($t = 65$). However, after approximately 125 time steps it starts deforming (Fig. 5b) until it turns into an in-phase synchronized pattern (Fig. 5a). This is also visible in the space-time plot (Fig. 5e).

To disclose the transition mechanism from amplitude chimera to the in-phase synchronized regime we analyze phase portraits in the amplitude chimera state (Fig. 6a) and during the transition (Fig. 6b) for three selected nodes of the network: one from the coherent part ($j = 50$) and the other two from the incoherent domain ($j = 99, 102$). The time ranges covered in Fig. 6a and b, respectively, are indicated by two green vertical lines in the space-time plot of Fig. 5e. In the regime of amplitude chimera the node from the coherent part ($j = 50$) oscillates with large amplitude around the origin, while the two nodes in the incoherent domain ($j = 99, 102$) oscillate with much smaller amplitudes around their own centers of mass shifted from the origin in two opposite directions (Fig. 6a). During the transition the amplitudes of the incoherent oscillations grow till they reach the value of the coherent oscillations. At the same time the centers of mass for nodes from the incoherent domain move towards the origin (Fig. 6b). Therefore, during the transition the phase trajectories for the nodes $j = 99, 102$ have spiral shapes (Fig. 6b) .

An important observation is that the transition mechanism described above depends on the initial conditions. For example, asymmetric initial conditions may lead to another scenario which involves an asymmetric state. In this case nodes from the incoherent part join one by one their neighboring synchronized state.

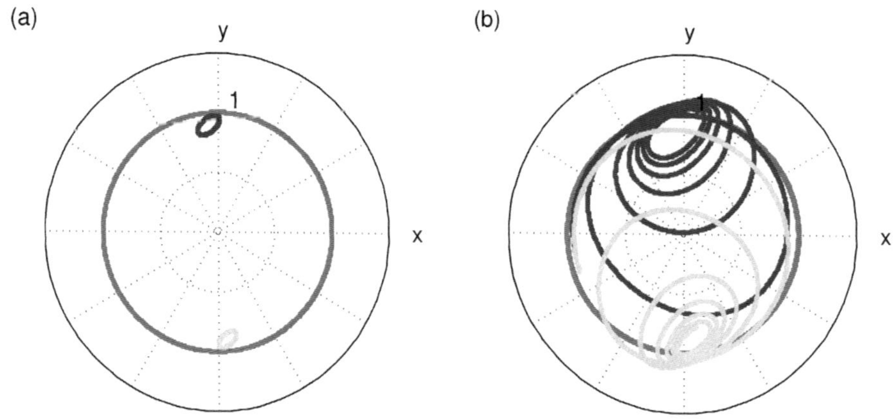

Figure 6. Phase portraits of three selected nodes of the network in the complex $z = x+iy$ plane for (a) amplitude chimera state and (b) during the transition to in-phase synchronization. Red color marks the trajectory of node $j = 50$ (from the coherent part). Dark and light blue colors show trajectories of nodes from the incoherent domain $j = 99$ and $j = 102$ correspondingly. Time ranges as marked in green in Fig. 5. Parameters: $N = 200$, $\lambda = 1$, $\omega = 2$, $P = 5$, $\sigma = 20$.

5. The role of initial conditions

The first initial condition we choose originates from the oscillation death problem and is typically used for the inhomogeneous steady state [21]. Specifically, the system is divided into two equal domains: one is located on the upper branch and the other is on the lower branch. In our particular example of $N = 200$ nodes this implies that one half of the network ($j = 1, ..., 100$) is on the upper branch $y^{*1} \approx +1$ while the other half is on the lower branch $y^{*2} \approx -1$ (Fig. 7a). In this case the lifetime of amplitude chimeras is $t < 200$ and decreases significantly when some small random shift from the branches is applied for every node.

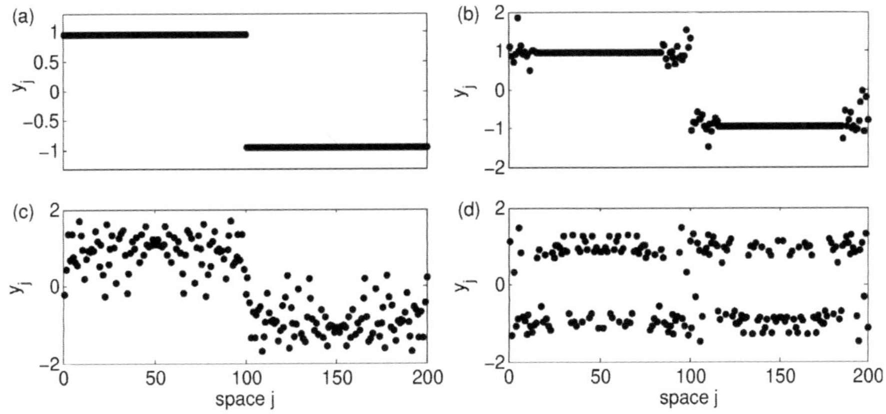

Figure 7. Initial conditions for amplitude chimeras used for $N = 200$, $\lambda = 1$, $\omega = 2$, $P = 4$ and $\sigma = 20$: (a) coherent distribution: one half of the nodes on the upper and one half on the lower branch of the inhomogeneous steady state (amplitude chimera lifetime $t = 130$); (b) combination of coherent and incoherent domains with no symmetries in the randomly defined incoherent part (amplitude chimera lifetime $t = 25$); (c),(d) random initial conditions with symmetries (optimal amplitude chimera lifetime $t > 3000$).

Another type of initial condition we have tested is the modification of the previously discussed set by adding two incoherent regions, where the nodes are distributed randomly (Fig. 7b). This leads to dramatically short lifetime ($t < 30$) of the amplitude chimera, which rapidly transforms into an in-phase synchronized state or a traveling wave.

Next we allow randomness in initial conditions for all nodes of the network, however, observing certain symmetries. In particular, in Fig. 7c the positions for 1/4 of the nodes $j = 1, ..., 50$ are chosen randomly around the upper branch, then these values are mirrored to the nodes $j = 51, ..., 100$ by setting $z_j = z_{1-j+N/2}$. Finally, the positions of the other half of the nodes $j = 101, ..., 200$ are obtained by applying a phase shift of π which we call "anti-phase partner" condition: $z_j = -z_{j+N/2}$ with $j \ mod \ N$. The initial condition scheme shown in Fig. 7d is constructed in a similar way as in Fig. 7c. The only difference is that the values for the first $j = 1, ..., 50$ nodes are chosen randomly around both branches of the inhomogeneous steady state (and not only around the upper branch like in Fig. 7c). This amendment does not have any impact on the lifetime of amplitude chimeras, which is on the other hand strongly affected by the presence of symmetries in the initial conditions. Our simulations suggest that amplitude chimeras tend to have very short lifetime if in the incoherent domain the initial positions of nodes near the upper branch are uncorrelated with those close to the lower branch. An increase of the amplitude chimera lifetime can be achieved by introducing symmetries in the initial distribution ("anti-phase partner" condition). Therefore, optimal results with $t > 3000$ are obtained for random initial conditions with symmetries (Fig. 7c,d). In conclusion, our numerical analysis shows that for random distributions without symmetries amplitude chimeras appear to be short transients towards in-phase synchronized regime, while symmetric conditions significantly increase their lifetime up to $t > 10^4$. Moreover, random initial distributions may also lead to asymmetric spatio-temporal patterns.

To provide a comprehensive view on the impact of initial conditions we compare the map of regimes calculated for a specially prepared initial set (Fig. 4) and for random initial conditions (Fig. 8). For the diagram shown in Fig. 4 we use as initial condition an amplitude chimera

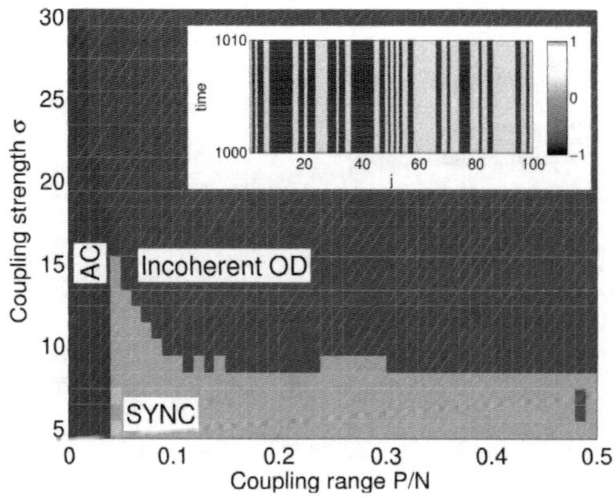

Figure 8. Map of dynamical regimes for $N = 100$, $\lambda = 1$, $\omega = 2$ in the plane of coupling range P/N and coupling strength σ for random initial conditions as in Fig. 7d): amplitude chimera (AC); in-phase synchronized oscillations (SYNC); incoherent oscillation death (Incoherent OD). The inset shows a space-time plot for the variable $y_j(t)$ for the coupling strength $\sigma = 30$ and coupling range $P/N = 0.5$.

profile at a fixed time (similar to Fig. 5c), which is obtained in the following way: For a fixed set of parameters in the amplitude chimera regime, the system is divided into two equal domains; half of the nodes are located on the upper branch and the other half is on the lower branch of the inhomogeneous steady state (similar to Fig. 7a); this initial condition then evolves into an amplitude chimera state which is used as initial condition for all other parameter values $(\sigma, P/N)$. The random initial distribution (similar to Fig. 7d) leads to a significantly different phase diagram (Fig. 8) in comparison with the specially prepared initial conditions (Fig. 4). The domain of the amplitude chimera is decreased for strong coupling if compared with Fig. 4. Moreover, the regime of chimera death obtained from carefully chosen initial conditions (Fig. 4) is replaced by the regime of incoherent oscillation death (inset in Fig. 8) which results from random initial conditions. It should be noted that the map of regimes shown in Fig. 8 is obtained for one single realization of random initial distribution and no averaging of initial conditions has been provided. Therefore, it is not statistically representative that chimera death is not observed in Fig. 8. This is merely due to the fact that the probability to achieve chimera patterns starting from one random realization of the initial condition is very low. However, the occurrence of chimeras for random initial conditions, in general, is not excluded.

6. Conclusions

In conclusion, we have provided a connection between two symmetry-breaking effects - chimera states and oscillation death. In particular, we uncover new spatio-temporal patterns, i.e., amplitude chimeras and chimera death. Amplitude chimeras represent a generalization of chimera behavior to amplitude dynamics, and chimera death generalizes chimeras to steady states. It is shown that different transition scenarios from amplitude chimeras to chimera death are possible. Chimera death patterns with different numbers of clusters in the coherent part exhibit a high degree of multistability, which is related to a variety of hysteretic scenarios.

Moreover, the transient dynamics of amplitude chimera is investigated disclosing the transition mechanism from amplitude chimera to the in-phase synchronized regime. Additionally, we find optimal initial conditions for amplitude chimeras showing that random distributions with particular symmetries essentially enlarge their lifetime.

Acknowledgments

This work was supported by DFG in the framework of SFB 910.

References

[1] A. Zakharova, M. Kapeller, and E. Schöll: *Chimera death: Symmetry breaking in dynamical networks*, Phys. Rev. Lett. **112**, 154101 (2014).
[2] Y. Kuramoto and D. Battogtokh: *Coexistence of Coherence and Incoherence in Nonlocally Coupled Phase Oscillators.*, Nonlin. Phen. in Complex Sys. **5**, 380 (2002).
[3] D. M. Abrams and S. H. Strogatz: *Chimera states for coupled oscillators*, Phys. Rev. Lett. **93**, 174102 (2004).
[4] G. C. Sethia, A. Sen, and F. M. Atay: *Clustered chimera states in delay-coupled oscillator systems*, Phys. Rev. Lett. **100**, 144102 (2008).
[5] C. R. Laing: *The dynamics of chimera states in heterogeneous Kuramoto networks*, Physica D **238**, 1569 (2009).
[6] A. E. Motter: *Nonlinear dynamics: Spontaneous synchrony breaking*, Nature Physics **6**, 164 (2010).
[7] E. A. Martens, C. R. Laing, and S. H. Strogatz: *Solvable model of spiral wave chimeras*, Phys. Rev. Lett. **104**, 044101 (2010).
[8] S. Olmi, A. Politi, and A. Torcini: *Collective chaos in pulse-coupled neural networks*, Europhys. Lett. **92**, 60007 (2010).
[9] G. Bordyugov, A. Pikovsky, and M. G. Rosenblum: *Self-emerging and turbulent chimeras in oscillator chains*, Phys. Rev. E **82**, 035205 (2010).
[10] J. H. Sheeba, V. K. Chandrasekar, and M. Lakshmanan: *Chimera and globally clustered chimera: Impact of time delay*, Phys. Rev. E **81**, 046203 (2010).
[11] M. Wolfrum and O. E. Omel'chenko: *Chimera states are chaotic transients*, Phys. Rev. E **84**, 015201 (2011).
[12] I. Omelchenko, Y. Maistrenko, P. Hövel, and E. Schöll: *Loss of coherence in dynamical networks: spatial chaos and chimera states*, Phys. Rev. Lett. **106**, 234102 (2011).
[13] C. R. Laing: *Fronts and bumps in spatially extended Kuramoto networks*, Physica D **240**, 1960 (2011).
[14] I. Omelchenko, O. E. Omel'chenko, P. Hövel, and E. Schöll: *When nonlocal coupling between oscillators becomes stronger: patched synchrony or multichimera states*, Phys. Rev. Lett. **110**, 224101 (2013).
[15] J. Hizanidis, V. Kanas, A. Bezerianos, and T. Bountis: *Chimera states in networks of nonlocally coupled hindmarsh-rose neuron models*, Int. J. Bifurcation Chaos **24**, 1450030 (2014).
[16] M. J. Panaggio and D. M. Abrams: *Chimera states: Coexistence of coherence and incoherence in networks of coupled oscillators*, Nonlinearity **28**, R67 (2015).
[17] K. Bar-Eli: *On the stability of coupled chemical oscillators*, Physica D **14**, 242 (1985).
[18] K. Tsaneva-Atanasova, C. L. Zimliki, R. Bertram, and A. Sherman: *Diffusion of calcium and metabolites in pancreatic islets: Killing oscillations with a pitchfork*, Biophys. J. **90**, 3434 (2006).
[19] E. Ullner, A. Zaikin, E. Volkov, and J. García-Ojalvo: *Multistability and clustering in a population of synthetic genetic oscillators via phase-repulsive cell-to-cell communication*, Phys. Rev. Lett. **99**, 148103 (2007).
[20] A. Kuznetsov, M. Kaern, and N. Kopell: *Synchrony in a population of hysteresis-based genetic oscillators*, SIAM J. Appl. Math. **65**, 392 (2004).
[21] A. Zakharova, I. Schneider, Y. N. Kyrychko, K. B. Blyuss, A. Koseska, B. Fiedler, and E. Schöll: *Time delay control of symmetry-breaking primary and secondary oscillation death*, Europhys. Lett. **104**, 50004 (2013).
[22] A. Koseska, E. Volkov, and J. Kurths: *Oscillation quenching mechanisms: Amplitude vs. oscillation death*, Phys. Rep. **531**, 173 (2013).
[23] D. P. Rosin, D. Rontani, N. D. Haynes, E. Schöll, and D. J. Gauthier: *Transient scaling and resurgence of chimera states in coupled Boolean phase oscillators*, Phys. Rev. E **90**, 030902(R) (2014).
[24] P. Ashwin and O. Burylko: *Weak chimeras in minimal networks of coupled phase oscillators*, Chaos **25**, 013106 (2015).
[25] A. M. Hagerstrom, T. E. Murphy, R. Roy, P. Hövel, I. Omelchenko, and E. Schöll: *Experimental observation of chimeras in coupled-map lattices*, Nature Physics **8**, 658 (2012).
[26] M. R. Tinsley, S. Nkomo, and K. Showalter: *Chimera and phase cluster states in populations of coupled chemical oscillators*, Nature Physics **8**, 662 (2012).

[27] E. A. Martens, S. Thutupalli, A. Fourrière, and O. Hallatschek: *Chimera states in mechanical oscillator networks*, Proc. Nat. Acad. Sciences **110**, 10563 (2013).

[28] L. Larger, B. Penkovsky, and Y. Maistrenko: *Virtual chimera states for delayed-feedback systems*, Phys. Rev. Lett. **111**, 054103 (2013).

[29] L. Schmidt, K. Schönleber, K. Krischer, and V. Garcia-Morales: *Coexistence of synchrony and incoherence in oscillatory media under nonlinear global coupling*, Chaos **24**, 013102 (2014).

[30] M. Wickramasinghe and I. Z. Kiss: *Spatially organized dynamical states in chemical oscillator networks: Synchronization, dynamical differentiation, and chimera patterns*, PLoS ONE **8**, e80586 (2013).

[31] M. Wickramasinghe and I. Z. Kiss: *Spatially organized partial synchronization through the chimera mechanism in a network of electrochemical reactions*, Phys. Chem. Chem. Phys. **16**, 18360 (2014).

[32] F. M. Atay: *Distributed delays facilitate amplitude death of coupled oscillators*, Phys. Rev. Lett. **91**, 094101 (2003).

[33] B. Fiedler, V. Flunkert, P. Hövel, and E. Schöll: *Delay stabilization of periodic orbits in coupled oscillator systems*, Phil. Trans. R. Soc. A **368**, 319 (2010).

[34] C. U. Choe, T. Dahms, P. Hövel, and E. Schöll: *Controlling synchrony by delay coupling in networks: from in-phase to splay and cluster states*, Phys. Rev. E **81**, 025205(R) (2010).

[35] Y. N. Kyrychko, K. B. Blyuss, and E. Schöll: *Amplitude and phase dynamics in oscillators with distributed-delay coupling*, Phil. Trans. R. Soc. A **371**, 20120466 (2013).

[36] I. Schneider: *Delayed feedback control of three diffusively coupled Stuart-Landau oscillators: a case study in equivariant Hopf bifurcation*, Phil. Trans. R. Soc. A **371**, 20120472 (2013).

[37] C. M. Postlethwaite, G. Brown, and M. Silber: *Feedback control of unstable periodic orbits in equivariant hopf bifurcation problems*, Phil. Trans. R. Soc. A **371**, 20120467 (2013).

[38] G. C. Sethia, A. Sen, and G. L. Johnston: *Amplitude-mediated chimera states*, Phys. Rev. E **88**, 042917 (2013).